JACOB BEKENSTEIN

The Conservative Revolutionary

JACOB BEKENSTEIN

The Conservative Revolutionary

Edited by

Lars Brink
Chalmers University of Technology, Sweden

Viatcheslav Mukhanov
Ludwig-Maximilians-Universität of Munich, Germany

Eliezer Rabinovici
Hebrew University of Jerusalem, Israel

K K Phua
Founding Director Emeritus
Institute of Advanced Studies
Nanyang Technological University, Singapore

World Scientific

NEW JERSEY • LONDON • SINGAPORE • BEIJING • SHANGHAI • HONG KONG • TAIPEI • CHENNAI • TOKYO

Published by

World Scientific Publishing Co. Pte. Ltd.
5 Toh Tuck Link, Singapore 596224
USA office: 27 Warren Street, Suite 401-402, Hackensack, NJ 07601
UK office: 57 Shelton Street, Covent Garden, London WC2H 9HE

British Library Cataloguing-in-Publication Data
A catalogue record for this book is available from the British Library.

The editors and publisher would like to thank the following publishers of the various journals for their assistance and permissions to include the selected reprints found in this volume:
The American Physical Society (Physical Review D);
Springer Nature (Lettere al Nuovo Cimento).

JACOB BEKENSTEIN
The Conservative Revolutionary

ISBN 978-981-120-395-4

For any available supplementary material, please visit
https://www.worldscientific.com/worldscibooks/10.1142/11373#t=suppl

Printed in Singapore

Contents

Preface

A theoretical discovery in physics is often made by one person, not by a collaboration, neither by a dedicated search from the society for new knowledge. This is why the history of physics is so much connected with famous names starting from Albert Einstein. One such name in the latter part of the list is Jacob Bekenstein, who unfortunately died far too early in 2015. He was a pioneer in black hole physics making seminal contributions already in his thesis. Even though there was much resistance and scepticism towards his work in the beginning he continued to pursue it. Today these are the starting points in black hole physics together with the discoveries of Stephen Hawking at the time. Bekenstein went on to make more discoveries and was a leading figure in this field that has grown to become one of the most important fields in modern fundamental physics. It was a great loss to the global physical society but also to the world at large when Jacob Bekenstein passed away.

One way to keep the memory of a loved and important person is to collect a memorial volume dedicated to him with contributions from his friends and the leading scientists in the field. For us editors of this volume it was natural and most wanted to collect these contributions to give a full picture of Jacob Bekenstein, his life and his science for future generations as well as for present day scientists. We were very happy that there was such a huge positive response from the physicists around the world. We believe that this volume will be a lasting memorial of Jacob Bekenstein as well as an important textbook for any student of quantum gravity and cosmology.

The Editors:

Lars Brink — Viatcheslav Mukhanov
Eliezer Rabinovici — Kok Khoo Phua

Jacob Bekenstein (1947–2015): A Conservative Revolutionary

Eliezer Rabinovici

Jacob (Yacov) Bekenstein, a faculty member of the Racah Institute of Physics of the Hebrew University, Jerusalem, Israel, passed away on 16 August 2015 while visiting the Department of Physics in Helsinki.

Jacob made seminal contributions to the understanding of the properties of black holes and information storage in quantum field theory and general relativity.

Each of these was revolutionary and some generated quite a large amount of resistance and strong opposition at the time they were introduced. As groundbreaking as they were they came to Jacob through very conservative considerations.

His powerful tool was the very old fashioned *Gedankenexperiments.* This approach (in English, "thought experiments") has played a very important role in the developments of new ideas in physics: Galileo envisaged falling bodies, Maxwell had demons, Einstein chased light rays and studied elevators and Schrödinger mulled over the fate of cats. Jacob Bekenstein took these experiments to a whole new level.

He revolutionised the understanding of black holes after performing a *Gedankenexperiment* upon the suggestion of his adviser at Princeton University, John Wheeler. In the experiment, the content of a teacup was poured into a black hole. Jacob was able to transform the implied threat of the reduction of entropy in this very esoteric process (in apparent violation of the second law of thermodynamics) into the totally unexpected realisation that black holes have thermodynamical properties. It was a triumph of the human spirit. Not only did he ascribe to black holes properties such as entropy, which reflects the number of degrees of freedom, but he also brought about a revolution in the deep understanding of thermodynamics and the structure of space as seen at high-enough energies, where black holes and their likes emerge and play a role.

Pioneering Study

For well over a century, textbooks have been teaching us that thermodynamical properties such as entropy are extensive, namely that they are proportional to the volume of the system under study. However, for reasons of consistency, Jacob decreed that black holes are objects whose thermodynamical properties must be proportional only to the area of the system. The body of work in which these ideas were developed and argued dates from 1972. As the years pass, and even though the full implications of this deep insight are yet to be understood, appreciation of this pioneering study is only increasing.

Since the beginning of his work, Jacob pointed out the crucial role played by the properties of processing information in unlocking the secrets of black-hole and quantum gravity. His research led him to suggest bounds on the amount of information that can be stored in a given volume of space–time. This bound is called the "Bekenstein Bound", and has direct implications on theories in which gravity seems to play no role. Jacob's work, based again on *Gedankenexperiments*, raised a great deal of interest as well as controversy. He published it in 1981, and it has since passed several refinements and generalisations.

The very special nature of black holes uncovered by Jacob increased the realisation that information, in certain interesting circumstances at the very least, is stored on the boundary of those systems. This general feature was named "holography" — a concept borrowed from quantum optics. In particular, in one set of cases, that information is encoded in the form of a quantum-field theory where gravity is not even present. This goes under the name of AdS/CFT correspondence. Issues related to AdS/CFT correspondence are still the subject of active research in 2015, more than 40 years after Bekenstein's seminal paper appeared.

Physics may be about uncovering the objective secrets of nature, but it is done by humans, and most humans appreciate due recognition and credit. Jacob's ideas in the 1970s encountered both significant resistance and attempts to diminish his role. Not least from Stephen Hawking, who briefly mentions Bekenstein's role in his bestselling book *A Brief History of Time*. He writes: "I was motivated partly by irritation with Bekenstein, who I felt misused my discovery." And then: "However, it turned out in the end that he [Bekenstein] was basically correct, though in a manner he had certainly not anticipated."

This reflects the enormous pressure that the 25-year old Jacob had to withstand, essentially all alone. Jacob who, while being rather shy was very opinionated, answers Hawking's remarks in his book *Of Gravity, Black Holes and Information*, where he writes: "I cannot, of course, do anything about another's anger but. . . Nobody can be held to blame for giving in a published result a novel interpretation of his own." He also mentions that "I almost lost confidence then." One can only imagine the strength he needed to stick to his intuition. The main argument against him at the time was the conventional wisdom that black holes should have zero temperature and thus supposedly no entropy. So much for conventional wisdom. Out of this cocktail of emotions that Bekenstein and Hawking brewed, one following the other, breakthroughs came in the understanding of the physical properties of black holes.

A Dedicated Teacher

As mentioned previously, Jacob was highly opinionated. In his book, one can find critical remarks on aspects of string theory, as well as a line such as "aspiring students, beware" with regards to string theory and "beware of cheap interdisciplinary fads". As I have been studying string theory for quite a few years and as I did not find it easy to interest Jacob in the field, I venture to say that in my opinion, some of this was self-critique, because both string theory and Jacob's breakthroughs relied on the consistency of physical structures and seem painfully far from any near-future experimental verifications. He was after all a conservative revolutionary. In recent years, Jacob actually searched for some experimental evidence for quantum aspects of gravity itself. He was an excellent and dedicated teacher, and as such he also encouraged students to study advanced mathematics and improve their mastery of the English language.

At the Racah Institute of Physics, the high-energy theorists sit on a higher floor; nevertheless, it seemed to me that Jacob seemed to have a quite guarded respect for the field. He was very interested in new discoveries coming (and not coming) out of the LHC.

Part of Jacob's personality was that he practised religion in a private manner. He was named after Jacob in the bible, about whom it is said that he struggled with a formidable stranger (an angel, according to some), and did not lose. Very few faced the formidable challenges of presenting a theory of gravity with as much success as Jacob Bekenstein. One is also told about the biblical Yacov that he dreamed about: a ladder grounded on Earth and reaching heaven on which there was motion upwards and downwards. Jacob always made every research effort to be grounded on the Earth, but he did not shy away from climbing the ladder and bringing back his insights.

Jacob (Yacov) David Bekenstein was born in Mexico City on 1 May 1947. His father, Joseph Bekenstein, worked as a carpenter, and his mother, Esther (Nee) Vladaslavotsky, was a homemaker. They were Jewish immigrants from Poland to Mexico. He went on to graduate from Princeton University, obtaining a PhD in 1972 under John Wheeler. After time as a postdoctoral fellow at the University of Texas in Austin, he moved to the Ben-Gurion University of the Negev in Beersheba, where he eventually became chairman of the astrophysics department. In 1990, he joined the faculty of the Hebrew University of Jerusalem.

His awards include the Rothschild, Israel, Wolf and Einstein prizes in Physics. He was a member of the Israeli Academy of sciences and humanities.

Jacob is survived by his wife Bilha, his daughter Rivka (a physicist) and his sons Yehonadav (a physicist) and Uriya (a biologist), as well as by his six grandchildren.

He is dearly missed and his ideas rest with us.

It is to his family that we dedicate this book.

In Loving Memory of Jacob Bekenstein

The Bekenstein Family

We, the Bekenstein family deeply thank all involved in bringing to light this wonderful memory of Jacob, our beloved husband, father and grandfather.

Jacob had a revolutionary mind and we were lucky to share our lives with the special man he was: not only a leader in science, but at the same time a truly modest person. In everyday family life, we experienced his unique personality, witnessing his resolute dedication to his scientific work, manifested in the small details of his daily habits, as in his uncompromising integrity.

As a father and an educator, Jacob was a role model. His life was a living example of what he valued: starting from his daily schedule and habit of *early to bed early to rise*, through his healthy life style, and his choice of being an ethical vegetarian, and finally by prioritizing his family needs and the education of his children.

Following his heart, Jacob had left leading academic centers in the United States and moved to start an independent scientific carrier In Israel, at the Ben-Gurion University of the Negev. Having the education of his children in mind paved the move of the family from Beer Sheba in the Negev to Jerusalem, which had better educational opportunities. Jacob was also dedicated to continuous self-education on a broad spectrum of topics including science history and Jewish studies. He would always "wind down" with a book in his hand.

One of us (Yehonadav) had the unique opportunity to take Jacob's course in advanced quantum mechanics during his graduate studies, providing a first-hand glimpse on Jacob as a physics lecturer. On top of being very organized and prompt, he would invest endless time perfecting his course notes. These notes are being used today by the Israeli Physics Society as an example of high-quality teaching material.

In addition to delivering the course material he found it important to share anecdotes on the 'heroes' (in his words) that have played a significant role in shaping and progressing our scientific understanding.

Receiving scientific recognition, as he did, never changed his attitude — he always remained humble and treated any student or person as an equal with a kind and welcoming smile regardless of stature.

For us the family which has lost Jacob in an instant and without a chance for a proper farewell it is heartwarming to see efforts in the form of books, talks, and prizes that are being dedicated to his memory.

We deeply thank his friends and colleagues and especially Lars Brink, Slava Mukhanov, Eliezer Rabinovici and K K Phua for their hard work, endless patience and dedication in putting together and editing this meaningful book.

Photos of Jacob Bekenstein

In front of the main chalet at school on black holes, in Les Houches, France (1972).

Taking a break with Teitelboim (left) and Quintana (right) at school on black holes, in Les Houches, France (1972).

On a trip to dead sea (Israel, 1979), holding his first born son Yehonadav.

Visiting his parents Joseph and Esther, with sons Yehonadav, Uriya, and daughter Rivka (Brooklyn, 1988).

Celebrating with PhD advisor John Archibald Wheeler (left) and wife Bilha Bekenstein (center), Physics Wolf Prize for Wheeler (Jerusalem, 1997).

Hiking with son Uriya (Iceland, 1999).

School of physics, Erice, Sicily (early 2000).

Receiving the 2012 Wolf Prize during the ceremony that took place in front of the Chagall wall tapestries at the Knesset. [This photo, also featured on the cover of the present book, is reproduced with the kind permission of the Wolf Foundation.]

The Early History as Seen from Some Who were Involved

"It can be seen that $\kappa/8\pi$ is analogous to temperature in the same way as A is analogous to entropy. It should however be emphasized that $\kappa/8\pi$ and A are distinct from the temperature and entropy of the black hole. In fact, the effective temperature of a black hole is absolute zero." This is a quote from the classical paper by J. M. Bardeen, B. Carter and S. Hawking, "The Four Laws of Black Hole Mechanics" published in *Commun. Math. Phys.* in 1973, a year after Jacob Bekenstein had announced his discovery of black hole entropy in *Lett. al Nuovo Cimento.* It is very clear from this quote the Bekenstein's idea was not accepted by the community straight away. Even more, it faced a strong resistance at the very beginning and one needs to have a character as strong as Jacob's not to give up and pursue in a subsequent paper what he believed is correct with the support of many beautiful gedanken experiments published in *Phys. Rev. D* in 1973. In fact, at the very beginning, it was not clear why the second law of thermodynamics should be legitimate for the external observer in the presence of a black hole. It might work or equally well it could fail. In fact, at this time it was already clear that black holes have no hairs and hence neither baryon nor lepton numbers are conserved from the point of view of the observer because they can simply disappear in the black hole without any trace. Why should entropy be exceptional? If a pedestrian turns around a corner he is not obliged to leave a message behind. Why could it not happen that entropy could just be hidden inside the black hole without leaving any trace to those who decided to stay outside? It could well happen and would not contradict any known fundamental physical law.

Only after the discovery of the Hawking radiation and resolving some extra puzzles it became clear that Jacob's intuition was absolutely correct, although it did not follow from any unbeatable logical arguments. This situation is not so unusual for many great discoveries in physics which sometimes are made contrary to common sense, and the courage to go against the common sense characterize really great physicists to whom undoubtedly Jacob Bekenstein belongs. This is why we believe it will be interesting to get the story of Jacob's discovery from the first hands.

In this section we have collected the memories of those — R. M. Wald, J. M. Bardeen, and G. W. Gibbons — who were present at these times and witnessed how it was happening.

Robert M. Wald was a contemporary graduate student of Jacob Bekenstein in Princeton in the early 1970s and he was sitting on the front bench observing and interfering with Jacob during the time when he formulated his proposal about black hole entropy and also what happened after this period.

James M. Bardeen was already a well-established relativist in the USA who got together with Hawking and Carter at a workshop in 1972 where they formulated their famous four laws. He describes this and follows up the history up to modern times.

Gary W. Gibbons was a graduate student in Cambridge when this history started working with his slightly older colleague Stephen Hawking and he describes the history from the Cambridge horizon and then also follows up the history to modern times.

Jacob Bekenstein and the Development of Black Hole Thermodynamics

Robert M. Wald
Enrico Fermi Institute and Department of Physics,
The University of Chicago, 5640 South Ellis Avenue,
Chicago, Illinois 60637, USA
rmwa@uchicago.edu

I give some personal reflections on Jacob Bekenstein's pioneering work on associating an entropy with a black hole proportional to its area and on the generalized second law of thermodynamics.

Jacob Bekenstein and I were classmates in graduate school in Princeton. I arrived in the Autumn of 1968. Bekenstein did not arrive until 1969, but he had already completed a master's degree, so we were at about the same stage in our knowledge of physics and general relativity when we both began to do research in early 1970. Both of us were primarily interested in the theory of black holes. We had desks near each other in the newly opened Jadwin Hall, and we interacted on nearly a daily basis. We both completed and defended our Ph.D. theses in the Spring of 1972.

General relativity was in the midst of a golden age while we were at Princeton. Penrose had proved the first singularity theorem a few years earlier, and the global methods that he thereby introduced were soon extended — particularly by him, Hawking, and Geroch — to obtain a large number of remarkable new insights into the theory. Israel's theorem on uniqueness of static black holes had very recently been proven. The remarkable global structure and other properties of the Kerr black hole had recently been fully understood, and the Penrose energy extraction process had just been proposed. Pulsars had very recently been discovered and associated with rotating neutron stars.

This golden age was reflected by the students attracted to general relativity at that time. Among the students at Princeton working on problems in general relativity with whom Bekenstein and I overlapped for at least a year were Demetrios Christodoulou, Marc Davis, Steve Fulling, Bei-Lok Hu, Bahram Mashhoon, Niall O'Murchadha, Stuart Shapiro, Claudio Teitelboim/Bunster, and Bill Unruh. I recall attending a "town hall" meeting of faculty and graduate students sometime in

1971–72 where the particle theory faculty decried the fact that they could hardly attract any students!

The atmosphere in relativity group at Princeton was dominated by John Wheeler. Although he was highly overcommitted and frequently out of town or otherwise unavailable, he was a wonderful advisor in that he generally treated his graduate students as his superiors, he was very supportive and encouraging, he was truly open to all ideas, and he had boundless energy and enthusiasm. At the time, however, I was not particularly enamored by his style of doing physics. To me, some of his grand ideas seemed to be over-reaches, and I preferred working on problems that were sufficiently well posed to have a well-defined mathematical resolution.

One of the problems of great interest to me and others at Princeton at the time was the idea that, in Wheeler's phrase, "black holes have no hair" — that a black hole, once formed or perturbed, will approach a stationary final state that is uniquely determined by its mass, angular momentum, and electric charge. It is here that Bekenstein made his first major contribution to general relativity: He showed that black holes cannot have "scalar hair"; i.e., there cannot exist a nonvanishing stationary Klein–Gordon scalar field around a stationary black hole.[1] Given the general difficulty of proving black hole uniqueness results, I was quite amazed by how simple and elegant Bekenstein's proof was. He realized that there was no need to use Einstein's equation; the Klein–Gordon equation alone in an arbitrary static or stationary-axisymmetric black hole background was sufficient to rule out time-independent scalar field solutions that are well behaved at infinity and at the horizon.

The "no hair" ideas had a very interesting implication: If one formed one black hole by the collapse of neutrons and another by the collapse of photons, the final state black holes would be exactly the same as long as they had the same total mass and angular momentum. Thus, the information about the original number of baryons present would be lost. This implied that the law of baryon conservation and, similarly, lepton conservation — which, at the time, were believed to be exact — would essentially become meaningless when black holes are present. Although no local violation of these laws would occur, one would have no way, in principle, of verifying that they were satisfied. Wheeler's term for this phenomenon was that in the presence of black holes, the laws of baryon and lepton conservation are "transcended."

Wheeler was quite comfortable with the idea of transcendence of the laws of baryon and lepton conservation. However, it is quite remarkable (particularly in retrospect) that he was not comfortable with a similar transcendence of the second law of thermodynamics: If one takes a "cup of tea" — the example Wheeler used — and throws it into a black hole, the entropy of that cup of tea will be lost, and total entropy of the universe will apparently decrease. One could maintain that the entropy is "really still there" inside the black hole — just as one could maintain that the baryons are "still there" — but, at best, the second law of thermodynamics would become observationally meaningless. My own view at the time was that the second law of thermodynamics is a statistical law, not a fundamental law, so its "transcendence" would be more palatable than the transcendence of an apparently fundamental law like baryon conservation. Thus, I was quite comfortable with the transcendence of the second law of thermodynamics. But Wheeler did not feel this way.

Wheeler suggested to Bekenstein that he look into ways that the second law might be rescued. I don't know whether it was Bekenstein or Wheeler who first thought of considering black hole area as possibly relevant to this issue, but it was a natural idea to consider at the time in this context: In 1970, Christodoulou had shown that one could assign an "irreducible mass" to a Kerr black hole that could not decrease in any process involving dropping particle matter into the black hole.[2] Very shortly thereafter, Hawking proved his famous "area theorem" that showed, in complete generality, that the surface area of the event horizon of a black hole cannot decrease with time.[3] Christodoulou's "irreducible mass" of a Kerr black hole was just its area. A non-decreasing quantity like black hole area would be a good place to look for something that might rescue the second law.

However, at the time, I felt that this was an utterly ridiculous project to work on. First, as already mentioned, I was not troubled by the apparent transcendence of the second law of thermodynamics. Second, the analogy between the second law and the area theorem seemed extremely artificial; it seemed quite unnatural to me to try to marry a statistical law with a mathematical theorem. But, most importantly, in the absence of a fully developed quantum theory of gravity, what could one possibly show and/or how could one possibly argue for the validity of any highly speculative ideas on black hole entropy that one might propose? I have a distinct memory of leaving a group meeting of Wheeler and all of his students shortly after Bekenstein had started working on this project, and seeing Bekenstein and Wheeler stay behind to talk further about the second law ideas. The thought that flashed through my mind at the time was "Boy, am I glad that I have some good problems to work on!"

In December 1971 — when Bekenstein was deeply involved in developing his ideas on area representing entropy and on the generalized second law — Bob Geroch visited Princeton to give a colloquium. This was the first time I had met Geroch, and I was immediately taken with his style of doing physics and his infinite supply of fascinating ideas. The colloquium he gave remains one of the best I have ever heard. It was a very wide ranging talk that touched on many different aspects of general relativity. Part of the talk discussed black holes, and he spent perhaps 10 minutes making the point that one could use a black hole to run a Carnot cycle with 100% efficiency. One could do this by lowering a box of high entropy matter all the way to the horizon of the black hole before dropping it in. All of the "heat" of the matter could thereby be converted to "work" in the laboratory from which one did the lowering. Thus, Geroch pointed out that, in this well-defined sense, black holes are systems at absolute zero temperature. I recall talking briefly with Bekenstein immediately after the colloquium. I was in high spirits about what a great talk it was. Bekenstein, of course, also thought it was a great talk, but he did not seem in as high spirits and, indeed, seemed a bit worried/concerned.

In retrospect, it seems clear that Bekenstein must have immediately realized that assigning an absolute zero physical temperature to a black hole would lead to severe consistency problems with black hole thermodynamics. In particular, Geroch's suggestion of lowering a box of matter containing entropy all the way to the horizon of a black hole could certainly be used to violate any proposal for a generalized second law, since, in this process, entropy would be lost, but the black hole would end up in the same state in which it began. Hence, whatever entropy, $S_{\rm BH}$, one might try to assign to the black hole, if one performed this process carefully enough, $S + S_{\rm BH}$ would necessarily decrease. Thus, Bekenstein had his work cut out for him if he was to salvage his ideas.

Bekenstein did salvage his ideas, using his considerable physical insight and understanding of what effects might be relevant to the issue. First, he argued that black hole entropy should be linearly proportional to area A (rather than being a nonlinear monotonic function of A). On dimensional grounds, the entropy then had to be *inversely* proportional to $\hbar$. This meant that the entropy of a macroscopic black hole would be absolutely enormous compared with the entropy of typical ordinary matter of comparable size and energy. But this meant, in turn, that to violate the generalized second law by lowering matter towards a black hole, one had to bring the matter *extremely* close to the horizon before dropping it in — so close that it was not clear that this would be possible to do this for physical matter. This basic idea to use limitations on the properties of physical matter to rescue the generalized second law was clearly present in Bekenstein's original papers.[4,5] It became more formalized in his 1981 paper as the "Bekenstein bound," $S/E \leq 2\pi R$, on the ratio of entropy, S, to energy, E, for matter confined to a region of size R.[6]

Nevertheless, black hole thermodynamics really is inconsistent if one assigns a finite entropy but a vanishing temperature to a black hole. This fact was sharpened in the remarkable paper of Bardeen, Carter, and Hawking, which showed that black holes obey precise analogs of the laws of thermodynamics.[7] The fact that black holes obey such laws was, in some sense, supportive of Bekenstein's thermodynamic ideas. However, by obtaining a precise first law of black hole mechanics, it was clear that if one wishes to assign a physical entropy to a black hole proportional to its area, A, then one must also must also assign to it a nonzero temperature proportional to its surface gravity, κ. But the fact that the physical temperature of a black hole vanishes then can be used to obtain contradictions with the generalized second law. In particular, suppose the black hole is placed in a thermal bath of radiation at a lower temperature than the assigned temperature of the black hole. Although the black hole may be smaller than typical wavelengths in this bath, it nevertheless, will still absorb some radiation. The generalized second law will then be violated, as heat will flow from the cooler to the hotter body. Indeed, Bardeen, Carter, and Hawking explicitly noted (using a Geroch-type box lowering argument rather than this thermal bath argument) that the horizon area of a black hole could not represent its physical entropy, and that the second law of thermodynamics is "transcended" in the presence of a black hole.

Sometime around 1973, I recall Bekenstein also telling me that his work had independently come under attack by a physicist for, I believe, the information theoretic ideas that he had used to assign an entropy to a black hole and to justify its thermodynamics. (I do not recall the name of this physicist or his specific criticism — he was not a general relativist and I had not heard his name before or since.) So, it certainly seemed that the main ideas that Bekenstein had worked on to date were not panning out very well! As already indicated above, I had not thought that assigning to a black hole a physical entropy proportional to its area made much sense in the first place — although I thought that Bekenstein had done a remarkably good job in advancing and defending this idea — and, as far as I was concerned, the paper by Bardeen, Carter, and Hawking provided a definitive refutation of the idea that the area of a black hole could represent its physical entropy.

Then, a miracle occurred! In 1974, Hawking calculated particle creation effects for a body that collapses to a black hole, and he made the amazing discovery that a distant observer will see a steady, thermal distribution of particles emerge at a temperature $T = \kappa/2\pi$.[8] So, a black hole truly has a nonzero *physical* temperature proportional to its surface gravity! Black hole thermodynamics now appeared to be entirely consistent. In particular, if one placed a black hole in a radiation bath of temperature $T_{\text{bath}} < \kappa/2\pi$ the black hole radiation would dominate over absorption, and the generalized second law would hold. The entropy $S_{\text{BH}} = A/4$ could now be interpreted as the *physical* entropy of the black hole — with the unknown constant in Bekenstein's original proposal now fixed by the value of the Hawking temperature. Bekenstein was right!

However, there remained a problem. Although the Hawking radiation by a black hole was sufficient to directly rule out any contradiction with the generalized second law if one immersed a black hole in a thermal bath of radiation, the Geroch-type box lowering argument still posed a potential difficulty for the generalized second law: As Geroch had originally argued, if one could lower a box arbitrarily close to the horizon, one should still be able get rid of the entropy in the box without increasing the black hole area. Here, it would seem, Hawking radiation or other quantum effects would not help, since one could perform this process for an arbitrarily large black hole, whose Hawking temperature is arbitrarily small. Thus, it appeared that one would have to rely on Bekenstein's proposed bound, $S/E \leq 2\pi R$, to prevent one from lowering the box close enough to the horizon to violate the generalized second law. However, I was unhappy with this possible resolution for the following two reasons: (i) It did not appear that this bound was sufficient to avoid a violation of the generalized second law; if one used, e.g., a rectangular box, it would be necessary for the quantity "R" in the bound to be the *shortest* dimension of the box, whereas the arguments in favor of the bound took R to be the *largest* dimension. (ii) It seemed to me that the consistency of black hole thermodynamics should not hinge on some property of matter that would not otherwise be needed for the consistency of thermodynamics.

Bill Unruh shared these concerns. Remarkably, we independently and simultaneously came up with a resolution in 1981 — one week before I visited him in Vancouver: Although quantum effects such as Hawking radiation are negligible for freely falling observers near a large black hole, they are not negligible for quasi-stationary bodies near the horizon, such as a box of matter being slowly lowered towards the horizon. Such bodies undergo a huge acceleration and therefore feel the effects of "acceleration radiation," which produces a buoyancy force. As we showed, as a consequence of this buoyancy force, the optimal place from which to drop a box of matter into the black hole is not the horizon but rather the "floating point" of the box. When the box is dropped from its floating point, the area increase of the black hole is exactly sufficient to compensate for the loss of the matter entropy, independently of whether or not any entropy bounds are satisfied by the matter.[9]

We immediately wrote to Bekenstein to tell him that the consistency of black hole thermodynamics did not rely on their being an entropy bound on ordinary matter. Bekenstein agreed with the validity of the buoyancy effect that we had found, but he felt that our approximations to describe it were not adequate and that some entropy bound on matter — although perhaps a weaker bound than the one he had previously proposed — would still be needed for the validity of the generalized second law. There followed a dueling series of papers[10–14] where Bekenstein gave examples where our approximations would not be adequate and argued that buoyancy would not be sufficient for the validity of the generalized second law, whereas we/I countered by showing that when the full calculation is done, buoyancy is sufficient. We never reached agreement on this issue.

Bekenstein and I remained friends throughout the years, but I saw him very infrequently after he moved to Israel in 1974. I was very pleased to come to Jerusalem in 2012 to celebrate the 40th anniversary of his original proposal that black hole area should represent its entropy. Although by then Bekenstein had won the Wolf Prize and many other high honors, he was as unassuming as he had been as a graduate student. He also was much more eager to discuss his new ideas than revel in past glories. I was also very pleased to be able to congratulate him in person at the American Physical Society meeting in 2015 upon his winning the Einstein Prize. It is tragic that he passed away only six months later.

Acknowledgments

This research was supported in part by NSF Grant PHY 15-05124 to the University of Chicago.

References

1. J. D. Bekenstein, *Phys. Rev. D* **5**, 1239 (1972).
2. D. Christodoulou, *Phys. Rev. Lett.* **25**, 1596 (1970).
3. S. W. Hawking, *Phys. Rev. Lett.* **26**, 1344 (1971).
4. J. D. Bekenstein, *Phys. Rev. D* **7**, 2333 (1973).
5. J. D. Bekenstein, *Phys. Rev. D* **9**, 3292 (1974).
6. J. D. Bekenstein, *Phys. Rev. D* **23**, 287 (1981).
7. J. M. Bardeen, B. Carter and S. W. Hawking, *Commun. Math. Phys.* **31**, 161 (1973).
8. S. W. Hawking, *Commun. Math. Phys.* **43**, 199 (1975).
9. W. G. Unruh and R. M. Wald, *Phys. Rev. D* **25**, 942 (1982).
10. J. D. Bekenstein, *Phys. Rev. D* **27**, 2262 (1983).
11. W. G. Unruh and R. M. Wald, *Phys. Rev. D* **27**, 2271 (1983).
12. J. D. Bekenstein, *Phys. Rev. D* **49**, 1912 (1994).
13. M. A. Pelath and R. M. Wald, *Phys. Rev. D* **60**, 104009 (1999)
14. J. D. Bekenstein, *Phys. Rev. D* **60**, 124010 (1999).

On the Microscopic Interpretation of the Bekenstein–Hawking Entropy

James M. Bardeen

Physics Department, Box 1560, University of Washington, Seattle, WA 98195-1560, USA

bardeen@uw.edu

As a graduate student Jacob Bekenstein constructed an ingenious argument for why a black hole should have an entropy proportional to the area of its event horizon, and came within a factor of 10 of effectively predicting the black hole temperature discovered shortly later by Hawking from a quantum field theory calculation. This Bekenstein–Hawking entropy has a clear interpretation as the macroscopic thermodynamic entropy associated with the black hole. However, its microscopic interpretation in terms of the quantum properties of the black hole has remained controversial for over 40 years. I argue that, for a generic vey large black hole, if the semi-classical approximation is valid, the Bekenstein–Hawking entropy may not impose *any* limit on the number of quantum degrees of freedom in the matter swallowed by the black hole or on the entanglement (von Neumann) entropy between the interior and exterior of the black hole. If there is such a limit, it must come from a breakdown of the usual semi-classical approximation in the vicinity of the horizon.

It is an honor to have been asked to contribute to this memorial to the life and work of Jacob Bekenstein. While I had relatively little direct contact with Jacob over the years, I have long admired how he was able to bring subtle physical reasoning to bear on difficult issues. Nowhere is this more evident than in his work on black hole entropy, begun as a graduate student at Princeton. He argued in his Ph.D. thesis and in a paper[1] published in early 1973, that the area increase theorem for classical black holes established by Hawking[2] suggested that black holes should have thermodynamic properties, with an entropy proportional to the area of the event horizon, and estimated the black hole entropy as

$$S_{\mathrm{bh}} \cong \frac{\ln 2}{2} \frac{A}{4\pi\hbar} \tag{1.1}$$

in units with $G = c = 1$. In the same paper Jacob proposed a generalized second law, that when matter is accreted, the ordinary entropy lost into the black hole is always compensated by an at least equal increase in the entropy of the black hole.

At about the same time, Bardeen, Carter, and Hawking published a paper[3] titled "The four laws of black hole mechanics," in which we pointed out the precise correspondence between the laws governing the properties and transformations of classical black holes with the four laws of thermodynamics. However, we concluded that this could *only* be an analogy, since black holes could not radiate and could not have a real temperature and a real entropy. In particular, the "first law" of black hole mechanics is

$$\delta M = \frac{\kappa}{8\pi}\delta A + \Omega_{\rm H}\delta J_{\rm H} + \text{matter terms}\,, \tag{1.2}$$

in which κ is the "surface gravity" of the black hole, the redshifted gravitational acceleration of a static observer at the horizon. We also argued against a generalized second law, claiming it was possible to add entropy to a black hole without any increase in horizon area, not being aware of the arguments to the contrary, that take into account the minimum radial extent of any matter injected into the black hole, made in Bekenstein's paper.

Jacob's insight was confirmed a year later, when Hawking published his famous result[4] that, when quantum field theory is taken into account, black holes do in fact radiate like a black body at the "Hawking" temperature

$$T_{\rm H} = \frac{\kappa\hbar}{2\pi}\,, \tag{1.3}$$

from which one immediately concludes that the black hole thermodynamic entropy, now called the Bekenstein–Hawking entropy, is

$$S_{\rm BH} = \frac{A}{4\hbar}\,, \tag{1.4}$$

about a factor of 10 larger than Bekenstein's initial estimate. The various ramifications of this identification have been reviewed by Wald.[5] The non-zero temperature implies a luminosity, which after taking into account spin-dependent grey-body factors, for a Schwarzschild black hole has the form

$$L_{\rm H} = \frac{4\pi}{245760\pi^2}\frac{\hbar}{M^2}k_s = 4\pi M^2\sigma T_{\rm H}^4 k_s\,, \tag{1.5}$$

where σ is the Stefan–Boltzmann constant. The coefficient k_s was calculated by Page[6] for spin 1 and spin 2 massless fields (photons and gravitons), with the results $k_1 = 6.4928$ and $k_2 = 0.7404$. The spin 0 calculation from Elster[7] and Taylor, Chambers and Hiscock[8] gave $k_0 = 14.36$.

The discovery of Hawking radiation raised profound questions about the fate of black holes, quantum field theory in curved spacetimes, and the nature of quantum gravity that have been vigorously debated in the literature for more than 40 years, with no end in sight, Jacob continued to make important and wide-ranging contributions, but I will in the rest of this paper focus on a few issues that I feel, in the spirit of Jacob's approach to physics, have not been dealt with properly by much of the community and take into account results for the semi-classical stress–energy tensor (SCSET) first obtained in the 1980's and 1990's[9] and recent calculations of Levi and Ori.[10]

First, there is the issue of how one should think about how the Hawking radiation is generated. A common assertion is that it is a pair creation process within a Planck length or so of the horizon (e.g., by Parikh and Wilczek[11]). The Hawking particle is supposedly created just outside the horizon and is paired with a "Hawking partner" particle just inside the horizon. Inside the horizon the partner can have a negative Killing energy, reducing the mass of the black hole, to balance the positive Killing energy of the Hawking particle that propagates away from the black hole. This is nonsense physically for a number of reasons. The Killing energy is irrelevant for a local pair creation process, since over scales very small compared with the curvature scale it is the energy and momentum in a local inertial frame that must be conserved. In a local free-fall frame straddling the horizon the Hawking particle must have a doubly enormous energy and radial momentum to compensate for two enormous redshift factors, one to go from the free-fall frame to a static frame just outside the horizon and the second to go from that static frame to an inertial frame at infinity in an asymptotically flat spacetime. A Hawking partner particle just insided the horizon must also have an enormously positive energy and radial momentum in the local inertial frame in order to have a negative Killing energy. There is no source for the enormous energy and momentum of the pair. What calculations of the SCSET show is an inflow of negative energy in a static frame near the horizon, *not* the outflow of positive energy required by the local particle creation scenario.

The explanation I consider reasonable was given early on in papers by Unruh[12] and Fulling.[13] Hawking radiation originates from vacuum modes, with zero energy, initially propagating very close to the horizon. The nonzero surface gravity of the black hole horizon means that these are redshifted until their wavelengths become comparable to M. The geodesic deviation associated with the spacetime curvature disrupts these modes, part of which propagate off to infinity, where relative to the vacuum of an observer in the asymptotically flat spacetime they are interpreted as particles with positive energy, and part of which propagate back into the black hole carrying negative energy, as is allowed locally in quantum field theory. A particle interpretation is not appropriate where the wavelength of the modes is comparable to the local curvature scale. The SCSET shows a gradual transition from ingoing negative energy to outgoing positive energy centered, at least for a conformally coupled scalar field in a Schwarzschild background, around $r = 3M$.

What is a bit troubling is that the modes giving rise to the Hawking radiation at later times must have had highly sub-Planckian wavelengths is a local freely-falling frames closer to the formation time of the black hole. This is not a problem in conventional local quantum field theory, which has modes with arbitrarily small wavelengths in an infinite Hilbert space. However, if one believes in a holographic property for quantum field theory in the presence of gravity, such as suggested by the form of the Bekenstein–Hawking entropy, the Hilbert space for quantum modes propagating close to the black hole horizon should be finite.

A more radical alternative was proposed by Bekenstein and Mukhanov.[14] They start from an assumption that the quantum black hole event horizon area is quantized, with $A = \alpha\hbar n$, n taking on positive integer values, and degeneracy $g(n)$ chosen to be consistent with the Bekenstein–Hawking entropy. The spacing of levels in energy is taken to correspond to a fixed angular frequency

$$\varpi = \frac{\ln 2}{8\pi M}\,. \tag{1.6}$$

implying a spectrum for the "Hawking" radiation quite different from the Planck spectrum. While on the whole I find the usual semi-classical picture more convincing, this idea could be checked if we are ever able to observe the evaporation of a primordial black hole.

Normally the value of the classical thermodynamic entropy of a system is thought to correspond to the total number of quantum degrees of freedom for the system. The quantum degrees of freedom are those representing excitations above the vacuum. For example, the entropy of a box filled with radiation in equilibrium at a certain temperature corresponds to the average number of photons present in the box, and is proportional to the volume of the box. However, S_{BH} is proportional to the surface area of the black hole. Due to the interior singularities that must be present, at least in classical general relativity, the volume of a black hole is not well defined.

What, then, can one say about the microscopic interpretation of the Bekenstein–Hawking entropy? At least roughly, S_{BH} corresponds to the maximum number of quanta that can collapse to form a black hole. If each quantum has energy ε the number of massless quanta to form a black hole of mass M is $N = M/\varepsilon$. But to end up in the black hole the wavelength associated with each quantum must be less than a few times the radius of the black hole, say $\lambda = 2\pi\hbar/\varepsilon < 6M$. The maximum number is then $N_{\mathrm{max}} \sim M^2/\hbar$. The particles that form a black hole will typically have much larger energies, so initially the non-gravitational contribution to the black hole quantum degrees of freedom will be much less than S_{BH}. Could quantum gravitational degrees of freedom saturate S_{BH}, as in the Bekenstein–Mukhanov model? Perhaps, but not if the only contribution is from gravitons treated semi-classically.

Of course, what has been the main focus of debate over the last 40+ years is the black hole information problem. In the semi-classical treatment of Hawking radiation, the Hawking quanta radiated are entangled with quantum excitations inside the black hole, whether these are considered "Hawking partners" or just vacuum excitations. The number of quantum degrees of freedom in the collapsing matter that forms the black hole, certainly in any astrophysical context, is many orders of magnitude smaller than S_{BH}. Assuming a pure state before the initial collapse (certainly not possible in the real world, but assumed as a matter of principle), it may or may be true that quanta swallowed up as the black hole forms are entangled with the outside world. In either case, the von Neumann entropy $S_{\rm vN}$ of the initial black hole is at most infinitesimal compared with $S_{\rm BH}$, after renormalization to remove the local entanglement of the vacuum across the boundary of the black hole.

However, as evaporation due to Hawking radiation proceeds, $S_{\rm vN}$ steadily increases and eventually, at the "Page time" of order $M^3/\hbar = M^3/m_{\rm p}^2$, $m_{\rm p}$ the Planck mass, becomes equal to $S_{\rm BH}$. At the Page time the black hole mass has decreased by about a factor of 2 from its initial value, so an initially large black hole ($M \gg m_{\rm p}$) is still very large. Naively the semi-classical approximation should still be valid and the emission of Hawking radiation should continue. Unless the quantum information trapped inside the black hole can start leaking out as subtle correlations in the Hawking radiation, reversing the increase in $S_{\rm vN}$, it would seem the endpoint is when the radius and mass of the black hole have shrunk down to near the Planck scale. If at that point the black hole simply disappears, leaving only the emitted Hawking radiation in a mixed state, the result would seem to be a violation of unitarity, one of the bedrock principles of quantum mechanics. Strictly speaking, this is not true, as re-emphasized recently by Unruh and Wald,[15] since a test of unitarity requires a complete Cauchy hypersurface, something that is not present after the black hole evaporates completely. A Planck scale remnant with an enormous entropy is also very unattractive. Assuming the late Hawking radiation is entangled with the early Hawking radiation, necessary if the final $S_{\rm vN}$ is to be small or zero, means that it cannot be entangled with Hawking "partners" inside the black hole, and Almheiri *et al.* have argued that this would imply singular flow of energy, a "firewall," just inside the horizon.

To address these issues, one must decide how much of local quantum field theory one is willing to abandon. In my discussion, I will insist that quantum information propagate causally in the macroscopic geometry of a large black hole in the vicinity of the horizon. Given this, I see no plausible mechanism for allowing significant quantum information associated with the generation of Hawking radiation to be preserved on the black hole horizon, to leak out at some later time as subtle correlations in the Hawking radiation. Outgoing null geodesics diverge away from the horizon, and quantum information even slightly inside the horizon ends up deep inside the black hole on a dynamical time scale. There is no local physics that identifies the precise location of an event horizon. It is locally like any other null hypersurface, and is locally indistinguishable from a Rindler horizon in Minkowski spacetime. One can make this explicit by a coordinate transformation.

Furthermore, whether a black hole even has a true event horizon, as opposed to just an apparent horizon, depends on knowing the entire future history of the black hole evolution. The classical theorems that prove the existence of an event horizon assume the energy–momentum tensor satisfies the null energy condition (NEC),

$$T_{kk} \equiv T_{\alpha\beta}k^\alpha k^\beta \geq 0\,, \tag{1.7}$$

with k^α an arbitrary null vector. The NEC can be violated locally in quantum field theory. It is in fact violated near the horizon of a black hole in the Unruh state. This has led to attempts to construct quantum null energy conditions (QNECs) that can be proven in quantum field theory and that have implications for the existence of black hole event horizons and singularities in black hole interiors.

The averaged null energy condition (ANEC) is the claim that

$$\int_{-\infty}^{\infty} d\lambda \langle T_{kk}(\lambda)\rangle \geq 0\,, \tag{1.8}$$

where the integral is along any complete achronal null geodesic with affine parameter λ. Achronal means no two points on the geodesic can be connected by a timelike curve. This has been proven for arbitrary states in Minkowski spacetime,[16] and is also valid for the null generators of a static bifurcate Killing horizon. Of course, an evaporating black holed does not have a static bifurcate Killing horizon.

The "half-ANEC," also proved in Ref. 13, relates the integral of the semi-classical null energy along the "outward" null geodesics orthogonal to a two-surface element ∂A to the rate of change of the entanglement entropy across the null 3-surface produced by deforming along the null geodesics,

$$2\pi \int_{\partial A}^{\infty} d\lambda \langle T_{kk}(\lambda, y)\rangle + \frac{\delta S_{\mathrm{vN}}(A)}{\delta\lambda} \geq 0\,. \tag{1.9}$$

This is the basis of a proof of a quantum version of the generalized second law by Wall.[17]

An essentially local QNEC has been proven in some generality by Balakrishnam *et al.*[18] This can be written somewhat schematically as

$$\langle T_{kk}(y)\rangle + \frac{1}{2\pi}\frac{\delta^2 S_{\mathrm{vN}}}{\delta\lambda^2} \geq 0\,. \tag{1.10}$$

It can be verified that the QNEC is indeed satisfied by the SCSET at the horizon of a Schwarzschild black hole for fields in the Unruh state using Page's[19] calculation of the evolution of the von Neumann entropy.

This QNEC was originally obtained as a consequence of the quantum focusing conjecture (QNC) formulated by Bousso *et al.*[20] The QNC is based on a "generalized entropy," which for a compact two-surface σ with area $A(\sigma)$ dividing a Cauchy hypersurface into an inner region and a non-compact outer region is

$$S_{\mathrm{gen}}(\sigma) - S_{\mathrm{out}} + \frac{A(\sigma)}{4G\hbar}\,, \tag{1.11}$$

where S_{out} is the von Neumann entropy of the exterior region and G is Newton's constant in the classical limit. While S_{out} is formally infinite without renormalization, S_{gen} is finite. Deforming σ along the "outward" null geodesics orthogonal to σ generates a sequence of two-surfaces $\sigma'(\lambda)$. The "quantum expansion" is $\Theta \equiv (4G\hbar/A)(dS_{\mathrm{gen}}/d\lambda)$. The QFC states that

$$\frac{d\Theta}{d\lambda} \leq 0\,. \tag{1.12}$$

Then if $dS_{\rm gen}/d\lambda < 0$ initially, as will be the case if σ is sufficiently inside the apparent horizon of the black hole, $S_{\rm gen}$ will keep decreasing. At a finite affine parameter, since $S_{\rm out} \geq 0$, either $A \to 0$ signaling the formation of a caustic or the geodesic cannot be extended because of a singularity. Starting from a non-compact Cauchy surface, the same arguments as used in the classical Penrose singularity theorem then imply a quantum singularity theorem.

Bousso *et al.*[19] argue that the QNC is not necessary to prove a quantum singularity theorem, since on a causal event horizon $dS_{\rm gen}/d\lambda \geq 0$ by the GSL. An initially quantum-trapped null hypersurface must be inside the causal event horizon and by definition stay inside. However, a black hole does not have a causal event horizon if quantum backreaction can eventually remove all trapped surfaces.

To sum up, I argue that almost all quantum information trapped by the black hole ends up in the deep interior, and as long as the quantum corrections to the geometry are very small in the general vicinity of the horizon, as is true semi-classically up to and well beyond the Page time for a large black hole, it would require a very serious violation of causality for this to appear as correlations in the Hawking radiation. I agree with Bousso and Porrati[21] that the suggestion of Hawking *et al.*[22] that information carried by zero modes could restore purity of the Hawking radiation is not viable. The zero modes share information with the Hawking radiation at large radii, but cannot extract information from the black hole.

The implication is that unless there is a major breakdown of the semi-classical approximation in the vicinity of the horizon by the Page time, the von Newmann entropy of the black hole will exceed the Bekenstein–Hawking entropy, eventually by a very large factor, and $S_{\rm BH}$ cannot be interpreted as the total number of quantum microstates associated with the black hole. This argument is hardly new. See a recent paper by Rovelli[23] for a discussion. In fact, one could imagine a cyclic process in which, after a long period of Hawking evaporation, the collapse of a shell of matter onto the black hole restores the original mass of the black hole, with $S_{\rm vN}$ and the entropy in the Hawking radiation increasing steadily through each cycle. To avoid all this, the semi-classical approximation must fail in a way related to the ratio of $S_{\rm vN}$ to $S_{\rm BH}$ even though, for a solar mass black hole, the semi-classical corrections to the geometry are roughly of order 10^{-76} around and outside the horizon.

One possibility is that, due to a finite Hilbert space, by the Page time there are no more available vacuum modes to be converted into Hawking quanta, and the Hawking evaporation just stops, leaving an inert massive remnant with $S_{\rm vN} \leq S_{\rm BH}$. I haven't seen this discussed much in the literature, and there may well be good reasons why it doesn't make sense.

Another possibility is that a deep connection between entanglement and geometry in quantum gravity generates a backreaction that strongly modifies the geometry in the vicinity of any compact two-surface with area A as the (renormalized) $S_{\rm vN}$

approaches $A/(4G\hbar)$. Also assume that quantum backreaction can prevent formation of a singularity in the deep interior, contrary to the QFC. If not, quantum information would seem to be irretrievably lost. In a spherically symmetric context, for simplicity, the absence of a singularity means that there must be an inner boundary to the region of trapped surfaces, i.e., an inner apparent horizon, at some radius r_-. As long as $r_- \ll 2M$ the Schwarzschild geometry may still be a good approximation around the outer apparent horizon. Causal propagation focuses quantum information close to r_- since outgoing radial null geodesics in the trapped surface region end up close to r_- on a dynamical timescale and ingoing radial null geodesics asymptote to r_- after passing through the origin. The $S_{\rm vN}$ associated with a sphere just outside r_- keeps increasing during Hawking evaporation, which by assumption can be accommodated only by increasing $A_- = 4\pi r_-^2$. This requires a substantial inflow of negative energy across r_-. Approaching the Page time, the inner and outer apparent horizons approach each other, conceivably to merge and disappear, releasing all trapped quantum information. I have constructed a very *ad hoc* toy model of this type.[24]

There have been a number of suggestions that spacetime singularities in black hole collapse are naturally avoided in loop quantum gravity (see e.g. Ref. 25). To resolve the black hole information problem and preserve unitarity the geometry cannot have a Cauchy horizon, a feature of most non-singular black hole models (see e.g. Ref. 26). Recently Rovelli and collaborators (Ref. 27) have proposed a model in which the interior of the black hole "tunnels" into the interior of a white hole. The evaporating black hole disappears and the white hole is created in a Planck scale region visible from infinity, without a Cauchy horizon and without any curvature singularities.

A much better understanding of quantum gravity will be necessary to settle the debate on the fate of black holes and the implications for fundamental issues in quantum field theory. It is unfortunate that Jacob is no longer able to contribute.

References

1. J. D. Bekenstein, *Phys. Rev. D* **7**, 2333 (1973).
2. S. W. Hawking, *Phys. Rev. Lett.* **26**, 1344 (1971); S. W. Hawking, in *Black Holes*, eds. B. DeWitt and C. DeWitt (Gordon and Breach, New York, 1973).
3. J. M. Bardeen, B. Carter and S. W. Hawking, *Commun. Math. Phys.* **31**, 161 (1973).
4. S. W. Hawking, *Nature* **248**, 30 (1974); *Commun. Math. Phys.* **43**, 199 (1975).
5. R. M. Wald, *Liv. Rev. Rel.* **4**, 6 (2001) [arXiv:gr-qc/9912119].
6. D. N. Page, *Phys. Rev. D* **13**, 198 (1976).
7. T. Elster, *Phys. Lett. A* **94**, 205 (1983).
8. B. E. Taylor, C. M. Chambers and W. A. Hiscock, *Phys. Rev. D* **58**, 044012 (1998) [arXiv:gr-qc/9801044].
9. D. N. Page, *Phys. Rev. D* **25**. 1499 (1982); K. W. Howard and P. Candelas, *Phys. Rev. Lett.* **53**, 403 (1984); B. P. Jensen and A. C. Ottewill, *Phys. Rev. D* **39**, 1130 (1989); B. P. Jensen, J. G. McLaughlin and A. C. Ottewill, *Phys. Rev. D* **43,** 4142 (1991).

10. A. Levi and A. Ori, *Phys. Rev. Lett.* **117**, 231101 (2016) [arXiv:1608.03806].
11. M. K. Parikh and F. Wilczek, *Phys. Rev. Lett.* **85**, 5042 (2000) [arXiv:hep-th/9907001].
12. W. G. Unruh, *Phys. Rev. D* **15**, 365 (1977).
13. S. A. Fulling, *Phys. Rev. D* **15**, 2411 (1977).
14. J. D. Bekenstein and V. F. Mukhanov, *Phys. Lett. B* **360**, 7 (1995); J. D. Bekenstein, *Phys. Rev. D* **91**, 123052 (2015) [arXiv:1505.03253].
15. W. G. Unruh and R. M. Wald, *Rep. Prog. Phys.* **80**, 092002 (2017) [arXiv:1703.02140].
16. T. Faulkner, R. G. Leigh, O. Parrikar and H. Wang, [arXiv:1605.08072].
17. A. C. Wall, *Phys. Rev. D* **85**, 104049 (2012) [arXiv:1105.3445].
18. A. Balakrishnan, T. Faulkner, Z. U. Khandker and H. Wang, [arXiv:1706.09432].
19. D. N. Page, *JCAP* **1309**. 028 (2013) [arXiv:1301.4995].
20. R. Bousso, Z. Fisher, S. Leichenauer and A. C. Wall, *Phys. Rev. D* **93**, 064044 [arXiv:1506.02669].
21. R. Bousso and M. Porrati, *Class. Quantum Grav.* **34**, 204001 (2017) [arXiv:1706.00436].
22. S. W. Hawking, M. J. Perry and A. Strominger, *Phys. Rev. Lett.* **116**, 231301 (2016) [arXiv:1601.00921].
23. C. Rovelli, [arXiv:1710.00218].
24. J. M. Bardeen, [arXiv:1706.09204].
25. S. Hossenfelder, L. Modesto and I. Premont-Schwarz, *Phys. Rev. D* **81**, 044032 (2010) [arXiv:0912.1823].
26. S. A. Hayward, *Phys. Rev. Lett.* **96**, 031103 (2006) [arXiv:gr-qc/05060126]; T. De Lorenzo, C. Pacilio, C. Rovelli and S. Speziale, *Gen. Rel. Grav.* 47:41 (2015) [arXiv:1412.6015].
27. E. Bianchi, M. Christodoulou, F. D'Ambrosio, C. Rovelli and H. M. Haggard, [arXiv:1802.1802.04264v2].

Remembering Jacob

G. W. Gibbons
D.A.M.T.P., Cambridge University,
Wilberforce Road, Cambridge, CB3 0WA, UK

This is a brief personal account of my memories of the early development of the subject of Black Hole Thermodynamics and of the influence of Jacob Bekenstein on it.

1. Introduction

At the request of the Editors for an article which provided some balance to the majority in this volume which are by those who knew and worked with Jacob directly, this contribution is intended somewhat in the spirit of one of the novels of Lawrence Durrell's tetralogy: *The Alexandrian Quartet* whose purpose was, to quote a paraphrase of the author's introduction "an exploration of relativity and the notions of continuum and subject–object relation, with modern love as the theme". However I shall say nothing here about "modern love". The basic idea of the tetralogy is that each novel recounts the same series of events taking place in Alexandra according to the memories of that time of four participants, *Justine, Balthazar, Mountolive* and *Clea.*

My intention therefore is to convey some idea of how the theory of black hole thermodynamics evolved in the eyes of one Cambridge based observer. It is not intended as a survey of the entire field, still less of the many achievements of Jacob Bekenstein. That will be dealt with by others in this volume. Perhaps by Durrellian observers from other universities in other continents. Consequently my bibliography is restricted to papers relevant to that task and many others of the time relevant to a more global account of that exciting era are not included. Since the dominant figure in Cambridge during the period considered was the late Stephen Hawking much of what I have to say concerns him. Sadly I have not been able to check with him personally that my reports of remarks he made are accurate and so they must be taken as recollections.

2. Recollections

I shall begin before the main action with the problem set me by my first supervisor Dennis Sciama which owed a great deal to an article of Freeman Dyson.[1] The problem arose because Joe Weber had just published claims to be detecting gravitational radiation coming from the galactic centre. The only plausible source was some sort of gravitational collapse involving what had become known as black holes. I knew next to nothing about black holes at that stage and after an initial unsuccessful skirmish, turned to working with Stephen Hawking on the theory of gravitational wave detectors. In the autumn 1970 I visited the U.S. to talk about our ideas, taking in a visit to Princeton. I recall meeting John Wheeler and possibly Demetrios Christodoulou. In December met up with Stephen in Austin for the Fifth Texas Symposium on Relativistic Astrophysics. At the initial cocktail party I vividly recall Remo Ruffini approaching Stephen to tell him about the results of Demetrios[2] and himself[3] on the concept of the *irreducible mass* $M_{\rm irr}$ of a black hole and Stephen claiming that the irreducible mass must be connected to the *area A of the event horizon* which he had just shown[4] to be non-decreasing. Indeed:

$$A = 16\pi M_{\rm irr}^2 \,. \tag{1}$$

The symposium was later beautifully summarised by Peter Hajicek.[5] The three papers[2–4] may be regarded as establishing the foundations of *classical* black hole thermodynamics. They certainly had a profound impact on Jacob, particularly the first two.

The main result of Ref. 4, which has come to be called the Second Law of Classical Black Hole Mechanics, rests on an important assumption which Roger Penrose christened *Cosmic Censorship.*[6] After returning to Cambridge in January 1971 I continued to work on gravitational waves detectors for a while but turned to looking at various possible counter-examples of Cosmic Censorship, some of which I was able to rule out.[7,8] Interest in Black Holes in Cambridge at that time really began to take off with the realisation that Black Holes must surely exist in the real world. I was privileged to attend lectures by Brandon Carter on the basic theory which was in effect a draft of the lectures he was to give later at Les Houches in 1972. Penrose was a frequent visitor and Hawking produced the long version of his Physical Review Letter[9] which would also be the basis for his Les Houches lectures. He and I even indulged in a little search for candidate black holes among spectroscopic binary star systems.[10]

A key feature of black holes, greatly stressed by John Wheeler for some time was their "No Hair" property which Wheeler claimed leads to the *Transcendence of Baryon Number Conservation* This stimulated Jacob's earliest papers on black holes[11–13] submitted on May 1971, Nov 1 1971 and Jan 11 1972 respectively and formed a substantial part of this thesis[14] recommended for acceptance in May 1972.[a]

[a]In fact, years later I was sent anonymously a 1982 facsimile made by University Microfilms International. It was possibly sent at the suggestion of Jacob after our first meeting in 1980.

I remember Hawking telling me that he felt upset that he hadn't been given some credit for this since his paper[9] contained a simple no hair theorem for Brans–Dicke scalars and he seemed to feel that Jacob's paper was essentially a simple generalisation of it. In fact Jacob did give one reference to Stephen in Refs. 11 and 12 as S.W. Hawking unpublished and this might have been to a lecture Stephen had given during a visit he had made to Princeton. In Ref. 13 he thanks Stephen among others for helpful suggestions.

Jacob's papers on the transcendence of baryon number conservation by black holes certainly influenced the future debate on the unitarity of quantum gravity when black holes are taken into account. In retrospect the concentration on *baryon conservation* as a sacred law and the use of no-hair results for meson field looks rather quaint. I suspect few today would subscribe to sacredness of Baryon number conservation. Of course the problem persists for *any* global symmetry and may merely be telling us that quantum gravity and global symmetries are incompatible because of anomalies. However it underscores an important point. Almost all work on black holes was strongly motivated by the conviction that they were very likely part of the physical world about us and were subject to the laws of physics as we knew them at the time. Moreover many believed that they would play a central role in astrophysics and cosmology. Reciprocally the questions asked of the theory were principally addressed to that end.

The intense activity and manifest progress in Black Hole Theory lead Cécile and Bryce DeWitt to organise a Summer School at Les Houches during August 1972.[15] Both Brandon Carter and Stephen Hawking were among the lecturers. But because I had heard all that they had to say and had a Ph.D. thesis to write up and submit, I stayed in Cambridge. Thus I missed out on all the discussions among the participants, which, according to the proceedings, included Jacob Bekenstein. I had to wait for Stephen's return to hear what was new. One thing that was new was Bardeen, Carter and Hawking's codification of the *Four Laws of Black Hole Mechanics*[16] which they had worked on during the school and which is also covered by their lectures in the proceedings.[15] Another piece of news was the discovery of the first genuinely convincing Black Hole candidate: the X-ray source Cygnus X1. I was fortunate enough to be asked to review the proceedings for Nature[17] and my review was enthusiastic. I remain as enthusiastic today. The proceedings laid down the basis for all subsequent work on classical astrophysical black holes. Despite over 50 years of intense activity, both theoretically and observationally, with some spectacular successes, the basic picture laid out in that volume remains valid.

This raised the question: what to do next? Hawking's view was that relativists would either go into astrophysical modelling or look at more fundamental, essentially quantum mechanical, issues. At some stage he mentioned the fact there had been a great deal of discussion, whether the Four Laws we merely analogous to the Four laws of thermodynamics, or whether there was some sort of correspondence between entropy S and horizon area A as had already been suggested by Jacob in a paper which had been submitted on 22 May 1972[18] in which he had proposed that

$$S = \eta k \frac{A}{L_p^2} = \eta k \frac{c^3 A}{G\hbar}, \tag{2}$$

where k as Boltzmann's constant, $L_p = \sqrt{\frac{G\hbar}{c^3}}$ the Planck length and η a dimensionless number of order unity.

The consensus among classical relativists at that time was that η was in effect infinite and hence by the four laws, the temperature T zero. In fact, according to Ref. 18, Bob Geroch at a Colloquium at Princeton in December 1971 had attempted to show by means of a thought experiment involving lowering objects into black holes at the end of ropes as envisaged earlier by Penrose,[19] that the second law of thermodynamics could be transcended if one posited something like (2). My reaction to this at the time was based on my growing distrust of thought experiments which although ubiquitous in undergraduate physics courses, seldom produce a conclusive answer or mathematical framework within which to establish anything reliable. In this spirit, and with the intention of getting into the subject I wrote a brief note arguing that any physical rope would break before reaching the horizon and submitted on the encouragement of Hawking to Nature on 13 September.[20] It was evidently read by Jacob since many years afterwards he found a mistake in my calculation[21]!

Hawking himself was more interested in the low mass primordial black holes whose existence he had suggested might form under the high pressure conditions prevailing in the early universe.[22] A similar suggestion had been made by Zel'dovich and Novikov.[23] The obvious question was how they might be detected? Despite speculations at the time that a black hole may have been the cause of the 1908 Tunguska event[24] it was clear that a mini black hole would pass straight through the earth leaving very little trace, particularly if it were electrically neutral. If however it had an electric or conceivably a magnetic charge it might show up in cosmic ray searches for charged particles or more speculatively searches then underway for magnetic monopoles. Now it soon became clear to us that a black hole of small size carrying just one unit of electric charge would, if described by the Riessner-Nordstrom solution generate an extremely high electric field, far greater than the critical value at which electron-positron pairs would be produced. This would lead to a rapid discharge and render the black hole neutral. Hawking and I began studying this process. It was clear that at the level of the charged Klein–Gordan or Dirac equations this phenomenon was closely related to the then well-known mechanism of super-radiant loss of angular momentum both stimulated and spontaneous[25] and our preliminary conclusions were announced at two I.A.U. Symposia in Poland celebrating the 500th anniversary of the birth of Copernicus.[26] The two symposia were held in Warsaw and in Krakow. Both Alexei Starobinsky and Yakov Zel'dovich were present at both and Starobinsky presented a report in Warsaw on his purely classical results with Zeldovich[27] on super-radiant scattering. Zel'dovich spoke about at the Krakow symposium on particle creation in cosmology.[28] My recollection is that neither Bill Unruh nor Jacob Bekenstein were present at either meetings.

On returning to Cambridge Hawking and I continued our work. I spent most of my time reading the literature including Schwinger's Compendium of selected papers in Quantum Electrodynamics. I vividly recall Stephen's scandalised reaction to the lack of mathematical rigour in Schwinger's derivation of his famous formula. Part of my difficulty was that almost all of the literature of the day focussed on deriving Feynman diagrams. In the situation in question this rather put the cart before the horse because these problems were *external field problems* with no a priori vacuum state around which to expand. However at this time there was growing interest in looking at particle creation in cosmology as pioneered for example by Leonard Parker[31] and by Zel'dovich[32] and his group. In fact Stephen was familiar with this work since his paper[33] had been criticised by Zel'dovich and Pitaevsky[34] for assuming classical energy conditions for the energy momentum tensor.

In any event he set about solving the problem by combining scattering theory in an asymptotically flat background spacetime with Bogoliubov transformations and by December of 1973 he had convinced himself that even neutral, non-rotating black holes of surface gravity κ would emit particles with a thermal spectrum and temperature

$$T = \frac{\kappa \hbar}{2\pi k c} . \tag{3}$$

After a conversation with Dennis Sciama urging him to publish his result as soon as possible he submitted a letter to Nature on 17 January of the following year.[35] Hawking ends by remarking

> Bekenstein[6] suggested on thermodynamic grounds that some multiple of κ should be regarded as the temperature of a black hole. He did not, however suggest that a black hole could emit particles as well as absorb them. For this reason Bardeen, Carter and I suggested that the similarity between κ and temperature was only an analogy. The present result however seems to indicate there may be more to it than this. Of course this calculation ignores the back reaction of the particles on the metric and quantum fluctuations on the metric. This might alter the picture.

Reference 6 was the longer the Physical Review paper that Jacob had written after his initial Nuovo Cimento letter.[36] Stephen then submitted a full version of his calculation to Communications in Mathematical Physics on 12 April 1974.[37] In fact the paper did not appear for over a year after receipt and when it did appear, much to Stephen's annoyance its receipt date was given as 12 April 1975.

After Stephen's paper our original problem was subsumed as a special case but Stephen graciously urged me to write it up in the light of his work. This I did and it appeared as Ref. 38. In fact Jacob had by then noticed the significance of super-radiance[39] and Zaumen of Schwinger's formula.[40] Neither had he given a fully fledged Quantum Field Theory in Curved Spacetime, or what we referred to as QFTICS, treatment. A more detailed order of magnitude analysis of the effects of electric charge were provided by Carter[41] who had recently left Cambridge for Paris. By then my hopes of finding primordial black holes using etched particle track techniques[42] or even those used on lunar samples brought back to earth on the Apollo 11 mission in magnetic monopoles searches[43] were fading rapidly. Later I realised there was a slight hope for black hole dyons.[44]

Stephen spent the academic year 1974–75 at CalTech as a Fairfield Scholar. This year was a very productive one for him and initiated some important new developments. To name just three:

- In June 1975 he submitted a paper entitled "Black Holes and Thermodynamics"[45] where he refined the suggestions of Bekenstein[36,46] by claiming that $\eta = \frac{1}{4}$ and suggested that that entropy of the black hole is "a measure of the information about the initial state which was lost in the formation of the black hole". He also introduced the idea of a black hole in a box in equilibrium with its radiation. By time reversal invariance he made the provocative claim that in such circumstances black holes and white holes are indistinguishable.
- On 25 August he followed this up with a more provocative paper[47] with the still contentious claim that his results implied that evolution in quantum gravity is non-unitary and is implemented by a a linear map on density matrices he called the super-scattering operator.
- While in California Stephen worked with his long-term collaborator Jim Hartle. On 19 November the submitted a paper which opened up yet another new avenue.[48] In that paper they constructed a Feynman propagator in the background of a black hole and used it to derive the Einstein relations between spontaneous and a stimulated emission for a system in equilibrium.

The discovery of *Black Hole Explosions* had greatly raised expectations of possible observational consequences. Stephen working with Don Page had worked out some of these[49] and so had I with Brandon Carter, Doug Lin and Malcolm Perry.[50] These and similar calculations, combined with observations of the cosmic X-ray and gamma-ray backgrounds provided very stringent limits on the present density of primordial black holes of masses around 10^{14}–10^{15} grammes. In fact less than about a millionth of the present density. On the big bang theory Stephen had used to predict the possibility that this implied that the density of pbh's early times was very very much less than the density of the radiation which was believed to dominate the density at early times. In other words the early universe according to the Big Bang theory must have been extraordinarily smooth compared with the turbulent conditions that many believed prevailed then, a point made by Novikov, Polnarev, Starobinsky and Zel'dovich.[51] This gave rise to a puzzle: why was this. In retrospect the obvious plausible explanation was inflation. Indeed this puzzle could well be added to the famous list of Alan Guth.[52]

On his return to Cambridge in the autumn of 1975, Stephen described to us his work with Jim Hartle. The boundary conditions used differed from those previously used in an unsuccessful attempt of Boulware[53] to discuss quantum field theory in the presence of event horizons. The difference, and their result, arose from following the lead of so-called Euclidean Quantum Field Theory and Wick rotating the formal Schwinger type path integral giving the putative Feynman integral to imaginary time.

They key property was its periodicity (at least for bosons) in imaginary time. This allowed contours for amplitudes to be moved from those for emission to those for absorption at the expense of a Boltzmann type factor. A puzzle was that this periodicity extended all the way to infinity. In other words they did not tend at infinity to the Feynman propagators considered by Boulware.

The clue to the question why this all worked struck me while attending a seminar given by Malcolm Perry on his work with John Collins on the implications for cosmology of quark confinement.[54] His talk made reference to what he called finite temperature Green's functions and I was stimulated to ask whether they were periodic in imaginary time. Consultation of the Ref. 55 revealed that quite generally periodicity in imaginary time with period $\beta = \frac{\hbar}{kT}$ is key to understanding thermal equilibrium at the quantum level and works as well for interacting systems as it does for the linear systems which had been used hitherto in the black hole case. In other words the periodicity of the Wick rotation of black holes with period $\frac{2\pi}{\kappa}$ implies almost all of black hole thermodynamics. Malcolm and I quickly wrote up the consequences of our epiphany. The first paper was a Physical Review Letter submitted on 1 December 1975.[56] The detailed follow up[57] took longer: it was submitted by Stephen to to the Royal Society on July 12 in the following year.

In the meantime on 4 March 1976 Stephen and I had submitted a paper on the extension of the Hartle–Hawking Gibbons–Perry theory to De-Sitter spacetime.[58] The point of the paper was largely to illustrate the *universality of the Hawking effect and its thermodynamic interpretation* rather than for any application to cosmology. The viewpoint adopted was that De-Sitter spacetime with its cosmological horizon behaves like an inside out black hole. It was quite clear then that if the cosmological constant were truly constant then its effective energy density would always be negligible compare with the radiation density of a hot big bang universe. Of course the assumption of constancy of the cosmological constant is dropped in inflationary theories and it came as a shock to me when it became clear that the effect we had discovered in that paper is, according to inflationary theory the origin of structure in the universe. Indeed it is to inflation we must look for any sort of observational verification of the theory of (inside out) black hole thermodynamics-dynamics.

I spent the academic year 1976–77 in Munich and just before I left Cambridge we finished a draft of the first paper in which we gave treatment of the subject at the level of *semi-classical but non-perturbative quantum gravity.* The main point was to shift from the idea that scattering amplitudes were the things to calculate to the calculation of partition functions for systems in equilibrium,[59] an idea already anticipated in Ref. 57. In fact the paper was submitted by Stephen after I had left Cambridge on 4 October 1976. At the time Stephen had been toying with the idea of virtual black holes but we knew of no exact asymptotically flat background solutions describing such things.

While in Munich I had realised that there was a potential problem with the non-positivity of the gravitational action[60] which we addressed in a paper with Malcolm.[61]

3. Meeting Jacob

So far in this narrative Jacob has been off stage so to speak. At least I personally had never met him. This changed when I visited Santa Barbara in 1980 in order to attend what I think was the second programme at what was then ITP, now KITP. It was organised by Bryce DeWitt and its subject was quantum gravity. I had many interesting and enjoyable discussions with Jacob during my stay.

He had come to believe from gedanken experiments involving lowering ordinary non-self-gravitating material objects into a black hole that the consistency of the generalised second law[4,36,46] required (in units in which $k = \hbar = c = G = 1$) that

$$\frac{S}{E} < 2\pi R \tag{4}$$

where E is the energy and R some effective radius. We spent a long time looking at examples and came to the conclusion that the inequality was most vulnerable at low energies. I suggested considering looking at the entropy of a quantum field in a cavity. This requires regularization of zero-point energies. We also tried to make a clear link between information and entropy, even to the extent of talking to Bremermann at Berkeley.[b]

Jacob was more optimistic than I that we had achieved much in the way of solid results and he prepared a draft for submission for publication. In the event we came to an amicable agreement that I should to be a co-author and Jacob went ahead and submitted the draft to Physical Review on 7 July 1980.[62]

Although widely cited I believe that Jacob's bound is still controversial. The challenge as always is to make precise what it means and what would constitute a proof and what would constitute a refutation. Ultimately one is up against the ultra-violet divergences of quantum field theory.

After 1980 my interests turned to the roles of black holes in Supergravity and later Superstring Gravity Theory. This is not an area in which Jacob has worked. My feeling is that while supersymmetry has received little support from experiment or observation it does offer the possibility to answer the many quantum mechanical questions that Jacob and others have raised in a controlled setting So far the the results have not encouraging.[63,64]

[b] Jacob's reference to Bremermann almost certainly arose because I possessed a copy of the *Encyclopaedia of Ignorance.*

Another super-link with Jacob's work concerns his very early attempt[65] to apply Bohr–Sommerfeld–Wilson type quantization to a Kerr–Newman black hole. He postulated that

$$
\begin{aligned}
J &= j\hbar\,, & j &= 0, \pm 1 \pm 2, \ldots \\
Q &= ke\,, & k &= j = 0, \pm 1 \pm 2, \ldots \\
M_{\rm irr}^2 &= \frac{1}{2} n\hbar & n &= 1, 2, \ldots
\end{aligned}
\tag{5}
$$

on the grounds that all three left-hand sides of (5) should be adiabatic invariants. He continued to be interested in this topic for many years.[66]

A similar idea has been proposed[67,68] for the persistence of the area product formula which appears first in Ref. 69 for the Kerr–Newman black holes

$$
\frac{A_+}{16\pi}\frac{A_-}{16\pi} = \frac{J^2}{4} + \frac{Q^4}{16} \tag{6}
$$

where A_+ and A_- are the areas of the inner and outer horizons under axi-symmetric deformations.

The presence of both inner and outer horizons is still mysterious and despite being widespread in supergravity solutions[70] its universal significance has been questioned.[71] Whatever the true significance it is yet one more example of the fecundity of Jacob's earliest suggestions.

This more or less ends my account. Apart from a brief meeting at the 32nd Jerusalem Winter School in Theoretical Physics in January 2015, was my only other meeting with Jacob that I recall. I was both shocked and immensely saddened by his death later that year. Like his hero Einstein, he kept asking deep questions and confronting our cherished beliefs up to the end.

In concluding I would like to thank Don Page for helpful discussions.

References

1. F. J. Dyson, "The efficiency of energy release in gravitational collapse", *Comments on Astrophysics and Space Physics* **1**, 75–80 (1969).
2. D. Christodoulou, "Reversible and irreversible transformations in black hole physics", *Phys. Rev. Lett.* **25**, 1596 (1970), doi:10.1103/PhysRevLett.25.1596.
3. D. Christodoulou and R. Ruffini, "Reversible transformations of a charged black hole", *Phys. Rev. D* **4**, 3552 (1971), doi:10.1103/PhysRevD.4.3552.
4. S. W. Hawking, "Gravitational radiation from colliding black holes", *Phys. Rev. Lett.* **26**, 1344 (1971), doi:10.1103/PhysRevLett.26.1344.
5. P. Hajicek, "Report on the Fifth Texas Symposium on Relativistic Astrophysics", *General Relativity and Gravitation* **2**, 173–181 (1971).
6. R. Penrose, "Naked singularities", in *Sixth Texas Symposium on Relativistic Astrophysics*, ed. D. J. Hegyi, *Annals of the New York Academy of Sciences*, **224**, 125 (1973).
7. G. W. Gibbons, "The time symmetric initial value problem for black holes", *Commun. Math. Phys.* **27**, 87 (1972), doi:10.1007/BF01645614.

8. G. W. Gibbons, "Collapsing shells and the isoperimetric inequality for black holes", *Class. Quant. Grav.* **14**, 2905 (1997), doi:10.1088/0264-9381/14/10/016 [hep-th/9701049].
9. S. W. Hawking, "Black holes in general relativity", *Commun. Math. Phys.* **25**, 152 (1972), doi:10.1007/BF01877517.
10. S. W. Hawking and G. W. Gibbons, "Evidence for black holes in binary star systems", *Nature* **232**, 465 (1971), doi:10.1038/232465a0.
11. J. D. Bekenstein, "Nonexistence of baryon number for static black holes", *Phys. Rev. D* **5**, 1239 (1972), doi:10.1103/PhysRevD.5.1239.
12. J. D. Bekenstein, "Transcendence of the law of baryon-number conservation in black hole physics", *Phys. Rev. Lett.* **28**, 452 (1972).
13. J. D. Bekenstein, "Nonexistence of baryon number for black holes. ii", *Phys. Rev. D* **5**, 2403 (1972), doi:10.1103/PhysRevD.5.2403.
14. David Jacob Bekenstein, "Baryon Number, Entropy and Black Hole Physics", Dissertation presented to the Faculty of Princeton University in candidacy for the degree of Doctor of Philosophy.
15. C. DeWitt and B. S. DeWitt (eds.), *Black Holes, Les Astres Occlus* (Gordan and Breach, 1973).
16. J. M. Bardeen, B. Carter and S. W. Hawking, "The four laws of black hole mechanics", *Commun. Math. Phys.* **31**, 161 (1973), doi:10.1007/BF01645742.
17. G. W. Gibbons "Serious black holes", *Nature* **247**, 494 (1974).
18. J. D. Bekenstein, "Black holes and the second law", *Lett. Nuovo Cim.* **4**, 737 (1972), doi:10.1007/BF02757029.
19. R. Penrose, "Gravitational collapse: The role of general relativity", *Riv. Nuovo Cim.* **1**, 252 (1969), [Gen. Rel. Grav. **34** (2002) 1141].
20. G. W. Gibbons "On lowering a rope into a black hole", *Nature Physical Science* **240**, 77 (1972).
21. I. Fouxon, G. Betschart and J. D. Bekenstein, "The bound on viscosity and the generalized second law of thermodynamics", *Phys. Rev. D* **77**, 024016 (2008), doi:10.1103/PhysRevD.77.024016 [arXiv:0710.1429 [gr-qc]].
22. S. W. Hawking, "Gravitationally collapsed objects of very low mass", *Mon. Not. Roy. Astron. Soc.* **152**, 75 (1971).
23. Ya B. Zel'dovich and I. D. Novikov, "The hypothesis of cores retarded during expansion and the hot cosmological mode", *Astronomicheskii Zhurnal* **43**, 758 (1966).
24. A. A. Jackson and M. P. Ryan, "Was the Tungus event due to a black hole?", *Nature* **245**, 88–89 (1973).
25. Ya, B. Zel'dovich, *ZhETP Pis, Red.* **14**, 270–272 (1971); English Translation "Generation of Waves by a Rotating Body", *JETP Letts.* **14**, 180–181 (1971).
26. S. W. Hawking and G. W. Gibbons, "Quantum Aspects of Accretion onto Black Holes in the Early Universe", in *Gravitational Radiation and Gravitational Collapse; Proceedings of the Symposium*, Warsaw, Poland, September 5-8, 1973, ed. C. Dewitt-Morette (D. Reidel Publishing Co., 1974) (IAU Symposium, No. 64), p. 185.
27. A. A. Starobinksy, "Amplification of waves reflected from Kerr black holes", in *Gravitational Radiation and Gravitational Collapse; Proceedings of the Symposium*, Warsaw, Poland, September 5-8, 1973, ed. C. Dewitt-Morette, Symposium sponsored by the International Astronomical Union Dordrecht (D. Reidel Publishing Co., 1974) (IAU Symposium, No. 64), p. 94.
28. Ia. B. Zeldovich, "Creation of particles in cosmlogy", in *Confrontation of Cosmological Theories with Observational Data; Proceedings of the Symposium*, Krakow, Poland, September 10-12, 1973, ed. M. S. Longair (D. Reidel Publishing Co., 1974), p. 329–333; Discussion, p. 333, 00/1974.

29. J. S. Schwinger ed. *Selected Papers in Quantum Electrodynamics* (Dover, 1958).
30. J. S. Schwinger, "On gauge invariance and vacuum polarization", *Phys. Rev.* **82**, 664 (1951), doi:10.1103/PhysRev.82.664.
31. L. Parker, "Particle creation in expanding universes", *Phys. Rev. Lett.* **21**, (1968) 562, doi:10.1103/PhysRevLett.21.562.
32. Y. B. Zeldovich, "Particle production in cosmology", *Pisma Zh. Eksp. Teor. Fiz.* **12**, 44 (1970).
33. S. W. Hawking, "The conservation of matter in general relativity", *Commun. Math. Phys.* **18**, 301 (1970), doi:10.1007/BF01649448.
34. Ya. B. Zeldovich and L. P. Pitaevsky "On the possibility of the creation of particles by a classical gravitational field", *Comm. Math. Phys.* **23**, 185 (1971).
35. S. W. Hawking, "Black hole explosions", *Nature* **248**, 30 (1974), doi:10.1038/248030a0.
36. J. D. Bekenstein, "Generalized second law of thermodynamics in black hole physics", *Phys. Rev. D* **9**, 3292 (1974), doi:10.1103/PhysRevD.9.3292.
37. S. W. Hawking, "Particle creation by black holes", *Commun. Math. Phys.* **43**, 199 (1975) [Erratum: *ibid.* **46** (1976) 206], doi:10.1007/BF02345020, 10.1007/BF01608497.
38. G. W. Gibbons, "Vacuum polarization and the spontaneous loss of charge by black holes", *Commun. Math. Phys.* **44**, 245 (1975).
39. J. D. Bekenstein, "Extraction of energy and charge from a black hole", *Phys. Rev. D* **7**, 949 (1973), doi:10.1103/PhysRevD.7.949.
40. W. T. Zaumen "Upper bound on the electric charge of a black hole", *Nature* **247**, 530–531 (1974).
41. B. Carter, "Charge and particle conservation in black hole decay", *Phys. Rev. Lett.* **33**, 558 (1974), doi:10.1103/PhysRevLett.33.558.
42. R. L. Fleischer, H. W. Alter, S. C. Furman, P. B. Price and R. M. Walker, "Particle track etching", *Science* **178**, 255 (1972), doi:10.1126/science.178.4058.255.
43. L. W. Alvarez, P. H. Eberhard, R. R. Ross and R. D. Watt, "Search for magnetic monopoles in the lunar sample", *Science* **167**, 701 (1970), doi:10.1126/science.167.3918.701.
44. G. W. Gibbons, "Black hole dyons need not explode", *Phys. Rev. D* **15**, 3530 (1977), doi:10.1103/PhysRevD.15.3530.
45. S. W. Hawking, "Black holes and thermodynamics", *Phys. Rev. D* **13**, 191 (1976), doi:10.1103/PhysRevD.13.191.
46. J. D. Bekenstein, "Black holes and entropy", *Phys. Rev. D* **7**, 2333 (1973), doi:10.1103/PhysRevD.7.2333.
47. S. W. Hawking, "Breakdown of predictability in gravitational collapse", *Phys. Rev. D* **14**, 2460 (1976), doi:10.1103/PhysRevD.14.2460.
48. J. B. Hartle and S. W. Hawking, "Path integral derivation of black hole radiance", *Phys. Rev. D* **13**, 2188 (1976), doi:10.1103/PhysRevD.13.2188.
49. D. N. Page and S. W. Hawking, "Gamma rays from primordial black holes", *Astrophys. J.* **206**, 1 (1976), doi:10.1086/154350.
50. B. Carter, G. W. Gibbons, D. N. C. Lin and M. J. Perry, "Black hole emission process in the high energy limit", *Astron. Astrophys.* **52**, 427 (1976).
51. I. D. Novikov, A. G. Polnarev, A. A. Starobinsky and Ta. B. Zeldovich, "Primordial black holes", *Astron. Astrophys.* **80**, 104–109 (1979).
52. A. H. Guth, "The inflationary universe: A possible solution to the horizon and flatness problems", *Phys. Rev. D* **23**, 347 (1981), doi:10.1103/PhysRevD.23.347.
53. D. G. Boulware, "Quantum field theory in Schwarzschild and rindler spaces", *Phys. Rev. D* **11**, 1404 (1975), doi:10.1103/PhysRevD.11.1404.

54. J. C. Collins and M. J. Perry, "Superdense matter: Neutrons or asymptotically free quarks?", *Phys. Rev. Lett.* **34**, 1353 (1975), doi:10.1103/PhysRevLett.34.1353.
55. A. L. Fetter and J. D. Walecka, *Quantum Theory of Many Particle Systems* (McGraw-Hill, 1971).
56. G. W. Gibbons and M. J. Perry, "Black holes in thermal equilibrium", *Phys. Rev. Lett.* **36**, 985 (1976), doi:10.1103/PhysRevLett.36.985.
57. G. W. Gibbons and M. J. Perry, "Black holes and thermal green's functions", *Proc. Roy. Soc. Lond. A* **358**, 467 (1978), doi:10.1098/rspa.1978.0022.
58. G. W. Gibbons and S. W. Hawking, "Cosmological event horizons, thermodynamics, and particle creation", *Phys. Rev. D* **15**, 2738 (1977), doi:10.1103/PhysRevD.15.2738.
59. G. W. Gibbons and S. W. Hawking, "Action integrals and partition functions in quantum gravity", *Phys. Rev. D* **15**, 2752 (1977), doi:10.1103/PhysRevD.15.2752.
60. G. W. Gibbons, "The Einstein action of Riemannian metrics and its relation to quantum gravity and thermodynamics", *Phys. Lett. A* **61**, 3 (1977), doi:10.1016/0375-9601(77)90244-4.
61. G. W. Gibbons, S. W. Hawking and M. J. Perry, "Path integrals and the indefiniteness of the gravitational action", *Nucl. Phys. B* **138**, 141 (1978), doi:10.1016/0550-3213(78)90161-X.
62. J. D. Bekenstein, "A universal upper bound on the entropy to energy ratio for bounded systems", *Phys. Rev. D* **23**, 287 (1981), doi:10.1103/PhysRevD.23.287.
63. G. W. Gibbons, M. J. Perry and C. N. Pope, "Bulk/boundary thermodynamic equivalence, and the Bekenstein and cosmic-censorship bounds for rotating charged AdS black holes", *Phys. Rev. D* **72**, 084028 (2005), doi:10.1103/PhysRevD.72.084028 [hep-th/0506233].
64. G. W. Gibbons, M. J. Perry and C. N. Pope, "Partition functions, the Bekenstein bound and temperature inversion in anti-de Sitter space and its conformal boundary", *Phys. Rev. D* **74**, 084009 (2006), doi:10.1103/PhysRevD.74.084009 [hep-th/0606186].
65. J. D. Bekenstein, "The quantum mass spectrum of the Kerr black hole", *Lett. Nuovo Cim.* **11**, 467 (1974), doi:10.1007/BF02762768.
66. J. D. Bekenstein and V. F. Mukhanov, "Spectroscopy of the quantum black hole", *Phys. Lett. B* **360**, 7 (1995), doi:10.1016/0370-2693(95)01148-J [gr-qc/9505012].
67. M. Cvetic, G. W. Gibbons and C. N. Pope, "Universal area product formulae for rotating and charged black holes in four and higher dimensions", *Phys. Rev. Lett.* **106**, 121301 (2011), doi:10.1103/PhysRevLett.106.121301 [arXiv:1011.0008 [hep-th]].
68. D. N. Page and A. A. Shoom, "Universal area product for black holes: A heuristic argument", *Phys. Rev. D* **92**(4), 044039 (2015), doi:10.1103/PhysRevD.92.044039 [arXiv:1504.05581 [hep-th]].
69. A. Curir and M. Francaviglia, "Isoareal transformations of the Kerr–Newman black holes", *Acta Phys. Polonica B* **9**, 3–10 (1978).
70. M. Cvetic, H. Lu and C. N. Pope, "Entropy-product rules for charged rotating black holes", *Phys. Rev. D* **88**, 044046 (2013), doi:10.1103/PhysRevD.88.044046 [arXiv:1306.4522 [hep-th]].
71. M. Visser, "Area products for stationary black hole horizons", *Phys. Rev. D* **88**(4), 044014 (2013), doi:10.1103/PhysRevD.88.044014 [arXiv:1205.6814 [hep-th]].

The Black Hole Entropy

Having discovered equations describing an aspect of nature one is inclined to trust all the regular solutions of these equations and devise experiments that could further validate the theory from which these equations were derived. The situation is not quite the same when the solutions following from the equations have singularities. Such solutions include an electron spiraling down on the nucleus according to the rules of Classical Electrodynamics, a universe expanding or collapsing according to the equations of classical General Relativity, and a black hole derived from the same equations.

As there are, as of now, no observed physical singularities in nature, the first response to the identification of a singular solution is to search for reasons to invalidate it or if needed the equations themselves. Quantum Mechanics served to remove the solution leading to the collapse of the hydrogen atom. However, observations do support an expanding universe with its potential big bang singularity, and indirect and direct evidence for the reality of black holes, containing a singularity or not, has been piling up from many sources.

As far as general relativity is concerned one does not set aside the singular solutions but rather one studies their consequences. A dramatic consequence was unraveled by Bekenstein, namely that one can associate thermodynamical properties to black holes. Bekenstein pointed out that a finite entropy could be associated with black holes and Hawking demonstrated that the same goes for the temperature. It was quantum mechanics that rendered these two quantities finite. In addition, Bekenstein was led to suggest that the entropy of the black hole was directly proportional to the area of its horizon, not its volume. It does not have the extensive property of low energy thermodynamics entropy. It is termed Holographic.

All these facts have thickened the plot. Given its thermodynamical features, should one consider a black hole as a fundamental object in some theory or is it actually/merely an effective description of many microstates in that theory? Does its thermal nature threaten the unitarity of a quantum theory of gravity? Does quantum mechanics come again to the rescue and heal the singularities? Can a physical system contain unresolved singularities as long as they are surrounded by horizons, this in a framework of a quantum theory of gravity? Or perhaps the theory of gravity is itself some effective description while the fundamental underlying structure is different.

These questions have surfaced to a large extent thanks to Bekenstein's work and in this chapter leading scientists working in the field offer their answers to these questions. Often their scientific elaboration is sprinkled with personal memories of their encounters with Bekenstein.

Valeri P. Frolov reviews the unfolding of the deep series of puzzles following from Bekenstein's discovery and suggests solutions to some.

Shahar Hod describes his struggles with the idea of the quantization of the horizon of a black hole and the insights and results he has gained from them.

Gerard 't Hooft suggests a couple of modifications to the laws of physics that enable one to expose and clearly define the microstates which constitute the black hole and its entropy. He evaluates how this impacts on the absence of a black hole information paradox. He also argues away the threat that singularities in the classical black hole configuration may pose.

Ted Jacobson critically discusses the evolution of the concepts of entropy and information from Carnot to Bekenstein with emphasis on the role of the thermodynamical way of approaching nature.

Juan Maldacena considers the developments in the understanding of the origins of the entropy of black holes. This is in the framework of string theory which enables one to count precisely the microstates constituting the entropy of special types of black holes and allows one a more detailed analysis when gravity/gauge duality is present. He also touches upon the Bekenstein bound which is the subject of a different chapter.

Viatcheslav Mukhanov takes the reader through the research track he followed together with Bekenstein in studying the idea of the quantization of the area of the black hole horizon and its consequences.

Andrew Strominger delves into the detailed mechanics which, with the help of a large number of conservation laws, could perhaps allow one to mine the complete pure state information from the radiation emanating out of a black hole.

Remarks on Black Hole Entropy

Valeri P. Frolov
Theoretical Physics Institute, University of Alberta,
Edmonton, Alberta, Canada T6G 2E1
vfrolov@ualberta.ca

The notion of black hole entropy was introduced by Jacob Bekenstein in 1972. During past 45 years this subject was in the center of interests of the modern theoretical physics. In this paper we briefly discuss "puzzles" of the black hole physics, connected with their entropy. We also demonstrate that when the standard energy conditions are violated entropy associated with the event horizon can have quite unexpected behavior.

1. Instead of Introduction

Jacob Bekenstein became famous after he published in 1972 three papers on black hole entropy and no-hair theorem.[1–3] He was only 25 years old at that time. His work was very close to my scientific interests and I followed all his publications. But for the first time I met Jacob only in the middle of 90th at the conference in Moscow. By that time he was already a world-wide recognized expert in black holes. What surprised me at our first meeting, was that this famous scientist turned out to be such a shy person, very friendly and widely open for discussions.

After this I have met Jacob at many conferences and had a lot of discussions with him. I remember that at one of the Quantum gravity meeting at Utrecht we were sitting together with Jacob. After my talk he mentioned that it was interesting but I spoke too fast. His talk was after mine. At the end of it because of the shortage of time he was speaking even faster than me. We laughed together when I mentioned this to him.

On many occasions I tried to convince Jacob, that black hole quantization, which he and Slava Mukhanov have advocated, not necessary means that the spectrum of black hole radiation should be discrete. My argument was that in their model the discrete levels have huge degeneracy and the interaction between these internal degrees of freedom can make the spectrum of radiation practically continuous. Jacob did not agree with me and proposed arguments, based on the gedanken experiments, supporting his point of view. Jacob could be a bit stubborn.

In 2016 there was a celebration of a hundred anniversary of black holes. We with Don Page were organizing a meeting "Black Holes' New Horizons" devoted to this event, which was held in May 2016 in Oaxaca (Mexico). We wanted very much to see Jacob at our meeting and sent him in the beginning of March of 2015 our invitation letter. A few days later (on March 6) I got the following answer from Jacob:

"Dear Valeri,

So kind of you to invite me to the workshop in Oaxaca. I would have liked to go, but I am, of late, skipping meetings which involve very long travel. And this would be a classic example of such. I imagine I will be missing a lot of fun; such is life. But I wish you and Don a successful organizing of what must be a very exotic meeting. I hope you enjoy it. My very best wishes,

Jacob"

Jacob Bekenstein passed away on August 16, 2015. To honor his memory we dedicated to him a special session at our Black Hole meeting. Bill Unruh and Barack Kol, who knew Jacob very well, spoke about their memories dedicated to Jacob's life.

2. Black Hole Entropy Puzzles

2.1. *Bekenstein-Hawking entropy of astrophysical black holes is huge*

In his groundbreaking papers[1,3] Jacob Bekenstein claimed that a black hole should have entropy proportional to its surface area. He arrived at this conclusion by considering gedanken experiments with black holes. His conclusion was that this entropy is $S_H = \beta A/l_{Pl}^2$, where the dimensionless coefficient β is of order of 1. Hawking calculations of the quantum radiation of black holes[4,5] demonstrated that it is thermal and the temperature of a black hole of mass M is

$$T_{BH} = \frac{\hbar c^3}{8\pi G M k_B}\,, \tag{1}$$

where k_B is the Boltzmann constant. This result allowed to fix the parameter β in the Bekenstein's expression for the entropy, $\beta = 1/4$. The exact expression for what is called now Bekenstein-Hawking entropy is[a]

$$S_{BH} = \frac{A}{4 l_{Pl}^2}\,. \tag{2}$$

Already in his first papers, Bekenstein estimated the black hole entropy for stellar mass black holes, and found that it is huge. In Ref. 3 he wrote:

---

[a]The standard subscript BH used in the expression has double meaning. It is used as an abbreviation for the "Black Hole". It can also be understood as the abbreviation for "Bekenstein-Hawking".

> entropy of a black hole is enormous. For example, a black hole of one solar mass would have $S_{bh} \approx 10^{60}$ erg °K^{-1}. By comparison the entropy of the sun is $S \approx 10^{42}$ erg °K^{-1}; those of a white dwarf or a neutron star of one solar mass even smaller. The large numerical value of black-hole entropy serves to dramatize the highly irreversible character of the process of black-hole formation.

Now we know that besides stellar mass black holes there also exist supermassive black holes with the mass $M \sim 10^6 - 10^{10} M_{\odot}$. The corresponding dimensionless entropy of these objects is $S_{BH} \sim 10^{89} - 10^{97}$.

The papers[6,7] contain estimations of the contribution of matter and black holes to the entropy of the Universe. The table below taken from the paper[6] summarizes its results.

Table 1. Entropies and energies. The table is taken from Ref. 6.

Objects	Entropy	Energy
10^{22} stars	10^{79}	$\Omega_{\text{stars}} \sim 10^{-3}$
Relic neutrinos	10^{88}	$\Omega_{\nu} \sim 10^{-5}$
Stellar heated dust	10^{86}	$\Omega_{\text{dust}} \sim 10^{-3}$
CMB photons	10^{88}	$\Omega_{\text{CMB}} \sim 10^{-5}$
Relic gravitons	10^{86}	$\Omega_{\text{grav}} \sim 10^{-6}$
Stellar BHs	10^{97}	$\Omega_{\text{SBH}} \sim 10^{-5}$
Single supermassive BH	10^{91}	$10^7 M_{\odot}$
$10^{11} \times 10^7 M_{\odot}$SMBH	10^{102}	$\Omega_{\text{SMBH}} \sim 10^{-5}$
Holographic upper bound	10^{123}	$\Omega = 1$

The main conclusion is that up to best of our present knowledge most of the entropy of our Universe is in the form of the black hole entropy. The following example also illustrates a peculiar property of black holes as thermodynamical systems. Let us assume that one kilogram of matter falls down into the black hole of mass $10^{10} M_{\odot}$. The change of the entropy in this process is as large as $\Delta S_{BH} \sim 10^{77}$ and it is comparable with the thermal entropy of a single star.

The concept of black hole entropy, introduced by Jacob Bekenstein, is now widely used. Google search shows that more than a half of million documents uses this notion. There are several puzzles of the black hole physics which are connected with this notion. Below we briefly discuss some of them.

One can illustrate the key problem connected with the notion of the black hole entropy as follows. Suppose one has a sphere of radius R. Its surface area is $A = 4\pi R^2$. Let us consider triangulation of the sphere and assume the area of a single triangle is $s = l^2$ (see Fig. 1). The total number of the triangles is $N = A/s$. One can use such a sphere to encode information, for example, by coloring triangles by black and white colors. Total number of possible combinations is 2^N, and the amount of information that one can transfer by using such a code is $\sim A/l^2$. The

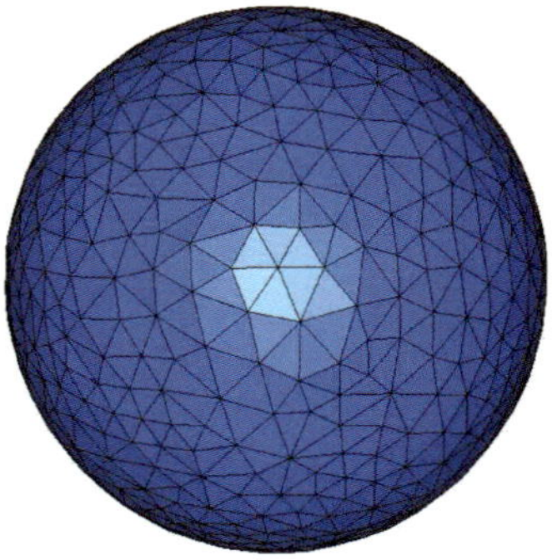

Fig. 1. Sphere triangulation.

entropy corresponding to the loss of this information is $S \sim A/l^2$. This simple observation shows that in order to obtain S_{BH} the size of elementary cells must be of order of the Planckian length. In other words, a plausible explanation of the statistical-mechanical origin of the black hole entropy should appeal to quantum gravity.

2.2. *Black hole entropy and entanglement*

In the interaction with surrounding matter a black hole behaves as a heated body with temperature T_{BH} and entropy S_{BH}. Black holes obey laws, similar to the laws of the standard thermodynamics[8] (for a comprehensive discussion see e.g. reviews[9,10]). A natural question which attracted a lot of attention is: Is the analogy of black holes with thermodynamical systems complete? Do there exist internal degrees of freedom responsible for the black hole entropy? One of the first attempts to answer this question was proposed by Bombelli, Koul, Lee and Sorkin (BKLS).[11] They propose to relate the black hole entropy to the entanglement entropy of zero point fluctuations. A similar idea was proposed by 't Hooft[12] in his famous brick wall model. The BKLS idea was rediscovered in Refs. 13 and 14, where the entanglement entropy of a free quantum field in the spacetime of a black hole was calculated. There were many interesting and important publications on this subject (see e.g. Refs. 15–18). A nice review, which contains a detailed analysis of these and other publications and describes the "state of art" of the black hole entropy problem by 1994 can be found in the paper by Jacob Bekenstein.[19]

The origin of such entanglement entropy can be understood as follows. The interior of the black hole is bounded by the event horizon. For an external observer a regular at the horizon (vacuum) state of a quantum field can be described in terms of Rindler particles (RP). Each of these RP has a partner particle (PP) located inside the horizon and which is strongly correlated with it. A superposition of RP and PP is a pure quantum state. Averaging over unobservable by an external observer PP states results in the density matrix $\hat{\rho}_{RP}$ for RP states. It describes the system entanglement. The entanglement entropy is defined as $S = -\mathrm{Tr}_{RP}(\hat{\rho}_{RP} \ln \hat{\rho}_{RP})$.

Some of the Rindler particles penetrate the potential barrier, surrounding the black hole. They form the Hawking radiation. However, the vast majority of Rindler particles are reflected by the potential barrier and fall down inside the black hole. At the same time new RPs are injected by the black hole. As the result, the distribution of the RP outside of the horizon is stationary. The RPs during their (short) lifetime outside the black hole contribute to the entanglement entropy. The corresponding density matrix $\hat{\rho}_{RP}$ is thermal. This property is closely related with the thermality of the Hawking radiation. Since the vast majority of the Rindler quanta that contribute to the entropy are located in the very close vicinity of the horizon, this entropy is proportional to the surface area of the black hole.

The problem of this and similar approaches to the black hole entropy is that this quantity is divergent. The divergence is a consequence of the assumption that the black hole surface is infinitely sharp. Certainly, this is an idealization. In fact, as a result of emission of the Hawking quanta, the horizon fluctuates. This effect of the horizon fluctuation was described in the remarkable paper by Jacob Bekenstein.[20] In terms of proper distance, the scale of such fluctuations is of the order of the Planck length, l_{Pl}. By using this cut-off one obtains the finite expression for the black hole entanglement entropy

$$S_{ent} = \beta S_{BH} \,, \tag{3}$$

where the dimensionless parameter β is of order of 1.[14]

There were several, more formal proposals, for making S_{ent} finite. One of the most popular attempts was some analogue of the renormalization procedure. It was proposed to prescribe to a black hole (infinite) negative, so-called "geometric", entropy, in such a way that after adding it to the (infinite) positive entanglement entropy one obtains a finite answer, which is identified with S_{BH}. The weak point of this approach is that it does not allow one to explain black hole entropy by statistical-mechanical counting of some black hole degrees of freedom. It introduces the notion of "geometric entropy", which does not have any statistic-mechanical meaning.

2.3. *Universality problem*

The above described approach, in which the entropy of the black hole is related to existing physical fields, has a fundamental problem. Each of the fields gives independent (additive) contribution to S_{ent}. Hence, even if the proper physical mechanism of the cut-off is found, the factor β, which enters (3), depends on the number and characteristics of the fields that contribute to S_{ent}. In particular, it depends on the spectrum of mass and properties of fundamental elementary particles, even of those that are not yet discovered.

This problem is closely related to another, even more fundamental problem: The universality of the black hole entropy S_{BH}. Consider a black hole with mass much larger than the Planckian mass. Such an object is classical and one can neglect possible quantum corrections. In other words, in order to find the Bekenstein-Hawking entropy of the black hole it is sufficient to use the corresponding solution of the classical Einstein equations. At the same time, one is going to obtain the same answer by counting some microscopic degrees of freedom of some version of quantum gravity. In order to obtain the required value of S_{BH} the corresponding constituents must be heavy, with mass comparable with the Planckian one. The calculations based on the fundamental theory should correctly reproduce the result of the low energy classical gravity. In other words, there must exist a mechanism, that guarantees that the calculations based on counting of states in the fundamental theory always give the correct answer, S_{BH}, which is known from the low energy gravitational theory. This is a so called universality problem.[21–25]

There are several approaches to counting the black-hole degrees of freedom and calculation of its entropy based on the some "fundamental" theory, such as string theory, loop gravity, induced gravity, etc. A discussion and comparison of these approaches can be found in comprehensive reviews.[26,27] Common feature of these approaches is that the low energy gravity is an emergent phenomenon generated by fundamental quantum heavy constituents (strings, loops, etc.). In such a case the black hole entropy is obtained by counting the number of states of such constituents.

2.4. *Information loss paradox*

Another black hole puzzle is a so-called information loss paradox. It can be formulated as follows. Suppose a black hole was formed as a result of the collapse of matter, which originally was in pure quantum state. For example, it was a coherent state of collapsing photons. After its formation black hole emits Hawking radiation and its mass decreases. If as result of this process the black hole disappears, the final state of the system would be radiation emitted by the black hole. In other words, a pure state is transformed into a state described by the density matrix with non-zero entropy. This implies the violation of unitarity of quantum mechanics.[28] Certainly, there is an option that the evaporation is not complete and some remnant would remain. If the initial mass of the black hole was large, this remnant must contain partners of all the particles emitted outside during Hawking evaporation. Since the initial state was pure, the entropy of these partners is the same as the total entropy of the emitted outside particles. In the classical description, this entropy is finally absorbed by the singularity inside the black hole. If in an UV complete version of Einstein gravity one or more new universes are created inside the black hole,[29,30] they "inherit" the corresponding entropy.

The information loss paradox arises if for some reason this "scenario" is considered as unsatisfactory, and one insists that the evaporation of the black hole is complete and no remnant is left. This point of view is adopted in the string theory. There exist large number of publications devoted to the unitarity problem in the black hole evaporation (see e.g. reviews[31–38]). For example, one can assume that

(1) the information "comes out" with the Hawking radiation,[32] or (2) the information is "stored" in soft-modes near the horizon,[39,40] or (3) the information "comes out" at the final state of the evaporation (see e.g. Refs. 37, 41 and 42 and references therein).

There is another interesting question which is directly connected with the information loss paradox. Let us consider two black holes, which at some moment of time T have same mass M. Let us assume that they were formed as a collapse of matter in the pure state. Let us also assume that one of these black hole was just formed, while the other one was formed long time ago, at time $T_0 < T$. Its initial mass was $M_0 > M$, and it lost the mass $\Delta M = M_0 - M$ as a result of the Hawking evaporation. Before the mass of the second black hole becomes M it emits radiation which has entropy $\sim \Delta S \sim M_0^2 - M^2$. For example, if $M_0 = \sqrt{2}M$, the emitted entropy ΔS is of the order of $S_{BH}(M)$. The time during which the surface area of the event horizon of an evaporating black hole is halved is often called "Page time".[43]

Evaporation of particle in the black hole exterior is accompanied by creation of its partner inside the black hole. By time T the entropy of such partners is the same, as the entropy of the emitted particles, ΔS. Thus the "internal states" of these two black hole, "young" and "old", with the same mass M at time T is different. If this information returns to the external observer during the last stage of their complete evaporation, this phase duration should be different for two black holes. A result reported by Gregory Vilkovisky some time ago[44–46] might be interesting in connection with this problem.

The number of papers on this subject is fast growing. This means that at the moment we do not have a solution of this puzzle, at least the one which satisfies most the gravity-string community.

2.5. *Where black hole entropy is located?*

In the discussion of black holes many different definitions of their horizon were proposed, such as: event horizon, apparent horizon, isolated horizon, etc. Most of the statistical-mechanical calculations of the black hole entropy was done under an assumption that the black hole is either static or stationary. In this case different definitions of black hole horizon agree. And it is natural to associate the black hole entropy with the event horizon. However, if the black hole is not stationary and its mass and shape depend on time the question where black hole entropy is located becomes non-trivial. Difference between different definitions of the horizon becomes very important, when the weak energy condition is violated. Such a violation is inevitable when quantum effects connected with the Hawking radiation are taken into account. For example, there may exist a closed apparent horizon, while there is no event horizon.[47] In the rest of the paper we discuss possible "paradoxical" behavior of the event horizon in the case when the energy conditions are violated.

3. Wormhole and Entropy of Rindler Horizon

If there exist matter violating the energy conditions, the Einstein equations allow solutions which describe traversable wormholes (a detailed discussion of this subject can be found in a nice book[48]). Such wormholes can be used as a device for "study" of black hole interiors.[49] When the mass and the size of mouths of the wormhole are small and the mass of the (non-rotating) black hole is large this effect allows a simple analytical description. We shall use an approach proposed by Emparan and collaborators for study of the event horizon for a small mass black hole falling into the very large black hole.[50–52] Namely, we approximate the event horizon of the black hole of large mass M by the Rindler horizon. This is a plane null surface in a flat spacetime. In the presence of a static wormhole, this plane first passes one of the mouths, and some time later it passes the other mouth. We are going to describe what happens in such a process with the event horizon. Here and later we consider a case when the number of spacetime dimensions is 4 or higher. This does not make the problem more complicated, but it is useful for a general discussion.

Let us consider $(D = n + 2)$-dimensional flat spacetime and denote by

$$X^a = (T, X, \vec{Z}), \qquad \vec{Z} = Z_1, \dots, Z_n\,, \tag{4}$$

Cartesian coordinates in it. Denote by $\gamma_\pm$ two straight lines defined by the equations

$$\gamma_\pm : X = \pm\frac{1}{2}L, \qquad Z_i = 0\,. \tag{5}$$

We assume that these two lines represent positions of the two spherical mouths of a traversable wormhole. We assume that the radii of the mouths are equal and very small, and that the time at the mouths obeys the relation $T_+ = T_-$. In such a model, any null ray that enters into γ_--mouth at time T comes out γ_+-mouth at the same moment of time.

We identify the unmodified Rindler horizon with a null plane Π: $T = X - L/2$. It crosses γ_+ at $T = 0$. We also denote by N the past null cone with the vertex at γ_--mouth at the moment of time $T = 0$ (see Fig. 1).

In our problem there exists only one dimensional parameter L which gives the length scale. It is convenient to use dimensionless coordinates $x^a = X^a/L = (t, x, \vec{z})$. We also denote $\xi = x + 1/2$. In these coordinates $(t, \xi, \vec{z})$ the equations of the past null cone N and the undeformed horizon Π are

$$\xi^2 + \rho^2 = t^2, \qquad t = \xi - 1\,, \tag{6}$$

where $\rho = \sqrt{\vec{z}^2}$. This null cone intersects the horizon Π at n-dimensional surface Γ, which is a simultaneous solution of the equations (6)

$$x = -\frac{1}{2}\rho^2, \qquad t = -\frac{1}{2}(\rho^2 + 1)\,. \tag{7}$$

We denote by N_+ a part of the cone N locates above the null plane Π, and by $\mathcal{H}$ a part of the plane Π located outside the surface Γ. It is easy to check that Rindler horizon, modified by the presence of the wormhole, is $H = N_+ \cup \mathcal{H}$. Really, an

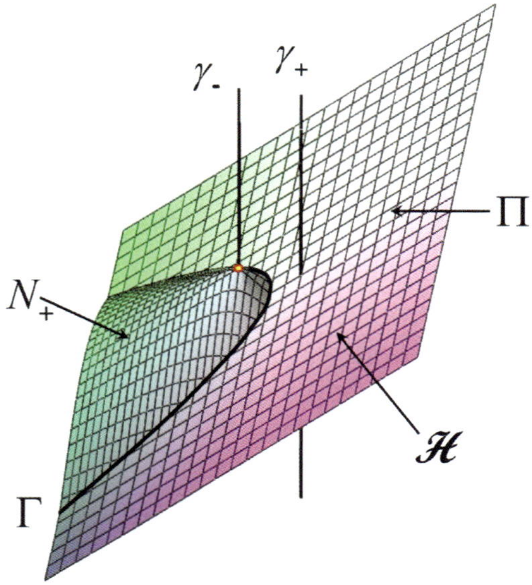

Fig. 2. This picture illustrates how the surface of Rindler horizon is modified in the presence of a static traversable wormhole. At this picture time changes in the vertical direction. Two solid lines $\gamma_\pm$ represent positions of the mouths of the traversable wormhole.

additional region lying between N_+ and Π is visible in the presence of the wormhole. The rays from this domain can pass through the wormhole and appear outside γ_+ mouth before the null plane Π crosses it.

Since interesting modification of the event horizon occurs at negative value of time t, it is convenient to define $\tau = -t$. We denote by $\mathcal{B}_\tau$ a slice of the distorted Rindler horizon at time $t = -\tau$. For $t \geq 0$ $\mathcal{B}_\tau$ is n-dimensional Euclidean spaces E^n. For $1/2 < \tau < 1$, $\mathcal{B}_\tau$ slice is a sum of 2 disconnected n dimensional spaces. One of them is E^n and the other is a round sphere S^n of radius τ. The surface area of a unit n-dimensional sphere S^n is

$$S_n = \frac{2\pi^{n/2}}{\Gamma(n/2)}\,. \tag{8}$$

In a flat space a sphere of radius τ contains a volume $V_{n+1} = \frac{1}{n}\tau^{n+1}S_n$.

In order to calculate the surface area of the horizon slices let us make some remarks. Let us denote by $d\sigma_n^2$ the line element of the surface of n dimensional unit sphere

$$d\sigma_n^2 = d\theta^2 + \sin^2\theta d\sigma_{n-1}^2\,. \tag{9}$$

To illustrate the properties of the horizon slices at different moments of time we consider a section $\theta_1 = \ldots \phi = 0$, and chose the parameter θ so that its value $\theta = 0$ corresponds to the lowest point of the sphere.

Let us denote

$$\cos\theta_\Gamma = 1 - \frac{1}{\tau}, \qquad \text{for } \tau > 1/2\,. \tag{10}$$

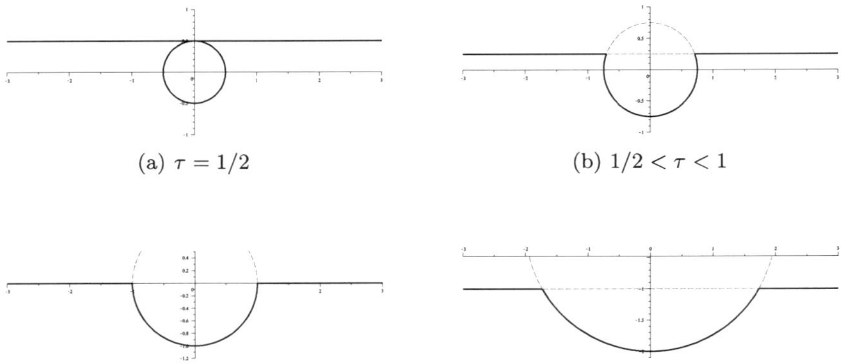

(a) $\tau = 1/2$ (b) $1/2 < \tau < 1$

(c) $\tau = 1$ (d) $\tau > 1$

Fig. 3. $t = -\tau =$ const slices of the horizon.

The angle θ_Γ determines a point where Γ curve intersects the slice $t = -\tau =$ const. In the interval $\tau \in [1/2, \infty)$, θ_Γ monotonically decreases from π at $\tau = 1/2$ to 0 when $\tau \to \infty$. For large τ, $\theta_\Gamma \sim (2/\tau)^{1/2}$. The structure of slices $\mathcal{B}_\tau$ for $\tau \geq 1/2$ is shown at Fig. 3. For $\tau = 1$, $\mathcal{B}_1$ is a half of the sphere S^n of radius 1 glued to a part of E^n (see Fig. 3(c)). For $\tau > 1$ the slice has the form presented at Fig. 3(d).

The line element of the spherical component of the event horizon at $\tau > 0$ is

$$dl^2 = \tau^2 d\sigma_n^2 \,. \tag{11}$$

The area of the distorted part of the event horizon is

$$A_n(\tau) = \tau^n S_{n-1} P_n, \qquad P_n(\theta_\Gamma) = \int_0^{\theta_\Gamma} d\theta \sin^{n-1}(\theta) \,. \tag{12}$$

The area of the part of the undistorted horizon located inside Γ at time τ is

$$A_n^0(\tau) = \frac{1}{n}\tau^n \sin^n(\theta_\Gamma) S_{n-1} \,. \tag{13}$$

Thus, the change of the area of the horizon at $\tau > 1/2$ is

$$\Delta A_n = A_n^0(\tau) - A_n(\tau) = S_{n-1}\Delta, \qquad \Delta = \frac{1}{n}(2\tau - 1)^{n/2} - \tau^n P_n(\theta_\Gamma) \,. \tag{14}$$

For the case $D = 4$ ($n = 2$) and $\tau \geq 1/2$, the area deficit ΔA_2 does not depend on time and it is

$$\Delta A_2 = -\pi \,. \tag{15}$$

For $\tau \in [0, 1/2]$

$$\Delta A_2 = -4\pi\tau^2 \,, \tag{16}$$

so that at $\tau = 0$, that is when the horizon is not distorted, ΔA_2 vanishes.

The evolution of the entropy of the Rindler horizon in the presence of the wormhole can be described as follows. The Rindler event horizon is defined by the following condition: At late time in future it coincides with the plane Π. The surface of this horizon in the past is slightly different from this plane. This difference asymptotically vanishes in the infinite past, however the size of domain where the horizon differs from Π, grows in the past. As a combination of these two effects, ΔA_2 remains the same. If we consider evolution of this "real" Rindler horizon, we come to the conclusion that it does not change up to the moment of time $t = -1/2$. At this moment a spherical part of the event horizon separates. Its initial radius is $1/2$ and its area is π. The area of the unperturbed horizon A_n^0 becomes less then the area of the initial horizon by the value π. Later in time the separated part of the horizon shrinks to zero. The net result of this evolution is that the area of the Rindler horizon at late time is less that its initial value by quantity π. If we restore the dimensionality one gets the following expression for the decrease of the entropy

$$\Delta S = S(t=\infty) - S(t=-\infty) = -\pi \frac{L^2}{4l_{pl}^2} \,. \tag{17}$$

If the number of spacetime dimensions is greater than four, the behavior of the entropy associated with the Rindler horizon is even more peculiar. For example for $D = 5$ ($n = 3$) one has

$$\Delta A_3 = 4\pi\Delta_3 \,, \tag{18}$$

$$\Delta_3 = \frac{1}{6}(\tau+1)(3\tau-2)\sqrt{2\tau-1} - \frac{1}{2}\tau^3 \arccos\left(\frac{\tau-1}{\tau}\right) . \tag{19}$$

For $\tau \to \infty$ ΔA_3 is negative and its absolute value infinitely grows

$$\Delta A_3 \sim \left[-\frac{2\sqrt{2}}{5}\tau^{1/2} + \frac{\sqrt{2}}{14}\tau^{-1/2}\right] S_2 \,. \tag{20}$$

Calculations for $D = 6$ give

$$\Delta A_4 = 2\pi^2\Delta_4, \qquad \Delta_4 = \frac{1}{4} - \frac{2}{3}\tau \,. \tag{21}$$

Similar results are valid for any $n \geq 3$: the corresponding ΔA_n is negative and divergent in the limit $\tau \to \infty$.

4. A Wormhole in a Black Hole Geometry

As we mentioned, the Rindler horizon is a certain idealization of the black hole horizon in the limit when the mass of the latter is very large. However, there exit two properties that differ the Rindler horizon from the horizon of the black hole: its topology differs from the spherical topology of the black hole, and its surface area is infinite. The first difference implies that the Rindler horizon is not a trapped surface. In this section we discuss how the black-hole event horizon evolves in the presence of a traversable wormhole, when one of its mouths falls into it.

4.1. *Null rays in the black hole interior*

We write the metric of $(D = n + 2)$-dimensional spherical black hole in the form

$$dS^2 = -f dV^2 + 2dV dr + r^2 d\omega_n^2 \,, \tag{22}$$

where $f = 1 - (r_g/r)^{n-1}$. In is convenient to use the dimensionless form of the coordinates and the metric. For this purpose we put

$$V = r_g v, \qquad r = r_g x, \qquad dS^2 = r_g^2 ds^2 \,, \tag{23}$$

where

$$ds^2 = -f dv^2 + 2dvdx + x^2 d\omega_n^2, \qquad f = -\frac{1 - x^{n-1}}{x^{n-1}} \,. \tag{24}$$

In the black hole interior $x \in (0, 1)$ the function f is negative.

We consider a situation when one of the mouths of a traversable wormhole falls down into the black hole, while the other mouth is kept at rest in its exterior. As earlier, we assume that the size of the mouths is very small. We also assume that the inner mouth reaches the singularity at the moment $v = 0$ of the advanced time. All the causal curves that enter the inner mouth before this would appear in the black hole exterior. We denote by N the past null cone with the vertex at $x = 0$, that is in the position of the inner mouth at $v = 0$. We also denote by Γ an intersection of N with the surface $x = 1$, which is a position of unperturbed horizon. We denote by N_+ a part of N with $x \leq 1$ restricted by Γ, and by $\mathcal{H}$ a part of the horizon H outside Γ, that is an undistorted part of the horizon. Thus the surface of the distorted horizon is $N_+ \cup \mathcal{H}$.

To find the n-dimensional surface Γ one needs to solve equations for null rays in the black hole geometry. Let us write the line element of a unit sphere in the form

$$d\omega_n^2 = d\theta^2 + \sin^2\theta d\omega_{n-1}^2 \,. \tag{25}$$

Because of the spherical symmetry it is sufficient to consider only null rays, propagating along the axis of the symmetry of S^{n-1}. In the coordinates (v, x, θ, θ_i), with $i = 1, \ldots, n-2$ one can put $\theta_i = 0$. Thus a null ray from the cone N can be specified by functions $(v(\nu), x(\nu), \theta(\nu))$, where ν is an affine parameter along the ray. The problem reduces to finding null rays in 3D spacetime with metric

$$ds^2 = -f dv^2 + 2dvdx + x^2 d\theta^2 \,. \tag{26}$$

It has two Killing vectors ∂_v and ∂_θ, hence one has

$$-f\dot{v} + \dot{x} = E \,, \tag{27}$$

$$x^2\dot{\theta} = L \,, \tag{28}$$

$$-f\dot{v}^2 + 2\dot{v}\dot{x} + x^2\dot{\theta}^2 = 0 \,. \tag{29}$$

The first two relations correspond to the conservation of the energy E and the angular momentum L. The last relation is a normalization condition for the null ray velocity. One can exclude the parameter E. It is sufficient to change the affine parameter $\nu \to \tilde{\nu} = E^{-1}\nu$. We use a dot to denote a derivative with respect to this new parameter.

To determine a position of the distorted horizon one needs to trace back in time the beam of null rays that arrive to the inner mouth of the wormhole at the moment when it reaches the singularity, that is a point $(v = 0, x = 0, \theta = 0)$. The null rays from such a beam differs by the value of the parameter $\lambda = L/E$. Inside the black hole one has $f = -|f|$ and the equations (27)–(29) take the form

$$\dot{x} = 1 - |f|\dot{v}\,, \tag{30}$$

$$\dot{\theta} = \frac{\lambda}{x^2}\,, \tag{31}$$

$$-|f|\dot{v}^2 + 2\dot{v} + \frac{\lambda^2}{x^2} = 0\,. \tag{32}$$

The last equation gives

$$\dot{v} = \frac{1 - \epsilon U}{|f|}\,, \qquad U = \sqrt{1 + \lambda^2 |f|/x^2}\,. \tag{33}$$

The parameter ϵ takes the values ± 1. For $\lambda = 0$ the equation (33) takes the form

$$\dot{v} = \frac{1 - \epsilon}{|f|}\,. \tag{34}$$

Hence for $\epsilon = +1$, $\dot{v} = 0$. Such ray propagates along fixed advanced time surface. For the other ray with $\lambda = 0$, one has $\dot{v} = 2/|f|$. Such a ray propagates along fixed "retarded time" surface. Only the rays with $\epsilon = +1$, after being traced back in time, cross the event horizon.

4.2. Γ-*curve*

A curve Γ is formed by intersection of the rays with $\epsilon = 1$ with the undistorted horizon. To find Γ one needs to solve the following equations

$$\dot{v} = \frac{1 - U}{|f|}\,, \quad \dot{x} = U, \quad \dot{\theta} = \frac{\lambda}{x^2}\,. \tag{35}$$

These equations imply

$$\frac{dv}{dx} = \frac{1 - U}{|f|U} \equiv -\frac{\lambda^2}{x^2 U(1 + U)}\,, \qquad \frac{d\theta}{dx} = \frac{\lambda}{x^2 U}\,. \tag{36}$$

The function U is singular at $x = 0$. The right-hand side of these equations can be written in a clearer form. Let us denote

$$U = \frac{\mathcal{U}}{x^{\frac{n+1}{2}}}\,, \qquad \mathcal{U} = \sqrt{\lambda^2(1 - x^{n-1}) + x^{n+1}}\,. \tag{37}$$

Then the equations (36) take the form

$$\frac{dv}{dx} = -\frac{\lambda^2 x^{n-1}}{\mathcal{U}(\mathcal{U} + x^{\frac{n+1}{2}})}\,, \qquad \frac{d\theta}{dx} = \frac{\lambda x^{\frac{n-3}{2}}}{\mathcal{U}}\,. \tag{38}$$

The quantity λ enters as a parameter in the right-hand side of these equations. It is easy to find solutions of (38) for special limiting value of this parameter. Namely for $\lambda = 0$ one has $v(x) = \theta(x) = 0$. For $\lambda \to \infty$, $\mathcal{U} \sim \lambda\sqrt{1-x^{n-1}}$ and equations (38) take the form

$$\frac{dv}{dx} = -\frac{x^{n-1}}{1-x^{n-1}}, \qquad \frac{d\theta}{dx} = \frac{x^{\frac{n-3}{2}}}{\sqrt{1-x^{n-1}}}. \tag{39}$$

The solution of the second equations is

$$\theta = \frac{2}{n-1}\arcsin(x^{-\frac{n-1}{2}}). \tag{40}$$

In particular, the value θ_∞ of the angle θ at the moment when a null ray with the impact parameter $\lambda = \infty$ reaches the unperturbed horizon $x = 1$ is

$$\theta^\Gamma_\infty = \frac{\pi}{n-1}. \tag{41}$$

The corresponding advanced time $v^\Gamma_\infty = -\infty$.

For a finite value of the impact parameter λ the coordinates $(v_\Gamma, \theta_\Gamma)$ of the points, where the ray crosses the unperturbed horizon $x = 1$, are

$$v_\Gamma(\lambda) = -\lambda^2 \int_0^1 dx \frac{x^{n-1}}{\mathcal{U}(\mathcal{U} + x^{\frac{n+1}{2}})}, \tag{42}$$

$$\theta_\Gamma(\lambda) = \lambda \int_0^1 dx \frac{x^{\frac{n-3}{2}}}{\mathcal{U}}. \tag{43}$$

These relations are nothing but parametric equations of the surface Γ.

4.3. *Shape and area of the distorted event horizon*

In order to determine a shape of the event horizon at some moment of time one needs to integrate backward in time v null ray equation for both subsets of the rays, with $\epsilon = \pm 1$. We need to find intersection of such rays with $v_0 = \text{const}$ surface, with $v_0 < 0$. For the rays from the subset $\epsilon = -1$ the corresponding range of λ is $\lambda \in [0, \infty)$. For the other subset of rays with $\epsilon = +1$, λ is in the interval $\lambda \in [\lambda_\Gamma(v_0), \infty)$. $\lambda_\Gamma(v_0)$ is the impact parameter of the ray that crosses Γ at $v = v_0$. To make the calculations easier we introduce a new parameter $\mu = \epsilon/\lambda$, and denote $\mu_\Gamma = 1/\lambda_\Gamma$. Then μ changes from $-\infty$ till μ_Γ. The negative μ corresponds to a subset with $\epsilon = -1$, while the positive μ correspond to the other subset. We also denote $y = x^n$. Then the corresponding set of null-ray equations takes the form

$$\frac{dy}{dv} = -nVW, \qquad \frac{d\theta}{dv} = -\frac{W}{y^{\frac{n+1}{2n}}}, \tag{44}$$

$$V = \sqrt{1 - y^{\frac{n-1}{n}} + \mu^2 y^{\frac{n+1}{n}}}, \qquad W = V + \mu y^{\frac{n+1}{2n}}. \tag{45}$$

The initial conditions at $v = 0$ are $y = \theta = 0$.

At $y = 0$ one has $V(0) = W(0) = 1$. The first, y-equation is regular at this point, while the second equation has integrable singularity. In order to treat this singularity it is sufficient to write $\theta = \tilde{\theta} + \Delta\theta(y)$ and to chose the (finite at $y = 0$) function $\Delta\theta$ so that the equation for $\tilde{\theta}$ becomes regular at $y = 0$. One has

$$\Delta\theta = \begin{cases} 2y^{1/4} + \frac{1}{3}y^{3/4}, & \text{if } n = 2\ , \\ \frac{2}{n-1}y^{\frac{n-1}{2n}}\ , & \text{if } n \geq 3\,. \end{cases} \tag{46}$$

The equation for $\tilde{\theta}$ is

$$\frac{d\tilde{\theta}}{dv} = \tilde{\Theta}, \qquad \tilde{\Theta} = \Theta - \frac{d\Delta\theta}{dy}Y\,. \tag{47}$$

For $n \geq 3$

$$\tilde{\Theta} = -\frac{W}{y^{\frac{n+1}{2n}}}(1 - V)\,. \tag{48}$$

It is easy to check that for $\tilde{\Theta} \sim y^{\frac{n-3}{2n}}$, so that that the right-hand side of (47) is finite at $y = 0$.

By integrating the equations (45) till the moment v_0 of the advanced time v, one obtains values $x_0(\mu) = x(v_0, \mu)$ and $\theta_0(\mu) = \theta_0(v_0, \mu)$. These expressions determine parametrically the shape of the $v = v_0$ slice of the deformed horizon. The full slice of the deformed horizon corresponds to the values of μ in the interval $\mu \in (-\infty, \mu_\Gamma]$. The point μ_Γ, where $x_0 = x_0(\mu_\Gamma) = 1$ and $\theta_0 = \theta_0(\mu_\Gamma)$, is a point of Γ where the slice v_0 intersects the unperturbed horizon.

The Fig. 4 shows the coordinate form of the slices $v = v_0$ of the event horizon for different values of the advanced time v_0. In these diagrams we used polar coordinates (x, θ). By rotation of the plots around the horizontal line one obtains a two-dimensional surface illustrating the form of the slice $v =$ const of the event horizon. It should be emphasized that the two-geometry induced on such a rotation surface by its embedding into three-dimensional flat space does not coincide with the two geometry of a slice of the distorted horizon. The latter is

$$d\omega^2 = x^2(\theta)(d\theta^2 + \sin^2\theta d\phi^2)\,. \tag{49}$$

Namely this metric determines the area element of the horizon slice, $d\sigma = x^2(\theta)\sin(\theta)d\theta d\phi$. The surface area of the deformed slice of the horizon is

$$A_n(v_0) = r_g^n S_{n-1} P_n[x], \qquad P_n[x] = \int_0^{\theta_\Gamma} d\theta \sin^{n-1}(\theta) x^n(\theta)\,. \tag{50}$$

The same expression with $x(\theta) = 1$ give the value of the surface area $A_n^0(v_0)$ of the unperturbed part $0 \leq \theta \leq \theta_\Gamma$ of the horizon. The difference

$$\Delta A_n(v_0) = A_n^0(v_0) - A_n(v_0) \tag{51}$$

determined the total difference of the areas of the undistorted and distorted horizons. The total area of undistorted horizon is $A_H^n = r_g^n S_n$.

The relative change of the horizon area $\Delta_n = \Delta A_n(v_0)/A_H^n$ is

$$\Delta_n = \frac{P_n[1] - P_n[x]}{B_n}, \qquad B_n = \int_0^{\pi} d\theta \sin^{n-1}(\theta)\,. \tag{52}$$

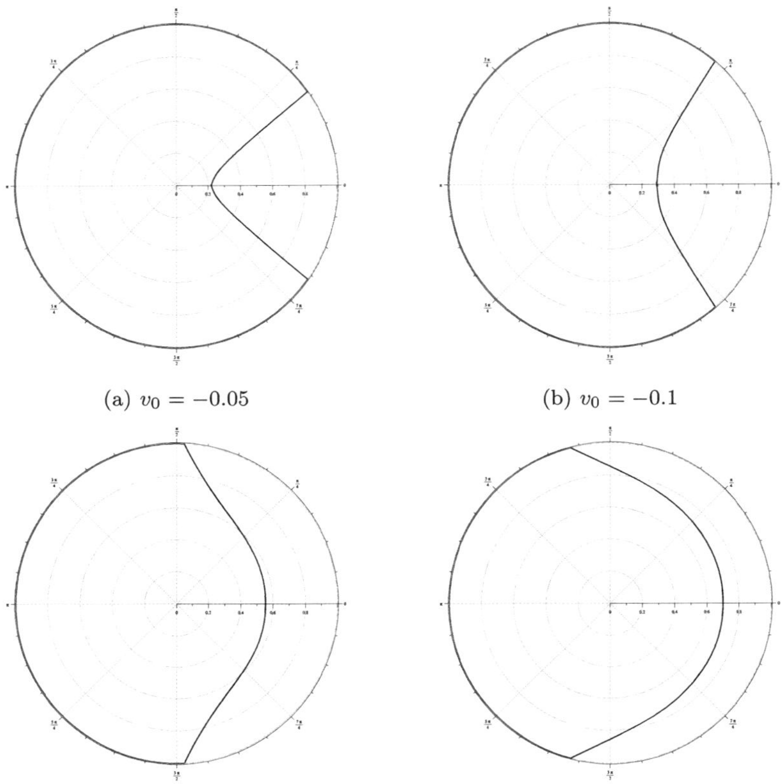

(a) $v_0 = -0.05$ (b) $v_0 = -0.1$

(c) $v_0 = -0.5$ (d) $v_0 = -1.0$

Fig. 4. Shape of the horizon slices at different moments of the advanced time v_0.

Omitting technical details, we just present here the results of the numerical calculations of the quantity Δ_n for the physically most interesting case $n = 4$ ($D = 4$). For a small absolute value of negative v_0 Δ_2 is an increasing function of $|v_0|$. Its values are: 0.2 for $v_0 = -0.01$; 0.35 for $v_0 = -0.05$; and 0.41 for $v_0 = -0.1$. Near $v_0 \approx -0.2$ Δ_2 reaches the maximum ≈ 0.43 and it further decreases again: 0.36 for $v_0 = -0.5$; 0.28 for $v_0 = -1.0$; and 0.1 for $v_0 = -2.0$.

If we relate the black hole entropy to its surface area according to (2), we would arrive to the following conclusion. In the above described gedanken experiment with a wormhole, the black hole entropy depends on the advanced time v. At first it decreases. At some moment $v < 0$ it reaches the minimal value and after this increases again. The black hole entropy returns back to its original value at the moment of advanced time $v = 0$, when the inner mouth of the wormhole is crashed in the singularity. The maximal decrease of the entropy reaches 43% of its original value. Similar results are valid for higher dimensional black hole.

We found the above described situation rather paradoxical. Even in the presence of a tiny mass wormhole, which practically does not change the mass and gravitational field of the black hole, there exists a process in which the black hole entropy decreases by 43% of its original value. The time scale of this process is $\sim r_g/c$. It is natural to relate this change of the entropy to the fact that during the time when the inner mouth is falling in the black hole interior, the information about the black hole interior becomes available to the external observer. At any rate, this dramatic decrease of the entropy is not expected in the "normal" thermodynamical system, without a dramatic change of their parameters. There might exist an interesting possible resolution of this "paradox". In 1990 Jacob Bekenstein discussed quantum limitations on the storage and transmission of information.[53] It might happen that the above assumption that by using of the wormhole with mouths of arbitrary small size one can extract all the information from a spacetime domain, which geometrically became available, is wrong. If the rate of the information extraction is limited by the geometric characteristics of the wormhole, this idealization would be oversimplified. An interesting question is how to measure the black hole entropy in a situation when the area of the black hole changes fast, and what are requirements on the devices used in such measurements.

Acknowledgments

The author thanks the Natural Sciences and Engineering Research Council of Canada (NSERC) and the Killam Trust for their financial support.

References

1. J. D. Bekenstein, *Lett. Nuovo Cim.* **4**, 737 (1972).
2. J. D. Bekenstein, *Phys. Rev. D* **5**, 1239 (1972).
3. J. D. Bekenstein, *Phys. Rev. D* **7**, 2333 (1973).
4. S. W. Hawking, *Nature* **248**, 30 (1974).
5. S. W. Hawking, *Commun. Math. Phys.* **43**, 199 (1975).
6. P. H. Frampton, S. D. H. Hsu, T. W. Kephart and D. Reeb, *Class. Quant. Grav.* **26**, 145005 (2009).
7. C. A. Egan and C. H. Lineweaver, *Astrophy. J.* **710**, 1825 (2010).
8. J. M. Bardeen, B. Carter and S. W. Hawking, *Commun. Math. Phys.* **31**, 161 (1973).
9. R. M. Wald, *Living Rev. Rel.* **4**, 6 (2001).
10. R. M. Wald, *NATO Sci. Ser. II* **60**, 477 (2002).
11. L. Bombelli, R. K. Koul, J. Lee and R. D. Sorkin, *Phys. Rev. D* **34**, 373 (1986).
12. G. 't Hooft, *Nucl. Phys. B* **335**, 138 (1990).
13. M. Srednicki, *Phys. Rev. Lett.* **71**, 666 (1993).
14. V. Frolov and I. Novikov, *Phys. Rev. D* **48**, 4545 (Nov 1993).
15. L. Susskind, L. Thorlacius and J. Uglum, *Phys. Rev. D* **48**, 3743 (1993).
16. C. G. Callan, Jr. and F. Wilczek, *Phys. Lett. B* **333**, 55 (1994).
17. C. Holzhey, F. Larsen and F. Wilczek, *Nucl. Phys. B* **424**, 443 (1994).
18. D. N. Kabat and M. J. Strassler, *Phys. Lett. B* **329**, 46 (1994).
19. J. D. Bekenstein, Do we understand black hole entropy?, in *Proceedings of the 7th Marcel Grossmann Meeting*, Stanford, USA, July 24–30, 1994 (World Scientific, 1996), pp. 39–58.

20. J. D. Bekenstein, Black hole fluctuations, in *Quantum Theory of Gravity*, ed. S. M. Christensen (Adam Hilger Ltd, 1984), p. 148.
21. V. P. Frolov, Black hole entropy and physics at Planckian scales, in *String Gravity and Physics at the Planck Energy Scale. Proceedings, NATO Advanced Study Institute*, Erice, Italy, September 8–19, 1995 (1995), pp. 187–207.
22. S. Carlip, *J. Phys.: Conf. Ser.* **67**, 012022 (2007).
23. S. Carlip, *Black Hole Entropy and the Problem of Universality*, in *Quantum Mechanics of Fundamental Systems: The Quest for Beauty and Simplicity: Claudio Bunster Festschrift* (2009), pp. 91–106.
24. A. Strominger, *Nucl. Phys. B - Proc. Suppl.* **192**, 119 (2009).
25. H. Saida, *J. Phys.: Conf. Ser.* **410**, 012139 (2013).
26. S. Carlip, *Lect. Notes Phys.* **769**, 89 (2009).
27. S. Carlip, Black hole thermodynamics, in *One Hundred Years of General Relativity: From Genesis and Empirical Foundations to Gravitational Waves, Cosmology and Quantum Gravity (Volume 2)* (World Scientific, 2017), pp. 415–465.
28. S. W. Hawking, *Phys. Rev. D* **72**, 084013 (2005).
29. V. P. Frolov, M. A. Markov and V. F. Mukhanov, *Phys. Rev. D* **41**, 383 (1990).
30. V. P. Frolov, M. A. Markov and V. F. Mukhanov, *Phys. Lett. B* **216**, 272 (1989).
31. J. Preskill, Do black holes destroy information?, in *International Symposium on Black Holes, Membranes, Wormholes and Superstrings*, Woodlands, Texas, January 16–18, 1992 (1992), pp. 22–39.
32. D. N. Page, *Int. J. Mod. Phys. D* **3**, 93 (1994).
33. T. Banks, *Nucl. Phys. Proc. Suppl.* **41**, 21 (1995).
34. L. Thorlacius, *Nucl. Phys. Proc. Suppl.* **41**, 245 (1995).
35. A. Strominger, Les Houches lectures on black holes, in *NATO Advanced Study Institute: Les Houches Summer School, Session 62: Fluctuating Geometries in Statistical Mechanics and Field Theory*, Les Houches, France, August 2–September 9, 1994 (1994).
36. S. D. Mathur, *Class. Quant. Grav.* **26**, 224001 (2009).
37. V. P. Frolov, *JHEP* **05**, 049 (2014).
38. W. G. Unruh and R. M. Wald, *Rept. Prog. Phys.* **80**, 092002 (2017).
39. S. W. Hawking, M. J. Perry and A. Strominger, *Phys. Rev. Lett.* **116**, 231301 (2016).
40. S. W. Hawking, M. J. Perry and A. Strominger, *JHEP* **05**, 161 (2017).
41. S. W. Hawking, arXiv:1401.5761 (2014).
42. J. M. Bardeen, arXiv:1406.409 (2014).
43. D. N. Page, *Phys. Rev. Lett.* **71**, 3743.
44. G. A. Vilkovisky, *Annals Phys.* **321**, 2717 (2006).
45. G. A. Vilkovisky, *Phys. Lett. B* **638**, 523 (2006).
46. G. A. Vilkovisky, *Phys. Rev. D* **77**, 124043 (2008).
47. V. P. Frolov and G. A. Vilkovisky, *Phys. Lett. B* **106**, 307 (1981).
48. M. Visser, *Lorentzian Wormholes: From Einstein to Hawking* (American Inst. of Physics, 1996).
49. V. P. Frolov and I. Novikov, *Phys. Rev. D* **48**, 1607 (1993).
50. R. Emparan and M. Martinez, *Class. Quant. Grav.* **33**, 155003 (2016).
51. R. Emparan and M. Martnez, *Int. J. Mod. Phys. D* **25**, 1644015 (2016).
52. R. Emparan, M. Martinez and M. Zilhao, *Phys. Rev. D* **97**, 044004 (2018).
53. J. D. Bekenstein and M. Schiffer, *Int. J. Mod. Phys. C* **1**, 355 (1990).

Bekenstein, I, and the Quantum of Black-Hole Surface Area

Shahar Hod

The Ruppin Academic Center, Emeq Hefer 40250, Israel
The Hadassah Institute, Jerusalem 91010, Israel

Professor Jacob Bekenstein was known not only for his brilliant and original physical ideas, but also for their clear presentation in his lectures and seminal research papers. I here provide a short review of Bekenstein's pioneering ideas about the quantization of black holes. I also describe my attempt, as a young and extremely naive student, to prove him wrong and how I got convinced in the correctness and utility of his deep physical intuition. Finally, my personal contribution to the ongoing attempts to understand the evenly spaced (discrete) area spectrum of quantized black holes, as originally suggested by Bekenstein in the early days of his scientific career, is described.

In memory of Prof. Jacob Bekenstein, whose insightful and elegant works have inspired my research

1. Preface

The remarkably deep and unique physical insights of Professor Jacob Bekenstein, that were also characterized by extremely beautiful mathematical elegancy, have inspired the imagination of many physicists, I among them, during the last decades. The sudden death of Jacob has left a huge hole in many hearts, but his original ideas still live with us. I am glad and honored for the invitation to contribute a personal essay for the special memorial volume for Jacob Bekenstein.

2. The Bekenstein Quantization of the Black-Hole Surface Area

The quantization of the black-hole horizon area was first discussed by Jacob Bekenstein, then a young and brilliant researcher in Princeton university,[a] more than four decades ago.[1] Bekenstein has based his original and ground breaking quantization argument on the intriguing physical fact discussed in Ref. 2 that, while the conserved charges of a Kerr–Newman black hole (mass, electric charge, and angular

[a]Bekenstein had received his Ph.D. from Princeton University at 1972 and then moved to the University of Texas at Austin.

momentum) may change due to the absorption of test particles, the surface area of the black hole may remain unchanged during an assimilation process of a point (structureless) particle if the later is absorbed at the black-hole horizon while being in a turning point[b] of its radial motion.

It was therefore suggested by Bekenstein that the black-hole surface area may serve as a classical adiabatic invariant. In particular, in the spirit of the Ehrenfest principle that has been formulated in the early days of quantum mechanics,[3] Bekenstein has boldly conjectured that the surface area of a black hole should have a discrete (quantized) spectrum of the form[1,c]

$$A_n = \gamma\hbar \cdot n \quad ; \quad n = 1, 2, 3, \dots , \tag{1}$$

where γ is a dimensionless constant.

Bekenstein has further suggested[1] to estimate the spacing of the black-hole area levels [that is, the value of the dimensionless physical coefficient γ in (1)] from a semi-classical gedanken experiment in which a black hole absorbs a finite-size neutral particle whose radial location and momentum are related by the quantum Heisenberg uncertainty principle.[4,d] In particular, Bekenstein has explicitly proved that, taking cognizance of quantum effects that influence the motion of test particles in black-hole spacetimes[1,4,d] the minimal increase in the black-hole horizon area is given by the simple expression

$$(\Delta A)_{\text{min}} = 8\pi\hbar\,. \tag{2}$$

Intriguingly, the minimal increase (2) in the black-hole surface area, as first suggested by Bekenstein in his seminal works,[1,4] is *universal* in the sense that it does not depend on the physical parameters of the black hole and the absorbed particle that have been used to derive it. This physically interesting observation naturally suggests that a lower bound of the form (2) is of fundamental importance in a quantum theory of gravity.[1,4]

3. Challenging (and then Confirming) the Bekenstein Area Quantization

In 1997, while I was investigating the inner structure of black holes under the insightful and enjoyable supervision of Professor Tsvi Piran at the Hebrew university of Jerusalem, I had a casual conversation with Dr. Avraham Mayo, then a student of Professor Bekenstein. The short and (for me) life changing conversation took place at the corridor of the physics department, few steps from Jacob's office. Until that moment I only knew Professor Bekenstein by his name. In particular, at that

[b] That is, for the classical absorption process to be an adiabatic one, the absorbed particle should be characterized by a zero radial momentum.

[c] We shall use natural units in which $G = c = 1$. In these units $\hbar$ has the dimensions of length^2.

[d] As emphasized by Bekenstein,[4] the Heisenberg uncertainty principle implies that, within the framework of semi-classical general relativity, a quantized (finite-size) particle cannot be absorbed precisely at the black-hole horizon from an exact radial turning point of its motion.

time I had no real information about Bekenstein's research interests nor about his incredible scientific achievements. Mayo had told me that his supervisor, Professor Bekenstein, is working on the quantization of black holes.[e]

The intriguing concept of bringing the ideas of quantum physics into the realms of black-hole physics and gravitation was totally new for me and it had immediately caught my imagination. I remember myself running, after the short conversation with Mayo, to the library in order to search for Bekenstein's works on the quantization of black holes.[1] I have read his seminal papers with much excitement and passion. The papers were written in a remarkably clear style (which, as I later learned, characterizes the entire set of Jacob's scientific works). In particular, the brilliant and (for me) new physical ideas were presented in these seminal papers in a mathematically elegant fashion.

In the first few days after reading, for the first time, Bekenstein's papers on the quantization of the black-hole horizon area, I was incredibly enthusiastic and happy about the beauty of physics that was reflected to me from his papers. In particular, the remarkably simple and mathematically elegant quantum area spectrum (1) that emerged from Bekenstein's deep physical insights[1] seemed to me as a new and exciting window into the elusive world of quantum gravity.

However, as a young, bold, and probably extremely naive student, I took it as a scientific, and even more, as a personal challenge to prove that the physically intriguing conclusion of Prof. Bekenstein, regarding the discrete quantization of the black-hole surface area, is wrong. In particular, I have noticed that the interesting analysis of Bekenstein presented in his seminal work[1] is restricted to the regime of *neutral* particles which are absorbed by black holes. I had the feeling that an analogous gedanken experiment, which involves the absorption of *charged* (rather than neutral) particles by charged black holes, may provide a physically interesting counter-example to the intriguing lower bound (2) derived by Bekenstein on the increase in black-hole surface area.

Analyzing the absorption of a charged particle of proper mass μ and electric charge q by a charged Reissner–Nordström black hole of mass M and electric charge Q, I have found that the minimally allowed increase in the black-hole surface area is given by the simple expression[5]

$$(\Delta A)_{\text{min}} = \frac{4\pi\mu^2}{qE_+}\,, \tag{3}$$

where $E_+ = Q/r_+^2$ is the electric field of the charged black hole at its surface $r_+ = M + (M^2 - Q^2)^{1/2}$.

Intriguingly, at first glance one may deduce from the analytically derived expression (3) that, for charged black holes, the minimal increase $(\Delta A)_{\text{min}}$ in surface area may become arbitrarily small in the $E_+ \to \infty$ limit. For two days I was very excited from the (obviously naive) thought that I, a young researcher, have found a counter-example to Bekenstein's well known lower bound (2).

[e]I would like to thank Dr. Avi Mayo for bringing Jacob's seminal works[1] on the quantization of black holes to my attention.

However, I have then realized that, within the framework of a quantum theory of gravity, the black-hole electric field cannot become arbitrarily large. In particular, vacuum polarization effects associated with the Schwinger pair production mechanism[6] set the upper bound[7]

$$E_+ \leq \frac{\pi\mu^2}{q\hbar} \tag{4}$$

on the strength of the black-hole electric field. Substituting (4) into (3), I have found the simple lower bound[5]

$$(\Delta A)_{\text{min}} = 4\hbar \tag{5}$$

on the increase in surface area of charged black holes due to the absorption of charged particles.

The fact that the lower bound (5), which I have derived for the minimal increase in the horizon area of charged black holes, and the lower bound (2) which, as originally proved by Bekenstein,[1] provides the minimal increase in black-hole surface area due to the absorption of neutral particles, are of the same order of magnitude has convinced me that there is something very deep and physically correct in Bekenstein's intriguing ideas about the quantization of black holes.

4. The Bohr Correspondence Principle and Black-Hole Area Quantization

The intriguing fact that both expressions (2) and (5) for the minimal increase in the black-hole surface area are universal (that is, independent of the physical parameters of the black hole and the absorbed particle) clearly indicates that these lower bounds may be of fundamental physical importance in black-hole physics and in the elusive quantum theory of gravity.

It should be realized, however, that the precise value of the fundamental dimensionless constant γ in the quantized Bekenstein area spectrum (1) cannot be determined by the gedanken experiments of Refs. 1 and 5. In particular, due to the inherent fuzziness of the Heisenberg uncertainty principle and due to the approximated nature of the expression (4) for the critical electric field, the estimated lower bounds (2) and (5) on the black-hole area increase can be challenged.

Furthermore, Bekenstein and Mukhanov[8,9] have correctly argued that the thermodynamic Bekenstein–Hawking area–entropy relation[1,10]

$$S_{\text{BH}} = \frac{A}{4\hbar} \tag{6}$$

for black holes, together with the Boltzmann–Einstein formula

$$g_n = e^S \in \mathbb{N} \tag{7}$$

for the number of microstates in statistical physics, imply that the value of the dimensionless physical constant γ in the quantum spectrum (1) should be of the simple functional form

$$\gamma = 4\ln k \quad \text{with} \quad k = 2, 3, \ldots . \tag{8}$$

Although physically convincing, the combined thermodynamic-statistical-physics argument presented in Refs. 8 and 9 could not fix uniquely the value of the integer k in the functional relation (8). As a researcher taking his first steps in the exciting academic world, I felt that it is a physically worthy challenge to try to determine the value of this mysterious integer.

In particular, one of the physical questions that most intrigued me as a young student was how the observed (classical) world emerges from the fundamental (quantum) description of nature. At that time, I already knew that, using the physical concept of coherent states, one can learn about the classical description of macroscopic phenomena from the underlying microscopic quantum description. But can one learn about the fundamental quantum world from its macroscopic coarse-grained classical description?

It took me a few days to realize that, luckily enough, the answer to the above stated question is 'Yes!'. In particular, the Bohr correspondence principle that was formulated in the early days of quantum mechanics[3] allows one to relate, in the regime of large quantum numbers, the fundamental discrete properties of a quantum physical system to the corresponding macroscopic (classical) properties of the system.

In particular, I have realized that the classical resonant oscillation frequencies of black-hole spacetimes can be related, through the Bohr correspondence principle, to the underlying quantum properties of these fundamental physical objects.[11,12] Taking cognizance of the Bekenstein quantized area spectrum (1) and the area-mass relation $A = 16\pi M^2$ for Schwarzschild black holes,[1] one finds the characteristic emission frequency

$$\hbar\omega^{\text{quantum}} = M_n - M_{n-1} = \gamma \cdot \frac{\hbar}{32\pi M} \tag{9}$$

which is associated with the quantum transition $n \to n-1$ of the black-hole area state in the asymptotic semi-classical $n \gg 1$ regime.

I have therefore conjectured, in the spirit of the Bohr correspondence principle,[3,4,11] that the characteristic classical resonant frequency[f] of these spherically symmetric black holes should have the simple functional relation [see Eqs. (8) and (9)]

$$\omega_{\text{R}}^{\text{classical}} = \frac{\ln k}{8\pi M}, \tag{10}$$

where k is the (yet unknown) integer parameter of Bekenstein and Mukhanov.[8,9]

[f]It is worth noting that black holes are characterized by an infinite set $\{\omega_n\}_{n=1}^{n=\infty}$ of damped gravitational quasinormal resonant modes. It is well known that, within the framework of classical general relativity, black holes do not emit radiation. Thus, in applying the Bohr correspondence principle to the classical oscillation frequencies of black holes, I have suggested[5,11] to focus on the highly damped ($n \gg 1$) black-hole resonant modes, which are characterized by infinitely small lifetimes (the characteristic zero lifetimes of these highly damped modes correspond to the no emission status of classical black-hole spacetimes).

The characteristic asymptotic resonant oscillation frequency of the Schwarzschild black hole was, at that time, known only numerically.[13] In particular, using sophisticated numerical techniques, it was computed with an accuracy of seven digits after the decimal point:[13]

$$M\omega_{\mathrm{R}}^{\mathrm{classical}}(n) = 0.0437123 \quad \text{for} \quad n \gg 1 \,. \tag{11}$$

It was Saturday morning, twenty years ago,[12] that I took my calculator and, with a shaky hand, calculated the numerical value of the expression $\ln 2/8\pi$ [see Eq. (10)]. To my deep disappointment, the answer I got did not agree with (11). But I did not lose heart; I turned to calculate the value of $\ln 3/8\pi \ldots$ and, to my excitement, the calculator returned the answer 0.0437123. It now agreed, digit-by-digit, with the numerically computed[13] characteristic black-hole resonant oscillation frequency (11).

As a young student, it was amazing to feel, for the first time in my (then) short scientific career, that I can count on my physical intuition. Moreover, I was thrilled to know that I have found the last missing piece in the black-hole quantization scheme of Bekenstein and Mukhanov. In particular, substituting $k = 3$ into Eqs. (1) and (8), one obtains the discrete Bekenstein area spectrum

$$A_n = 4\hbar \ln 3 \cdot n \quad ; \quad n = 1, 2, 3, \ldots \tag{12}$$

of quantum Schwarzschild black holes.

It is worth emphasizing that the evenly spaced quantum black-hole area spectrum (12), as originally formulated by Bekenstein[1,4] [see Eq. (1)], is consistent both with the thermodynamic entropy-area relation (6) of black holes, with the Boltzmann–Einstein relation (7) of statistical physics, and, as I have explicitly shown in Ref. 11, with the Bohr correspondence principle of quantum physics.

5. Summary

Jacob was known for his original and highly insightful physical ideas, but not less for the beautiful and elegant ways he was using to present them in his lectures and research papers. I, as many others, benefited a lot from reading Jacob's seminal works on the physics of black holes. His ingenious insights have inspired the imagination of many physicists, I among them. Jacob's pioneering ideas about the integration of quantum effects into black-hole physics continue to inspire the physics community in its ongoing search for a unified and self-consistent quantum theory of gravity and thermodynamics.

As a final personal note, I would like to comment that, unfortunately, I have never asked Jacob what he thought about my idea to use the Bohr correspondence principle and the asymptotic resonant frequencies of black holes in order to determine the value of the dimensionless physical parameter γ in his black-hole quantum area spectrum (1). Now, that it is too late, I can only regret for not doing so.

However, as fate would have it, many years after the appearance of my paper[11] with the suggested black-hole quantization spectrum (12), I came across a nice photo of Prof. Bekenstein that was taken in his office in the occasion of him winning the 2012 Wolf Prize in physics.[14] In the background of that photo I saw Jacob standing near his famous blackboard, on which I could vividly see the fundamental factor of $4 \ln 3$.[14,15]

Acknowledgments

This research is supported by the Carmel Science Foundation. I would also like to thank Yael Oren, Arbel M. Ongo, Ayelet B. Lata, and Alona B. Tea for stimulating discussions.

References

1. J. D. Bekenstein, *Phys. Rev. D* **7**, 2333 (1973); J. D. Bekenstein, *Lett. Nuovo Cimento* **11**, 467 (1974).
2. D. Christodoulou, *Phys. Rev. Lett.* **25**, 1596 (1970); D. Christodoulou and R. Ruffini, *Phys. Rev. D* **4**, 3552 (1971).
3. M. Born, *Atomic Physics* (Blackie, London, 1969).
4. J. D. Bekenstein, *Cosmology and Gravitation*, M. Novello, ed. (Atlantisciences, France, 2000) [arXiv:gr-qc/9808028].
5. S. Hod, *Phys. Rev. D* **59**, 024014 (1999); S. Hod, *Gen. Rel. Grav.* **31**, 1639 (1999).
6. J. Schwinger, *Phys. Rev.* **82**, 664 (1951).
7. L. Parker and J. Tiomno, *Astrophys. J.* **178**, 809 (1972); T. Damour and R. Ruffini, *Phys. Rev. Lett.* **35**, 463 (1975).
8. V. Mukhanov, *JETP Lett.* **44**, 63 (1986).
9. J. D. Bekenstein and V. F. Mukhanov, *Phys. Lett. B* **360**, 7 (1995).
10. S. W. Hawking, *Commun. Math. Phys.* **43**, 199 (1975).
11. S. Hod, *Phys. Rev. Lett.* **81**, 4293 (1998).
12. See also https://birthofidea.tecnico.ulisboa.pt/#hod .
13. H. P. Nollert, *Phys. Rev. D* **47**, 5253 (1993); N. Andersson, *Class. Quantum Grav.* **10**, L61 (1993).
14. See http://support.huji.ac.il/news-events/news/bekenstein_wolf_prize2012/
15. See also https://www.nytimes.com/2015/08/22/science/space/jacob-bekenstein-physicist-who-revolutionized-theory-of-black-holes-dies-at-68.html

Black Hole Microstates

Gerard 't Hooft

Institute for Theoretical Physics, Utrecht University,
Postbox 80.089, 3508 TB Utrecht, the Netherlands
http://www.staff.science.uu.nl/~hooft101/
g.thooft@uu.nl

Usually, it is assumed that black hole microstates can only be constructed by modifying the known laws of physics. Here it is argued that basically two modifications are needed. One of these concerns the topology of the space-time continuation of the Schwarzschild metric, that must be modified such that there is only one asymptotic region rather than two. The other concerns the information contents of Fock Space for the Standard Model, for which a holographic constraint appears to be needed. A qualitative description is given of the microstates one then gets, how they cure the black hole information paradox, and the way they add to the total mass of a black hole. Thus, the quantum black hole serves as a 'theoretical laboratory', assisting us in our search for internal logic in quantum gravity theories. A striking observation is that (super)string theory is not among the modifications needed.

Dedicated to the memory of Jacob D. Bekenstein

1. Introduction

Black hole physics carries a profound set of implications for the physics of the subatomic particles. J. D. Bekenstein[1–3] was one of the first to realise this. Even without asking questions about the quantum mechanical properties of black holes, one can speculate on the significance of the fact that the black hole parameters appear to obey relations that look exactly like the thermodynamical variables in samples of ordinary matter. Thus, black holes appear to have not only energy, charge and angular momentum, but also a temperature and an entropy.

A black hole isn't black. It is not exactly black. It couldn't be, because for something to be exactly black, one needs to say that it is surrounded by an exact vacuum, but to define an exact vacuum, one needs an infinite space. The black hole has only a finite size, and this is why it must be expected to emit some amount of light.

The first to actually compute the temperature of a black hole was Hawking.[4] Since this temperature is usually quite low, and its entropy quite big, these features tend to be hidden, so that there is no direct clash with other branches of physics, including the predictions from usual General Relativity about the way in which material is sucked in, accelerates, and disappears from view.

Our next primary observation is that black holes cannot respect additive conservation laws such as the conservation of baryon number or lepton number, because they can absorb arbitrary amounts of neutrons and proton–electron pairs, while keeping their mass to baryon number ratio much smaller that that of single particles. This became even more evident when it was realised that if indeed the black hole temperature does not vanish, black holes will radiate away energy, so that the ratio between energy and baryon or lepton number can shrink to zero.

The exception to this rule is the case where charges are coupled to Abelian fields such as electro-magnetism. In that case electric charge is manifest from the asymptotic values of the electric field strength, while any violation of electric charge conservation would clash with gauge-invariance, which would cause inconsistence with unitarity and/or locality for these gauge theories.

Thus, one might be inspired to suspect that black holes will have something to tell us about the fundamental interactions between elementary particles, normally laid down in the Standard Model.[a] On the other hand, the fact that black holes may have properties in common with ordinary matter may suggest that there exist domains of physics where the fundamental distinction between black hole physics and particle physics may vanish. Thus, anyone interested in unified theories of physics should be interested in black holes; they are not just curious little puzzles of logic, they can teach us important, and highly non-trivial, lessons.

When the physics community was striving at understanding the synthesis of quantum mechanics with *special* relativity, we had the luxury of a plethora of experimental approaches and observations that revealed to us how Nature works in this domain. Initially there were many competing theories, of which some were more severely criticised than others, just because they seemed to by logically imperfect. One theory that originally seemed to be logically deficient, *quantised field theory*, emerged victoriously, just because experimental observations pointed us the way; the logical imperfection of this approach turned out not to lead to insuperable practical barriers, while the theory did allow us to construct a complete scheme of all possibilities.[5] What is called the Standard Model now, turned out to be just one of these possibilities. Some parameters of the theory, notably a handful of masses and coupling strengths, had to be determined by careful measurements, after which many other features of the theory were accurately predicted, enabling us to scrutinise the accuracy of the model in meticulous detail.

[a]With 'Standard Model' we here refer to whatever quantum field theory that is needed to describe elementary particles and their interactions at energy scales reaching up to the Planck scale.

What was left was the need to generalise these results such that *General* Relativity would be included. Here, we again have various contenders for the position of a unifying theory. But this time, understanding of the interplay of quantum mechanics and General Relativity, is being hampered by the unavailability of direct experimental tests. This drawback is used by some to argue that string theories are completely useless, while others, notably string theorists themselves, tend to downplay this difficulty as being merely a technical hindrance.

Yet we have seen how important the role of experiments has been in arriving at the Standard Model, and how easy it is to make severe errors of judgment when we ask for a valid theoretical scheme.

It turns out to be instructive now to regard quantised black holes as a *theoretical laboratory*.[6] The theoretical laboratory is not quite as good and impartial as the experimental one. We can only perform *thought experiments*. Our point is that, the quantum black hole behaviour depends in an essential way on the physical properties of its horizons. The black hole horizon acts as an *infinite microscope*: particles that emerge from the past horizon experience the effects of the particles that entered into the future horizon some time earlier, as if these have been magnified by almost infinite factors, and this way they fill a gap that cannot be reached by conventional particle accelerators.

Of course, the theoretical lab does not give us unequivocal results on the outcome of experiments, but it does give us something: the outcome of the theoretical experiments must, at all times, agree with some form of internal logic. This logic includes the laws for unlimited magnifications at the horizons, and it is non-trivial.

We demand this logic to cover some essential features:

- A black hole must behave exactly as the 'classical' or, 'unquantised' black hole, in the limit where the black hole radius $R_{\rm BH}$ is large compared to the Planck length, or, the black hole mass, $M_{\rm BH}$, is much larger than the Planck mass.
- The quantum black hole should allow a description in terms of quantum states, or energy levels, in terms of which one can describe its evolution by imposing some Schrödinger equation. It is these quantum states that are often referred to as *black hole microstates*. There exists a quantum Hamiltonian H for the black hole, that accurately dictates its evolution. Unitarity demands that this Hamiltonian be hermitian.

Strictly speaking, unitarity does not hold for systems with an asymptotic domain, but this merely implies that we have to split off the effects from particles and other features that are far separated from the black hole; imagine some boundary condition far from the black hole, in the form of a box that keeps matter particles from escaping to infinity. In particular, massless particles must be handled with care, but all of this should not lead to any difficulties of principle.

- We demand a notion of *locality*. In some treatises, it looks as if a quantum black hole has direct communication lines open with some other black hole at some far distance, or even in some other universe. This could easily lead to direct interactions between such distant partners. Locality forbids this (we return to this point in Sec. 5).

These principles may seem to be logical and inescapable, but nevertheless papers were published purportedly showing that they are mutually incompatible.[7] This is because the first and the second demand above appear to imply that an observer falling into a black hole cannot pass through the horizon. This would be contradictory, because one can also conclude from classical general relativity that the exact location of the horizon cannot be determined by Cauchy data alone; one needs to know what happens in the infinite future, thus the fate of an observer cannot depend on the position of a horizon at any finite time.

The resolution of this problem includes a postulate concerning space-time topology,[8,9] a possibility that was overlooked until recently. In such a space-time, locality does not forbid a direct interaction between a point on the horizon and its antipode.

Just because they are almost, but not quite, mutually incompatible, our logical constraints turn out to be powerful and restrictive, so that they provide important new information. As we shall see at the end, Sec. 5, one more assumption will be needed, but it is important to emphasise that we neither postulate nor exclude the validity of string theory. The point to remember here is, that it is physics at a lower energy scale where we encounter sufficient internal structure for characterising the microstates. What we do see is that the emergent mathematics concerns the horizon, which resembles a string worldsheet. For that reason, it is conceivable that a modified form of string theory might emerge.[10,11]

2. Replacing Fock Space by Microstates

To formulate any theory for a physical system, one must first specify what the fundamental degrees of freedom are. In most treatments addressing the quantisation of gravity, one assumes some space-time that exhibits curvature way beyond any perturbative description. The problem with this is that one then looses control over what one might expect to happen. Are there asymptotic states at all? Is there a vacuum state? To what extent may one allow for non-trivial topologies? And so on. The beautiful (and important) feature of black holes is that one might limit oneself to states that do allow for perturbative descriptions. This immediately opens the possibility of a systematic treatment of 'all microstates'. We claim that, in describing a simple, semiclassical black hole, these microstates do form a complete set, and that this set does admit total control. It is difficult to overemphasise the importance of this. Usually investigators refer to their system as being 'chaotic', a euphemism for ignorance and lack of understanding.

Our discussion in this paper is mainly a qualitative one. More technical details are to be found in Refs. 12 and 13. First, we distinguish what we call *hard* particles from *soft* particles. Hard particles are particles whose mass and/or momentum exceed the Planck value, so that their effects on space-time curvature cannot be handled perturbatively. Soft particles are all other particles. We don't ask to *ignore* the effects of soft particles on space-time curvature; rather, we assume that the effects soft particles have on space-time, can be handled perturbatively. Thus, there will be gravitons in our soft-particle spectrum, but these are handled as in non-Abelian gauge theories.

Next, the stage on which the microstates are defined will be represented as a Cauchy surface in space-time, and this space-time will be represented as a Penrose diagram for the *eternal* black hole, while all points in region *II* of this diagram are *identified* with the antipodes of their mirror images in region *I* (this will be referred to as the 'antipodal identification).

To many colleagues, it sounded surprising, or even objectionable, that, on this space-time metric, neither the effects of the imploding matter that was responsible for the black hole being formed, nor the effects of the evaporating particles at later times, are taken care of — the metric assumes the black hole background to be strictly static. This however is for a very good reason: we *only* include the soft particles as quantum particles on our Cauchy surface. Since the hard particles are as yet omitted, also their effects on the metric are not shown. How to include hard particles will be considered later (Sec. 4). Note that, for characterising microstates, only the Cauchy surface is needed (solid line in Fig. 1), and this does not cross parts of space-time in the future or in the past; only the present counts, so it should not matter what hard particles would be doing in the early past or the late future.

Regions *III* and *IV* in Fig. 1 are not needed. The *black hole singularity* (wavy lines in Fig. 1) is always in the *infinite future* and the *infinite past.*[b]

It is easy to follow the evolution of particles in this space-time during a short time-interval, where the time for the external observer stays within the range

$$|t| \ll \mathcal{O}(M_{\rm BH} \log M_{\rm BH}) \ , \tag{2.1}$$

in Planck units. During this period the black hole metric stays the same, apart from small perturbative changes due to the soft particles present. All in- and out-going particles will be soft ones, and their effects can be calculated using 'Standard Model' physics, combined with perturbative (quantum) gravity.

All out-going particles will be kept in the domain of soft particles. In principle, we could extend our analysis to include hard particles sent in from the outside, such as other black holes. This is known to occur in the world of black holes, but we can omit such initial states from the beginning. Also, out-going particles

[b]The "black hole singularity" is often thought to reside in "the black hole center", but this is an illusion: there is nothing in the black hole center. We explain this issue in the Appendix.

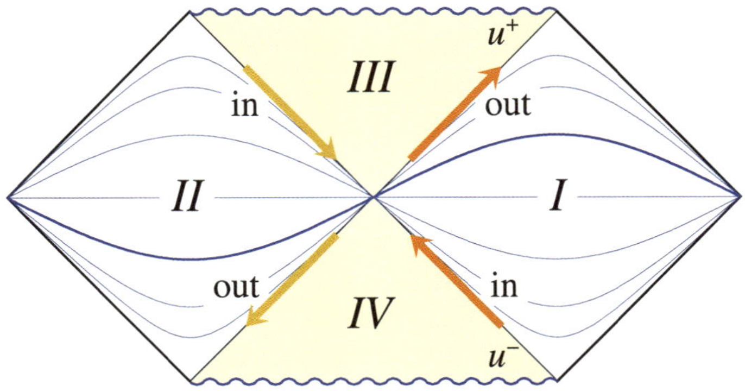

Fig. 1. Penrose diagram for (Schwarzschild) black hole with the effects of in- and out-going matter omitted. Curved lines are equal-time lines.

that are as hard as black holes, will be a rarity that is not expected to contribute significantly to the unitarity constraint. These would be genuine tunnelling events that are exponentially suppressed.

3. Time Dependence

The two significant coordinates u^+ and u^- in the Penrose diagram are light-cone coordinates. The two angular coordinates are less significant for our main arguments and are usually suppressed. Let t be the time coordinate for the external observer, then it is convenient to use a scaled time coordinate τ defined as

$$\tau = t/(4GM_{\rm BH}) \; . \tag{3.1}$$

If we perform a time translation, then, near the origin, the Penrose coordinates at time τ are related to the ones at time 0 as follows:

$$u^{\pm}(\tau) = u^{\pm}(0)\, e^{\mp\tau} \; . \tag{3.2}$$

This means that, as time t flows, the positions of objects are continuously Lorentz-transformed. This also holds for the momenta of in- and out-going particles: if p^- is the in-going momentum and p^+ the out-going one, then, near the origin,

$$p^{\pm}(\tau) = p^{\pm}(0)\, e^{\mp\tau} \; . \tag{3.3}$$

We observe that the in-momentum p^- at late times inflates exponentially with time, while the point u^+ where an in-particle crosses the future event horizon, shrinks exponentially with time. An out-going (Hawking) particle leaves the past horizon at a point u^- with momentum p^+. This momentum explodes exponentially at large negative time, while u^- expands rapidly for large positive times.

At any time, all in-going particles enter through the future event horizon and all out-going particles emerge from the past horizon. It is important to emphasise this, since it is often wrongly stated that late Hawking particles, somehow, manage to enter the scene through the future event horizon. The correct thing to say is that *black hole dynamics* describes how out-going particles emerge at the past horizon in quantum states that depend on how in-going particles enter in the future event horizon. The ensuing 'violation of causality' here is only apparent; in reality there will be a time delay that is large, typically

$$t_{\text{out}} = t_{\text{in}} + \mathcal{O}\big(4GM_{\text{BH}} \log(M_{\text{BH}}/M_{\text{Pl}})\big) \ , \tag{3.4}$$

which means that the link between out-going and in-going material is always made at values of the product u^+u^- near the Planck scale.

4. Evolution Over Longer Time Periods

In the previous section, we limited ourselves to evolution periods small compared to the intervals (2.1), for a very good reason. Following 3.3 we see that the momenta p^- of the in-going particles rapidly increase in time, while the momenta p^+ of the out-going particles rapidly increase when we follow them backwards in time. Thus, soft particles become hard very quickly, and, as soon as we encounter hard particles, we can no longer ignore their effects on the metric.

It is not hard to suggest a flaw in our treatment so-far: we ignored the effects of the earliest in-going particles and the late out-going ones; both will severely affect the metric of space-time. Should we not replace the Penrose diagram of Fig. 1 by one where these modifications have been taken into account? Vaidya[14] and Hawking[15,16] proposed such modifications.

This, however, would not concur with our philosophy. If we allow for these hard particles, we will very quickly be confronted with particles whose local momenta will be inflated way beyond the energy contents of the entire universe; they cannot possibly all represent existing microstates. A cut-off will be needed in any case. A much better strategy can be set up:

The only hard particles that we should expect to encounter are the ones that are squeezed either against the past horizon (the in-particles), or the future horizon (the out-particles). We can calculate the effect these will have on the metric.[9,17] *They cause a shift* $\delta u^\pm(\theta,\varphi)$ *in the coordinates* $u^\pm(\theta,\varphi)$ *that is proportional to the momentum distribution* $p^\pm(\theta',\varphi')$, and this can be calculated precisely. We refer to this shift as the *gravitational footprint* of the in-going particles, left behind on the out-going ones.

We can adjust this result somewhat: since the earliest in-going particles give the biggest effects, and these have been long forgotten anyway, we can safely replace $\delta u^\pm(\theta,\varphi)$ by the values $u^\pm(\theta,\varphi)$ themselves. Thus, the gravitational footprint is now taken to represent the entire spectrum of the out-going particles. The out-particles *are* the gravitational footprints of the in-particles.

What is nice about this procedure are two things: one, while p^- increases exponentially with time τ, the footprint $u^-(\theta,\varphi)$ expands. It represents the wave functions of the out-particles. Since these wave functions are spreading, the out-particles rapidly turn into soft particles. This means that we successfully replaced hard particles by soft ones, while keeping all information about the spectrum intact.

Secondly, we are forced to avoid double counting. As soon as we realise that the out-particles contain all relevant information, we can discard the hard particles. Since momentum is replaced by position, the relation between in- and out-particles is a Fourier transformation. The Fourier transformation is a unitary process. This information is preserved, using soft particles only.

The procedure can now be repeated as often as we want. This way we reach all epochs in time. At all moments, we have only soft particles living in a Penrose diagram for eternal black holes.[c]

5. Miscellaneous

It is important to note that, in the spectrum of the out-particles, their positions $u^-(\theta, \varphi)$ in the Penrose diagram can have any sign. It is an essential feature of the Fourier transform that unitarity is only maintained if both in $p^\pm$ space and in $u^\pm$ space, the wave functions are allowed to extend from $-\infty$ to $+\infty$. This means that both regions *I* and *II* are involved.

Now in the conventional Schwarzschild metric, all points (r, t, θ, φ) are represented twice in the Penrose diagram. Thus it would appear as if we are describing two black holes rather than one, *and these two black holes are interacting.* It was proposed to regard these two black holes as an "entangled" EPR pair.[18] This is untenable, since the two regions, *I* and *II* now interact. For a non-locally separated EPR pair it is essential that they should not interact directly. As in quantum field theory, one must impose that, when two space-time points x_1 and x_2 are space-like separated, or more generally, sufficiently far apart, then all operators $\phi(x_1)$ defined at x_1 and $\phi(x_2)$ defined at x_2, must commute,

$$[\phi(x_1), \phi(x_2)] = 0 , \tag{5.1}$$

which is not the case[9] when x_1 is in region *I* and x_2 is in region *II.*

As we indicated in Sec. 2, the only correct way to proceed is by demanding that region *II* refers to the *antipodes* of region *I.* This was extensively discussed and motivated in Refs. 9 and 13: the simplest way to characterise Schwarzschild space-time (and its extensions to Kerr and Kerr–Newman) is by folding the Kruskal–Szekeres coordinates such that (x, y, θ, φ) is identified with $(-x, -y, \pi - \theta, \varphi + \pi)$. This is singularity-free at all points with $r > 0$. The extra singularity at $r \to 0$ is hidden far away in the unphysical region. We do see that this redefines neighbourhoods of points in *I* to include points in *II*, so that the concept of locality is affected. This is

[c]From here on, there is no need to restrict ourselves on the eternal black hole, since only the data in regions *I* and *II* are considered. We may need an infinitesimal extension into regions *III* and *IV* only to check how the Cauchy surface continues from *I* into *II.*

important, since often lack of locality is brought forward as an argument why there should be a "black hole information problem".[7,19–25] With antipodal identification, the Hartle–Hawking state no longer acts as an element of a density matrix (a mixed state), but it now describes Hawking particles, in terms of a pure state of the black hole. This, we claim, is a fundamental improvement of the internal logic of the system.

There are several problems still wide open. First, using the principles described in this paper, an evolution operator for the Schwarzschild black hole can be derived, but the dynamical variables that are used are not the elements of Fock space, adapted to the curved black hole background.

The fundamental parameter, instead, is the total momentum density entering the black hole at every point on the future event horizon, and it is linked to a position operator for the out-going matter at the past event horizon; since the latter is the Fourier transform of the former, and provided we use the antipodal identification constraint, the evolution operator we found is indeed unitary.

What is missing however is a unitary mapping of the Fock space elements onto the momentum density states. This may be related to another deficiency: a cut-off will be needed for the values of the ℓ parameter of the partial waves used. In principle, it should be possible to arrive at a unitary relation with a judicious choice of this cut-off, but what we are really searching for is a more complete set-up of Nature's physical degrees of freedom. This is our still open question. One might expect some version of the holographic principle to apply here: a one-to-one mapping is searched for that maps the fields describing Standard Model states on a three-dimensional Cauchy surface onto the eigenvalues of momentum operators that live on the two-dimensional horizon.

It is instructive to note how the black hole mass evolves in our procedure. In-particles arrive from far away, and out-particles leave the scenery far away. What is conserved during all stages is the total mass. As soon as an out-particle or an in-particle is sufficiently far separated from the horizon, in spite of the particle being soft, its contribution to the mass must be subtracted from the total mass, to arrive at the mass of the remaining black hole. This is a small modification in the metric of what was supposed to be the eternal black hole. Near the horizon, particles blending with the horizon continue to contribute to the total black hole mass; only when their footprints leave the black hole, their contributions to the total mass is to be subtracted, as just described. Note that the exponential inflation of p^- of a particle entering at the horizon, has no effect on the total mass at all.

Finally, we stress the importance of investigations of this nature. As our 'theoretical laboratory', we employ black holes to check the internal logic of quantum gravity theories. Our present research established that space and time possess a non-trivial topology: the Schwarzschild metric is first re-written in Kruskal–Szekeres coordinates (compressed to form Penrose diagrams), but then we observe the necessity to fold this up exactly once, such that every space-time point in this frame is identified with a point at its antipode:

$$(x, y, \theta, \varphi) \equiv (-x, -y, \pi - \theta, \varphi + \pi) . \tag{5.2}$$

This mapping, necessary for the restoration of unitarity, inverts the arrow of time.

Acknowledgments

The author acknowledges discussions with P. Betzios, G. Dvali, A. Franzen, N. Gaddam, S. Giddings, S. Hawking, J. Maldacena, S. Mathur, S. Mukhanov, O. Papadoulaki, F. Scardigli, L. Susskind, and W. Vleeshouwers among others.

Appendix A. The Black Hole Singularity

In popular treatises of black holes, authors sometimes mention the black hole singularity as being something very mysterious, and "not understood".[26] It is presented as a region of the black hole, residing at its centre, where "the laws of nature become divergent". Indeed, the Schwarzschild metric diverges at $r = 0$. The mystery is then: how could such a singularity "pop up" when a black hole is formed?

In reality, the region $r < 2GM_{\rm BH}$ is not at the (space-like) centre at all, it is at the infinite future, while what we see when we look at a black hole horizon, we see the infinite past. The r coordinate is a time coordinate there. The classical Schwarzschild black hole has one singularity at $r = 0$, and if an observer would approach this point, his time coordinate puts him beyond the infinite future for all outside observers, at all times. So we do not have to worry about singularities popping up, they emerge long after the universe ceases to exist, so that there is no clash with ordinary physics at all.

It makes more sense to compare the region behind the horizon, $r < 2GM_{\rm BH}$, region *III* in Fig. 1, with the analytic extension of some coordinate in the complex plane. In the complex plane many models of physical systems may generate singularities such as poles and branch cuts without any alarming physical consequences. The reader is encouraged to investigate the elliptical orbit of a gravitating planet in the complex time plane, to note that the planet reaches singularities there. Astrophysicists ignore such singularities simply because the time coordinate never becomes complex. Similarly, in the case of black holes, we do not have to worry about the time-like coordinate r in the region $r < 2GM_{\rm BH}$ just because these values for the time coordinate are unphysical.

In the present work, we also have a singularity at $r \to 0$ that corresponds to the "infinite past". It is region *IV*, which is equally unphysical. What happened here is that we (deliberately) ignore the imploding matter that actually formed the black hole, in a distant but finite past. Ignoring that, we get an artificial singularity at the infinite past. Physically, it is of no concern. Mathematically, this allows us to regard black holes that are stationary for a sufficient amount of time in order to study its actual time dependence, as explained further in the text of this paper.

For our mathematics, we merely need the domain $r > 2GM_{\rm BH} - \varepsilon$, for small enough $\varepsilon > 0$, in order to find the smooth extension of our metric towards the (much more important) region *II* of the Penrose diagram.

References

1. J. D. Bekenstein, "Generalised second law of thermodynamics in black hole physics", *Phys. Rev. D* **9**, 3292–3300 (1974), DOI: 10.1103/PhysRevD.9.3292.
2. J. D. Bekenstein, "Quantum Black Holes as Atoms", in *Proceedings of the Eight Marcel Grossmann Meeting*, eds. T. Piran and R. Ruffini (World Scientific, Singapore, 1999), pp. 92–111; arXiv:gr-qc/9710076.
3. J. D. Bekenstein, "Quantum Information and Quantum Black Holes", in *Advances in the Interplay between Quantum and Gravity Physics*, eds. P. G. Bergmann and V. de Sabbata (Kluwer, Dordrecht, 2002), pp. 1–26.
4. S. W. Hawking, "Particle Creation By Black Holes", *Commun. Math. Phys.* **43**, 199 (1975) [Erratum-*ibid.* **46**, 206 (1976)].
5. G. 't Hooft, *In Search of the Ultimate Building Blocks* (English translation of: "Bouwstenen van de Schepping"), (Cambridge Univ. Press, 1997), ISBN 0 521 550831/550833.
6. A. Zichichi, Erice lecture notes, 2018.
7. A. Almheiri, D. Marolf, J. Polchinski and J. Sully, "Black Holes: Complementarity or Firewalls?", *JHEP* **1302**, 062 (2013), arXiv:1207.3123 [hep-th].
8. G. 't Hooft, "Black hole unitarity and antipodal entanglement", *Found. Phys.* **49**(9), 1185–1198 (2016), DOI: 10.1007/s10701-016-0014-y; arXiv:1601.03447v4 [gr-qc] (2016).
9. G. 't Hooft, "The firewall transformation for black holes and some of its implications", *Foundations of Physics* **47**(12), 1503–1542 (2017), DOI: 10.1007/s10701-017-0122-3, e-Print: arxiv:1612.08640 [gr-qc] (2016).
10. G. 't Hooft, "Strings from gravity", in *Unification of Fundamental Interactions*, Proceedings of Nobel Symposium **67**, Marstrand, Sweden, June 2–7, 1986, eds. L. Brink *et al. Physica Scripta* **T15**, 143–150 (1987).
11. G. 't Hooft, "The black hole interpretation of string theory", *Nucl. Phys. B* **335**, 138–154 (1990).
12. G. 't Hooft, "Diagonalising the Black Hole Information Retrieval Process", arXiv:1509.01695.
13. G. 't Hooft, "Virtual Black Holes and Space-Time Structure", *Foundations of Physics* (2018), DOI: 10.1007/s10701-017-0133-0.
14. P. C. Vaidya, "The external field of a radiating star in general relativity", *Current Science* **12**, 183 (1943); P. Vaidya, "The gravitational field of a radiating star", *Proc. Indian Acad. Sci. A* **33**, 264 (1951).
15. S. W. Hawking, "The unpredictability of quantum gravity", *Comm. Math. Phys.* **87**, 395 (1982).
16. A. Ashtekar and M. Bojobald, "Black hole evaporation: A paradigm", arXiv:gr-qc/0504029v2.
17. T. Dray and G. 't Hooft, "The gravitational shock wave of a massless particle", *Nucl. Phys. B* **253**, 173 (1985).
18. J. Maldacena and L. Susskind, "Cool horizons for entangled black holes", *Fortsch. d. Phys.* **61**, 781 (2013), arXiv:1306.0533 [hep-th]; L. Susskind, "Copenhagen vs Everett, Teleportation, and ER=EPR", arXiv:1604.02589v2 [hep-th].

19. S. W. Hawking, "Breakdown of predictability in gravitational collapse", *Phys. Rev. D* **14**, 2460 (1976).
20. S. B. Giddings, "Quantum mechanics of black holes", arXiv:hep-th/9412138; "The Black hole information paradox", arXiv:hep-th/9508151.
21. J. Polchinski, "The Black Hole Information Problem", arXiv:1609.04036 [hep-th].
22. S. B. Giddings, "Black holes, quantum information, and unitary evolution", *Phys. Rev. D* **85**, 124063 (2012), arXiv:1201.1037 [hep-th].
23. S. B. Giddings, "(Non)perturbative gravity, nonlocality, and nice slices", *Phys. Rev. D* **74**, 106009 (2006), hep-th/0606146.
24. S. B. Giddings and M. Lippert, "The Information paradox and the locality bound", *Phys. Rev. D* **69**, 124019 (2004), hep-th/0402073.
25. S. B. Giddings, "Models for unitary black hole disintegration", arXiv:1108.2015 [hep-th].
26. D. Cossins, "Working out what goes on inside black holes could mean ripping up all existing theories of reality", *New Scientist*, June 27, 2018.

Entropy from Carnot to Bekenstein

Ted Jacobson

Maryland Center for Fundamental Physics,
University of Maryland, College Park, MD 20742, USA

The second half of this essay tells the story of the genesis and early development of the notion of black hole entropy, in the style of an after dinner talk (apart from a long technical footnote). The first half sketches the development of the concept of entropy, beginning with Carnot, passing via Thomson, Clausius, Boltzmann, Planck, von Neumann, and Shannon on the way to Bekenstein, and the essay ends with a retrospective on Lemaître. The central theme is the fault-tolerant way that profound insights have emerged from simple (yet subtle) thermodynamic reasoning.

1. Introduction

Thermodynamics is a mysterious subject. Powerful conclusions can materialize out of scant inputs, and it is forgiving of fundamental misconceptions about the underlying details. Before recounting the tale of Jacob Bekenstein's discovery of black hole entropy and the generalized second law, I want to take a few steps further back, to review the origins of thermodynamics and the concept of entropy, and how it led to quantum physics and quantum field theory. This will help to place in perspective the sort of process by which radical insights into new physics can come about through the study of thermodynamics.

2. Heat Engines: Carnot, Thomson and Clausius

To begin with, it is amazing that Sadi Carnot in 1824 was able to infer that there is a universal upper bound to the efficiency of a heat engine, even while using an incorrect concept of heat and lacking the concept of total energy conservation.[a] A heat engine produces a quantity of work when drawing heat out of a thermal reservoir and depositing heat into a colder thermal reservoir. Carnot reasoned that the most efficient such engines could in principle come arbitrarily close to being

[a] A brief historical account of the work of Carnot, and its influence on the later developments of Thomson and Clausius is given in Ref. 1.

reversible.[b] He argued that all reversible heat engines, operating between the same two temperatures, must produce the *same* amount of work from the *same* amount of heat flow out of the hotter reservoir. Were that not the case, some of the work extracted from the more efficient engine could be injected to run the less efficient engine in reverse, while retaining the balance of the work for other purposes. If heat is a conserved "caloric" fluid, then when the less efficient engine is run backwards, the caloric flow from the colder to the hotter reservoir exactly cancels the flow that occurred in the more efficient engine. The result, as Carnot put it, would be "not only perpetual motion, but an unlimited creation of motive power without consumption either of caloric or of any other agent," which he rejected as being "inadmissable".

While Carnot's conclusion was correct, his argument contained a single deep flaw: heat is *not* by itself conserved! More heat flows out of the hot reservoir than flows into the colder reservoir, the difference being the work extracted. And less heat flows into the cold reservoir with the more efficient engine than with the less efficient engine. But now, if the less efficient engine is run backwards, the cold reservoir is no longer restored to its initial state: more heat is drawn out than went in. The leftover work, then, is not produced from *nothing*, but rather from the heat drawn out of the *colder reservoir*. While not as inadmissable as Carnot's result, this is nevertheless inadmissable. Its impossibility is Kelvin's version of the second law of thermodynamics. Alternatively, *all* of the work from the more efficient engine could be used to run the less efficient engine backwards, in which case the net result would be spontaneous (but engineered) heat flow from the colder reservoir to the hotter one, in violation of Clausius' version of the second law.

William Thomson (later Lord Kelvin) proposed in 1848 an absolute thermometric scale based on Carnot's theory, still using the concept of heat as a conserved caloric fluid that could not be converted to mechanical energy.[c] After being convinced of the mechanical equivalence of heat by James Joule's experiments, Thomson in 1851 revised his proposal. The new version was based on the fact that, for reversible heat engines, the ratio Q_h/Q_l of the heat Q_h drawn from the hotter reservoir to the heat Q_l deposited in the colder reservoir must be not only independent of the heat engine for a given Q_h (as required by the corrected Carnot argument), but also independent of Q_h. (Were the latter not the case, one could divide the operation into sub-processes and obtain reversible engines with different efficiencies.) The ratio Q_h/Q_l is therefore a property of the two reservoirs, independent of the construction of the engine, and so it serves to define in an absolute fashion the ratio of the temperatures of the two reservoirs, $T_h/T_l := Q_h/Q_l$. This is the

[b]This requires in particular that the two reservoirs are not in thermal contact.

[c]Thomson wrote: "The characteristic property of the scale which I now propose is, that all degrees have the same value; that is, that a unit of heat descending from a body A at the temperature T° of this scale, to a body B at the temperature $(T-1)^\circ$, would give out the same mechanical effect, whatever be the number T." As Klein explains: "The temperature T on the earlier absolute scale is a linear function of $\log\theta$, where θ is the corresponding temperature on the later absolute scale."

absolute thermodynamic temperature scale, which is defined up to an overall arbitrary constant multiple.

When re-expressed as $Q_h/T_h = Q_l/T_l$, Thomson's definition of temperature is directly connected to the concept of entropy which Rudolph Clausius developed over the decade from 1854 to 1865, when he finally gave entropy its name. If the entropy extracted from the hotter reservoir is defined as Q_h/T_h, and the entropy added to the colder reservoir is defined as Q_l/T_l, then a reversible engine can be described as one for which there is no net change of entropy: the entropy extracted from the hotter reservoir is exactly equal to that deposited in the colder one. At one level, this statement is virtually empty, being nothing but a restatement of the *definition* of thermodynamic temperature in different words. However, because the absolute temperature can be shown to be equivalent to the ideal gas thermometer scale, the criterion of zero entropy change has nontrivial implications for the macroscopic state variables. Moreover, this definition of entropy change can be integrated along a sequence of equilibrium states, to determine the difference in entropies of the initial and final equilibrium states. The result is independent of the sequence of states chosen to connect the two equilibria, because if it were to depend on the sequence, one could contrive a means to pump heat from a colder to a hotter reservoir, without the input of any net work, in violation of the second law. A concept of the entropy of an equilibrium state is thus defined, up to an overall additive constant, and up to the multiplicative constant inherent in the definition of absolute temperature. In a *non*-reversible engine, the ratio Q_l/Q_h is larger, so that more entropy enters the colder reservoir than leaves the hotter one, and thus entropy irreversibly increases.

3. Atoms and Radiation: Boltzmann and Planck

Clausius' entropy clearly had some relation to an underlying mechanical disorder, a relation to which Ludwig Boltzmann devoted three decades of thought, beginning at age 22 in 1866.[d] By 1871 he had obtained the expression $-\int \rho \ln \rho$ for the entropy of a dynamical system (not necessarily in equilibrium) as a phase space integral, the equilibrium distribution $\rho \propto \exp(-\beta H)$ being the probability density that the Hamiltonian has the value H at the temperature β^{-1}. The logarithm is defined only up to an additive constant, since a unit of phase space volume must be chosen in order to render the argument of the log dimensionless. This ambiguity was no cause for concern, because in any case Clausius' entropy was defined only up to an additive constant.

Boltzmann introduced in 1877 the notion that the probability of a macrostate of a gas could be identified with the number of compatible microstates, equilibrium being the configuration that maximizes this number, and entropy being its logarithm. To count the microstates he divided the velocity space of a molecule into small cells

[d]For an authoritative history of Boltzmann's work see *Atoms, Mechanics, and Probability. Ludwig Boltzmann's Statistico-Mechanical Writings. An Exegesis*, by Olivier Darrigol (Oxford: Oxford University Press, 2018).

of uniform volume, and described a microstate, or "complexion", by an assignment of cells to the atoms. He argued that the cell size is unimportant, as long as it can be chosen large enough to contain the velocities of a large number of atoms, while at the same time being small enough so that the relative variation of this number from one cell to the next is small. The cell size then affects significantly only the overall additive constant in the entropy, which plays no role in thermodynamics.

At this point the stage is set for the work of Max Planck.[e] Planck set out in 1894 to establish the second law of thermodynamics as an exact statement, like the conservation of energy. He initially thought the absorption and emission of electromagnetic radiation would be a fundamentally irreversible process that could account for an absolute formulation of the second law. At some point he realized this was simply inconsistent with classical electrodynamics, but nevertheless this is the concern that led him to the blackbody problem. Despite what we are told in textbooks and colloquia, the "ultraviolet catastrophe" was *not* the problem he set out to solve. (That phrase was introduced in 1911 by Paul Ehrenfest who was, by the way, a student of Boltzmann.) Following Boltzmann, Planck believed that entropy is well-defined out of equilibrium, and that it is maximized in equilibrium, which led him to the problem of understanding the blackbody spectrum from first principles.

Planck developed over a period of five years the theory of oscillators ("resonators") driven by a stochastic electromagnetic field and damped by radiation reaction. One product of this work was the equilibrium relation $\rho_\nu = 8\pi\nu^2 U_\nu/c^3$, between the energy density per unit frequency interval ρ_ν of the radiation field, and the average energy U_ν of a single damped oscillator with frequency ν. This reduced the problem to the one of finding the equilibrium distribution of U_ν values.

Wien had introduced a theoretical distribution law for ρ_ν which fit the experiments reasonably well at the time, and in 1899 Planck set out to account for Wien's distribution using thermodynamics. Taking Wien's law as given, Planck inferred the corresponding temperature dependence of U_ν, and from that he inferred that the entropy $S(U)$ of one resonator, as a function of its energy U, must satisfy the relation $S''(U) \propto -1/U$. He further showed that this relation is implied by the entropy maximization principle, Wien's displacement law, and the assumption that the entropy of a collection of N resonators of a given frequency depends only on their total energy, and not on N. This was all classical physics.

Soon came more data, however, extending the spectrum further into the infrared, and differing from the Wien distribution. This led Planck to realize that his assumption of the N-independence of the entropy was unjustified, and he quickly found the simplest modification of his formula for $S''(U)$ that would agree

[e]The motivation and evolving reasoning of Planck, as well as the reactions of his contemporaries to his work, was penetratingly analyzed by Thomas Kuhn in his masterpiece, *Blackbody Theory and the Quantum Discontinuity, 1894–1912*. The following comments are based on Kuhn's analysis, and on the forthcoming book by Anthony Duncan and Michel Janssen, *How We Got to Quantum Mechanics*.

with the observed fact that, for each frequency, at sufficiently high temperature $T(= 1/S'(U))$, the radiance is proportional to temperature. This led him directly to $S''(U) = -a/(U(1 + bU))$, which produced a spectral density that fit all of the data very well; in fact it produces precisely the Planck distribution! The equilibrium distribution of radiation was thus determined by two fundamental physical constants, a and b.[f] A giant step had been taken, but Planck as yet had no derivation of the new entropy formula from first principles.

For this he turned to Boltzmann's notion that entropy corresponds to the logarithm of the number of microstates compatible with a macrostate, and that entropy is maximized in equilibrium. Adapting Boltzmann's combinatorial counting of the microstates of gases, Planck set out to enumerate the number of ways that a given total energy can be distributed among a collection of resonators of different frequencies. The distribution of energy with respect to frequency that maximizes the total number of such "complexions" would then correspond to the equilibrium distribution. To this end it was necessary to adopt what could be called an energy 'bin size' (though Planck certainly did not refer to it this way), in order to discretize the counting. To maintain agreement with Wien's displacement law, he took the bin size to be proportional to the natural frequency ν of the resonators, $\Delta E = h\nu$, which required him to introduce a proportionality factor h with dimensions of action. This quickly led to the desired result for the entropy, at the end of 1900. As Kuhn convincingly argues,[g] contrary to what we are taught in school, and to what his early readers like Lorentz and Ehrenfest thought he was doing, Planck was most certainly *not* assuming that the energy of a resonator was itself quantized. In fact, he did not initially think that a break with classical physics was required.

Although Planck of course recognized that the observed radiation distribution depended on the constant h he had introduced to specify the bin sizes, he resisted the conclusion that classical physics was unable to account for that formula, despite the fact that the classical physics was complete without that constant. Eventually, he was convinced in 1908, by Lorentz (who had himself just come around, and who was not the only nor the first to note), that classical electromagnetic theory

[f]The same two constants had already entered the Wien distribution, which is after all just the high frequency limit of the Planck distribution, and Planck had enthusiastically noted that, together with Newton's constant and the speed of light, these allowed the construction of a natural system of units. Two years later in 1901, after having found his statistical derivation of the Planck distribution, he showed that consistency of equilibrium between radiation and ideal gas implied that these radiation constants must also determine the Avogadro constant N_A and the gas constant R (or Boltzmann's constant $k = R/N_A$), and vice versa. This allowed him to express the elementary electronic charge, which is related to N_A and R via electrochemical measurements, in terms of the measured radiation constants, and he found agreement with the available estimates.

[g]To all the evidence discussed by Kuhn, and by Duncan & Janssen, I would add Planck's Nobel Prize lecture in 1920, `https://www.nobelprize.org/prizes/physics/1918/planck/lecture/`. In that lecture, Planck nowhere mentions having quantized the energies of the oscillators. Rather, he refers to " 'elementary regions' or 'free rooms for action' ". He does describe the difficulties fitting his derivation into classical physics, but there he is (I believe) referring to subsequent developments.

cannot produce a sensible equilibrium distribution. In the meantime, only a few people, beginning with Ehrenfest and Einstein, saw that a break with classical physics was required. And since classical equipartition of electromagnetic energy, or maximization of the field entropy, led inevitably not to the Planck distribution but to the Jeans distribution at all frequencies, there must have been a flaw in Planck's reasoning if taken classically. In fact, Einstein had first noted the problem in 1906: the Planck distribution entailed a significant variation of the probability across one energy bin when $h\nu/kT$ is not much smaller than unity. By the end of the decade the conclusion was widely appreciated: only by quantizing the energy of the modes of the radiation field, and maximizing their entropy at fixed energy, could the Planck distribution be consistently derived. To account for a sensible theory of thermal equilibrium of electromagnetic radiation, quantum mechanics and quantum field theory had to be discovered! Yet, while the thermodynamics of radiation provoked the appearance of Planck's constant, it took another quarter century of effort, driven primarily by the challenge of understanding atomic spectra, before the complete theory of quantum mechanics and a rudimentary quantum electrodynamics were at hand.

4. Quantum Mechanics and Information Theory

Along with quantum mechanics came a radical shift in the nature of the probabilities that figure in the statistical definition of entropy. Quantum mechanics demanded a new role for probability in physics: that of governing indeterminism.[h] Classical states were replaced by quantum state vectors, whose squared amplitudes give fundamental probabilities (whatever those are). Moreover, because the state spaces of composite systems combine by tensor product, a generic quantum subsystem cannot be assigned a definite quantum state vector. This phenomenon of nonseparability, or entanglement of bipartite quantum states, was noted very early on by Lev Landau in 1927 (at the ripe old age of 19), in a paper on radiation damping of an electric dipole.[3] The opening lines of the paper read: "In wave mechanics, a system cannot be uniquely defined; we always have to do with a probabilistic ensemble (statistical conception). If the system is coupled with another one, its behavior has a double indeterminacy." Landau then proceeds to explain algebraically what is meant. The same year, John von Neumann introduced the entropy $S = -\mathrm{Tr}\rho \ln \rho$ associated with the density matrix ρ of a quantum system, in his paper "Thermodynamics of Quantum Mechanical Ensembles".[4] Five years later, in *The Mathematical Foundations of Quantum Mechanics*, von Neumann attributed to Landau's paper the notion that, in quantum mechanics, a subsystem of a pure quantum state can have entropy.

[h]In his Nobel Prize lecture [`https://www.nobelprize.org/prizes/physics/1954/born/lecture/`] Max Born gives an overview of the emergence of the statistical interpretation of quantum mechanics. For an in-depth historical look, focusing on the approaches of Jordan and von Neumann, see Ref. 2.

Our last station on the road to black hole entropy is information theoretic entropy, introduced by Claude Shannon in 1948.[5] For practical reasons, it was of interest to Shannon to quantify the information that can be transferred via a communication channel. The connection to thermodynamic entropy is that the microstate of a physical system can be viewed as a message whose information content is greater when the entropy of the corresponding macrostate is greater. Although quantum mechanics played no role in Shannon's considerations, the shift of focus from the state of the system to the state of knowledge of the communicator is in a sense required by quantum mechanics, at least if quantum probabilities are conceived as referring to information possessed by an observer. But the information theoretic interpretation of entropy further widens the application of this concept by detaching it from the particulars of the thermodynamics, phase space, or Hilbert space of a specific physical system, and focusing on the information acquisition of an agent who interacts with that system. As a result, the concept of entropy can be applied even when the nature of the system is only partially known. It is presumably for this generality, as well as its relation to the information acquired by an observer of a quantum system, that Jacob Bekenstein in 1972 reached for the information theoretic definition of entropy when trying to determine the entropy that should be assigned to a black hole.

5. Black Holes and Bekenstein

The genesis of black hole entropy has been traced to the famous story of John Wheeler and his graduate student, Jacob, discussing teacups and black holes. As Wheeler recalled it, the conversation took place in his office. In his 1990 book, *A Journey into Gravity and Spacetime*, he writes, "I told him [Jacob] of the concern I always feel when a hot cup of tea exchanges energy with a cold cup of tea."

I wanted to get to the bottom of this tea thing.[i] It was kind of confusing to me thinking of two cups of tea exchanging heat with each other. I mean what would be the arrangement to ensure good thermal contact? And anyway, how often do you find a cold cup of tea so close to a hot one? And why not just talk about the heat going from the tea into a single teacup when it was poured?

I found another reference, Wheeler's 1998 book: *Geons, Black Holes, and Quantum Foam: A Life in Physics*, written with Ken Ford. An Amazon.com search in the book yielded 8 passages in which the word tea appears — one being the index entry for the Fine Hall tea at Princeton. The first passage was about breaking for tea after discussions with Bohr, and mentioned that Bohr would lift the edge of the

---

[i]The following part of this article is closely based in part on an after-banquet speech given on the occasion of the conference, *40 Years of Black Hole Thermodynamics*, held in honor of Jacob Bekenstein at the Institute for Advanced Studies, Jerusalem, September, 2012. Here I have maintained the tone and style that were suited to that occasion. In particular, some of the text is written in first person.

rug in his office and kick the chalk bits under the carpet, to avoid being scolded by the janitor. Several were about tea in Fine Hall, and discussions there with physics luminaries. Wheeler says, "Bohr and I were regulars at the afternoon teas." He quotes Oppenheimer as saying that "Tea is where we explain to each other what we do not understand." Another entry mentions that Einstein invited Wheeler's general relativity class to tea at his house, during the last two years of Einstein's life. So it is clear that tea played a very important role in Wheeler's scientific life...

And then there is the legendary discussion with Jacob. Wheeler writes:

> *The idea that a black hole has no entropy troubled me, but I didn't see any escape from this conclusion. In a joking mood one day in my office, I remarked to Jacob Bekenstein that I always feel like a criminal when I put a cup of hot tea next to a glass of iced tea and then let the two come to a common temperature. My crime, I said to Jacob, echoes down to the end of time, for there is no way to erase or undo it. But let a black hole swim by and let me drop the hot tea and the cold tea into it. Then is not all evidence of my crime erased forever? This remark was all that Jacob needed. He took it seriously and went away to think about it.*

There are several notable and, frankly, puzzling things about this passage. First, Wheeler's 1990 "concern" has escalated into "feeling like a criminal" because of his facilitation of an increase of entropy. I read this passage to my wife, who is not a scientist. She asked me why did he feel like a criminal, if he was not breaking any laws? Good question. In fact, the second law of thermodynamics was fully upheld when the heat flowed from hot to cold, and the entropy increased. The crime, if there were one, would have been dropping the teacups into the black hole — *then* he might be breaking a law, namely the second law. Next, notice that the cold cup of tea has now become a glass of iced tea, further complicating the situation. I mean, wouldn't the iced tea really absorb more heat from the room temperature air than from the hot tea? From a physics point of view, however, the most interesting thing to me is that Wheeler said that he and others thought at the time that, since a black hole has zero temperature, it must have zero entropy. I guess Wheeler was thinking a zero temperature black hole was like a zero temperature material system which, if it had a unique ground state, would have zero entropy. But, unlike a material system, the black hole could absorb energy and still remain in equilibrium at zero temperature. One should therefore have thought it would have *infinite* entropy, because any time it absorbed heat Q its entropy would rise by the infinite amount $Q/0$.

Had I been there with Wheeler in his office that day, I think I would have said "What's the problem? The entropy went down into the black hole, so what?" And I may have tried to argue that since the black hole has zero temperature, it has infinite entropy, and therefore the second law, including the black hole entropy, is upheld. But, as Jacob appreciated, the problem is deeper than that: what good is the second law if you can't use it? It presumably applies to the accessible world, and nothing that goes on inside the black hole is accessible to those who remain outside. Moreover, Jacob believed that, like all other entropy, black hole entropy must be a measure of missing information. If a black hole really had infinite changes of entropy, then in any process an infinite amount of information would be lost, which sounds absurd. It should be possible to lose a finite amount of information, that is, to add a finite amount of entropy. This argument suggests that a black hole therefore must have a nonzero temperature. But Jacob didn't make the argument this way.

In an interview in the Israeli newspaper Haaretz [from June, 2012] Jacob spoke of his feelings about the "laws and order of physics". He is quoted there as saying: "I get a sense of security that not everything is random and that I can actually understand and not be surprised by things." Thermodynamics is the order that emerges from underlying randomness. I guess that in seizing upon Wheeler's question, Jacob was on a quest to tame the randomness of the world. He insisted on the second law, even when black holes are present, somehow knowing in his heart that the law must be truly fundamental.

The only hope for saving the second law in the presence of a black hole was to assign a finite entropy to the black hole itself. Roger Penrose had shown in 1969 that rotational energy can be extracted from a black hole,[6,7] and together with R. M. Floyd found in 1971 that, exploiting the fissioning of a body into two fragments, one with negative Killing energy and the other with positive Killing energy, such energy extraction always results in an increase of the horizon area.[8] In parallel, Demetrios Christodoulou had shown in 1970 that in such processes a quantity he called the *irreducible mass* — which is in fact proportional to the horizon area — cannot decrease.[9] And in 1971 Stephen Hawking proved a general theorem demonstrating that the horizon area cannot decrease.[10]

The obvious analogy of these results with the second law of thermodynamics suggested to Jacob that a black hole be assigned an entropy proportional to its horizon area.[11] But entropy is dimensionless (when temperature is assigned units of energy), so it was necessary to divide the horizon area by a universal constant with dimensions of area, and for this only one candidate presented itself: the (tiny) squared Planck length, $\hbar G/c^3$, which is equal to about $(10^{-33}\text{ cm})^2$. Jacob remarked that the appearance of $\hbar$ in the entropy "is not totally unexpected", since "the underlying states of any system are always quantum in nature", and he proposed the black hole entropy formula

$$S = \eta A/(\hbar G/c^3)\,, \tag{1}$$

where η is a dimensionless proportionality constant. As to the value of η, he wrote that

> *...it would be somewhat pretentious to attempt to calculate the precise value of the constant η without a full understanding of the quantum reality which underlies a "classical" black hole.*

Instead, he provided an *estimate* of η, using an argument which also provided an independent derivation of the entropy formula (1), using quantum mechanics and the information theoretic interpretation of entropy.

Jacob reasoned this way: The entropy of a black hole should correspond to the information it is hiding. When one additional bit of information is added, the entropy should increase by $\ln 2$, and there must therefore be a smallest nonzero amount of area that can be added *with certainty* to a black hole horizon, corresponding to one bit of information. In classical physics there is no lower limit. But he showed that, in view of the interplay of the gravitational redshift and the Heisenberg uncertainty relation, there is indeed in quantum physics a smallest area that can be added with certainty, and it has always the universal value $\sim \hbar G/c^3$, regardless of the mass, spin and charge of the black hole. Having thus determined that one bit of additional entropy corresponds to one square Planck length of area, he had established (1) from "first principles", and deduced that η is a number of order unity. He could then compute, using the Clausius relation, the "effective" temperature that should be assigned to a black hole. Were it not for quantum mechanics, the entropy would have been infinite and the temperature would have been zero. With quantum mechanics, the entropy was finite but huge, and the temperature was non-zero but tiny — indeed proportional to $\hbar$ (and in fact equal to the Hawking temperature up to an unknown order unity factor).[j]

With this understanding in hand, Jacob proposed the generalized second law (GSL) of thermodynamics: the sum of the ordinary entropy outside the black hole and the black hole entropy (1) can never decrease. To support this proposal, he

[j]Let me briefly explain in this technical footnote the argument just described. It made use of the first law of black hole mechanics which, as far as I know, was first derived by Jacob. He found that law simply by varying the mass, angular momentum, and charge parameters in the formula for the horizon area of a Kerr–Newman black hole. The result, with zero charge variation, can be written as $\delta A = (8\pi G/\kappa)(\delta M - \Omega_H \delta J)$, in units with $c = 1$. Here M is the mass, J is the angular momentum, Ω_H is the angular velocity of the horizon, and κ is the surface gravity. Jacob had a formula for κ in terms of the mass, spin and charge of the black hole, but he did not then know that this quantity is the surface gravity, which is an "intensive variable" that is constant on the horizon. That was shown by Bardeen, Carter and Hawking in 1973,[12] who called it the "Zeroth law". To minimize δA, one should thus minimize $\delta M - \Omega_H \delta J$, which is the variation of the conserved quantity corresponding to the horizon-generating Killing vector $\chi = \partial_t + \Omega_H \partial_\phi$. I'll call this conserved quantity the "boost energy", because on the horizon this Killing vector behaves like a Lorentz boost on a Rindler horizon in Minkowski spacetime. Jacob thus considered a particle of mass m outside the horizon, with 4-velocity parallel to χ, where its boost energy is equal to $|\chi|m$, with $|\chi|$ the norm of χ. This vanishes if the mass vanishes, or if the particle sits on the horizon, where $|\chi| = 0$. But the uncertainty relation precludes using vanishing boost energy to make δA vanish: If the particle is localized a distance d from the horizon (along the hypersurface orthogonal to χ), then $d \gtrsim \hbar/m$. If the mass is very small, then it is totally delocalized and cannot be known to have entered the black hole. On the other hand, if $\hbar/m$ is much smaller than the black hole, we can use the approximation $|\chi| \approx \kappa d$, so that the conserved quantity is $(\kappa d)m \gtrsim \hbar\kappa$. Referring back to the first law, this implies that $\delta A \gtrsim (8\pi G/\kappa)(\hbar\kappa) = 8\pi\hbar G = 8\pi(\text{Planck length})^2$. The smallest area increase is thus independent of the mass and spin of the black hole. The black hole entropy is therefore given by $S \sim A \ln 2/8\pi\hbar G$, corresponding to $\eta \sim \ln 2/8\pi$ in (1), and the temperature is $T_{\rm bh} \sim \hbar\kappa/\ln 2$ (which is $2\pi/\ln 2$ times the Hawking temperature, $T_H = \hbar\kappa/2\pi$). Notice that the minimal boost energy that can be added with certainty to the black hole coincides with $\sim T_{\rm bh}$.

argued that the black hole entropy should represent the inaccessible information about how the black hole might have formed, so that if a body with common entropy S_c is dropped into the black hole, S_{bh} must increase by at least S_c, implying that the generalized entropy must increase. This argument makes sense provided its premise, that S_{bh} represents the inaccessible information, is indeed valid. Of course he had no justification for that premise other than his deep insight and intuition that black hole thermodynamics should make sense. But he then tested the GSL, without assuming that premise. The first test involved a nonrelativistic oscillator enclosed in a box at finite temperature and dropped into a black hole, and he showed that the generalized entropy increases. The second test involved a beam of thermal radiation aimed at the black hole.

The story now takes a surprising twist. Of course the black hole absorbs radiation from the thermal beam and grows, increasing its entropy. But there is a problem: If the beam temperature is lower than T_{bh}, then heat flows from cold to hot! The GSL is violated because the beam entropy decreases by more than the black hole entropy increases.

In retrospect, we can see that to save the GSL there was only one way out: to suppose that the "effective" temperature T_{bh} of the black hole is in fact an *actual temperature*, so that the black hole must emit thermal radiation, and therefore entropy. However, Jacob did not make that argument. In his PRD paper he instead pointed out that, if the bath were colder than the black hole, the entropy-decreasing process of absorbing radiation would involve stochastic fluctuations, because the wavelength of the thermal radiation would be much larger than the black hole. And then came a crucial error: he asserted that in the domain of such fluctuations "the second law is irrelevant." His deeply felt, and correct, conviction — that the generalized second law must hold — led him to dismiss a counter example using a spurious argument.

Why did Jacob not instead infer that a black hole *must* have a real temperature, proportional to $\hbar$? I am fascinated and puzzled by this. I'm sure he understood these matters very well. While the second law does not hold at the level of individual fluctuations, it holds on average, as he himself stated in the very same paper. So, why did he not predict black hole radiation in order to save the GSL?

Jacob stated in the PRD article that attributing a *real* temperature to a black hole "can easily lead to all sorts of paradoxes, and is thus not useful"... but he doesn't say what those paradoxes might be. He states in his book, *Buchi Neri* (apparently published only in Italian), that he wrote up the article in between lectures at the famous Les Houches summer school in 1972. He recalls a long talk he had with Brandon Carter there, writing:

> *He stressed that since a black hole absorbs radiation perfectly, there can be no option but to assign it zero temperature, which contradicted my claim that the black hole has a real temperature. I almost lost my confidence then, and this accounts for the somewhat weak statement about black hole temperature I make in my paper...*

But while it is perhaps true that nothing can escape from a black hole, there is another, overlooked possibility: radiation could escape from the *outside*, with no causality violation, as Hawking discovered in 1974.[13]

It seems that what was missing from Jacob's thoughts at the time is just one fundamental point: even in "empty space," a black hole is immersed in the vacuum of quantum fields. I conjecture that, although he thought a lot about field quanta falling into a black hole, he never thought about the quantum field *vacuum* falling into a black hole. Had that thought crossed his mind, one day in 1972, I venture to guess that he would have immediately realized that the vacuum fluctuations give the black hole a *real* temperature, which could uphold the GSL in all settings, even when fluctuations dominate. I'd also guess that Jacob would have gone on to conceive the information-theoretic notion that black hole entropy arises from quantum entanglement of those vacuum fluctuations on either side of the horizon, an idea only suggested ten years later by Rafael Sorkin.[14] As it happened, Hawking discovered this real temperature two years later while investigating other matters entirely. Because Hawking's calculation doesn't involve "the quantum reality which underlies a classical black hole," but rather only the quantum reality of test fields on a classical black hole background, it turned out that Jacob's pessimism regarding the possibility of computing the proportionality constant η was unfounded. Indeed, Hawking's calculation showed that $\eta = 1/4$.

6. Cosmology and Lemaître

This story of horizons and the entropy of quantum fluctuations can actually be traced back an additional forty years before Jacob's paper, to the work of Georges Lemaître, a physicist, mathematician, and Roman Catholic priest from Belgium. Although is it not commonly known, in 1933 Lemaître was the first person to understand the nature of a black hole horizon, long before it was called a black hole. He realized that a black hole horizon is locally equivalent to a cosmological de Sitter horizon.[15,16] Other than Felix Klein, who had pointed to the globally nonsingular de Sitter hyperboloid, Lemaître was the first to understand that one can fall freely without disaster across a de Sitter horizon.[k] He showed this using a coordinate transformation, and he applied the same type of transformation to understand the black hole horizon. As later shown by Gary Gibbons and Hawking,[18] de Sitter horizons have an entropy and a temperature, much as do black hole horizons.

Lemaître was ahead of his time in addressing seriously, with all the currently available physics, the questions raised by cosmology. He deduced the distance-redshift relation, and using available data inferred a value for the constant in "Hubble's law" for the universal expansion before Hubble did,[19–22] argued that without a cosmological constant the universe would be too young to accommodate the geological age of the earth (given the universal expansion rate, which at that time was

[k] de Sitter, Hermann Weyl, and Einstein all initially thought de Sitter space was somehow singular at the horizon.[17]

incorrectly estimated to be around seven times its actual value), and formulated a theory of the origin and growth of structure. Noting that entropy increases according to the second law as the universe evolves, Lemaître believed that there must have been an initial state of zero entropy. He wrote a visionary paper about this in 1931.

Sir Arthur Eddington had mentioned this issue of zero initial entropy, in a 1931 address to the Mathematical Association, titled "The End of the World: from the Standpoint of Mathematical Physics," and published in the scientific journal Nature. As the title suggests, the main subject was the infinite future and the so-called heat death of the universe. But he also discussed the beginning of the universe, saying: "We have come to an abrupt end of space-time — only we generally call it the beginning"... and he opined, "Philosophically, the notion of a beginning of the present order of Nature is repugnant to me."

Lemaître responded to this in a brief letter to Nature (May 9, 1931), less than half a page long. On the same page of the journal with a letter from British Columbia on the chemistry of cheese ripening, and another letter with a photograph of insects found in the gut of a cobra in Malaysia, is Lemaître's letter, titled: "The Beginning of the World from the Point of View of Quantum Theory". In the letter he traces the universe back to the ground state of a primeval atom. He writes

> *...it may be that an atomic nucleus must be counted as a unique quantum, the atomic number acting as a kind of quantum number. If the future development of quantum theory happens to turn in that direction, we could conceive the beginning of the universe in the form of a unique atom, the atomic weight of which is the total mass of the universe. This highly unstable atom would divide in smaller and smaller atoms by a kind of super-radioactive process.*

Lemaître goes on to suggest that "Some remnant of this process might, according to Sir James Jeans's idea, foster the heat of the stars until our low atomic number atoms allowed life to be possible." We now know that in fact the elements evolved in the *opposite* direction: the actual source of stellar energy is not radioactive fission, but rather the fusion of light elements into heavier ones. And life could evolve only once stars that forged these elements exploded in supernovae, scattering their nuclear progeny for future chemistry. But the notion that the universe we see originated from the decay of a symmetrical, zero entropy quantum state is perfectly in line with our best current understanding.

Looking back at the beginning of time, Lemaître squarely faces what could be called the cosmic information problem: the initial, pure quantum state of his primordial atom would be unique, with no entropy. Quantum mechanics would preserve this purity, so from whence could the variety of all creation be found? Lemaître had no hesitation answering this question. He wrote:

> *Clearly the initial quantum could not conceal in itself the whole course of evolution; but, according to the principle of indeterminacy, that is not*

> *necessary. [...] the whole story of the world need not have been written down in the first quantum like a song on the disc of a phonograph. The whole matter of the world must have been present at the beginning, but the story it has to tell may be written step by step.*

To my eye, this bears an uncanny resemblance to the modern account of primordial cosmology. The primeval atom of Lemaître's intuition can be identified with the unstable vacuum of inflationary cosmology. The early universe, according to inflation, was a symmetrical, pure quantum state, the de Sitter vacuum, which spontaneously decayed into a plasma of matter and radiation. The structure we see today emerged from that zero entropy vacuum as a result of quantum indeterminism. It arose via gravitational instability, seeded by tiny deviations from equilibrium which had descended from nothing but primordial vacuum fluctuations.

Acknowledgments

I am grateful to B. Banihashemi, S. De Baerdemaeker, J. D. Dorfman, J. Uffink, and especially M. Janssen for comments, suggestions, and help with historical accuracy. This work was supported in part by NSF grant PHY-1708139.

References

1. M. J. Klein, Carnot's contribution to thermodynamics, *Physics Today* **27**, 23 (1974).
2. A. Duncan and M. Janssen, (Never) Mind your p's and q's: Von Neumann versus Jordan on the foundations of quantum theory, *Eur. Phys. J. H* **38**, 175–259 (2013).
3. L. Landau, Das Dämpfungsproblem in der Wellenmechanik, *Zeitschrift fur Physik* **45**, 430–441 (1927).
4. J. von Neumann, Thermodynamik quantenmechanischer gesamtheiten, *Nachrichten von der Gesellschaft der Wissenschaften zu Gättingen, Mathematisch-Physikalische Klasse* **1927**, 273–291 (1927).
5. C. E. Shannon, A mathematical theory of communication, *SIGMOBILE Mob. Comput. Commun. Rev.* **5**(1), 3–55 (2001).
6. R. Penrose, Gravitational collapse: The role of general relativity, *Nuovo Cimento Rivista Serie*, 1 (1969).
7. R. Penrose, "Golden Oldie": Gravitational collapse: The role of general relativity, *Gen. Rel. Grav.*, **34** (2002).
8. R. Penrose and R. M. Floyd, Extraction of rotational energy from a black hole, *Nature Phys. Sci.* **229**, 177–179 (1971).
9. D. Christodoulou, Reversible and irreversible transformations in black-hole physics, *Phys. Rev. Lett.* **25**, 1596–1597 (1970).
10. S. W. Hawking, Gravitational radiation from colliding black holes, *Phys. Rev. Lett.* **26**, 1344–1346 (1971).
11. J. D. Bekenstein, Black holes and entropy, *Phys. Rev. D* **7**, 2333–2346 (1973).
12. J. M. Bardeen, B. Carter and S. W. Hawking, The four laws of black hole mechanics, *Commun. Math. Phys.* **31**, 161–170 (1973).
13. S. W. Hawking, Black hole explosions? *Nature* **248**, 30–31 (1974).
14. R. D. Sorkin, 1983 paper on entanglement entropy: "On the entropy of the vacuum outside a horizon", arXiv:1402.3589.

15. G. Lemaître, L'Univers en expansion, *Annales de la Société Scientifique de Bruxelles*, **53** (1933).
16. G. Lemaître, The expanding universe, *Gen. Rel. Grav.* **29**, 641–680 (1997) [*Annales Soc. Sci. Brux. Ser. I Sci. Math. Astron. Phys. A* **53**, 51 (1933)].
17. J.-P. Luminet, *L'invention du Big Bang*, 2nd edn. (Le Seuil, Paris, 2014).
18. G. W. Gibbons and S. W. Hawking, Cosmological event horizons, thermodynamics, and particle creation, *Phys. Rev. D* **15**, 2738–2751 (1977).
19. G. Lemaître, Un Univers homogène de masse constante et de rayon croissant rendant compte de la vitesse radiale des nébuleuses extra-galactiques, *Annales de la Société Scientifique de Bruxelles* **47**, 49–59 (1927).
20. G. Lemaître, Republication of: A homogeneous universe of constant mass and increasing radius accounting for the radial velocity of extra-galactic nebulae, *Gen. Rel. Grav.* **45**, 1635–1646 (2013).
21. J.-P. Luminet, Lemaître's big bang, arXiv:1503.08304.
22. M. Livio, Lost in translation: Mystery of the missing text solved, *Nature* **479**, 171–173 (2011).

Developments in Black Hole Entropy

Juan Maldacena
Institute for Advanced Study,
Princeton, NJ 08540, USA

We give a brief overview of black hole entropy, covering main developments since Bekenstein's original proposal.

1. Black Hole Entropy

The idea that black holes have an entropy proportional to the area has been a remarkable realization.[1] It is a surprising confluence of statistical mechanics, quantum mechanics, special relativity and gravity. This can be seen clearly by restoring all the constants normally set to one in the black hole entropy formula[1,2]

$$S = \frac{(\text{Area})_{\text{horizon}}}{4l_{\text{Planck}}^2} \to S = k\frac{(\text{Area})c^3}{4G_N\hbar} \tag{1}$$

The formula is most remarkable because a black hole is just a solution of the classical equations. This formula, together with Einstein equations is consistent with the first law of thremodynamics[3]

$$\delta M = T\delta S \tag{2}$$

where M is the mass of the black holes and T is the temperature of the black hole, given by Hawking's formula.[2] The fact that the area of the horizon increases[4] is consistent with the 2nd law of thermodynamics.[5] It also shows that (1) is a coarse grained entropy, which is the kind of entropy that increases under the second law.

These results show that black holes behave as ordinary thermal systems for an observer who remains outside the black hole. They further suggest that in the full quantum theory, an outside observer, could view the black hole as an ordinary quantum system. Finding ways to realize this idea was a central topic of research in the past forty years.

Bekenstein proposed a generalized second law for the total entropy, given by the area of the horizon plus the entropy outside the horizon[5]

$$S_{\text{total}} = \frac{(\text{Area})}{4l_{\text{Planck}}^2} + S_{\text{matter outside}} \tag{3}$$

The validity of this generalized second law seems to imply some constraints on the entropic content of matter.[6] This arises from a thought experiment involving matter falling into a black hole. If we call S_m the entropy of the matter, we find that the entropy of the outside decreased by S_m. However, the area of the black hole increased, as implied by (2), $\delta S_{BH} = \delta E_m/T$. So the generalized second law of thermodynamics would be obeyed if this later quantity is larger than S_m. This implies that there should be an upper bound on the entropy content in terms of the energy. Matter with some entropy should necessarily carry some energy. Now, this should hold for matter that is close to the black hole horizon. An important property of black holes is that the asymptotic time translation symmetry acts as a boost in the near horizon region. Namely, near the horizon we can define coordinates $x^\pm$ where the radial and time direction has a usual flat space geometry $ds^2 = -dx^+dx^- + \cdots$. The horizons are at $x^\pm = 0$ (past and future horizons). The asymptotic time translation symmetry that shifts the asymptotic time t acts as a boost in the $x^\pm$ plane. We can normalize the boost generator so that x^+ has eigenvalue one. In terms of Rinder coordinates, $x^\pm = \pm re^{\pm\tau}$ the boost generator is conjugate to τ time translations. The time τ is related to the asymptotic time t by

$$t = \frac{1}{2\pi T}\tau \tag{4}$$

where T is the Hawking temperature. The precise rescaling is obtained by the full black hole geometry and (4) would just be the expression for the Hawking temperature. In other words, in terms of the time τ the inverse Hawking temperature is 2π, so that the physical value measured at infinity comes purely from the rescaling of the two time coordinates. The same rescaling factor appears in the energy measured with respect to time t versus the boost energy measured with respect to time τ

$$\frac{E}{T} = \frac{E_t}{T} = 2\pi B \tag{5}$$

Then the necessary condition for obeying the second law is that

$$S_m \leq 2\pi B_m \tag{6}$$

where B_m is the boost eigenvalue of the matter. Bekenstein's formulation of the bound involved bounding the entropy in terms of the energy and the size R of the system.[6] However, (6) is a more precise form of the bound, it is what can be proven and it is also what we need for the second law.

It was suspected for many years that (6) would imply some restriction on the types of QFT that can be coupled to gravity. However, it was shown by Casini,[7] bulding on previous ideas by Ref. 8, that a version of (6) is actually true in any relativistic quantum field theory.

Before discussing the proof, it is necessary to notice that (6) has an $\hbar$ on the left hand side. So the bound is trivially obeyed in the classical limit. In order to get close to saturating the bound it is necessary to consider quantum effects. In relativistic quantum field theory we cannot localize particles or excitations. Therefore, in order to talk about a bound like (6) we need to define things more precisely. In fact, the proper quantum version involves the von Neumann entropy of the quantum field theory state restricted to the right Rindler wedge. This von Neumann entropy is sometimes called "entanglement" entropy since it arises because we are dividing a pure state into two parts. We can compute this entropy for the state in question, containing some matter, and we subtract the same quantity for the vacuum to form the difference ΔS. Then the bound is[7]

$$\Delta S \leq 2\pi \langle \Delta B \rangle \tag{7}$$

where ΔB is the expectation value of the Boost generator in the state in question minus its expectation value in the vacuum. There are divergencies when we compute the entropies but these cancel when we compute the differences. There can be some remaining ambiguities in the definition of the entropy but these are the same as the ambiguities defining the precise form of the boost generator on the half space. It turns out that (7) is the same as the positivity of relative entropy.[7] Equivalently, we can think of the vacuum as thermal state on the Rindler wedge. Then (7) reduces to the statement that the difference in free energy between the state in question and the vacuum is positive, which is true since the thermal equilibrium state minimizes the free energy.

In recent years the discussion of black hole thermodynamics has been extended to the semiclassical domain, where we consider also quantum fields and include their entropy. In that case, a more precise version of black hole entropy includes also the von Neumann entropy of the quantum fields outside the black hole horizon[9]

$$S_{\text{total}} = \frac{(\text{Area})}{4G_N} + S_{V.N.} + \cdots \tag{8}$$

where the dots are some extra pieces that we will not discuss in detail here, having to do with extra Wald terms[10] and counterterms. This a more precise version of (3), which is the correct expression up to order G_N^0 in the semiclassical expansion. Note that while (1) depends only on gravity, (8) depends also on the type of matter fields that we have in the theory. Using the monotonicity of relative entropy, A. Wall, has proven[11] that the second law of black hole thermodynamics holds for (8). This shows that no new inequality is needed from matter other than the ones that follow from the fact that matter is described by relativistic quantum field theory. In particular, the precise form (7) of the Bekenstein bound is automatically obeyed.

The fact that black hole entropy exists, and obeys the second law, can be viewed as evidence that black holes should have a unitary description as viewed from the outside. Otherwise, if unitarity were preserved only after adding a second asymptotic region beyond the singularity, then why couldn't entropy simply disappear into that region?

2. The Search for Black Hole Microstates

We can think of the formula for black hole entropy as a kind of "experimental" result, analogous to the entropy that an experimentalist would measure on a real world material by staring from zero temperature and putting in energy slowly to get it to a desired final sate, and computing the entropy from the first law. Since the work of Boltzmann we expect that entropy should be associated to the number of microstates of the system. So the question is: what are the microstates that give rise to the entropy of black holes? One would expect that these microstates should be a fairly universal since, at leading order, their number is independent of the details of the theory.

The most naive picture for these microstates comes from simply computing the von Neumann entropy of the quantum fields outside the black hole. This gives a UV divergent quantity going like[9]

$$S_{V.N.} = \frac{(\text{Area})}{\epsilon^2} + \cdots \tag{9}$$

where ϵ is the UV cutoff. In the semiclassical approximation we should always choose the cutoff ϵ to be larger than l_{Planck}. Therefore (9) is smaller than (1). However, this has not prevented speculation that perhaps by some suitable principle we would find a cutoff procedure that would reproduce the black hole entropy. The most precise ideas in this direction involve using a Pauli–Villars-type regulator for the fields and viewing the Newton constant as arising as induced gravity.[12] This has the drawback that the UV sensitive terms come from ghosts. A more correct and solid point of view is to think of (9) as part of a correction to black hole entropy and view the divergent term as a term that is canceled by a similar renormalization of the Newton constant, which leads to a second area term included in the dots in (8) that cancels the divergent terms in (9). This cancellation is automatic when we compute the free energy and the entropy using Euclidean methods.[13] But this still leaves us without a concrete picture for the microstates. Nevertheless it suggests that they are related to states that exist near the UV of the bulk. These are states that do not have an explicit, calculable description, within the regime that gravity is a good approximation. Of course, the black hole entropy formula still "knows" about them.

2.1. *Black hole entropy in string theory*

The above discussion suggests that in a UV finite theory of gravity perhaps the von Neumann entropy is finite and accounts for the full entropy of the black hole. String theory is definitely UV finite.[14] Unfortunately, it is not known how to compute the Von Neumann entropy of a subregion. It had been speculated that perhaps the black hole entropy is related to the entropy of open strings ending on the horizon.[15] Though this is an attractive idea, to the best of my knowledge, there is no precise computation of the entropy along these lines. However, there are two more indirect approaches to compute black hole entropy in string theory which we discuss below.

2.1.1. *Black hole entropy for supersymmetric black holes*

A very fruitful set up involves the computation of the entropy of extremal black holes in supersymmetric gravity theories coming from string theory compactifications.[16] Extremal black holes are a type of charged black hole. A non-singular charged black hole obeys the condition $M \geq Q$ (in some units). An extremal black hole obeys $M = Q$. In sufficiently supersymmetric theories this sometimes coincides with a BPS bound.[17] A BPS bound is a bound that arises from the supersymmetry algebra, which also has the form $M \geq Q$. States saturating this bound are specially protected, in the sense that supersymmetry implies that their number does not change when we change the coupling constants of the theory. By changing the coupling we can start with a black hole solution and turn it into in a weakly coupled collection of strings and D-branes.[18] We will not describe in detail what these objects are. The important point is that they obey clear and simple rules. These objects can be assembled in a number of ways, given the initial charges. For large charges, this number agrees precisely with the one expected from the Bekenstein-Hawking black hole entropy formula.[16]

This counting has been done with increasing degree of precision. Quantum corrections as in (8) have been matched. Most recently it was understood how to perform the gravity computation in a non-perturbative fashion in order to reproduce the precise counting of microstates. See Ref. 19 for a review and further references. Of course, such computations depend on the details of the theory because the entropy of black holes does indeed depend on the details when once we go beyond the area term.

This has been a great success and it gives us confidence that string theory is a fully consistent theory of quantum gravity.

2.1.2. *Black hole entropy from the gauge/gravity duality*

Black hole entropy can also be studied from the point of view of the gauge/gravity duality.[20–22] This is a duality that equates gravity in asymptotically anti-de Sitter (AdS) boundary conditions with a strongly coupled conformal field theory. A black hole in AdS is equated with a thermal state in the conformal field theory. This state is a hot fluid of strongly interacting particles. The entropy of the black hole is equated to the entropy of this fluid in the quantum field theory. In general, this is difficult to compute because the field theory is strongly coupled. A particularly simple case is the three-dimensional case, AdS_3. Here the boundary is $1+1$-dimensional and the powerful conformal symmetry can be used to compute the thermal entropy as[23]

$$S = \frac{\pi}{3} c L T \tag{10}$$

where L is the size of the boundary region, T is the temperature and c is the so-called central charge of the two-dimensional conformal field theory. This central charge can be computed also in the gravity theory,[24] $c = 2\frac{R_{AdS_3}}{2G_{N\,3}}$. So, one can start from the gravity theory, compute the central charge, assume that the theory has a unitary realization as a CFT and then compute the black hole entropy.[25]

Another approach has been to compute the entropy (or energy) numerically in the quantum mechanical description.[26] This was done for an interacting matrix quantum mechanics theory[27] that has a gravity dual.[28] This has the advantage of being a direct computation, without any reliance on conformal symmetry. In fact, this quantum mechanical model is not conformal invariant.

3. From Coarse Grained Entropy to Fine Grained Entropy

As we mentioned before, the entropy computed by the Bekenstein–Hawking formula should be interpreted as a coarse grained entropy, since it grows under time evolution, obeying the second law of thermodynamics. One can wonder whether there is any formula that computes the full von Neumann entropy of the microscopic density matrix. In fact, such a formula was proposed by Ryu and Takayanagi[29] and Hubeny–Rangamani–Takayanagi[30] (HRT). Their prescription is to find an extremal surface in the whole geometry. A surface which has minimal area on a given spacelike slice but has maximal area among all possible spacelike slices.[31] This prescription becomes particularly interesting for spacetimes containing a black hole which is connected through the interior to a second asymptotic region. An example is the standard eternal Schwarzschild black hole, which is the maximal analytic extension of the Schwarzschild solution containing two asymptotic regions, see Fig. 1(a). These configurations can be viewed as entangled black holes.[32,33] In this case the extremal surface is the bifurcation surface and it coincides with the horizon. A more interesting case arises when we consider a deformation of this geometry obtained by adding matter that falls into the two black holes see Fig. 1(b). In this case the Penrose diagram becomes more elongated and the extremal surface area is smaller than the area of the horizon. Furthermore, the extremal surface is the same at all times, a fact which is related to the constancy of the fine grained entropy of the system under unitary evolution. An interesting aspect of the extremal surface (or HRT) surface is that it typically sits behind the event horizon, see Fig. 1(b).

The HRT formula is even more remarkable that the Bekenstein–Hawking formula because it knows about the microscopic entropy of the system. In other words, one might think that gravity is "just" a hydrodynamic or thermodynamic approximation to the exact system. However, the HRT formula shows that gravity knows about the microscopic von Neumann entropy of the system. In the context of holography, the HRT formula can be used to compute the entanglement entropy of subregions of the boundary quantum field theory. This has been useful for two reasons. First, it has provided a tool for studying the entanglement patters of strongly coupled field theories. The lessons from these holographic theories serve as interesting examples in order to understand the problem in more general theories. Second, the fact that the entropy is given by a simple geometric quantity suggests that entanglement plays a crucial role in determining the geometry of spacetime.

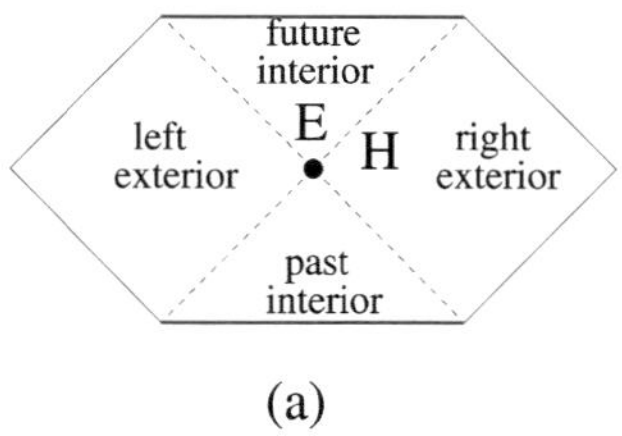

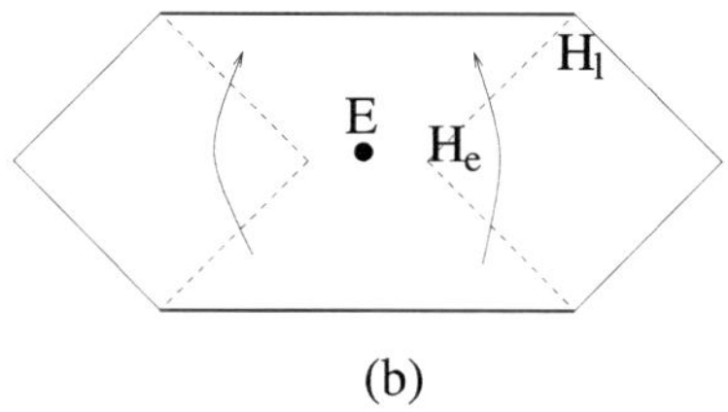

Fig. 1. (a) The Penrose diagram of the eternal Schwarschild solution. It has the geometry of a non-traversable wormhole connecting two regions that look like the exterior geometry of a black hole. These regions are sharing the interior regions. In this case, the external surface E coincides with the event horizon H, whose area is time independent. (b) We consider a configuration where we add extra matter falling into the black holes, represented here by the arrows. Now the event horizon area is a minimum at H_e and grows towards its late times H_l, with $A_{H_l} > A_{H_e}$ as implied by the area theorem. The extremal surface is inside the wormhole and has a smaller area $A_E < A_{H_e} < A_{H_l}$.

4. Conclusions

Black hole entropy has been a central object of interest in the exploration of quantum aspects of gravity. The generalized entropy was defined in the quantum theory and the generalized second law was proven. This proof works for any matter theory described by relativistic quantum field theory, no further constraint on matter was necessary. In particular, a version of the Bekenstein bound (7) is automatically obeyed for any quantum field theory. Through string theory we can now identify the black hole microstates. There are two broad ways to do this. One involves supersymmetric black holes, where one can count the states by going to weakly coupled limit without losing any states. The other involves black holes in special spaces, such as anti-de Sitter spacetimes, where one can find a dual formulation in terms of a strongly interacting theory on the boundary. In this case, the black hole looks like a hot fluid and the entropy is the entropy of this fluid. This describes the entropy of both the black hole and the spacetime around it. This boundary description is manifestly unitary. So, if we believe the duality we conclude that black holes preserve information. An important problem that remains is to describe the black hole interior in the full quantum theory. In other words, we would like to be able to describe the interior using the same variables in which we describe the microstates with a unitary dynamics. In fact, several paradoxes arise in some approaches to the problem.[34,35]

The entropy defined in terms of the area of the event horizon is a coarse grained entropy. A fined grained notion of entanglement entropy was also recently defined and has proven also to be very useful.

References

1. J. D. Bekenstein, *Phys. Rev. D* **7**, 2333 (1973), doi:10.1103/PhysRevD.7.2333.
2. S. W. Hawking, *Commun. Math. Phys.* **43**, 199 (1975), doi:10.1007/BF02345020,

10.1007/BF01608497 [*Euclidean Quantum Gravity*, eds. G. W. Gibbons and S. W. Hawking (World Scientific, 1973), pp. 167–188].
3. J. M. Bardeen, B. Carter and S. W. Hawking, *Commun. Math. Phys.* **31**, 161 (1973), doi:10.1007/BF01645742.
4. S. W. Hawking, *Phys. Rev. Lett.* **26**, 1344 (1971), doi:10.1103/PhysRevLett.26.1344.
5. J. D. Bekenstein, *Phys. Rev. D* **9**, 3292 (1974), doi:10.1103/PhysRevD.9.3292.
6. J. D. Bekenstein, *Phys. Rev. D* **23**, 287 (1981), doi:10.1103/PhysRevD.23.287.
7. H. Casini, *Class. Quant. Grav.* **25**, 205021 (2008), `arXiv:0804.2182 [hep-th]`, doi: 10.1088/0264-9381/25/20/205021.
8. D. Marolf, D. Minic and S. F. Ross, *Phys. Rev. D* **69**, 064006 (2004), `arXiv:hep-th/0310022 [hep-th]`, doi:10.1103/PhysRevD.69.064006.
9. L. Bombelli, R. K. Koul, J. Lee and R. D. Sorkin, *Phys. Rev. D* **34**, 373 (1986), doi:10.1103/PhysRevD.34.373.
10. R. M. Wald, *Phys. Rev. D* **48**, R3427 (1993), `arXiv:gr-qc/9307038 [gr-qc]`, doi: 10.1103/PhysRevD.48.R3427.
11. A. C. Wall, *Phys. Rev. D* **85**, 104049 (2012), `arXiv:1105.3445 [gr-qc]`, doi:10.1103/PhysRevD.85.104049 [Erratum *ibid.* **87**, 069904 (2013), doi:10.1103/PhysRevD.87.069904].
12. J.-G. Demers, R. Lafrance and R. C. Myers, *Fields Inst. Commun.* **15**, 227 (2001), `arXiv:gr-qc/9507042 [gr-qc]`.
13. G. W. Gibbons and S. W. Hawking, *Phys. Rev. D* **15**, 2752 (1977), doi:10.1103/PhysRevD.15.2752.
14. E. Witten (2012), `arXiv:1209.5461 [hep-th]`.
15. L. Susskind and J. Uglum, *Phys. Rev. D* **50**, 2700 (1994), `arXiv:hep-th/9401070 [hep-th]`, doi:10.1103/PhysRevD.50.2700.
16. A. Strominger and C. Vafa, *Phys. Lett. B* **379**, 99 (1996), `arXiv:hep-th/9601029 [hep-th]`, doi:10.1016/0370-2693(96)00345-0.
17. R. Kallosh, A. D. Linde, T. Ortin, A. W. Peet and A. Van Proeyen, *Phys. Rev. D* **46**, 5278 (1992), `arXiv:hep-th/9205027 [hep-th]`, doi:10.1103/PhysRevD.46.5278.
18. J. Polchinski, *Phys. Rev. Lett.* **75**, 4724 (1995), `arXiv:hep-th/9510017 [hep-th]`, doi:10.1103/PhysRevLett.75.4724.
19. A. Sen, *Gen. Rel. Grav.* **46**, 1711 (2014), `arXiv:1402.0109 [hep-th]`, doi:10.1007/s10714-014-1711-5.
20. J. M. Maldacena, *Int. J. Theor. Phys.* **38**, 1113 (1999), `arXiv:hep-th/9711200 [hep-th]`, doi:10.1023/A:1026654312961,10.4310/ATMP.1998.v2.n2.a1, [*Adv. Theor. Math. Phys.* **2**, 231 (1998)].
21. S. S. Gubser, I. R. Klebanov and A. M. Polyakov, *Phys. Lett. B* **428**, 105 (1998), `arXiv:hep-th/9802109 [hep-th]`, doi:10.1016/S0370-2693(98)00377-3.
22. E. Witten, *Adv. Theor. Math. Phys.* **2**, 253 (1998), `arXiv:hep-th/9802150 [hep-th]`, doi:10.4310/ATMP.1998.v2.n2.a2.
23. J. L. Cardy, *Nucl. Phys. B* **270**, 186 (1986), doi:10.1016/0550-3213(86)90552-3.
24. J. D. Brown and M. Henneaux, *Commun. Math. Phys.* **104**, 207 (1986), doi:10.1007/BF01211590.
25. A. Strominger, *JHEP* **02**, 009 (1998), `arXiv:hep-th/9712251 [hep-th]`, doi:10.1088/1126-6708/1998/02/009.
26. E. Berkowitz, E. Rinaldi, M. Hanada, G. Ishiki, S. Shimasaki and P. Vranas, *Phys. Rev. D* **94**, 094501 (2016), `arXiv:1606.04951 [hep-lat]`, doi:10.1103/PhysRevD.94.094501.
27. T. Banks, W. Fischler, S. H. Shenker and L. Susskind, *Phys. Rev. D* **55**, 5112 (1997), `arXiv:hep-th/9610043 [hep-th]`, doi:10.1103/PhysRevD.55.5112 [435 (1996)].

28. N. Itzhaki, J. M. Maldacena, J. Sonnenschein and S. Yankielowicz, *Phys. Rev. D* **58**, 046004 (1998), `arXiv:hep-th/9802042 [hep-th]`, doi:10.1103/PhysRevD.58.046004.
29. S. Ryu and T. Takayanagi, *Phys. Rev. Lett.* **96**, 181602 (2006), `arXiv:hep-th/0603001 [hep-th]`, doi:10.1103/PhysRevLett.96.181602.
30. V. E. Hubeny, M. Rangamani and T. Takayanagi, *JHEP* **07**, 062 (2007), `arXiv:0705.0016 [hep-th]`, doi:10.1088/1126-6708/2007/07/062.
31. A. C. Wall, *Class. Quant. Grav.* **31**, 225007 (2014), `arXiv:1211.3494 [hep-th]`, doi: 10.1088/0264-9381/31/22/225007.
32. W. Israel, *Phys. Lett. A* **57**, 107 (1976), doi:10.1016/0375-9601(76)90178-X.
33. J. M. Maldacena, *JHEP* **04**, 021 (2003), `arXiv:hep-th/0106112 [hep-th]`, doi:10. 1088/1126-6708/2003/04/021.
34. S. D. Mathur, *Class. Quant. Grav.* **26**, 224001 (2009), `arXiv:0909.1038 [hep-th]`, doi:10.1088/0264-9381/26/22/224001.
35. A. Almheiri, D. Marolf, J. Polchinski and J. Sully, *JHEP* **02**, 062 (2013), `arXiv:1207.3123 [hep-th]`, doi:10.1007/JHEP02(2013)062.

Quantum Black Holes

Viatcheslav Mukhanov
Theoretical Physics, Ludwig Maxmillians University,
Theresienstr. 37, 80333 Munich, Germany
viatcheslav.mukhanov@physik.lmu.de

In 1974 Jacob Bekenstein has noticed that the area of the black hole behaves as adiabatic invariant and hence most likely must be quantized in integers of the Planckian area. In these notes I will present the arguments which led him and myself to such a conclusion and will discuss the consequences of the black hole quantization.

1. Historical Introduction

In 1985 I was puzzled by the question why the energy (mass) of a black hole in quantum gravity must be continuous instead of being quantized? In fact, in many respects a black hole behaves as a bound system, which in quantum theory has generically a discrete energy spectrum. Assuming that the mass is quantized I first used an analogy with the hydrogen atom to obtain its spectrum. The spectrum and the width of the spectral lines for the hydrogen atom can be derived even without knowing too much about full quantum theory by just using the Einstein's idea of the quanta of radiation and the correspondence principle. Treating black hole as "an atom" which emits typical quanta of the Hawking radiation as a result of the transitions between quantum levels, I found that the black hole mass must be proportional to the square root of the integer number and hence its area is quantized in integers of the Planckian area. Around the same time Yan Kogan came to a similar conclusion studying "the condition for a consistent description of the motion of a test string in the external field of a black hole".[7] At the beginning we were planning to write paper together but finally decided to split because neither of us had any reasonable explanation for the degeneracy of quantum levels needed to explain the black hole entropy. First, noting that the mass spectrum of the excited string behaves similar to the black hole spectrum we have tried to "explain" the entropy as originated from the exponential degeneracy of highly excited string states. Obviously, this did not work straightforwardly because the corresponding string entropy is proportional to the string mass, while the black hole entropy must grow as mass squared. An attempt to relate the string mass to the squared mass of the black hole using gravitational mass defect was not quite successful. Therefore, I suggested to Yan to publish our findings separately. In the meantime I decided to try to find more "down to earth" explanation for the exponential degeneracy of quantum levels. One of the ideas, widely discussed in the literature, was to explain the entropy as originated from the different internal geometries of the black hole, which are indistinguishable from the point of view of an external observer, for whom this black hole is entirely characterized by its mass M, angular momentum J and charge Q. To simplify the formulae I will consider below only the Schwarzschild black hole although the results below can be easily generalized for the rotating charged black hole. At this time (not anymore[a]) I thought that the number of different internal microstates must be in one to one correspondence with the number of ways to build black hole of mass M. For example, one can produce it out of two photons with frequencies ω_1 and ω_2, satisfying $\omega_1+\omega_2 = M$ (everywhere I use Planck units, setting $c = \hbar = G = k_B = 1$). Another possible way is to form black hole, for example, from three quanta with frequencies ω_1, ω_2 and ω_3, with $\omega_1+\omega_2+\omega_3 = M$, etc. If no restrictions on the frequencies are imposed, the number of possible ways to make the black hole is infinite and hence the corresponding entropy must also be infinite. The only way to rescue the entropy definition above is to assume that the black holes are quantized. This imposes constrains on the frequencies making the number of ways to form black hole to be finite. These arguments have led me to the same quantization law I obtained before. These two completely different ways of reasonings for the area quantizations I published first in Lebedev Institute preprint.[8] However, in the published version of the paper[9] I skipped the derivation by analogy with the hydrogen because of rather severe restrictions on the number of pages in JEPT letters. The preprint I have sent to the only person outside of the Soviet Union, to Jacob Bekenstein. Living inside of, as I thought, "an eternal black hole" called Soviet Union I did not care too much about making my results known to the "external observers." However, I could not resist sending paper to someone who originated the whole concept. Moreover, Ya. Zeldovich was strongly opposing this idea claiming that the levels will be split and degeneracy completely removed as it happens for usual thermodynamical systems in equilibrium with the thermal bath. Although I tried to argue that the black hole cannot be treated as a normal thermodynamical system because of its negative heat capacity, Zeldovich was insisting that my arguments are silly. To my great surprise I got an answer from Jacob, where he wrote that he completely agrees with me on all points including

[a]In fact, for the given mass M the number of so-called internal geometries must be infinite because, for example, we can first form a black hole with an arbitrary large mass and then wait until this mass decreases to M as a result of Hawking evaporation. It is clear that the internal geometry in this case must be different from those one which we would have if the black hole was originally formed with mass M. Moreover, "inside" nonsingular black hole, in absolute future with respect to the external observer we can have different kind of universes. The only way to avoid this problem is to relate entropy with degeneracy due to the unknown way to form the black hole from the matter of mass M.

the degeneracy issue. He also pointed out that he came to the conclusion that the area must be quantized long ago basing on observation that it behaves like adiabatic invariant.[1] I was extremely happy. This meant that I am not alone and there is another person in the world who also likes this "silly" idea.

After collapse of Soviet Union I went to ETH (Zurich) and in 1993 wrote to Jacob "inviting myself" to Jerusalem. His reaction was positive and for the first time in my life I went to the homeland of all religions where I finally met Jacob personally. Someone who has already overlapped with him before told me that Jacob is extremely religious and not very "communicative." This impression happened to be completely wrong. First of all Jacob was just moderately religious, perhaps a bit selective in a choice of discussion partners, but I did not find on his side even a slightest trace of self-conceit or, so wide spread in our community, "inferiority or superiority complexes" (often hardly distinguishable from outside). One could feel immediately his strong character. Jacob never cared too much about "main stream" research or sociology and was entirely occupied just with those ideas which he was considering to be interesting irrespective of the "community fashion." His judgement of people was rather sharp, but he was keeping it mainly for himself and only in rare cases was sharing with the close friends. Jacob's opinions on questions of science and politics might be sounded not always "politically correct" in a modern useless sense of this notion, but were exactly to the point and very deep. He was a person of value and cared more about the essence than about making impression on outside. In spite of the fact that sometimes Jacob looked as a very "soft person" he could fight to the bitter end if he thought that something is wrong and worth fighting. This does not mean at all that he was stubborn and I witnessed how in couple of cases Jacob changed his mind without any problem just under the pressure of facts. Many of these observations I made much later but from the very beginning we found common language without any difficulties (might be because on many relevant issues we were sharing the same opinion). Therefore my interactions with Jacob were always extremely friendly and fruitful and in a short time we became real friends.

I was invited to Jacob's place and introduced to his wife Bilha and kids, Yehonadav, Uriya and Rivka, who by now have grown to the worthy people and I recall that Jacob was always proud of them. Although during my first visit to Jerusalem my touristic preferences prevailed over scientific interests we, nevertheless, found time for discussions and as a result in a year have published the paper with detailed derivation of the quantum black hole spectrum.[3] Working on this paper and facing problems I was ready (many times) to give up. However, it was not so easy to do so working with Jacob. Finally, I realized what Einstein had really meant when he wrote "It is not that I am so smart, it is just that I stay with problems longer". Having this experience with Jacob and after I heard his story how he got an idea about black hole entropy I realized that this Einstein's remark is entirely applicable to Jacob. Being on the track he was infinitely persistent, might be not always extremely quick, but was never giving up if he had a feeling that there remain questions to be understood and clarified. Jacob intuition was impressive and his understanding of statistical physics was the deepest one I ever met.

After our paper was finished we started to look for an operator algebra which would allow us to formalize the description of quantum black holes. With this purpose Jacob visited me in Zurich. I was impressed when after discussing the properties of the algebra, Jacob mentioned that Bryce DeWitt told him that everything about algebra is in Jacobi identities, went to his office and soon was back with Virasoro algebra about which he has completely forgotten (if he ever knew) and rediscovered by himself with a half an hour. For obvious reasons this algebra did not work because degeneracies were not large enough. Another interesting memory to some extent characterizes the Jacob's attitude about life preferences. I recall that when someone complained that during two weeks he could not find enough time to talk to Jacob because he was too busy, Jacob replied with unforgettable ironic smile under his mustache "If somebody is too busy this just means that he is doing too many irrelevant things." Sometimes he could be really sharp.

Since I moved to Germany in 1998, our worldlines were often crossing either in Jerusalem or on some "neutral territory" as, for instance, in IHES in Bures-sur-Yvette or in Princeton. In particular, in 1998 we have continued the search for the right algebra during our visit to IHES. Unfortunately, without final success. These attempts Jacob has summarized in Refs. 4 and 5. Every time when I was coming to Jerusalem we were discussing all possible questions from science to politics and life. Jacob was an exceptional human. Several times when I had really serious problems in my life only Jacob, as nobody else, could find absolutely right words to support me. Last time we met in Jerusalem Winter School in 2015 and arranged our next meeting to IHES in October. Just a couple of months before Jacob passed away...

Below I describe simple reasonings in favor of the black hole quantization. First, I will present never before published derivation of the black hole spectrum basing on analogy with the hydrogen atom. Forgetting about quantum mechanics I will consider the simplest model for the classical atom and with the minimal assumption about quanta of radiation will show how the Bohr spectrum can be derived. Moreover just using correspondence principle I will estimate the width of the levels and show that the result obtained is in agreement with full quantum theory. We will see that even for the highly excited states the classical physics is not reproduced precisely in apparent conflict with naive interpretation of the correspondence principle. I hope that this consideration will convince reader that the derived by analogy black hole quantization with its unexpected consequences for the Hawking radiation might make sense in spite of the fact that we do not have yet a consistent theory of non-perturbative quantum gravity. In the remaining part of the paper I will recall the justification for the entropy of quantum black hole and show that the modified thermal spectrum is just a consequence of the exponential degeneracy of quantum levels.

2. Quantization

Hydrogen. Let us consider the hydrogen atom and verify whether we can derive its energy levels and their width only by using (a) the Einstein quanta of radiation and (b) the correspondence principle. The quantized energy E depends on the integer number n and our task is to determine $E_n = E(n)$ neglecting all other quantum numbers, such as angular momentum and spins. Moreover we simplify the classical picture assuming that the electron is moving on circle orbit of radius r around the proton at rest. According to Einstein the quantum transition from n-th to $n-1$ level must be accompanied by the emission of photon with frequency

$$\omega_{n,n-1} = E_n - E_{n-1} \, . \tag{1}$$

In its minimal interpretation the correspondence principle would mean that for large n this frequency must coincide with the frequency of classical radiation which is about the frequency of rotation of the electron around center. On a bound orbit of radius r the kinetic energy of the electron, $T = mv^2/2$, is twice less than the magnitude of the negative potential energy $U = -\alpha/r$, where $\alpha = e^2$ is the fine structure constant. Hence, the total energy is

$$E = T + U = -\frac{\alpha}{2r} = -\frac{m\omega^2 r^2}{2} \, , \tag{2}$$

from where it follows that

$$\omega = \left(\frac{\alpha}{mr^3}\right)^{1/2} = \left(-\frac{8}{m\alpha^2}E^3\right)^{1/2} \, . \tag{3}$$

Taking into account that $E_n - E_{n-1} \simeq dE_n/dn$ for $n \gg 1$ and equating $\omega_{n,n-1}$ from (1) to ω we obtain the following equation for E_n,

$$\frac{dE_n}{dn} \simeq \left(-\frac{8}{m\alpha^2}E_n^3\right)^{1/2} \, , \tag{4}$$

solving which gives us the spectrum

$$E_n = -\frac{m\alpha^2}{2n^2} \, , \tag{5}$$

which is valid even for small n although the derivation above is justified only for $n \gg 1$. For large n the distance between the nearby levels is about

$$\Delta E_n = E_n - E_{n-1} \simeq \frac{m\alpha^2}{n^3} \simeq \omega_n \, . \tag{6}$$

One could wonder whether for $n \gg 1$ we must obtain exactly classical behavior or there will be still essential differences between the classical and quantum hydrogen even in this case? To answer this question we must determine the width of the highly excited levels and compare it with the distance between the nearby levels. With this purpose let us use the correspondence principle in its minimal formulation by assuming that the overall intensity of the emitted radiation by quantum atom must be in agreement with the result obtained in the classical theory. The classical radius of the electron orbit corresponding to energy $E_n = -\alpha/2r_n$, given in (5), is

$$r_n = \frac{n^2}{m\alpha}. \tag{7}$$

According to the Maxwell theory the intensity of the emitted radiation is given by the second time derivative of the dipole momentum, that is,

$$I \simeq (e\ddot{r})^2 \simeq \alpha\omega^4 r^2, \tag{8}$$

or substituting here ω and r from (6) and (7) we obtain

$$I_n \simeq \frac{m^2\alpha^7}{n^8}. \tag{9}$$

Thus, the time interval τ_n during which atom emits the energy needed for transition from level n to $n-1$, given in (6), is equal to

$$\tau_n \simeq \frac{\Delta E_n}{I_n} \simeq \frac{n^5}{m\alpha^5} \simeq \frac{n^2}{\alpha^3}\omega_n^{-1}. \tag{10}$$

Correspondingly the width of the level n is

$$W_n \simeq \frac{1}{\tau_n} \simeq \frac{m\alpha^5}{n^5} \simeq \frac{\alpha^3}{n^2}\Delta E_n, \tag{11}$$

and it gets smaller compared to the distance between nearby levels as n grows. Hence the emitted lines of the isolated hydrogen atom become sharper and the quantum nature of the hydrogen become even more distinctive as n grows. The detailed picture of emission by the quantized atom is very different from the classical picture. According to classical theory the transverse part of the vector potential A at distance $R \simeq \omega_n^{-1}$ is

$$A = \frac{\dot{d}}{R} \simeq \frac{e\omega_n r_n}{R} \simeq \frac{e^3}{n}\frac{1}{R}, \tag{12}$$

and by factor e^3/n smaller than the level of inevitable quantum fluctuations in scales R in conflict with uncertainty relation for radiation. According to classical theory the electron continuously emits radiation during the time interval τ_n changing energy from E_n to E_{n-1}. In quantum theory a quantum of radiation of frequency ω_n is emitted randomly during the time interval ω_n^{-1} per n^2/α^3 rotations. From the formulae above we see that the quantum features of the isolated hydrogen atom become more distinctive for large n. For instance, in a hypothetical case $e \gg 1$ the discrete energy levels make sense only for $n > e^3$ because otherwise their width is larger than the distance between the nearby levels.

Black Hole. Now I will apply a similar line of reasoning for a black hole of mass M assuming that the mass is a function of the integer number n. It is not excluded that this assumption might find a justification in some unknown yet nonperturbative quantum gravity. However, in the absence of such a theory we can only try to guess what could be the spectrum of the black hole in a way similar to how we did it above for the hydrogen atom. Although the consideration below can be easily generalized for a charged, rotating black hole[9] to simplify the formulae I will consider here only the Schwarzschild black hole, which from the point of view of external observer is even more simple than hydrogen because it is entirely characterized only by its mass. Once again I will use the correspondence principle and the quanta of Hawking radiation (instead of Einstein's photons). For quantum black hole this radiation can be treated as spontaneous radiation due to the transition from n to one of the nearby levels, for instance, to level $n-1$. The typical frequency of the Hawking quanta is of order of black hole temperature,

$$T_H = \frac{1}{8\pi M} . \tag{13}$$

According to the correspondence principle, for large n

$$\frac{dM_n}{dn} \simeq M_n - M_{n-1} \simeq \gamma T_H , \tag{14}$$

where γ is a numerical coefficient of order unity, and hence

$$\frac{dM_n}{dn} \simeq \gamma \frac{1}{8\pi M_n} . \tag{15}$$

Integrating this equation we obtain the mass spectrum

$$M_n \simeq \sqrt{\frac{\gamma}{4\pi} n} , \tag{16}$$

and therefore the area of black hole A_n must be quantized in terms of integers of the Planck area,

$$A_n = 16\pi M_n^2 = 4\gamma n . \tag{17}$$

We will show below that the numerical coefficient γ in (14) is most likely equal to $\ln 2$ and hence the distance between the nearby levels is

$$\Delta M_n = M_n - M_{n-1} \simeq \frac{\ln 2}{8\pi M_n} . \tag{18}$$

To estimate the width of the levels we use the correspondence principle assuming that the total flux of Hawking radiation from the classical black hole must be about the same as the flux from quantum black hole (similar how we did it for the hydrogen). According to Ref. 6

$$\frac{dM}{dt} \simeq -\frac{9N}{10\pi \cdot 8^4 \cdot M^2} , \tag{19}$$

where N is the number of distinct massless quanta in Nature. On the other hand the rate of emission for the highly excited quantum black hole is

$$\frac{dM_n}{dt} \simeq -\frac{2\Delta M_n}{\tau_n} \simeq -\frac{2\ln 2}{8\pi M_n \tau_n} , \tag{20}$$

where τ_n is the lifetime on level n. The factor 2 here accounts for the possibility of direct transitions to levels $n-2$, $n-3$, etc.[3] Comparing (19) and (20) we find

$$\tau_n \simeq \frac{5 \cdot 8^4 \cdot \ln 2 \cdot M_n}{18N} , \tag{21}$$

and correspondingly the width of level nth is

$$W_n \simeq \frac{1}{\tau_n} \simeq \frac{18N}{5 \cdot 8^4 \cdot \ln 2 \cdot M_n} \simeq 4.5 \cdot 10^{-2} N \, \Delta M_n \,. \tag{22}$$

For $N < 20$ the level width is less than the distance to nearby level. Therefore, the spectrum of the quantum black hole can be very distinctive from the usual Planckian spectrum even for the large black holes. As it was shown in,[3] the resulting spectrum is perfectly consistent with the thermal spectrum of photons in a finite box with reflecting boundaries. In this latter case the frequencies are proportional to the integers of the inverse size of the box and the spectrum is not continuous but consists of lines with intensities obeying the Planckian distribution. Similar for the black holes instead of continuous spectrum we have the thermal spectrum consisting of rather sharp lines.

Thus, we have found that the spectrum of radiation from the large quantum black holes can be very different from the continuous Hawking spectrum which he found considering quantum fields in a given classical background. It is not as surprising as it looks at the first glance if we recall the situation with the highly excited hydrogen states where the line structure of emitted radiation becomes even more profound for large n. Therefore, the correspondence principle in both cases above should be understood in a rather restrictive way. Namely, at very large n only the averaged characteristic of the emitted radiation, but not its detailed features, must be in agreement with the classical theory. If the consideration above is correct, this opens experimental possibility to check the effects of unknown yet nonperturbative quantum gravity observing radiation from the exploding primordial black holes.

2.1. *Entropy*

To define the statistical entropy of quantized black hole we assume that it is due to the lost information about the way how the black hole at level n was formed. Because of the event horizon this information can never be recovered by the external observer. The number of different ways to form the quantum black hole from a given matter is finite. For example, one possible way is first to form a smallest black hole at the first level and after that go step by step to the level n. The other possibility is to jump directly to level n without intermediate steps etc. The total number of ways to get to the level n is obviously equal to the number of subdivisions of the integer number n into ordered sums (for instance, for $n = 3$ we have $3 = 1 + 1 + 1, 3 = 1 + 2, 3 = 2 + 1, 3 = 3$ where one has to distinguish $1 + 2$ from $2 + 1$). The result for the level n is

$$\Gamma(n) = 2^{n-1} \tag{23}$$

and the black hole entropy is equal

$$S_n = \ln \Gamma(n) = (n-1) \ln 2 \,. \tag{24}$$

It is proportional to the "elementary" entropy $\ln 2$, corresponding to one bit of information. Assuming that the relation between entropy and area

$$S = \frac{1}{4}A + \text{const}\,, \tag{25}$$

still holds we obtain the following quantization rule

$$A_n = 4\ln 2 \cdot n\,, \tag{26}$$

in full agreement with (17) if we set $\gamma = \ln 2$ and take the constant of integration in (25) to be equal $-\ln 2$, so that $n = 0$ describes the state without black hole. The minimal increase of the area, $\Delta A = 4\ln 2$, is in agreement with the earlier result by Bekenstein[1] (see also Ref. 6) and it was obtained before he suggested the area quantization in Ref. 2.

As it follows from (23) each level of the black hole is exponentially degenerate. At the first glance one could expect the splitting of these highly degenerate levels after black hole gets in equilibrium with an infinite thermal bath which we need to define its macrocanonical entropy. However, the black hole has a negative heat capacity and therefore macrocanonical definition is not applicable here. One has to use the microcanonical ensemble instead. In this case there is no any reason to expect the splitting of the levels. Moreover, as it was shown in Ref. 3 the perfectly thermal properties of the radiation are entirely due to the exponential degeneracy of the levels. To give an idea why this happens let us consider the large black hole at level $n \gg 1$. Then all energy levels between n and $n - m$ for $m \ll n$ are nearly equidistant with the energy difference between nearby levels corresponding to the fundamental frequency

$$\bar{\omega} = \frac{\ln 2}{8\pi M_n} = \ln 2 \cdot T_H\,. \tag{27}$$

It is clear that the black hole can directly go from the state n to any nearby lower states and, for instance, going to the level $n - 2$ emits quantum of radiation with the frequency $2\bar{\omega}$ or with the frequency $\omega = m\bar{\omega}$ for the direct transition from level nth to $n - m$. For the external observer the black hole is even more simple than hydrogen atom because it has no hairs and entirely characterized by its mass and entropy. Therefore one can expect that the matrix elements for all transitions to the nearby levels are the same and the probability for direct transition from level n to $n - m$ with emission of quantum of frequency $\omega = m\bar{\omega}$ is simply proportional to the degeneracy of level $n - m$, that is,

$$P(\omega) \propto 2^{n-m} \propto \exp\left(-\ln 2 \cdot \frac{\omega}{\bar{\omega}}\right) = \exp\left(-\frac{\omega}{T_H}\right). \tag{28}$$

It was shown in Ref. 3 that probabilities for all possible sequences of quanta are precisely as for the thermal radiation with temperature T_H in a finite box of size $\bar{\omega}^{-1}$. It is rather remarkable that all thermal properties of the Hawking radiation follow from the degeneracy (entropy) of quantized black hole under assumption of equal matrix element for the transitions to close levels. Although the spectrum of the quantum black holes is thermal it nevertheless different from the continuous spectrum derived by Hawking. It must consist of lines with the width of few percent of the distance between nearby lines. Moreover, the minimal frequency which black hole can emit is $\bar{\omega} = \ln 2 \cdot T_H$. The probability of the emission of quanta with the wavelength much larger than the size of the black hole is strongly suppressed.

References

1. J. Bekenstein, Generalized second law of thermodynamics in black-hole physics, *Phys. Rev. D* **9**(12), 3292 (1974).
2. J. Bekenstein, The quantum mass spectrum of the Kerr black hole, *Lett. al Nuovo Cimento.* **11**(9), 467 (1974).
3. J. Bekenstein and V. Mukhanov, Spectroscopy of the quantum black hole, *Phys. Lett. B* **360**, 7 (1995).
4. J. Bekenstein, Quantum black holes as atoms, in *Proceeding of 8th Marcel Grossman Meeting*, Jerusalem, Israel, 1997, eds. T. Piran and R. Ruffini (World Scientific, 1999).
5. J. Bekenstein, The case for discrete energy levels of a black hole, *Int. J. Mod. Phys. A* **17**(S1), 21 (2002).
6. B. DeWitt, Quantum field theory in curved space, *Phys. Rep.* **19**(6), 296 (1975).
7. Ya. Kogan, Quantization of the mass of a black hole in string theory, *JETP. Lett.* **44**(4), 267 (1986).
8. V. Mukhanov, Evaporation and entropy of quantized black hole, Lebedev Institute preprint N 163, 1986.
9. V. Mukhanov, Are black holes quantized?, *JETP. Lett.* **44**(1), 63 (1986).

Black Hole Information Revisited

Andrew Strominger
Center for the Fundamental Laws of Nature, Harvard University, Cambridge, MA 02138, USA

We argue that four-dimensional black hole evaporation inevitably produces an infinite number of soft particles in addition to the thermally distributed 'hard' Hawking quanta, and moreover that the soft and hard particles are highly correlated. This raises the possibility that quantum purity is restored by correlations between the hard and soft radiation, while inclusive measurements which omit the soft radiation observe the thermal Hawking spectrum. In theories whose only stable particle is the graviton, conservation laws are used to argue that such correlations are in principle sufficient for the soft gravitons to purify the hard thermal ones.

Dedicated to the memory of Jacob D. Bekenstein

1. Introduction

The Bekenstein–Hawking area law for black hole entropy

$$S_{BH} = \frac{\text{Area}}{4G} \tag{1.1}$$

in my view stands alongside the Einstein and Schroedinger equations as one of the most interesting equations of twentieth-century physics. Unlike the latter two equations however, it remains both mysterious and paradoxical. One form of the paradox is how the information content of a black hole implied by (1.1) is released when a black hole evaporates and disappears. In this contribution we shall point out that soft particles, ignored in most discussions of the topic, carry enough information to play a central role in the maintenance of quantum purity.

Quantum black holes have been argued[1] to emit a nearly thermal spectrum of radiation. This Hawking radiation is 'hard' in the sense that thermal emission probabilities vanishes as the energy $\omega \to 0$. From a variety of viewpoints it is known that such a process is impossible in 4D: hard radiated quanta are always accompanied by an infinite cloud of tightly correlated soft quanta. In this note we conjecture that the full evaporation process is unitary, but that the inclusive cross section traced over all final soft quanta reproduces the Hawking result at leading order. Put differently, a detector with an infrared (IR) cutoff would confirm Hawking's prediction. This proposed unitarity restoration process is an alternative to that of Page[2] in which quantum purity is restored by correlations between early and late time hard Hawking quanta. We then show that, for the special case of pure 4D gravity, supertranslation and superrotation conservation laws do imply that tracing over the outgoing soft quanta fully decoheres the hard emissions. Similar conclusions were independently reached in the very recent work of Carney *et al.*[3] who showed, using rather different quantum information-theoretic methods, that for a generic scattering process (not restricted to black holes) tracing over soft quanta greatly and sometimes fully decoheres the final state.

This paper does not attempt to explain the dynamical process by which soft/hard correlations arise, much as Page's work does not attempt to explain the dynamical origin of late/early hard correlations. Rather an alternate possibility, in which soft quanta play a central role, is presented herein for the final outcome of a unitary theory of black hole dynamics.

Reversing the perspective, whether or not soft quanta do fully restore unitary in black hole evaporation, they cannot be ignored in the discussion. All nontrivial scattering processes produce an infinite number of them, and they can store a large amount of information with a very small energy cost. Any process which excludes them — even for a burning lump of coal — will violate unitarity.

This work is a logical outgrowth of the observation[4] that supertranslation charge conservation constrains black hole evaporation, the analysis of near-horizon soft hair in Refs. 5 and 6, the reinterpretation of IR divergences as imposing conservation laws given in Ref. 7 and more generally recent insights into symmetries, conservation laws and soft theorems reviewed in Ref. 8. The importance of soft contributions to gravitational entanglement entropy was discussed in Ref. 9 and more generally to black hole information in Ref. 10.

2. Soft Particle Creation by Black Holes

Consider a black hole which is formed by gravitational collapse and then eventually evaporates and disappears. According to Hawking's semiclassical computation, the outgoing Hawking radiation has a nearly thermal (up to greybody factors and sub-semiclassical corrections) blackbody spectrum, in which the probability per unit time of emitting a boson of energy ω is approximately

$$N(\omega) \sim \frac{\omega^2}{e^{\omega/T_H} - 1} \,. \tag{2.1}$$

Here the Hawking temperature for a black hole of mass M is

$$T_H = \frac{1}{8\pi GM} \,. \tag{2.2}$$

After the black hole has evaporated, Hawking argued,[11] the final outgoing state is a density matrix

$$|\Psi_{\rm in}\rangle \rightarrow \rho_{\rm Hawking} = \sum_{\alpha} \rho_\alpha |H_\alpha\rangle\langle H_\alpha| \,. \tag{2.3}$$

Here the outgoing states $|H_\alpha\rangle$ have particle occupation numbers roughly of the form (2.1) and the index α runs over all the states in the thermal radiation ensemble. Since (2.3) entails a breakdown of determinism, this conclusion is hard to swallow.

One might expect instead[2] that there is some yet-to-be-found flaw in the reasoning of Ref. 11 and that there is a unitary process of the form

$$\mathcal{S}|\Psi_{\rm in}\rangle = \sum_{\alpha} c_\alpha |H_\alpha\rangle \,, \tag{2.4}$$

for some constants c_α which we do not yet understand how to compute. In fact, the literal form (2.4) is generically impossible for a simple reason. Picture the black hole as some kind of unitary black box which spews out quanta from the horizon in a pure state which is some element in the ensemble (2.3). These quanta will have IR divergent exchanges of soft gravitons which occur far outside the horizon and are insensitive to the details of the black box. Exponentiating these exchanges will give $e^{-\infty}$ and set the amplitude to zero, just as soft photon exchanges do for $e^+e^- \rightarrow e^+e^-$ (with no external photons) in QED. The amplitude to produce any state with quanta distributed as in (2.1) will vanish. Nonvanishing amplitudes must have poles in the number of soft quanta as $\omega \rightarrow 0$. This conclusion can alternately be viewed as a consequence of supertranslation charge conservation[7,12] as we review below.

Put another way, due to IR divergences, Hawking's leading order semiclassical result for unaccompanied hard radiation is not close to the exact answer. It is corrected by an infinite amount.

This type of IR problem is familiar in quantum field theory. The usual fix is to compute inclusive cross sections with an IR cutoff. This is fine if we are computing collider cross sections since real detectors have an intrinsic IR cutoff. But subtle issues of unitarity cannot be addressed in the context of inclusive cross sections. For that we need an $\mathcal{S}$-matrix.

In QED a second type of fix for this type of IR problem, which does involve an $\mathcal{S}$-matrix, was given long ago by Faddeev and Kulish (FK).[13] The FK construction was recently extended to gravity in Ref. 14. All asymptotic hard particles are dressed with a cloud of soft gravitons involving $\omega \rightarrow 0$ poles. The IR hard-particle pair exchange divergences are cancelled by divergences in exchanges between hard particles and soft graviton clouds. $\mathcal{S}$-matrix elements of the dressed hard particles are IR finite.

There is a large ambiguity in the FK choice of the dressing. IR finiteness only fixes the coefficients of the $\omega \rightarrow 0$ poles in the dressing factors. For the purposes of IR safety, one can take the dressing to be entirely composed of gravitons below an arbitrarily low energy scale Λ_{IR}.

This suggests a natural modification of (2.4). We propose it be modified to

$$\mathcal{S}|\Psi_{\rm in}\rangle = \sum_\alpha c_\alpha |H_\alpha\rangle|S_\alpha\rangle = |\Psi_{\rm out}\rangle\,, \tag{2.5}$$

where we have decomposed the Hilbert space into hard and soft factors, which are separated by a scale $\Lambda_{IR} \ll T_H$. The Hawking radiation $|H_\alpha\rangle$ is in the hard part and the soft part $|S_\alpha\rangle$ contains the dressing with the soft poles required for IR finiteness. This simply states that thermally-distributed hard particles are accompanied by a nonthermal cloud of soft particles. While we know that *some* soft dressing must be present, without a complete theory of black hole dynamics we do not know how to compute the exact soft spectrum.

We now conjecture that the soft factors associated to every distinct hard factor are orthogonal:

$$\langle S_\alpha|S_\beta\rangle = \delta_{\alpha\beta}\,. \tag{2.6}$$

Physically this means that there is a unique soft emission spectrum for every hard state. Since there is an infinite degeneracy of soft states this is not *a priori* ruled out. We will make plausibility arguments for (2.6) in a restricted context in the next section but for now we just regard (2.6) as a conjecture. It follows from (2.6) that detectors which only observe the hard Hawking radiation, and miss the soft dressings, will observe a density matrix

$$\rho_H = \sum_\alpha c_\alpha c_\alpha^* |H_\alpha\rangle\langle H_\alpha|\,. \tag{2.7}$$

Taking

$$c_\alpha = \sqrt{\rho_\alpha}e^{i\theta_\alpha}, \quad |\Psi_{\rm out}\rangle = \sum_\alpha \sqrt{\rho_\alpha}e^{i\theta_\alpha}|H_\alpha\rangle|S_\alpha\rangle \tag{2.8}$$

where ρ_α is defined in (2.3), we find

$$\rho_H = \rho_{\rm Hawking}\,. \tag{2.9}$$

This motivates the conjecture that black hole evaporation is a unitary process of the form (2.5) with the magnitude of c_α given by (2.8) at leading semiclassical order. A detector which cannot capture the soft radiation will see the Hawking spectrum.[a] In particular there need be no correlations between Hawking quanta before and after the Page time as put forth in Ref. 2.[b] Rather the information is stored in correlations between the hard and soft quanta.[c]

[a]Confirming unitarity would require at a minimum measuring the soft pole using gravitational memory detectors.[4]

[b]The conjecture of this paper makes no assertion about the time at which the information comes out, only that it all comes out in the infinite future. We avoid doing so because it is unclear if this time can be well-defined in the presence of soft particles: see Refs. 10 and 9 for some discussion. However if the time can be well-defined it is possible that early hard particles dressed with soft quanta could be correlated with late hard particles dressed with soft quanta and in this sense our conjecture is a refinement of Ref. 2.

[c]As it was recently shown in Ref. 3 that soft and hard quanta are generically highly entangled in any scattering process, even for a burning lump of coal some of the information is contained in such correlations.

The Page proposal (2.4) did not provide any algorithm for how the constants c_α determining the out states are computed, and in the intervening decades no such algorithm has clearly emerged. In contrast in the present proposal the magnitude of c_α in (2.5) is given at leading order by the Hawking computation as $\sqrt{\rho_\alpha}$, but no algorithm is provided here for determining the relative phases θ_α of the outgoings states. That would require insights and methods beyond those of the present work.

3. A Restricted Example

In the previous section we made a rather general proposal that thermal Hawking radiation is purified by correlations with soft quanta. In this section we put flesh on this proposal and show that the key conjecture (2.6) holds for a simplified 4D setting in which gravitons are the only stable particles. In this case, there should be a unitary $\mathcal{S}$-matrix with purely gravitational in and out states. We do not know of a concrete 4D example of such a theory, but IIB string theory is a 10D example in which the only stable particles are the graviton and its superpartners. In the last section the generalization to theories with multiple stable massive or massless particles is briefly discussed.

We seek more detail about the soft clouds $|S_\alpha\rangle$ arising from a given $|H_\alpha\rangle$. The crucial pole structure can be determined from supertranslation charge conservation[12] following a similar analysis given for QED in Ref. 7. There is one conserved supertranslation charge for every angle $(z, \bar{z})$ on the celestial sphere at null infinity. In a convenient basis the outgoing charge can be decomposed into soft and hard pieces as

$$\begin{aligned} Q^+(z,\bar{z}) &= Q^+_S(z,\bar{z}) + Q^+_H(z,\bar{z}) - \frac{M\gamma_{z\bar{z}}}{4\pi}\,, \\ Q^+_H &= \frac{\gamma_{z\bar{z}}}{4\pi G}\int_{\mathcal{I}^+} du T_{uu}\,, \\ Q^+_S &= -\frac{\gamma_{z\bar{z}}}{16\pi G}\int_{\mathcal{I}^+} du \left(D_z^2 N^{zz} + D_{\bar{z}}^2 N^{\bar{z}\bar{z}}\right)\,, \end{aligned} \tag{3.1}$$

where in the first line we have subtracted off the zero mode proportional to the total mass which is not our interest here. In the conventions of Ref. 8 u is retarded time, D_z is the covariant derivative on the unit sphere with metric $\gamma_{z\bar{z}}$, $T_{uu}(u, z, \bar{z})$ (and T_{uz} below) is a coefficient of the r^{-2} term in the expansion about $\mathcal{I}^+$ of the gravity wave stress tensor and

$$N_{zz}(u, z, \bar{z}) = \partial_u C_{zz}(u, z, \bar{z}) \tag{3.2}$$

is the Bondi news with C_{zz} the leading transverse perturbation of the asymptotically flat metric. Q^+_H is the hard energy flux through the angle $(z, \bar{z})$. Q^+_S is a soft graviton zero mode at the same angle. Supertranslation charge conservation equates the total Q^+ to a similar expression for the incoming charge Q^- on $\mathcal{I}^-$. However for simplicity we restrict to black holes especially prepared in a 'neutral' eigenstate with[d]

$$Q^-|\Psi_{\rm in}\rangle = 0\,. \tag{3.3}$$

It follows that $Q^+|\Psi_{\rm out}\rangle = 0$ or

$$Q^+_H(z,\bar z)|\Psi_{\rm out}\rangle = -Q^+_S(z,\bar z)|\Psi_{\rm out}\rangle\,. \tag{3.4}$$

The hard part of the outgoing state $|H_\alpha\rangle$ is a collection of hard gravitons. Ignoring, for the rest of this paragraph, the spin of the graviton, such states can be characterized by the positions z^k_α where the outgoing Hawking quanta exit through $\mathcal{I}^+$ with energies E^k_α. Such momentum eigenstates diagonalize the hard supertranslation charge

$$Q^+_H(z,\bar z)|H_\alpha\rangle = \sum_k E^k_\alpha \delta^2(z-z^k_\alpha)|H_\alpha\rangle\,. \tag{3.5}$$

Since black holes tend to emit at low angular momenta, it might seem more natural to use an angular momentum basis. However such bases do not diagonalize Q^+_H and therefore do not obey a simple relation of the form (3.5). It then follows from (3.4) that the soft modes must obey

$$Q^+_S(z,\bar z)|S_\alpha\rangle = -\sum_k E^k_\alpha \delta^2(z-z^k_\alpha)|S_\alpha\rangle\,. \tag{3.6}$$

Soft clouds which obey such relations were constructed (for QED) in Ref. 7, and necessarily produce $\omega \to 0$ poles in field expectation values. Up to the spin, the eigenvalue in (3.5) almost uniquely determines the hard state. A counterexample is the case of two gravitons at *exactly* the same z_k each with energy $\frac{1}{2}E_k$. This is not an issue in the present context because the probability of producing two *exactly* collinear gravitons is zero in the Hawking formula. Since eigenstates of a hermitian operator with distinct eigenvalues are orthogonal we may then normalize so that the soft states $|S_\alpha\rangle$ obey (2.6).

Now let us account for spin by adding an extra label $s^k_\alpha = \pm 2$ to each hard graviton. Two states $|H_\alpha\rangle$ which differ only in the spins of one or more gravitons will have the same eigenvalue of Q^+_H in (3.5). However the soft state is further constrained by conservation of superrotation charge $Q^\pm_z$ at each angle on the celestial sphere.[15,16] We now show that these charges do discern the spin and force the soft clouds to be orthogonal if the hard graviton spins are not aligned.[e] As in (3.3) we restrict to

$$Q^-|\Psi_{\rm in}\rangle = 0 = Q^+|\Psi_{\rm out}\rangle\,. \tag{3.7}$$

The outgoing superrotation charge can be decomposed into soft and hard pieces (in the conventions of Ref. 8) as

---

[d]Asymptotic FK dressed particles states are constructed so as to commute with the supertranslation charges and hence convenient for making such a neutral black hole.

[e]This was suggested in Ref. 3.

$$Q_z^+(z,\bar z) = Q_{Sz}^+(z,\bar z) + Q_{Hz}^+(z,\bar z)\,,$$
$$Q_{Hz}^+ = \frac{1}{8\pi G}\int_{\mathcal{I}^+} du(T_{uz} - uD_zT_{uu})\,, \tag{3.8}$$
$$Q_{Sz}^+ = -\frac{1}{16\pi G}\int_{\mathcal{I}^+} duD_z^3C^{zz}\,.$$

Q_{Hz}^+ is essentially the 'twist' (the hard angular momentum plus boost weight $E\partial_E$) radiated through $(z,\bar z)$:[f]

$$Q_{Hz}^+|H_\alpha\rangle = \frac{i}{2}\sum_k D^{\bar z}\delta^2(z-z_\alpha^k)(s_\alpha^k - 2 - E_\alpha^k\partial_{E_\alpha^k})|H_\alpha\rangle\,. \tag{3.9}$$

Since our interest here is in determining the effect of the spin on $|S_\alpha\rangle$, we construct a conserved charge involving only the spin by integrating an appropriate linear combination of Q_{Hz}^+ and $Q_{H\bar z}^+$ over a small region R_α^k surrounding any point z_α^k. Let us define

$$Q_H^+(z_\alpha^k) = -i\int_{R_\alpha^k} d^2z\gamma_{z\bar z}\big((z-z_\alpha^k)Q_{Hz}^+ - (\bar z - \bar z_\alpha^k)Q_{H\bar z}^+\big)\,, \tag{3.10}$$

with a similar expression for $Q_S^+(z_\alpha^k)$. We then have

$$Q_H^+(z_\alpha^k)|H_\alpha\rangle = s_\alpha^k|H_\alpha\rangle\,, \tag{3.11}$$

and from (3.7)

$$Q_S^+(z_\alpha^k)|S_\alpha\rangle = -s_\alpha^k|S_\alpha\rangle\,. \tag{3.12}$$

This constrains the behavior of the soft cloud at subleading order in the soft expansion,[g] and implies that all (noncollinear) hard graviton states appearing in the evaporation of an initially uncharged black hole are accompanied by a different soft state. This is then enough information to conclude that the soft graviton clouds obey (2.6).

4. Comments

In this section we mention some unresolved issues for further exploration.

More massless particles, beyond the graviton, produces degeneracies in the supertranslation and superrotation charges, which can no longer on their own distinguish all states in the Hawking ensemble. At the same time they also introduce new infinities of conserved charges, possibly even in the case of scalars,[17] which may further correlate the hard and soft factors. Moreover Section 3 employed only half (rotations, and not boosts) of the correlations imposed by superrotation charge conservation, and there are perhaps further unexplored correlations from conservation laws associated to the sub-subleading soft theorem.[15,18–21] In the real world

[f] Here we have taken $Y^z = \gamma^{z\bar z}\delta^2(z-z_\alpha^k)$, $Y^{\bar z} = 0$ in Eq. (5.4) of Ref. 16 and integrated by parts.
[g] Note that there is no IR divergence associated to superrotations. Constraints of this type on the soft state might generalize the present paper above $D = 4$.

there are stable massive particles such as the electron. Their effects on the present analysis might best proceed via the hyperbolic resolution of future timelike infinity initiated in Ref. 22, which brings in concepts from AdS_3/CFT_2. It is certainly possible however that counterexamples to the central conjecture (2.6) can be found beyond the simple example of the preceding section.[h] If so, one might impose (2.6) as a consistency condition restricting sensible quantum theories of gravity. Another open issue is the extension to higher dimensions. Soft theorems and their associated conservation laws still exist and correlate the soft state accompanying any hard emission, but the structure of IR divergences is very different.

The analysis of Section 3 makes a number of simplifying restrictions including to incoming states with no supertranslation or superrotation charges. Although highly nongeneric, the information paradox is still present and can be addressed with these restrictions. It would be interesting to extend the analysis to generic incoming states.

We do not have a proposal here for a complete calculation of $|\Psi_{\text{out}}\rangle$ including soft modes, which would include a determination of the phases in (2.8). Rather, we have presented an alternative pattern for information return which relies on the nonthermal soft graviton production in the black hole evaporation process which is in any case required by the soft theorems. Observers equipped with only hard particle detectors see the mixed thermal radiation state predicted by Hawking. Corrections to the Hawking calculation are of a very minimal form and arise only in the regime in which we already know the calculation must break down due to IR divergences.

Acknowledgments

I am grateful to T. Dumitrescu, S. Giddings, D. Harlow, S. Hawking, D. Kapec, P. Mitra, S. Pasterski, M. Perry, A. Raclariu, G. Semenoff and S. Zhiboedov for useful conversations. This work was supported by NSF grant 1205550 and the John Templeton Foundation.

References

1. S. W. Hawking, "Particle Creation by Black Holes," *Commun. Math. Phys.* **43**, 199 (1975) [Erratum: **46**, 206 (1976)].
2. D. N. Page, "Is Black Hole Evaporation Predictable?," *Phys. Rev. Lett.* **44**, 301 (1980).
3. D. Carney, L. Chaurette, D. Neuenfeld and G. W. Semenoff, "Infrared quantum information," arXiv:1706.03782 [hep-th].
4. A. Strominger and A. Zhiboedov, "Gravitational Memory, BMS Supertranslations and Soft Theorems," *JHEP* **1601**, 086 (2016).

[h]Certainly theories with exact global symmetries will violate (2.6), since two different hard states related by the symmetry will produce the same soft state. However it is believed such global symmetries cannot arise in quantum gravity.

5. S. W. Hawking, M. J. Perry and A. Strominger, "Soft Hair on Black Holes," *Phys. Rev. Lett.* **116**(23), 231301 (2016) [arXiv:1601.00921 [hep-th]].
6. S. W. Hawking, M. J. Perry and A. Strominger, "Superrotation Charge and Supertranslation Hair on Black Holes," *JHEP* **1705**, 161 (2017) [arXiv:1611.09175 [hep-th]].
7. D. Kapec, M. Perry, A. M. Raclariu and A. Strominger, "Infrared Divergences in QED, Revisited," arXiv:1705.04311 [hep-th].
8. A. Strominger, *Lectures on the Infrared Structure of Gravity and Gauge Theory* (Princeton University Press, 2018), arXiv:1703.05448 [hep-th].
9. D. Kapec, A. M. Raclariu and A. Strominger, "Area, Entanglement Entropy and Supertranslations at Null Infinity," *Class. Quant. Grav.* **34**(16), 165007 (2017) [arXiv:1603.07706 [hep-th]].
10. T. Banks and W. Fischler, "Holographic Theory of Accelerated Observers, the S-matrix, and the Emergence of Effective Field Theory," arXiv:1301.5924 [hep-th]; T. Banks, W. Fischler, S. Kundu and J. F. Pedraza, "Holographic Space-time and Black Holes: Mirages As Alternate Reality," arXiv:1401.3341 [hep-th]; T. Banks, "The Super BMS Algebra, Scattering and Holography," arXiv:1403.3420 [hep-th]; T. Banks and W. Fischler, "Holographic Space-time, Newton's Law and the Dynamics of Black Holes," arXiv:1606.01267 [hep-th].
11. S. W. Hawking, "Breakdown of Predictability in Gravitational Collapse," *Phys. Rev. D* **14**, 2460 (1976).
12. A. Strominger, "On BMS Invariance of Gravitational Scattering," *JHEP* **1407**, 152 (2014) [arXiv:1312.2229 [hep-th]].
13. P. P. Kulish and L. D. Faddeev, "Asymptotic conditions and infrared divergences in quantum electrodynamics," *Theor. Math. Phys.* **4**, 745 (1970) [*Teor. Mat. Fiz.* **4**, 153 (1970)].
14. J. Ware, R. Saotome and R. Akhoury, "Construction of an asymptotic S matrix for perturbative quantum gravity," *JHEP* **1310**, 159 (2013) [arXiv:1308.6285 [hep-th]].
15. F. Cachazo and A. Strominger, "Evidence for a New Soft Graviton Theorem," arXiv:1404.4091 [hep-th].
16. D. Kapec, V. Lysov, S. Pasterski and A. Strominger, "Semiclassical Virasoro symmetry of the quantum gravity $\mathcal{S}$-matrix," *JHEP* **1408**, 058 (2014).
17. M. Campiglia, L. Coito and S. Mizera, "Can scalars have asymptotic symmetries?," arXiv:1703.07885 [hep-th].
18. M. Campiglia and A. Laddha, "Sub-subleading soft gravitons: New symmetries of quantum gravity?," *Phys. Lett. B* **764**, 218 (2017) [arXiv:1605.09094 [gr-qc]].
19. M. Campiglia and A. Laddha, "Sub-subleading soft gravitons and large diffeomorphisms," *JHEP* **1701**, 036 (2017) [arXiv:1608.00685 [gr-qc]].
20. E. Conde and P. Mao, "BMS Supertranslations and Not So Soft Gravitons," *JHEP* **1705**, 060 (2017) doi:10.1007/JHEP05(2017)060 [arXiv:1612.08294 [hep-th]].
21. A. Laddha and A. Sen, "Sub-subleading Soft Graviton Theorem in Generic Theories of Quantum Gravity," arXiv:1706.00759 [hep-th].
22. M. Campiglia and A. Laddha, "Asymptotic symmetries of QED and Weinberg's soft photon theorem," *JHEP* **1507**, 115 (2015) [arXiv:1505.05346 [hep-th]].

The Bekenstein Bound

Endowing black holes with entropy has led to the examination of the manner in which information can be stored in space time as well as to the question of whether there is a bound on the amount of information that can be contained within a region possessing a given energy. Jacob Bekenstein suggested such a bound using black holes to obtain a bound which itself did not contain Newton's constant. The bound raised as many questions as it attempted to answer. This chapter contains a description of attempts to sharpen the concepts involved in the bound and ways of generalizing them to space times including cosmologies.

Tom Banks and Willy Fischler's contribution surveys in detail the cosmological generalizations and implications of the Bekenstein bound; from the early works of Fischler–Susskind and Bousso to their own joint work on the Holographic Space time.

Raphael Bousso provides a critical review of the formulations and derivations of the Bekenstein bound and of its various profound off-springs, this covering his own contribution to its generalization to general space-times.

Don N. Page's contribution begins by providing counter examples to the Bekenstein bound that highlight the need to sharpen the definition of the basic concepts such as energy, volume and entropy which appear in the various forms of the bound. He describes his own attempts and finishes with those of H. Casini.

Gabriele Veneziano discusses the evolution of the understanding of the Bekenstein bound and its implications in more general setups. These include string inspired cosmologies, in particular some that possess a pre-Big Bang era.

Erik P. Verlinde uses the Bekenstein bound and the black hole entropy formula as an opportunity to reconsider what are the basic concepts that underlie the phenomena of gravity. He is led to restore the basic role of the concept of inertia as well as to M. Milgrom's way to describe galaxy dynamics.

Cosmological Implications of the Bekenstein Bound

Tom Banks
Department of Physics and NHETC,
Rutgers University, Piscataway, NJ 08854, USA
banks@physics.rutgers.edu

Willy Fischler
Department of Physics and Texas Cosmology Center,
University of Texas, Austin, TX 78712, USA
fischler@physics.utexas.edu

A brief review of "Holographic Space-Time" in the light of the seminal contributions of Jacob Bekenstein.

1. Introduction

Jacob Bekenstein's revolutionary discovery of the entropy of black hole horizons has important implications for cosmology and, in the opinion of the present authors, may provide the key clue to a general quantum theory of gravity. The present article will review attempts to extend Bekenstein's ideas to cosmological, and more general space-times. The logical line of argument runs from Bekenstein and Hawking,[1] through the Gibbons–Hawking discovery of de Sitter temperature,[2] the fundamental results of Jacobson,[3] and the initial cosmological conjecture of Susskind and one of the present authors,[4] which led to the general conjecture of Bousso.[6] We view our own work on Holographic Space-time (HST) to be the natural extension of these ideas.[8] There have been other attempts to generalize Bekenstein's results to cosmological space-times,[7] but they appear to lie outside the general pattern that emerges from this historical development.

Bekenstein's assignment of entropy to black holes implied a non-zero temperature and Hawking's revolutionary calculation of the thermal spectrum of black holes was verification of Bekenstein's conjecture. This is sometimes taken to imply that quantum field theory must be a good description near the horizon of a large black hole, but as shown by the work of Ref. 17 the key to the thermal nature of black holes is the periodic identification of the Euclidean time coordinate in the non-singular continuation of black hole metrics to Euclidean signature. This implies thermal behavior of any quantum system using that time coordinate. The latter argument is the simplest way to understand the Gibbons–Hawking temperature of de Sitter space.[a]

The first clue about extending Bekenstein's conjecture to general space-times came more than two decades later, in a paper by T. Jacobson.[3] In our opinion, the seminal work of Bekenstein, and its elaboration by Hawking, Gibbons and Jacobson, will eventually be seen as laying the foundations of a general quantum theory of gravity. Jacobson considered a spacetime whose radii of curvature are larger than all microscopic scales, and argued that there was consequently a large region of spacetime which was approximately flat. He then makes three postulates (one only implicit)

- One can localize the energy and entropy density in small regions and use hydrodynamic (localized thermodynamic) equations.
- Bekenstein's law relating area to entropy is valid for infinitesimal changes in area, in particular those which are described by Raychauduri's equation and related to the Einstein curvature.
- Unruh's relation between acceleration and temperature in flat space is valid in the limit of infinite acceleration. There is a lot hidden in this assumption. It implies in particular that the entropy that appears in thermodynamic relations is the maximal entropy allowed for the system. In the context of quantum mechanics this means that the entropy should be interpreted as the logarithm of the dimension of a localized quantum system.

The thermodynamic relation

$$dE = TdS\,, \tag{1}$$

can be written in terms of the change in area of a causal diamond as a consequence (for example) of the change in proper time between its tips,

$$4dEL_P^2 = \int_{\mathbf{H}} \theta d\lambda d\mathbf{A}\,. \tag{2}$$

We are thinking about an Unruh trajectory that is near the boundary of the diamond, with its inflection point nearest to a point p on the boundary. χ^a is the approximate boost Killing vector (here we assume low curvature) that generates this trajectory: $\chi^a = -\kappa\lambda k^a$, where κ is the acceleration of the trajectory where the norm of χ^a is 1, k^a is the tangent to the generators of the local Rindler horizon, and λ is an affine parameter that vanishes at p and is negative to the past of p. $d\mathbf{A}$ is the change in area on the horizon. We take κ to be very large, so that

[a]Since there is no general argument that different dS spaces can be thought of as states of a single system, it is unclear if one should write the thermodynamic relation $dE = TdS$ in this context, but if one does, and uses the Bekenstein relation between horizon area and entropy, one deduces the correct energy density of dS space.

the trajectory stays near the point p. That means that we are taking the infinite temperature limit since $T = \kappa/2\pi$. In this limit κ appears linearly on both sides of the first law of thermodynamics since we also have

$$dE = -\kappa \int_{\mathbf{H}} \lambda T_{ab} k^a k^b d\lambda d\mathbf{A} \,. \tag{3}$$

Note that this is proportional to the time dependent Hamiltonian for motion along the trajectory. Since p is on the maximal area surface at the boundary of the diamond the expansion and shear vanish at p and the Raychauduri equation is approximately

$$d\theta = -R_{ab} k^a k^b d\lambda \,, \tag{4}$$

which can be integrated to find θ near p. Thus we get

$$8\pi L_P^2 \int_{\mathbf{H}} T_{ab} k^a k^b d\lambda d\mathbf{A} = \int_{\mathbf{H}} R_{ab} k^a k^b d\lambda d\mathbf{A} \,. \tag{5}$$

If we impose covariant conservation of the stress tensor, then this can only be true if Einstein's field equation holds for every point in space time, when dotted twice into every local null vector at that point. Consequently, one gets Einstein's equation with an undetermined cosmological constant.

Our interpretation of the latter fact is that the cosmological constant is *not* a local energy density. We conjectured long ago that the c.c. was instead an asymptotic boundary condition. A causal diamond has two large scale geometric properties: the proper time of the geodesic between its tips, and the area of its holographic screen. The c.c. controls the relation between the two limits where these quantities become large. For positive Λ the area remains finite for proper time going to infinity. For negative Λ the area goes to infinity at finite proper time. For vanishing Λ we have $A \sim t^{d-2}$ as $t \to \infty$.

Jacobson's result shows that Einstein's field equations are the hydrodynamic equations of the BHJFSB entropy law for causal diamonds. String theory tells us that we can think of all fields in physics as originating from the supersymmetric partners of and other features of the metric in high enough dimension. Thus, all of field theory is hydrodynamics. This is a clue to the nature of quantum gravity. In familiar physics, one quantizes the equations of hydrodynamics only near the ground state of a system where all low energy excitations are quantized hydrodynamic fields (including the fermions of Fermi liquid theory). These are linear field theories, with small perturbative corrections. The classical non-linear equations of hydrodynamics are valid instead in high entropy situations where the microscopic quantum states have nothing to do with phonons. Entropy appears in these equations, but is not explained by them. Similarly we expect that the entropy of horizons will come from quantum systems very different from bulk quantum field theory.

We also reiterate that Jacobson's use of infinite Unruh temperature systems to derive his result, implies that the entropy law applies to the full dimension of the quantum Hilbert space, not some less random density matrix.

2. The Fischler–Susskind Bound

In the early spring of 1998, Fischler and Susskind were wondering whether the spectacular discovery of Jacob Bekenstein in the context of black holes could hold for general space-times. Their prejudice at the time was that indeed this new paradigm should apply more generally. In particular, cosmological space-times seemed a fertile and especially useful arena in the search for a first step in generalizing Bekenstein's idea. This led them to consider flat FRW universes as simple geometries where one might be able to extend the Bekenstein entropy bound.

At that time we had learned from Matrix Theory and what later became known as the AdS-CFT connection, that the holographic principle, requires that the degrees of freedom of a spatial region reside not in the interior as in an ordinary quantum field theory but on the surface of the region. Furthermore it requires the number of degrees of freedom per unit area to be no greater than 1 per Planck area. As a consequence, the entropy of a region must not exceed its area in Planck units. The task at hand was how to generalize this principle to cosmology. What follows is a brief summary of the Fischler–Susskind paper on the holographic principle in cosmology.[4]

The FRW metric has the usual form

$$ds^2 = dt^2 - a^2(t)dx^i dx^i \,. \tag{6}$$

The number of spatial dimensions will be kept general so that i runs from 1 to d. The cosmological assumptions are the usual ones including homogeneity, isotropy and in the late universe constant entropy density in comoving coordinates.

A first attempt was very quickly ruled out: the entropy in any region of coordinate size $\Delta x \sim R$ never exceeds the area. Since our assumptions require the entropy density to be constant, the entropy in a region is proportional to its coordinate volume R^d. Furthermore in flat space the surface area of the region grows with R like $[Ra(t)]^{d-1}$. Obviously when R becomes large enough, the entropy exceeds the area and the principle is violated.

A more sophisticated version goes as follows. Consider a spherical spatial region Γ of coordinate size R with boundary B. Now consider the light-like surface L formed by past light rays from B toward the center of Γ. One can consider then three situations. If R is smaller than the coordinate distance to the cosmological horizon R_H then the surface L forms a light cone with its tip in the future of the singularity at $t = 0$. If R is the size of the horizon then L is still a cone but with its tip at $t = 0$. However if $R > R_H$ the surface is a truncated cone. Now consider all the entropy (particles) which pass through L. For the first two cases this is the same as the entropy in the interior of Γ at the instant t. But in the last case the entropy within Γ exceeds the entropy passing through L. The proposed holographic principle is that the entropy passing through L never exceeds the area of the bounding surface B. It is not difficult to see that for the homogeneous case, this reduces to a single condition:

The entropy contained within a volume of coordinate size R_H should not exceed the area of the horizon in Planck units. In terms of the (constant) comoving entropy density σ

$$\sigma R_H^d < [aR_H]^{d-1} \tag{7}$$

with every thing measured in Planck units. Note that both R_H and a are functions of time and that this equation must be true at all time.

Let us first determine whether the bound is true today. The entropy of the observable universe is of order 10^{86} and the horizon size (age) of the universe is of order 10^{60}. Therefore the ratio of entropy to area is much smaller than 1. Now consider whether it will continue to be true in the future. Assume that $a(t) \sim t^p$. The horizon size is determined by

$$R_H(t) = \int_0^t \frac{dt}{a(t)} \sim t^{1-p} \,. \tag{8}$$

Thus in order for this equation to continue to be true into the remote future we must satisfy

$$p > \frac{1}{d} \,. \tag{9}$$

In other words there is a lower bound on the expansion rate.

The bound on the expansion rate is easily translated to a bound on the equation of state. Assume that the equation of state has the usual form

$$P = \gamma\epsilon \tag{10}$$

where P and ϵ are pressure and energy density. Standard methods yield a solution of the Einstein equations

$$a(t) \sim t^{\frac{2}{d(1+\gamma)}} \,. \tag{11}$$

Thus $p = \frac{2}{d(1+\gamma)}$ and the inequality for the equation of state becomes

$$\gamma < 1 \,. \tag{12}$$

This bound is well known and follows from entirely different considerations. It describes the most incompressible fluid that is consistent with special relativity. A violation of the bound would mean that the velocity of sound exceeds the velocity of light. We will take this agreement as evidence that our formulation of the cosmological holographic principle is on the right track.

Although violating the bound on the equation of state is impossible, saturating it is easy. We will give two examples. In the first example the energy density of the universe is dominated by a homogeneous massless minimally coupled scalar field. It is well known that in this case the pressure and energy density are equal.[b]

Another example involves flat but anisotropic universes, for example the anisotropic[5] universes with metric:

$$ds^2 = dt^2 - \Sigma_i \, t^{2p_i} dx_i{}^2 \,. \tag{13}$$

[b]In HST cosmology this equation of state describes the early stage of the universe.[9]

The ratio entropy/area, S/A, for this case is easily evaluated to be:

$$S/A = \Pi_i \; R_{H,i}/[\Pi_j \; t^{p_j} t^{1-p_j}]^{d-1/d} \tag{14}$$

where $R_{H,i} = t^{1-p_i}$ is the coordinate size of the horizon in direction i. The denominator on the left-hand side of this equation is the proper area of the horizon.

It follows that:

$$S/A = t^{\;1-\Sigma_i \; p_i} \, . \tag{15}$$

The conditions on the exponents of the Kasner solutions are obtained by using the Einstein equations.The exponents satisfy the following equations:

$$\Sigma_i \; p_i = 1 \, , \tag{16}$$

$$\Sigma_i \; p_i^2 = 1 \, . \tag{17}$$

It follows from these equations that S/A for these flat anisotropic universes is constant in time.Thus depending on the boundary condition the holographic bound may be saturated by these universes. Note that in these "models" we do not really explain the entropy of the system. The scalar field/Kasner metric acts like a hydrodynamic variable, which encodes the entropy but does not explain it.[c]

Having established that the holographic principle will not be violated in the future we turn to the past. First consider the entropy–area ratio ($\rho = S/A$) at the time of decoupling. Standard estimates give a ratio which is about 10^6 times bigger than todays ratio. Thus at decoupling

$$\rho(t_d) \sim 10^{-28} \, . \tag{18}$$

During a radiation dominated era it is easily seen that ρ is proportional to $t^{-1/2}$,

$$\rho = 10^{-28} \left[\frac{t_d}{t} \right]^{1/2} \, . \tag{19}$$

Remarkably, in Planck units $t_d^{1/2} \sim 10^{28}$ so that $\rho < 1$ for all times later than the Planck time. The entropy in the universe is as large as it can be without the holographic principle having been violated in the early universe!

When Fischler and Susskind tried to extend their approach to closed FRW universes they found that their arguments led to a violation of the bound for such cosmologies. This puzzle was resolved, and a much more general formulation of a Bekenstein like bound discovered, in the work of Bousso.[6]

[c]Indeed, we get the same equation of state as these models in a kinetic dominated, or mixmaster universe, which has gravitational/scalar fields that depend on position but have no spatial gradient terms in their action. These models have *too much* entropy to be compatible with the cosmological version of Bekenstein's bound.

3. The Bousso Bound

The work by Fischler and Susskind inspired Bousso to generalize the bound and express it in a covariant language. The result was Bousso's magnificent general conjecture, which he called the Covariant Entropy Bound. A detailed description of this can be found in Bousso's contributions to this volume. We will only mention a few differences of emphasis, which have developed in our extensive use of Bousso's work in our own.

Bousso's bound applies to a general space-like two surface and is described in terms of conditions on the light sheets emanating from it. A particular case of such a two surface is associated with a causal diamond. The boundary of the diamond is the intersection of two light-cones, which form a continuous, but not differentiable manifold. On each light-cone there is a metric of the form

$$ds^2 = (du - A_i(\mathbf{x}, u)dx^i)B_j(\mathbf{x}, u)dx^j + g_{ij}(\mathbf{x}, u)dx^i dx^j \,. \tag{20}$$

The area of the space-like two surfaces at fixed u increases as one moves away from the tips of the diamond. By continuity, there is a maximum value of the area for some value of u, which might be at the joining point of the two diamonds. The area of that maximal area surface satisfies the Covariant Entropy Bound of Bousso. Bousso's conjecture says that it bounds four times the entropy of states localized in the causal diamond. In our work, we have used the phrase "holographic screen of the diamond," to refer to this maximal area surface. In Bousso's work the name *holographic screen* refers to the full null history of this 2-surface.

There is a well-known theorem in Lorentzian geometry, which says that a metric is completely determined in terms of properties of its collection of causal diamonds. For any pair of diamonds, there is a maximal area diamond in its intersection, and the causal structure of the manifold is completely determined by specifying the collection of diamonds and their intersections. The conformal factor is determined in terms of the areas of all the diamonds. If, in addition, we use the Bousso–Jacobson generalization of Bekenstein's area law to arbitrary causal diamonds, and assume that the entropy bound is in fact saturated for the maximally uncertain density matrix associated with a causal diamond, then we can translate a Lorentzian geometry into a statement about quantum Hilbert spaces.

Indeed, we have long known how to translate causal relations into properties of a quantum operator algebra.[18] Bekenstein's seminal work, as generalized by Jacobson, Fischler–Susskind, and Bousso, completes the process of translation. The more refined version of Bousso's conjecture, which insists that the entropy bound is saturated, is best justified by Jacobson's use of infinite temperature Unruh trajectories to derive Einstein's equations from hydrodynamics.

It is an immediate consequence of these considerations that the ancient idea of quantum gravity as a theory of fluctuating geometries is not correct. Jacobson's arguments view the geometry of space-time as a hydrodynamic variable. The more refined quantum version of this idea that the present authors have proposed in the theory of Holographic Space-time (HST) views geometry as an emergent description relating the dimensions of Hilbert spaces nesting inside each other. There is no quantum interference involved.

One can understand how these ideas are consistent with string theory's match onto low energy effective field theory by recalling that in ordinary condensed matter physics, hydrodynamic equations are useful in two quite different regimes. For systems with a ground state, hydrodynamic variables associated with conserved symmetry currents often give a complete quantum mechanical description of the low energy excitations of the system. They give rise to a set of gapless boson and fermion fields, which are weakly interacting. The dominant longwavelength interactions are described by generalized Navier–Stokes equations because those equations follow from symmetries and the derivative expansion. Recently, two groups have presented general effective field theory descriptions of the hydrodynamic modes of relativistic quantum field theories.[10]

The traditional use of hydrodynamics however is in a completely different regime, high in the energy spectrum, where the density of states is very large. In this regime, hydrodynamic variables satisfy non-linear classical equations, and the quantum fluctuations around those solutions are exponentially small as a function of the entropy density. The microscopic quantum fluctuations of the system take place on a wide variety of time scales, between the natural microscopic scales τ and $e^S\tau$. They have nothing to do with quantum fluctuations of the hydrodynamic fields, but rather with short wavelength fluctuations, which do not change the hydrodynamic variables by perceptible amounts, because the latter are coarse grained volume averages.

Jacobson's derivation of the Einstein's equations takes place in this high entropy regime. The novel feature of theories of gravity are connected to the fact that when the c.c. is non-negative, this high entropy regime is one of extremely low energy. For a causal diamond of radius R, for positive c.c. or $R \ll R_{AdS}$ for negative c.c., typical states contribute to the Hamiltonian an amount of order R^{-1}. In the limit $R/L_P \to \infty$, in flat space, these states become "soft gravitons". They are topological degrees of freedom, which decouple from the scattering operator of finite energy states.

The traditional methods of string theory hide these states in two different ways. In perturbative string theory in Minkowski space, amplitudes for producing large numbers of soft gravitons in processes initiated by a few finite energy particles, are very small at low energy.[d] At energies of order string scale, long strings, the precursors[12] of black holes, are produced even in few particle collisions. These large meta-stable objects emit infinite numbers of soft gravitons in their evolution and

[d]Here we are implicitly assuming a Minkowski space of sufficiently high dimension. It is well known that in four dimensions the gravitational scattering operator has vanishing matrix elements between Fock space states even in perturbation theory. We believe that in all dimensions Fock space is a subspace of a representation space of an asymptotic current algebra, generated by the action of currents that are delta functions in angle, and that the scattering operator is not unitary on this subspace.[11]

decay. The non-perturbative approach to string theory in AdS spaces, is a system with a gap, and the asymptotic soft graviton states are simply banished from the Hilbert space. High entropy is encountered, as in non-gravitational physics, only far above the gap. As a consequence[13] gravity in AdS space is a poor guide to local physics and cosmology.

Thus, traditional string theory deals only with the low entropy part of the spectrum of a quantum gravitational Hamiltonian, where quantized hydrodynamics is a good approximation to much of the physics. Classical gravity, like the Navier–Stokes equation, is still a good description of coarse grained properties of high entropy states, but it is a mistake to quantize it in those situations. The theory of Holographic Space Time (HST) takes the JFSB generalization of Bekenstein's law as a fundamental principle and gives a transparent model of the physics of horizons and inflation. We turn next to a brief description of the principles of HST and its application to inflationary cosmology.

4. The HST Formalism

Consider a causal diamond D_1 in a general space-time, embedded in a larger causal diamond, D_2 whose holographic screen has finite area. The BJFSB principle implies that each of these diamonds is associated in the quantum theory with a finite dimensional operator algebra. The notion of causality implies that the smaller algebra $\mathcal{A}_1$ has a nontrivial commutant inside the larger one. Otherwise we would have a genuine EPR paradox, with quantum information being transmitted faster than the speed of light. For finite dimensions, this implies that

$$\mathcal{A}_\infty \otimes \bar{\mathcal{A}}_1 , \tag{21}$$

and that the Hilbert space has a similar factorization. This generalizes to diamonds that are not nested, but share an intersection region (Fig. 1). That region contains a maximal area causal diamond, which is represented in quantum mechanics as a

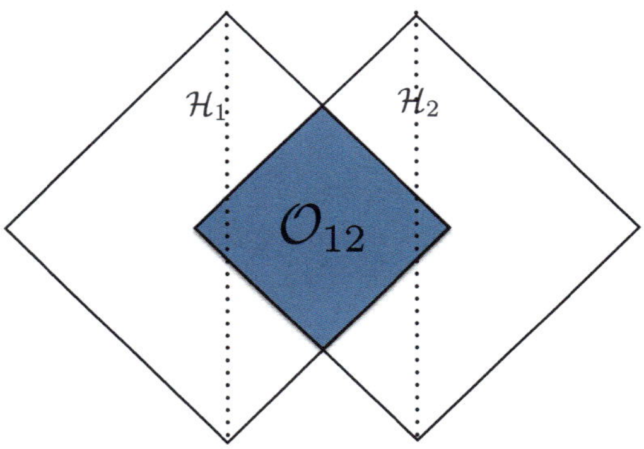

Fig. 1. Intersecting diamonds.

tensor factor in the individual diamond Hilbert spaces. Consistency of the two diamonds' descriptions of the intersection implies that the state on individual diamond Hilbert spaces always gives a density matrix for the intersection diamond D_{12} with the same eigenvalue spectrum. This is the quantum analog of Einstein's general principle of relativity, as we will see in a moment. The unitary transformations relating the two descriptions of the density matrix are quantum analogs of general coordinate transformations between different physical gauges. We therefore call these consistency requirements *Quantum General Coordinate Invariance* (QGCI).

A causal diamond in space-time is characterized by two intrinsic geometrical parameters, the maximal proper time between its tips, and the maximal area on its boundary. In HST, the latter is encoded in the dimension of a Hilbert space, while the former is directly connected with the time evolution in the quantum system. We will have a different quantum description for every time-like trajectory (though we will stick to geodesics in this article). Choosing a set of nested proper time intervals along a trajectory is equivalent to choosing a nested set of causal diamonds, and part of the geometrical information is encoded in the relation between the proper time dependence of the Hamiltonian and the growth of the area. In order to be consistent with QGCI the time evolution over any interval I, must factorize into $U_{\rm in}U_{\rm out}$ where $U_{\rm in}$ is constructed from operators in $\mathcal{A}(I)$ and $U_{\rm out}$ from operators in $\bar{\mathcal{A}}(I)$. Thus, the Hamiltonian *must* be time dependent. This is a reflection of the geometrical fact that any time slicing that remains within a causal diamond has a generator which is not a Killing vector of space-time, even in space-times with time-like Killing vectors. Figure 2 shows the relation between the causal coordinates of HST and the standard FRW coordinate systems of cosmology. Note that in any such coordinate system there is a large redshift between the proper time along a geodesic and the local times of degrees of freedom that are localized on the boundary. This is incorporated in the Hamiltonian in a relative factor of $1/N$ between terms

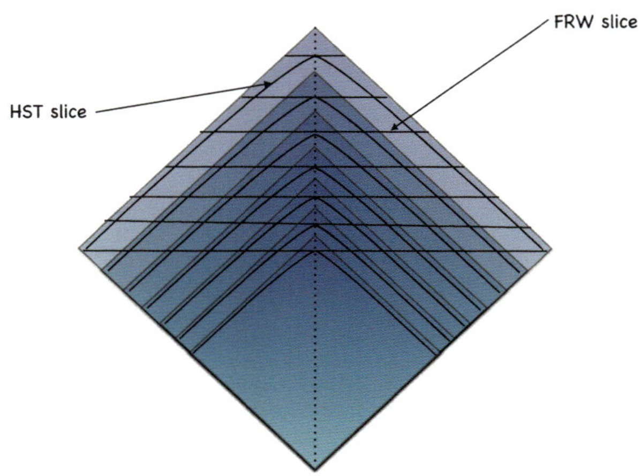

Fig. 2. HST and FRW coordinates.

describing the boundary and terms describing excitations localized in the bulk. N scales like $A^{1/2}/L_P$ when that quantity is large. Note finally that associating the Hamiltonian with proper time along particular trajectories completely resolves the "Problem of Time" in GR. Relativity is imposed as consistency conditions between quantum systems with different Hamiltonians, rather than insisting that any particular description be invariant.

Although this is not the historical line of argument, one can motivate the explicit models we study, by two observations about black holes. The first is that inserting a black hole into de Sitter (dS) space *reduces* the area of the horizon and thus the entropy of the system. In fact, the Schwarzschild dS metric gives a completely classical (except for its use of Bekenstein's quantum identification of entropy and area) derivation of the Gibbons–Hawking temperature of this space-time. The value of that temperature suggests that the states responsible for the entropy of empty dS space contribute to the Hamiltonian of a static geodesic via an operator $1/NH_B$, where H_B has an N independent bound. Objects localized near the geodesic will have energies that are $o(N^0)$. One is also led to describe localized states by a number of constraints of order EN where each constraint operator acts on a Hilbert space whose size is independent of N. Indeed a randomly chosen state will satisfy this many constraints with quantum probability e^{-cEN}. E should thus be thought of as proportional to the energy of the state in Planck units and must be $\ll N$ if the state is to be meta-stable.

The second important property of black holes that we will use is that they are fast scramblers.[14] Information localized on the horizon is scrambled in a time of order the logarithm of the entropy in horizon radius units. This is a property of all black holes in Minkowski or dS space, but in AdS space, the scrambling is only ballistic (a power of the entropy) when the black hole radius exceeds the AdS radius. This follows both from the nature of the quasi-normal modes[15] and from the fact that the quantum theory is a local field theory.[e]

To motivate the choice of variables and Hamiltonians in HST models, we think about scattering theory in asymptotically flat space (AfS). It is convenient to describe the conformal boundary of AfS by two copies of the momentum light cone $P^2 = 0$, one at each angle for massless momenta penetrating the boundary and one for momenta flowing along the boundary. We need the latter to describe currents carried by massive particles. We can describe everything at infinity by currents carrying quantum numbers through the sphere at infinity. These are generalized functions $J_i(P)$. Our own world contains particles of half integer spin, so among the these there must be spinors $\mathcal{Q}^i_\alpha(P)$, which are null plane spinors for P and perhaps also null plane spinors for the reflected null vector $\tilde{P}$. The spin statistics connection implies that the Q operators at different points must anti-commute. In fact, at coincident points they satisfy a generalization of the Awada–Gibbons–Shaw

[e]More precisely, the scrambling is fast out to the AdS radius, and then becomes ballistic at larger distances. Note that the infall time to the singularity is always the AdS radius and does not increase with the Schwarzschild radius as it does for non-negative c.c.

algebra,[16] but we will not have to use that. The Q^i_α carry quantum numbers sufficient to describe everything in space-time so it is reasonable to assume that they generate the entire operator algebra.

Now imagine a finite causal diamond associated with a finite proper time interval along a particular geodesic. We have to find a finite dimensional subalgebra of variables compatible with rotation invariance around that geodesic, and the only way to do that is to impose an angular momentum cutoff. That is, we take the operator algebra of the diamond to be generated by $q^i_{ab} = -q^i_{ba}$, where the indices run from 1 to something of order $N = R/L_P$.[f] Rotation invariance of the commutation relations implies that $[q^i_{ab}, q^j_{cd}]_+ = (\delta_{ac}\delta_{bd} - \delta_{ad}\delta_{bc})B^{ij}$. For fixed values of the indices the representation space is the fundamental representation of $SU(K/L)$ for finite $K + L$. Given these variables we can construct matrices

$$M^{ij}_{ab} \equiv q^i_{ac}q^j_{cb}\,, \tag{22}$$

and take the Hamiltonian to be the trace of a polynomial in these matrices. The matrices transform like anti-symmetric tensors of any rank and the trace projects out the two forms. Any such Hamiltonian is invariant under $U(N)$ transformations of the indices, which are the "fuzzy" analog of area preserving maps. It is well known that under such a map, a small spherical cap is mapped into an "amoeba" of arbitrary shape with the same area. Thus, the fact that a region is small just means that it takes time to scramble information injected into a small number of degrees of freedom. Each action of the Hamiltonian scrambles some finite number of variables, and thus the number of variables reached is exponential in the number of time steps. There is no suppression of interaction between "distant" regions, because the Hamiltonian does not know about the geometry of the sphere but only its measure. The time scale of horizon dynamics $\sim N$ is reproduced if we write

$$H_{\text{in}} = \frac{1}{N}\text{Tr } P(M/N)\,, \tag{23}$$

where we have invoked 't Hooft scaling of matrix models.

We now want to ask about constraints that could correspond to localized jets of particles penetrating the boundaries of the diamond. Such jets should be non-interacting if N is large, and the obvious (only?) way to do that is to constrain the initial states such that all of the matrices M are block diagonal, with a number of blocks of size $E_k \sim 1$ and one block of size $\sum E_k N$. This reproduces (in the sense of scaling with N the dS black hole entropy formula if $\sum E_k$ is identified with a multiple of the asymptotically conserved (in flat space) energy. Of course in dS space energy is conserved only over times of order R. In fact, the generic Hamiltonian of this form leads to an asymptotically conserved energy if we allow $N \sim t$ as $t \to \infty$ and reproduces the gross thermal properties of dS space if N is kept large but finite as t is taken to infinity.

We do not have space here to describe the details, but these models define a scattering theory that is manifestly compatible with space-time translation invariance,

[f]This gives us all half integer representations of $SO(3)$ up to the cutoff.

unitarity and causality. Its asymptotic Hilbert space has a description as a representation of the Poincare invariant (generalized) AGS algebra. The constraint of Lorentz invariance, which follows from the HST consistency conditions for geodesics in relative motion, has not yet been implemented. Experience with string theory leads us to expect that most of the models we have described will not pass this test.

All of the models have meta-stable excitations with all of the scaling properties of entropy, temperature, energy and scrambling properties that we expect.[19] The conditions for black hole formation also meet semi-classical expectations: a black hole is formed when the energy in Planck units inside a causal diamond is of order the Schwarzschild radius in Planck units. The models all resolve the firewall paradox[20]: when a system falls onto a black hole the entropy of the relevant Hilbert space increases by a large amount because of the off-diagonal matrix elements. The latter are however initially frozen and the time that it takes to equilibrate the infalling system is a time over which that system behaves as if it were in flat space.

5. Holographic Cosmology

The HST model of cosmology proposes new answers to some of the oldest questions in the subject: how the initial singularity is avoided and why the universe started out in a low entropy state. There is nothing in the quantum mechanics of time dependent Hamiltonians that insists that any finite point in time must be infinitely far away from past or future time-like infinity. In HST, merely positing an initial time at some finite time in the past, tells us that if we choose the nested diamonds to all begin at the initial point, then $H_{\rm in}(t_1)$ acts on a Hilbert space of very low dimension. It cannot resemble the Hamiltonian of a quantum field theory and no features of it exhibit classical hydrodynamic flow. There is nothing singular about the beginning of the universe, it is just not described by the classical or quantum Einstein equations.

The Boltzmann–Penrose question has a more interesting resolution. One can construct a one parameter set of models in which each causal diamond along each geodesic, is always in a maximal entropy state. The Hamiltonians are simply chosen to be those of the previous section, but we start in a generic state so there are no terms in the Hamiltonian in constrained sectors. If we allow this evolution to go on forever, but insist that at large time t the Hamiltonian approaches that of a $1+1$ dimensional chiral CFT, with central charge $\sim t^2$, and UV cutoff $\sim 1/t$, living on an interval of size Kt with $1 \ll K \ll t$. Since the comoving volume of a flat FRW model scales like t^3, the comoving entropy and energy densities scale like

$$\sigma \sim 1/t \qquad \rho \sim 1/t^2 . \tag{24}$$

We automatically get the Friedman relation $\rho \sim 1/t^2$ and the characteristic energy/entropy density relation $\sigma \sim \sqrt{\rho}$ of the $p = \rho$ equation of state, which[4] showed saturated the early universe entropy bound. The large number K is related to the range of the indices i, while the t^2 central charge should be thought of as originating from t^2 field q_{ab} in the CFT.

If we modify the model so that after a certain time N, the number of q_{ab} does not increase then we get a flat FRW model that interpolates between $p = \rho$ and $p = -\rho$, with no localized excitations at any intermediate time. These models are completely formulated as consistent quantum theories of gravity. From the point of view of the asymptotic future dS space of these models, any localized excitation is a constrained state, with the approximately conserved energy E determining the number of constraints. For fixed E, the most probable state to find localized in the bulk is a single black hole of mass E. If we require that the universe have a radiation dominated era, this initial condition is not acceptable.

In HST, the global structure of the universe is not an initial condition. Homogeneity is a choice of *model*: we insist that we have a set of identical quantum systems, which can be associated with different geodesics in the universe. The Hilbert space of a given system is $\mathcal{H}(t, \mathbf{x})$ and its Hamiltonian $H(t, \mathbf{x})$, and both have the tensor factorization property implied by a nested sequence of causal diamonds corresponding to the nested sequence of intervals $[0, t]$. An asymptotic dS future implies that as $t \to \infty$, the dimension of the Hilbert space asymptotes to e^{cN^2}.

Now, fix the radius N of the asymptotic dS space and insist that, following the initial, generic, $p = \rho$ beginning, the Hilbert space remains at finite size $1 \ll n \ll N$ for N_e e-folds of proper time. That is, we have an inflationary era, with inflationary Hubble parameter n^{-1} in Planck units. We then let the horizon expand gradually, by changing the size of the Hilbert space on which $H_{in}(t)$ acts. This is what corresponds to the slow roll era. As the horizon expands, we must satisfy the QGCI constraints. Geodesics originally outside the inflationary horizon volume enter into it (Fig. 3). This occurs on FRW time slices prior to the end of slow roll. Let us label one particular geodesic as the origin, $\mathbf{0}$. During the period up to horizon entry, $H_{\text{out}}(t, \mathbf{0})$ along one geodesic must coincide with $H_{\text{in}}(t, \mathbf{x})$ for the trajectory just entering the horizon, for all t prior to horizon entry. Since t is in the inflationary regime, the latter operator acts on a Hilbert space of entropy n^2 and the state is random. This means that the initial conditions in the out part of the Hilbert space $\mathcal{H}(t, \mathbf{0})$ must be such that there is a tensor factor of entropy n^2 which has not interacted with the rest of the system. Given the form of the Hamiltonian this implies that the initial conditions are constrained, and there is a localized object of entropy n^2 coming into the horizon. Since the *model* is defined by a homogeneous and isotropic collection of quantum systems, we get a homogeneous and isotropic gas of black holes on FRW slices of the space-time. Homogeneity and isotropy cannot be exact however, because the HST consistency conditions only constrain the density matrix in the intersection of two diamonds. There are thus statistical fluctuations, in the entropy of the black holes and in their angular momentum. These are, by the usual rules of statistical mechanics, primarily Gaussian, with a size $\delta H/H \sim 1/n$, the inverse square root of the entropy. Classical cosmological perturbation theory gives the conventional gauge invariant measure of fluctuations

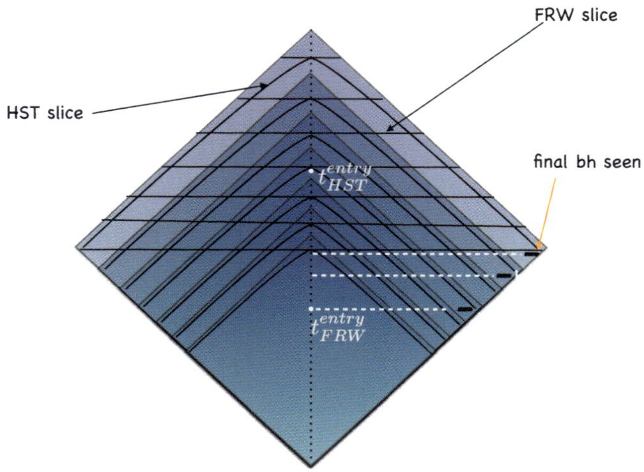

Fig. 3. Black holes entering horizon.

as $\zeta = \delta H/H\epsilon$, where $\epsilon = \dot{H}H^2$ is the slow roll factor. Non-Gaussianity is down by inverse powers of n.

The detailed form[21] of the predictions of this model is quite different than that of conventional inflation, particularly for tensor fluctuations and non-Gaussianities, but for scalar fluctuations the predictions depend on the slow-roll factor $\epsilon(t)$. The only observational evidence we have constraining $\epsilon(t)$ comes from scalar CMB fluctuations, so it is obvious that this model can fit the existing data. Nominally, the ratio of tensor to scalar fluctuations is of order $r \sim \epsilon^2$, and this suggests it is very small, but we have not yet been able to calculate the "$o(1)$" coefficients in this ratio. So far we have found no theoretical constraint on the form of $\epsilon(t)$ except for the fact that, as is clear from Fig. 3, inflation lasts half the conformal time of the universe (which is bounded because we have an asymptotic dS space). This gives a total number of e-folds as ~ 80. Approximate $SO(1,4)$ invariance of the fluctuations is obtained by combining the approximate $SL(2,R)$ of individual horizon volumes with the $SO(3)$ of any given geodesic.

6. Conclusions

Bekenstein's revolutionary association of entropy with black hole area provides the kernel, in our opinion, for the entire quantum theory of gravity. Quantum field theory calculations by Hawking and others, which validate Bekenstein's idea, do not provide us with a model for the microscopic degrees of freedom responsible for that entropy. The success of those calculations is based on the geometric fact that the analytic continuation of a horizon to imaginary time contains a one cycle of fixed length at infinity. Unruh's result on accelerated trajectories in flat space-time uses the same geometric observation and suggests that Bekenstein's idea applies to more general regions in space-time, and that states that quantum field theory treats as unique vacuum states, actually have large entropy. The work of Jacobson, Fischler–Susskind, and Bousso, gives us a natural extension of Bekenstein's conjecture to arbitrary causal diamonds in arbitrary space-times. Jacobson's work in particular shows that most of the states associated with the BHJFSB entropy, have very small energy when viewed from the geodesic coordinate system in a given causal diamond. The Schwarzschild de Sitter entropy formula then shows us that states localized in a causal diamond in de Sitter or asymptotically flat space, are constrained states of the fundamental degrees of freedom, with the conventional energy equal to the number of constraints divided by the dS radius.

The HST formalism is a general proposal for quantum theories of gravity in arbitrary space-times, based on the ideas outlined in the previous paragraph. A classical space-time gives rise to a set of causal diamonds. HST is based on combining the field theoretic idea that any such diamond corresponds to a Murray–von Neumann factor in the complete operator algebra of space-time. The BHJFSB entropy formula implies that most Hilbert spaces involved in quantum gravity are finite dimensional. The exceptions among "commonly encountered" space-times are the asymptotic Hilbert spaces of asymptotically flat and AdS space-times. These are two, quite different limits of the physics in finite causal diamonds. We have argued elsewhere[13] that the CFT boundary Hamiltonian of AdS space-times does not describe time evolution in causal diamonds much smaller than the AdS radius, except for a small set of states, in perturbation theory. Minkowski space physics is a straightforward extension of the physics of localized objects in finite causal diamonds and is approximately the same as the de Sitter physics of such objects over time scales short compared to the dS radius.[g]

Much of the work on HST, including the cosmological models of the penultimate section, does not depend heavily on the precise form of the variables or Hamiltonians that we use to describe the quantum systems underlying classical space-times. This is because of Bekenstein's lesson that the situations in quantum gravity that are poorly described by quantum field theory, have a coarse-grained thermodynamic description, which is adequate to predict most of what we need to account for current observations. It would not be amiss to designate the early universe in the HST model as a *Bekensteinian Cosmology*, dominated by hydrodynamics and the statistical fluctuations of systems with large but finite entropy.

Acknowledgments

The research of WF is based upon work supported by the National Science Foundation under Grant Number PHY-1620610.

[g]For bound clusters of objects, which have "decoupled from the Hubble flow" until they collapse into black holes and decay, the approximation remains valid for a much longer time. Deviations from it come only from thermal "accidents" in dS space, and from jets of radiation emitted from the cluster, which enter the horizon after a de Sitter time.

References

1. J. D. Bekenstein, "Black holes and entropy," *Phys. Rev. D* **7**, 2333 (1973); S. W. Hawking, "Particle creation by black holes," *Commun. Math. Phys.* **43**, 199 (1975) [Erratum *ibid.* **46**, 206 (1976)].
2. G. W. Gibbons and S. W. Hawking, "Cosmological event horizons, thermodynamics, and particle creation," *Phys. Rev. D* **15**, 2738 (1977), doi:10.1103/PhysRevD.15.2738.
3. T. Jacobson, "Thermodynamics of space-time: The Einstein equation of state," *Phys. Rev. Lett.* **75**, 1260 (1995).
4. W. Fischler and L. Susskind, "Holography and cosmology," hep-th/9806039.
5. E. Kasner, *Am. J. Math* **48**, 217 (1921).
6. R. Bousso, "A covariant entropy conjecture," *JHEP* **9907**, 004 (1999), hep-th/9905177; R. Bousso, "The holographic principle," *Rev. Mod. Phys.* **74**, 825 (2002), hep-th/0203101, and references therein.
7. R. Brustein and G. Veneziano, "A causal entropy bound," *Phys. Rev. Lett.* **84**, 5695 (2000), hep-th/9912055, doi:10.1103/PhysRevLett.84.5695.
8. T. Banks, "Holographic space-time: The takeaway," arXiv:1109.2435 [hep-th]; T. Banks, "TASI lectures on holographic space-time, SUSY and gravitational effective field theory," arXiv:1007.4001 [hep-th]; T. Banks and W. Fischler, "Holographic inflation revised," arXiv:1501.01686 [hep-th]; T. Banks and W. Fischler, "The holographic approach to cosmology," hep-th/0412097; T. Banks and W. Fischler, "Holographic cosmology," hep-th/0405200; T. Banks and W. Fischler, "Holographic cosmology 3.0," *Phys. Scripta T* **117**, 56 (2005), hep-th/0310288; T. Banks and W. Fischler, "An Holographic cosmology," hep-th/0111142; T. Banks and W. Fischler, "M theory observables for cosmological space-times," hep-th/0102077.
9. T. Banks and W. Fischler, "Holographic inflation revised," in *The Philosophy of Cosmology*, eds. K. Chamcham, J. Silk, J. D. Barrow and S. Saunders (Cambridge University Press, 2017), pp. 241–262, doi:10.1017/9781316535783.013, arXiv:1501.01686 [hep-th].
10. P. Glorioso and H. Liu, "Lectures on non-equilibrium effective field theories and fluctuating hydrodynamics," (and references therein), *PoS TASI* **2017**, 008 (2018), doi:10.22323/1.305.0008, arXiv:1805.09331 [hep-th]; F. M. Haehl, R. Loganayagam and M. Rangamani, "The fluid manifesto: Emergent symmetries, hydrodynamics, and black holes," *JHEP* **1601**, 184 (2016), doi:10.1007/JHEP01(2016)184, arXiv:1510.02494 [hep-th]; F. M. Haehl, R. Loganayagam and M. Rangamani, "Schwinger-Keldysh formalism. Part I: BRST symmetries and superspace," *JHEP* **1706**, 069 (2017), doi:10.1007/JHEP06(2017)069, arXiv:1610.01940 [hep-th]; F. M. Haehl, R. Loganayagam and M. Rangamani, "Schwinger-Keldysh formalism. Part II: Thermal equivariant cohomology," *JHEP* **1706**, 070 (2017), doi:10.1007/JHEP06(2017)070, arXiv:1610.01941 [hep-th].
11. T. Banks, "Current algebra on the conformal boundary and the variables of quantum gravity," arXiv:1511.01147 [hep-th]; T. Banks, "The temperature/entropy connection for horizons, massless particle scattering, and the origin of locality," *Int. J. Mod. Phys. D* **24**(12), 1544010 (2015), doi:10.1142/S0218271815440101, arXiv:1505.04273 [hep-th]; T. Banks, "The super BMS algebra, scattering and holography," arXiv:1403.3420 [hep-th].
12. L. Susskind, "Some speculations about black hole entropy in string theory," hep-th/9309145, in *The Black Hole, 25 Years After*, eds. C. Teitelboim and J. Zanelli (World Scientific, 1998), pp. 118–131; G. T. Horowitz and J. Polchinski, "A correspondence principle for black holes and strings," *Phys. Rev. D* **55**, 6189 (1997),

doi:10.1103/PhysRevD.55.6189, hep-th/9612146; J. L. F. Barbon and E. Rabinovici, in *From Fields to Strings: Circumnavigating Theoretical Physics Vol. 3*, M. Shifman, A. Vainshtein and J. Wheater (World Scientific, 2005), pp. 1973–2008, doi:10.1142/97898127753440048 [hep-th/0407236].
13. T. Banks and W. Fischler, "Holographic space-time models of anti-de Sitter space-times," arXiv:1607.03510 [hep-th]; "Soft gravitons and the flat space limit of anti-de Sitter space," arXiv:1611.05906 [hep-th].
14. P. Hayden and J. Preskill, "Black holes as mirrors: Quantum information in random subsystems," *JHEP* **0709**, 120 (2007), doi:10.1088/1126-6708/2007/09/120, arXiv:0708.4025 [hep-th]; Y. Sekino and L. Susskind, "Fast scramblers," *JHEP* **0810**, 065 (2008), arXiv:0808.2096 [hep-th].
15. G. T. Horowitz and V. E. Hubeny, "Quasinormal modes of AdS black holes and the approach to thermal equilibrium," *Phys. Rev. D* **62**, 024027 (2000), doi:10.1103/PhysRevD.62.024027, hep-th/9909056.
16. M. A. Awada, G. W. Gibbons and W. T. Shaw, "Conformal supergravity, twistors and the super-BMS group," *Annals Phys.* **171**, 52 (1986), doi:10.1016/S0003-4916(86)80023-9.
17. J. B. Hartle and S. W. Hawking, "Path integral derivation of black hole radiance," *Phys. Rev. D* **13**, 2188 (1976), doi:10.1103/PhysRevD.13.2188; G. W. Gibbons and S. W. Hawking, "Action integrals and partition functions in quantum gravity," *Phys. Rev. D* **15**, 2752 (1977), doi:10.1103/PhysRevD.15.2752; G. W. Gibbons and S. W. Hawking, *Euclidean Quantum Gravity* (World Scientific, 1993).
18. K. Fredenhagen, "An introduction to algebraic quantum field theory," in *Advances in Algebraic Quantum Field Theory*, eds. R. Brunetti *et al.* (Springer, 2015), pp. 1–30, doi:10.1007/978-3-319-21353-8_1.
19. T. Banks and W. Fischler, "Holographic space-time, Newton's law and the dynamics of black holes," arXiv:1606.01267 [hep-th]; T, Banks, W. Fischler, S. Kundu, and J. F. Pedraza, "Holographic space-time and black holes: Mirages as alternate reality," arXiv::1401.3341 [hep-th].
20. S. L. Braunstein, S. Pirandola and K. Zyczkowski, "Better late than never: Information retrieval from black holes," *Phys. Rev. Lett.* **110**, 101301 (2013), arXiv:0907.1190 [quant-ph]; A. Almheiri, D. Marolf, J. Polchinski and J. Sully, "Black holes: Complementarity or firewalls?", arXiv:1207.3123 [hep-th].
21. T. Banks and W. Fischler, "Holographic space-time, Newton's law and the dynamics of black holes," arXiv:1606.01267 [hep-th]; T. Banks and W. Fischler, "The holographic space-time model of cosmology," arXiv:1806.01749 [hep-th]

Black Hole Entropy and the Bekenstein Bound

Raphael Bousso

Center for Theoretical Physics and Department of Physics,
University of California, Berkeley, CA 94720, USA and
Lawrence Berkeley National Laboratory, Berkeley, CA 94720, USA
bousso@lbl.gov

I share some memories and offer a personal perspective on Jacob Bekenstein's legacy, focussing on black hole entropy and the Bekenstein bound. I summarize a number of fascinating recent developments that grew out of Bekenstein's pioneering contributions, from the Ryu–Takayanagi proposal to the Quantum Null Energy Condition.

1. Personal Memories

Jacob Bekenstein entered my life as a formula; he left it as a cherished friend.

I was Stephen Hawking's student, so I soon learned that black holes have entropy, according to Bekenstein's famous formula.[1–3] Stephen and I studied the pair creation of black holes in cosmology. We used Euclidean methods to compute the (negligible) rate at which this process would happen. One expects configurations with less entropy to be more suppressed, so the Bekenstein–Hawking entropy of de Sitter and black hole horizons, though not central to our project, played a role as a consistency check.

I was surprised to receive a postdoc offer from Bekenstein. We had not met personally, and none of my thesis work bore a close relation to his own interests at the time. My future plans leaned more towards cosmology than to the quantum aspects of black holes. My hope had been to spend time at Stanford and work with Andrei Linde. But I knew enough to understand that an offer from Bekenstein was something to be proud of. Also, I was relieved and grateful, because it was my only postdoc offer.

There was a problem: I was born in Israel, and though I had never lived there and do not speak Hebrew, I would have been subject to military service upon moving to Jerusalem. But Jacob was ready to help. After some bureaucratic wrangling, he obtained a permission for me to spend two years in Israel without being conscripted. Just then, with terrible timing, I learned that I had won a personal postdoctoral fellowship (by a German organization), with Stanford as the host institution. It was too late: I had already accepted Jacob's offer, and he had worked hard on my behalf to make it happen.

But when Jacob found out, he told me without hesitation that I should go to Stanford. I learned later that I was far from the only beneficiary of Jacob's habitual generosity and selflessness. He leaves behind a trail of gratitude.

Jacob's judgement was right. The physics questions he was grappling with at the time were profound, particularly the problem of a precise formulation of Bekenstein's nongravitational entropy bound, which I will discuss in Sec. 4. But it would take more than a decade for a significant breakthrough to come about. At Stanford, I was able to make contributions on a shorter time-scale. I worked on cosmology with Linde, as I had hoped. I could easily visit Joe Polchinski in (relatively) nearby Santa Barbara. And from Lenny Susskind and a fascinating talk by Willy Fischler,[4] I learned about the problem of formulating a holographic entropy bound for cosmological spacetimes, which led to Refs. 5 and 6. Ironically, Jacob had sent me straight into the beautiful web his work had spun, the fascinating connections between geometry and quantum information.

Though Jacob was personally kind and gentle, intellectually he was uncompromising and sometimes fierce. The question of whether nontrivial entropy bounds apply in general spacetimes was quite controversial in 1999. I was confident that I had resolved the issue, but it took some time for the debate to play out. Jacob was privately and publicly supportive, helping my work gain acceptance. His encouragement grew into a friendship and a source of strength over many years.

Jacob's colleagues did not allow his modesty to stand entirely in the way of proper recognition. It is a consolation to think back of the wonderful "fest" that his colleagues in Jerusalem insisted on organizing, allowing us to celebrate Jacob and his work just a few years before his untimely death. (He reluctantly agreed, but insisted that it be billed as a workshop.) Jacob received his share of prestigious prizes and awards. Yet, the full import of his most important insight, that black holes have entropy, is only beginning to dawn.

2. Black Hole Entropy

Bekenstein's 1972 proposal that black holes carry entropy[1] takes up four very small pages in a now extinct journal. It ranks as one of the most important physics papers of the 20th century. With 6 equations and less than 1500 words, it makes for an exceptionally quick and rewarding read.

Writing in concise, elegant prose, the author poses a puzzle: in the Second Law of thermodynamics, what accounts for the entropy of matter systems that disappeared into a black hole? In one sentence, he offers a proposal: black holes themselves must have some amount of entropy, such that the total (matter plus black hole) entropy according to an outside observer will never decrease. In the next sentence, he makes the proposal quantitative: a black hole's entropy should be given by its horizon area in Planck units, up to an unknown factor of order unity. He then verifies that this proposal upholds the Second Law in a simple example. In four final lines he explains the signficance of what he has done: a counting of the quantum states of spacetime. Bekenstein's 1972 paper marks the first-ever quantitative statement intrinsically about quantum gravity.

The most remarkable aspect of this revolutionary work (aside from the fact that it was written by a graduate student) is the premise from which it sets out. Bekenstein argues that we should *worry* about the loss of matter entropy into a black hole, because it would make it impossible to *verify* that the Second Law of thermodynamics still holds.[a]

A lesser mind might have been content with making excuses along the following lines. *Well, bad things happen all the time. I'm sorry that your box of gas fell into a black hole. It may be a bit awkward to visit your box there and to see how its entropy is doing. But this is an inconvenience, not a fundamental problem that needs solving. Probably the entropy ends up at the singularity inside the black hole, and we do not have a theory yet that describes the singularity. But if we had a full-fledged quantum gravity theory, we could account for the entropy in this way.* Bekenstein, by contrast, stood firm in insisting that the laws of physics continue to make sense in our laboratories, whether or not a black hole forms in one of the test tubes.

An observer who chooses to remain outside cannot access matter inside the black hole. And by the definition of "black hole," she cannot see the matter or receive signals from it while she remains outside. General relativity further dictates that by jumping into a black hole, she cannot "catch up" with the matter that formed it. Indeed, she cannot even receive a signal from any matter that entered the black hole more than a time $(R/c)\log(R/L_P)$ before she does. Here times are measured at some distant radius from which the infall starts; R is the radius of the black hole and $L_P = (G\hbar/c^3)^{1/2}$ is the Planck length, about 10^{-33} cm.[b]

It was a major leap for Bekenstein to elevate these obstructions to a crisis for the Second Law of thermodynamics. In insisting that the Second Law should retain operational meaning, Bekenstein was perhaps guided by the fact that black holes do carry gauge charges such as energy or electric charge. Their conservation can be verified at all times from the outside. He was familiar, moreover, with Hawking's recent proof of a theorem in general relativity: with reasonable (though ultimately invalid) assumptions on the positivity of the energy of matter, the area of all event horizons cannot decrease. There is a superficial analogy between this area theorem and the Second Law, which states that the total entropy cannot decrease.

---

[a]This problem was first posed to Bekenstein by his advisor, John Wheeler.[7] For a fascinating first-hand account of these events, see Bob Wald's contribution to this memorial volume.[8]

[b]In the remainder of this article, I will set $c = 1$ but keep Newton's constant G and Planck's constant $\hbar$ explicit.

But the Second Law is just a statistical probability, whereas the area theorem, within classical general relativity, was a mathematical certainty. This did not deter Bekenstein: his proposal was that the analogy is an identity. He thus asserted that the area of a geometric surface is more fundamentally a statistical quantity. Nearly half a century later, this remains the single deepest insight we have gained into the fabric of space and time. It has proven extraordinarily fruitful and lies at the core of today's most promising avenues for studying quantum gravity.

Whereas Bekenstein advocated that the area of a black hole should be treated as a real contribution to the total entropy, it is interesting that he did not go further and ascribe a temperature to black holes. In fact, his proposal implied (by the First Law of thermodynamics) that black holes have a temperature and hence must radiate. Rather ironically, this implication prompted Hawking to reject the proposal at first. Classically, nothing can come out of a black hole, so it seemed absurd to think that black holes could radiate. But of course, Bekenstein's proposal is quantum mechanical in nature, so a purely classical argument against its implications carries no weight. In 1974 Hawking discovered black hole radiation as a quantum effect, in an independent calculation. Working back via the First Law, this fixed the proportionality factor in Bekenstein's proposal to be 1/4.

3. Generalized Entropy and Generalized Second Law

The terms "generalized entropy" and "generalized second law" first appear in Bekenstein's 1973 paper.[2] We now think of them as fixed names for important concepts, but in this first instance, the word "generalized" is merely descriptive, a convenient way of referencing the notions introduced in the 1972 paper. Bekenstein distinguishes the "common" entropy of the matter systems outside the black hole from the "generalized" entropy. The latter includes both the black hole contribution and the matter entropy:

$$S_{\text{gen}} = \frac{A}{4G\hbar} + S_{\text{out}} \,. \tag{1}$$

The generalized second law (GSL) is Bekenstein's conjecture that the generalized entropy satisfies a Second Law of thermodynamics:

$$dS_{\text{gen}} \geq 0 \,. \tag{2}$$

It took nearly four decades for the GSL to be proven in a wide regime. Using methods from algebraic quantum field theory, and exploiting specific properties of null quantization, Wall[9] proved the GSL perturbatively in $G\hbar$, for free fields. The proof extended automatically to interacting fields, after it was established that their modular Hamiltonian takes the same form on such null hypersurfaces.[10,11] Beyond the perturbative expansion in $G\hbar$, the GSL remains a conjecture.

Thanks to the success of Bekenstein's proposal, the modifier "generalized" is now outdated.[c] The theoretical evidence that black holes have entropy is overwhelming. It would be silly to exclude this contribution from the total entropy and expect the Second Law to remain valid. Thus, generalized entropy is simply the entropy.

But there are reasons to keep the name. The sum $S_{\rm gen}$ is actually better defined than either of its two separate terms on the right hand side of Eq. (1). In quantum field theory, $S_{\rm out}$ can be defined as the von Neumann entropy of the quantum state restricted to the exterior of the horizon. This has divergences; the leading divergence is proportional to the area in units of the UV cutoff. Newton's constant G also has a nontrivial renormalization group flow. There is strong evidence that these flows cancel, so that the sum $S_{\rm gen}$ is finite and well-defined. (See the appendix of Ref. 12 for a detailed discussion and extensive references.) There are subleading divergences that cancel as well.[13–16]

Moreover, there are a number of theorems in general relativity which hold only if one assumes the null energy condition (NEC) on the stress tensor, that $T_{ab}k^a k^b \geq 0$ for any null vector k^a. Hawking's area theorem is an example. But in fact, the null energy condition can be violated by quantum fields. An important example of this is Hawking radiation: the black hole shrinks, in violation of the area theorem.

The generalized entropy can be viewed as a quantum-corrected area, since the area term dominates by one power of $\hbar$ over the matter entropy term. And unlike the area, the generalized entropy *does* increase during black hole evaporation: it can be shown that the entropy of the Hawking radiation more than compensates for the lost area.[17] Indeed, this behavior is required by the GSL.

We see that there are really two ways of looking at Bekenstein's generalized entropy. The first is the way he discovered it, as a black hole's contribution to the total entropy, and thus as a way to save thermodynamics. But we can also think of generalized entropy as a kind of quantum-corrected area that "saves" a theorem of general relativity. Hawking's area theorem fails in the presence of quantum effects, but the analogous statement for generalized entropy survives.

Moreover, $S_{\rm gen}$ can be associated not only to slices of a black hole event horizon, but to arbitrary spatial surfaces that split a Cauchy surface into two parts. For example, we can define a generalized entropy of the Earth's surface, given by its area plus the von Neumann entropy of its exterior (or interior). This broader notion of generalized entropy was not envisaged in Bekenstein's original proposal. But in recent years, it has proven exceptionally powerful.

Consider any GR theorem that involves the area of surfaces or some derivative thereof. Suppose, moreover, that the theorem relies on the NEC and hence breaks down in NEC-violating quantum states. Since we can view the generalized entropy as a quantum-corrected area, it is natural to ask whether we can "save" the GR theorem in the presence of quantum effects, by replacing area with generalized

[c]I thank Stefan Leichenauer for frequently complaining about this.

entropy in the statement of the theorem. This trick works for Hawking's area theorem, so why not apply it in other contexts?

For example, Penrose's theorem[18] guarantees a singularity in the presence of a trapped surface, assuming the NEC. But the NEC and hence Penrose's theorem can fail. It is thus not clear whether the inevitability of singularities and the associated breakdown of general relativity is merely an artifact of a classical limit. But Wall[19] used Eq. (1) to define the notion of a "quantum-trapped" surface. He further showed that the GSL implies a singularity in the future of any such surface. Thus, Penrose's singularity theorem can be transferred to firmer ground, onto the foundation laid by Bekenstein in 1972.

Another example is the area theorem for holographic screens,[20,21] which applies to surfaces in cosmology and inside black holes. By judiciously replacing area with generalized entropy in the analysis, this result can be elevated to a GSL for "Q-screens,"[22] and thus to a well-defined statement of the Second Law in cosmology.

Even the classical focussing theorem, which governs the evolution of cross-sectional area spanned by light-rays, can be elevated to a "quantum focussing conjecture," again by appropriately replacing area with generalized entropy. The QFC turns out to be a very powerful statement, to which we will return later.

4. Bekenstein Bound

4.1. *From the GSL to* $S \leq 2\pi ER/\hbar$

If black holes did not carry entropy, then as Bekenstein noted, it is inevitable that the Second Law of thermodynamics can be violated from the point of view of an external observer, because matter entropy can be lost into the black hole.

However, even if black holes do carry entropy as Bekenstein proposed, $S_{BH} = A/4G\hbar$, it is not immediately clear that the Second Law is saved. This would appear to depend on how the black hole responds to the addition of matter systems that carry entropy. The question is whether the area of a black hole increases *by enough* to make up for the lost entropy, or not.

The growth of the horizon area is controlled by the Einstein equation of general relativity, which involves geometric quantities and the stress-energy tensor of matter. Thus the area increase will manifestly depend on the size and energy density of the matter system, and on its path across the horizon.

From this viewpoint, it is completely unclear why the horizon area increase would be sensitive to the entropy of the matter system, i.e., to the number of its possible quantum states. But the GSL demands this. Thus the GSL carries within it a fascinating, deep connection between quantum information on the one hand, and energy and geometry on the other.

A stationary black hole is entirely characterized by its conserved charges: its electric charge, its angular momentum, and most importantly (since it limits the previous two), its mass. The area of its event horizon, therefore, is a function of only these three quantities. It has no manifest dependence on the entropy of the matter that entered the black hole (that is, on the log of the number of quantum states that this matter could have been in). So how can the area possibly "know" that it should increase enough to compensate for this lost matter entropy?

To make this concrete, suppose that we add a neutral matter system with no angular momentum to an unspinning neutral black hole. Thus, the black hole is described by the Schwarzschild solution with mass M before and mass $M + \delta M$ after the addition of the system. We can take the black hole to be large, so that $M \gg \delta M$. The Schwarzschild radius is proportional to the mass, so the area of the horizon increases in this process, by $\delta A = 32\pi G^2 M \delta M$. Hence, the entropy of the black hole will increase by $\delta S_{BH} = 8\pi G M \delta M/\hbar$.

In order to violate the GSL in this example, we would have to arrange for δS_{BH} to be smaller than the entropy S of the matter system we are adding to the black hole. This means we have to keep δM small. That is, we would like the matter system to add as little energy as possible to the black hole. This is best accomplished by extracting work from the entropy-carrying matter system, by lowering it slowly towards the horizon of the black hole (using a string, say), and only then dumping it in. By this trick, we effectively lower the mass of the system by redshifting it into a gravitational well.

It might seem that we can lower the system all the way to the horizon, where the redshift diverges. If this was possible, then we could extract all of its rest mass as work. Then the black hole mass would not increase at all once the matter system is finally dropped in: $\delta M = 0$. We could violate the GSL: the generalized entropy would decrease, because the black hole area would remain fixed same whereas the matter entropy would have disappeared.

But Bekenstein realized[3,23] that this is impossible: any matter system will have finite size, so we cannot lower its center of mass arbitrarily close to the horizon without tidal disruption of the system. For concreteness, consider a spherical box with radius R and roughly homogeneous mass distribution. We must be content to lower its center of mass to a proper distance R from the horizon. There is no more work we can extract, without having parts of the system already enter the black hole.

A bit of algebra shows that in this optimal case, $\delta M = ER/4GM$. (See, e.g., Ref. 24 for a brief derivation.) Thus, the black hole area will increase by $\delta A = 8\pi GER$, and the black hole entropy will increase by $\delta S_{BH} = 2\pi ER/\hbar$. The GSL requires that $\delta S_{\text{gen}} \geq 0$ and hence that $\delta S_{BH} > -\delta S_{\text{out}} = S$. By insisting on the validity of the GSL, we are thus led to the Bekenstein bound:

$$S \leq 2\pi ER/\hbar\,, \tag{3}$$

where S, E, and R are the entropy, energy, and radius of the matter system.

It is very significant that Newton's constant G does not appear in this final result. Neither do any parameters that describe the black hole. The Bekenstein bound is purely a nongravitational statement, and it is intrinsically about the matter system itself: the first universal entropy bound.

4.2. *Controversy*

The Bekenstein bound, Eq. (3), was at the center of a heated controversy in the 1980s and 1990s. Reference 25 has many references and a nice summary of the back and forth, which traces back at least to Ref. 26. (To my knowledge, the GSL itself was not controversial.) The debate centered largely on two questions: (i) Is Eq. (3) indeed necessary for the GSL? That is, does the Bekenstein bound follow rigorously from the GSL, rather than just being inspired by the handwaving arguments of the previous subsection? (ii) Is Eq. (3) true? Or is it possible to find matter systems whose entropy exceeds the Bekenstein bound?

Both were valid questions. The thought-experiment of lowering the system slowly towards the horizon, which links the GSL to the Bekenstein bound, is beset by subtleties and potential loopholes. Thus, it appeared logically possible for the bound to fail even if the GSL holds, or for the bound to be true but not to be implied by the GSL.

Concerning (i), Unruh and Wald[26] led the charge in arguing that known properties of quantum field theory were sufficient to protect the GSL. Thus the GSL would not impose new constraints such as Eq. (3). (See also Ref. 8.) I will comment on this viewpoint further at the end of Sec. 5.4.

Regarding (ii), there seemed to be fairly obvious counterexamples to the Bekenstein bound. For example, certain quantum fields have negative Casimir energy in certain cavities. Thus $E < 0$ and $S = 0$ in the ground state of the system.

In hindsight, the question we should have been focussing on was: (iii) Exactly how should one define S, E, and R in Eq. (3)? In fact, this question did receive some attention, but until 2008, there were no satisfactory proposals.

Bekenstein argued that E must include all necessary parts of the system, by which he meant a "complete system" that "could be dropped whole into a black hole."[27] This gets rid of the Casimir counterexample, because the rest mass of the walls of the cavity overcompensates for the negative energy of the ground state. However, in forcing us to include the rest energy of enclosures, Bekenstein risked trivializing his bound, by making it virtually impossible to find a system that would come close to saturating it.

Bekenstein also insisted that for nonspherical systems, R should be defined as the radius of the smallest circumscribing sphere. This helped fend off potential counterexamples, but it weakened the bound for no principled reason. In the thought-experiment that led to the bound, in Subsec. 4.1, we are free to orient a thin slab so that its long side is parallel to the horizon. Thus, the shortest dimension determines how close we can get it to the horizon before dropping it in. This yields a bound of the form

$$S \leq \pi E w/\hbar\,, \tag{4}$$

where w is the smallest separation of two infinite parellel plates between which we can fit the system. As pointed out to me by Casini, the "necessary parts" argument alone would suffice to eliminate the apparent counterexamples to this stronger bound, so it was not clear why Bekenstein went with the weaker version.[d]

4.3. *Relation to the covariant entropy bound*

In 2002, I was able to show[28] that the covariant entropy bound,[5] in its generalized form,[29] implied the "tight" version of Bekenstein's bound, Eq. (4), in a weakly gravitating setting. The generalized covariant entropy bound states that

$$S_{\text{matter}} \leq \frac{A - A'}{4G\hbar} . \tag{5}$$

Here S_{matter} is the entropy of the matter systems crossing a light-sheet of an arbitrary spatial surface Σ. A light-sheet is a family of non-expanding null geodesics (light-rays) that emanate orthogonally from Σ. A and A' are the areas of Σ and of the final surface on which the light-sheet is terminated, respectively.

The general argument that Eq. (5) implies Eq. (4) is quite simple. For this brief summary, let us make it even simpler by restricting to an approximately spherical matter system of radius R, and neglecting factors of order unity.

Consider a congruence of parallel light-rays in Minkowski space, such as would emanate from lasers mounted on a flat wall at $x = 0$ that all flash for an instant at $t = 0$. These light-rays occupy the null plane $x = t$. (Here, light-rays does not mean actual light, which might be blocked or deflected. We consider null geodesics, which pass through any matter system like X-rays and which are only affected by gravity.)

The matter system presents a cross-section of area $\sim R^2$ to the light-rays. In passing through the matter system, the geodesics will be focussed by an angle $\phi \sim GE/R$. (This is the same gravitational effect that is responsible for the famous bending of light by the sun.) This angle will be small by the assumption that gravity is weak, i.e., because our matter system is something like planet Earth, not a neutron star or an expanding universe. To first order in G, therefore, the cross-section spanned by the light-rays will have decreased by an annulus of width $\sim \phi R$ and hence of area

$$\Delta A \sim \phi R^2 \sim GER . \tag{6}$$

The generalized covariant entropy bound states that the entropy of the matter system satisfies $S \leq \Delta A/4G\hbar$, which yields the Bekenstein bound. A more careful analysis produces the tighter version of the bound, Eq. (4), including the prefactor π.

This was a pleasing result. It meant that Bekenstein's original 1981 bound did not stand logically apart from the "holographic" entropy bound I had found much

[d]In the conversations we had about this, Jacob was characteristically low-key, offering only that he intended the bound as a "rule of thumb." It was an interesting contrast to the intensity with which he pursued these debates in scientific publications.

later. Instead, it corresponded to a particular weakly gravitating limit of a more general, gravitational bound.

But this result also meant that the generalized covariant entropy bound was subject to the same questions as Bekenstein's bound. What about apparent violations such as Casimir energy? In some cases ΔA is better defined than E and R separately, but it remained unclear what exactly constitutes an "isolated" and "complete" matter system. Where does it begin and end?

We now know that this was the wrong question. One should simply pick a (null) region that will be considered, and compute its generalized entropy,[12] or in the nongravitational limit, the vacuum-subtracted von Neumann entropy of the quantum field theory state restricted to this region.[30,31] But it did not occur to me to use the generalized entropy for non-horizon surfaces, and I was deterred by the fact that the von Neumann entropy is divergent, due to the entanglement entropy across the boundary of the region. So, in a set of deservedly forgotten papers, I pursued various other approaches to sharpening the formulation of the Bekenstein bound.[32,33]

My goal was to answer question (iii), and to formulate the Bekenstein bound in a way that was neither wrong, nor trivial, nor ill-defined. The key challenge I focussed on was to define the size of a system, given that quantum wave functions are usually not sharply localized. I felt I was making progress, but I did not reach a fully satisfactory solution, and in fact I was on the wrong track.

But the problem did have a solution. Its discovery by Casini would form the seed of several recent breakthroughs that are significantly deepening our understanding of the connection between quantum information, energy, and spacetime.

4.4. *Marolf, Minic, and Ross*

In 2003 Marolf, Minic, and Ross[34] (MMR) pointed out a curious property of the entropy one can ascribe to a matter system. As they put it, it is "observer-dependent." Their insight built on earlier work by Marolf, Sorkin, Unruh, Wald, and others, as cited in Ref. 34.

Suppose that a field excitation with fixed energy E and spatial size R can be in a large number n of different quantum states. This would be possible, for example, in a theory with n different species of photon. We can then prepare a maximally mixed state of all possible species. That is, we know that a photon wavepacket is present, but not which species it was made from. The entropy of this wavepacket is $\log n$. If n is large enough, so that $\log n > 2\pi ER$, then the entropy will exceed the Bekenstein bound.

This "species problem" had long been recognized as presenting a challenge. A common response was to note that the actual number of light species in Nature cannot be freely chosen, and that it happens to be quite small. One might reason that this fact is no accident, but the result of the Bekenstein bound constraining the number of species.

But MMR noticed that the entropy of such a wave packet would not in fact grow like $\log n$, from the point of view of an accelerated observer who had access only to one side of a Rindler horizon. This was significant, since an observer hovering near the horizon of a black hole is locally like a Rindler observer.

MMR computed the entropy only in a certain limit of small excitations above the vacuum. But the calculation was done in full quantum field theory: they considered the actual density operators that describe the vacuum and the excited state, restricted to the Rindler patch ($x > 0$ at $t = 0$). The von Neumann entropy of either state diverges, due to entanglement across the boundary at $x = 0$. As usual, one would like to say that this divergent part is not important, and that instead that one should focus on the entropy of the excitation. But what formula should one attach to these words?

MMR reasoned that one could simply compare the divergent entropies with and without the excitation. The difference would capture the entropy that can be attributed to the excitation. Indeed, for ordinary matter systems localized well inside the Rindler region, this quantum field theory prescription reproduces the standard thermodynamic entropy (say, of a box of gas).

Written as a density operator, the global vacuum is the state $|0\rangle\langle 0|$. Restricting the vacuum state to the right Rindler patch ($x > 0$) means tracing out the left patch:

$$\sigma = \mathrm{Tr}_L \, |0\rangle\langle 0| \, . \tag{7}$$

Similarly, one can chose a global excited state ρ_g and define its restriction to the Rindler patch as

$$\rho = \mathrm{Tr}_L \, \rho_g \, . \tag{8}$$

The (divergent) von Neumann entropies of these states are $-\mathrm{Tr}\sigma \log \sigma$ and $-\mathrm{Tr}\rho \log \rho$, respectively. MMR's analysis was restricted to states such that $\rho = \sigma + \delta\rho$, with $\delta\rho \ll 1$. Working to first order in $\delta\rho$, the difference between the entropies is

$$\Delta S = -\mathrm{Tr}\left[\delta\rho \, (1 + \log \sigma)\right] = -\mathrm{Tr}[\delta\rho \, \log \sigma] \, . \tag{9}$$

Crucially, MMR found that the entropy difference in the Rindler patch, Eq. (9), does not grow as $\log n$ indefinitely, as it would if we computed it from the global quantum state. In the limit of a large number n of internal states, at fixed global energy and size of the system, the Rindler-patch entropy difference *saturates* at

$$\Delta S = 2\pi \, \delta E_R / \hbar \, , \tag{10}$$

where

$$\delta E_R = \int dx \, dy \, dz \, x \, T_{tt} \, , \tag{11}$$

is the Rindler energy and T_{tt} is the energy density.

In the special case where the system is sharply localized compared to its distance R from the Rindler horizon, the Rindler energy is simply $\delta E_R \approx ER$, where $E =$

$\int dx\, dy\, dz\, T_{tt}$ is the global energy. The result of MMR, Eq. (10), thus implies that in the limit of diverging species number n, the Bekenstein bound becomes saturated but not violated. This holds at least in the limit studied by MMR, if the entropy is defined as they proposed, from the viewpoint of the Rindler observer.

But the Rindler horizon could secretly be a black hole horizon, just as the Earth's surface looks flat to us. Expressed in terms of asymptotic energy and black hole temperature, the MMR result is that in the large n limit, the entropy of an excitation just outside a black hole saturates at $\delta E/T_{BH}$. By the First Law, this is precisely the amount by which the black hole entropy will increase if we drop in the system. Thus, MMR found that the GSL would be satisfied, and saturated, in this limit.

4.5. *Casini's formulation and proof of the Bekenstein bound*

For the results of MMR, described above, it is crucial that the entropy be defined as a vacuum-subtracted entropy. This is the difference between the von Neumann entropy of the excited state restricted to the Rindler region, ρ (or to the outside of the black hole), and that of the vacuum state so restricted, σ. In 2008, Casini[30] considered the vacuum-subtracted entropy of arbitrary states in the Rindler region. (Thus he went beyond a first-order analysis of states that were close to the vacuum.) He used tools from algebraic quantum field theory to state and prove a rigorously defined version of Bekenstein's bound.

Casini pointed out that

$$S_{\rm rel}(\rho|\sigma) = \Delta K - \Delta S\,, \tag{12}$$

which follows immediately from the definitions of the relative entropy,

$$S_{\rm rel} = \operatorname{Tr}\rho\log\rho - \operatorname{Tr}\rho\log\sigma\,, \tag{13}$$

the vacuum-subtracted entropy

$$\Delta S = -\operatorname{Tr}\rho\log\rho + \operatorname{Tr}\sigma\log\sigma\,, \tag{14}$$

and the modular energy ΔK,

$$\Delta K = \langle\rho|K_\sigma|\rho\rangle - \langle\sigma|K_\sigma|\sigma\rangle\,. \tag{15}$$

Here the modular Hamiltonian K_σ of the vacuum state σ is defined via the operator equation

$$\sigma = \frac{e^{-K_\sigma}}{\operatorname{Tr} e^{-K_\sigma}}\,. \tag{16}$$

In general, the modular Hamiltonian is not an integral of a local density. But in the case at hand, σ is the vacuum restricted to the Rindler patch. This implies that ΔK is proportional to the Rindler energy[35]:

$$\Delta K = \frac{2\pi}{\hbar}\int dx\, dy\, dz\, x\, T_{tt}\,. \tag{17}$$

As noted earlier, for objects that are well-localized at a distance R from the Rindler horizon, this implies that $\Delta K \approx 2\pi ER/\hbar$.

These observations motivated Casini to reformulate the Bekenstein bound as the claim that

$$\Delta S \leq \Delta K\,. \tag{18}$$

This statement is well defined for all finite-energy quantum states, not just those for which we can give an intuitive definition of E and R. For the latter, it reduces to $\Delta S \leq 2\pi ER/\hbar$. Thus Casini solved the decades-old problem of giving a sharp formulation of the Bekenstein bound.

Casini then proved this bound, in one line: Eq. (18) follows from Eq. (12) and the known property that the relative entropy is nonnegative, $S_{\rm rel} \geq 0$.

5. Recent Developments

Bekenstein's notion of generalized entropy lies at the heart of several important recent developments. All of them are explore the concept beyond its original setting, applying it to any surface Σ that splits the world[e] into two sides. It does not matter whether Σ lies on a black hole event horizon. As noted in Sec. 3, in this context it can be useful to think of generalized entropy as the "quantum-corrected area" of the surface Σ:

$$A_Q = A + 4G\hbar\, S_{\rm out}\,, \tag{19}$$

where

$$S_{\rm out} = -\mathrm{Tr}\,\rho \log \rho \tag{20}$$

is the von Neumann entropy of the state ρ of the quantum fields restricted to one side of the surface, and ρ is given by a partial trace of the global state over the other side as in Eq. (8). More refined algebraic definitions of $S_{\rm out}$ deal with the divergent entanglement structure[36,37] and with subtleties arising from gauge constraints.[38–40]

5.1. *Ryu–Takayanagi proposal*

In the context of the AdS/CFT correspondence,[41] Ryu and Takayanagi proposed in 2006 that the von Neumann entropy of any boundary subregion is given by the area of the minimal bulk surface Σ homologous to this region[42]:

$$S_{\rm bdy} = \frac{A[\Sigma]}{4G\hbar}\,. \tag{21}$$

Hubeny, Rangamani and Takayanagi soon extended the proposal from static to general settings, where there exist no minimal surfaces. Instead, the area of the

[e]By "the world," I mean a Cauchy surface.

smallest stationary[f] homologous bulk surface[43] computes the entropy. A useful "maximin" reformulation of this proposal was given by Wall.[44]

Faulkner, Lewkowycz, and Maldacena[45] developed the proposal further, going beyond leading order in $1/N$ (or equivalently, in G), where N is the rank of the gauge group of the boundary field theory. They argued that the boundary entropy is more accurately computed up to $O(N^0)$ by adding the ordinary matter entropy in the bulk to the area. Interestingly, this was not formulated explicitly as the generalized entropy in the sense of Bekenstein (Sec. 3).

This connection was made clear, and the proposal extended to all orders in $1/N$, by Engelhardt and Wall.[46] They suggested that the relevant surface Σ should be defined as the "quantum extremal surface" (really, quantum stationary surface), for which A_Q is locally stationary. The bulk generalized entropy A_Q then computes the boundary entropy:

$$S_{\text{bdy}} = \frac{A_Q[\Sigma]}{4G\hbar}\,. \tag{22}$$

These investigations helped resolve the problem of subregion–subregion duality.[47,48] References 44, 49, 50 argued that the dual to any boundary region is the entanglement wedge, i.e., the region enclosed by the boundary region and the bulk surface that computes the entropy. This is now well established.[51,52] It has led to beautiful insights concerning the role of quantum error correction in the emergence of the bulk spacetime.[53,54]

5.2. *Cosmological area theorem and second law*

The notion of a holographic screen was first made precise in Ref. 6. Given an arbitrary spacetime, a holographic screen can be non-uniquely constructed by choosing a null foliation of the spacetime. The screen is the hypersurface consisting of a one-parameter family of marginally trapped or anti-trapped surfaces called leaves. (In some cases, like asymptotically flat or AdS spacetime, the screen can reside on the conformal boundary.)

This definition was chosen so that the covariant entropy bound,[5] applied to each leaf, constrains the entropy on the entire null slice that it is embedded in, in terms of the area of the leaf. In other words, holographic screens make manifest the idea[55,56] that the world is a hologram. In asymptotically AdS spacetimes, we have an actual quantum gravity theory supporting this conjecture. But the covariant entropy bound applies to any spacetime, including the universe we live in, suggesting that the holographic nature of quantum gravity is universal. For a review see Ref. 24.

[f]By "stationary," I mean a surface whose area does not change to first order in small deformations. In the AdS/CFT literature this is more frequently referred to as an "extremal" surface. But "extremal" means minimal or maximal. In a Lorentzian geometry, stationary spatial surfaces can only be saddle points of the area functional.

In Refs. 20 and 21, Engelhardt and I distinguished between past and future holographic screens, which consist respectively of marginally trapped and marginally antitrapped surfaces. (These objects are similar to but more general than "dynamical horizons."[57]) We further proved that any past or future holographic screen satisfies a nontrivial area theorem: the area of the leaves increases monotonically along the screen. This is distinct from Hawking's area theorem, which applies to causal horizons. Holographic screens exist generically, including in our universe, and regardless of the presence of black hole or de Sitter horizons.

Our proof, like Hawking's for his theorem, relied on the Null Energy Condition, that the stress tensor satisfies $T_{ab}k^a k^b \geq 0$ for any null vector k^a. As noted earlier, this condition can be violated in quantum field theory. Thus, it is natural to propose that the quantum-corrected area would satisfy a more robust monotonicity property. We called this conjecture the generalized second law for cosmology[22]:

$$dS_{\text{gen}} > 0. \tag{23}$$

(The strict inequality arises from a genericity assumption that simplifies the analysis.)

This is the first rigorous and covariant statement of a Second Law in cosmology. It does not, for example, rely on symmetries like homogeneity to subdivide a portion of the universe and consider its comoving entropy. Such a definition would at best be approximate and would not apply in general spacetimes.

Yet, our Second Law remains somewhat mysterious, in that we know of no thermodynamic argument for why it should hold. This is in marked contrast to Bekenstein's GSL, which was arrived at not by quantum-correcting Hawking's area theorem, but by demanding that the laws of thermodynamics hold in the exterior of black holes. (See Secs. 2 and 3 above.) It would be nice to fill this gap.

5.3. *Quantum Focussing Conjecture*

A consequence of Einstein's equation, Raychaudhuri's equation

$$\theta' = -\frac{1}{2}\theta^2 - \varsigma_{ab}\varsigma^{ab} - 8\pi G\, T_{ab}k^a k^b \tag{24}$$

describes the evolution of a family (congruence) of null geodesics that form a null hypersurface.[58,59] Here θ is the expansion, i.e., the logarithmic derivative of a small area element transported along the congruence. θ is defined in terms of the trace of the null extrinsic curvature, whose symmetric part gives the shear, ς_{ab}. (We consider surface-orthogonal null congruences so there is no antisymmetric part or "twist.") The prime refers to a derivative with respect to the affine parameter along the geodesics.

Assuming the null energy condition, $T_{ab}k^a k^b \geq 0$, Eq. (24) implies

$$\theta' \leq 0\,. \tag{25}$$

This is the classical focussing property of light: matter focusses but never antifocusses light-rays. For example, initially parallel light-rays grazing the Sun are bent inwards, not outwards by the Sun's gravity.

If the null energy condition is violated, for example by Casimir energy or near a black hole horizon, then light-rays can antifocus, i.e., one can have $\theta' > 0$. One can ask, however, whether a quantum-corrected version of Eq. (25) remains valid.

To find an appropriate formulation,[12] we appealed to Bekenstein's generalized entropy, interpreting it as a quantum-corrected area $A_Q = 4G\hbar S_{\text{gen}}$. For any two-dimensional surface Σ that divides the world into two sides, we defined the "quantum expansion" Θ analogous to θ, as the response of the generalized entropy of Σ to local deformations of the Σ along an orthogonal light-ray.[g] It was then natural to conjecture that

$$\Theta' \leq 0\,. \tag{26}$$

This is the *Quantum Focussing Conjecture* (QFC). See Ref. 12 for precise definitions.

Some evidence for the validity of this conjecture comes from the fact that it reproduces other well-tested or proven statements as limiting cases. For example, if Σ has $\Theta \leq 0$ initially, then Eq. (26) implies the generalized covariant entropy bound,[5,29] Eq. (5). In particular, the QFC thus implies the Bekenstein bound[23] in the strong form of Eq. (4).[60]

The formulation of the QFC in terms of S_{gen} provides a more rigorous statement of both bounds: the generalized entropy of the initial surface is greater than that of the final surface. (There exists another quantum version of the covariant bound which is defined only in the weakly gravitating limit.[31,61] In this regime, both versions can be applied, and there appears to be no logical relation between them. It would be nice to understand this better. Perhaps there exists a stronger statement that implies both?)

A further consequence of the QFC will be discussed in the next subsection.

5.4. *Quantum Null Energy Condition*

The QFC led to a new result in quantum field theory, the *Quantum Null Energy Condition.*[12,62] The QNEC states that the local energy density of matter is bounded from below by an information-theoretic quantity:

$$T_{ab}k^a k^b \geq \frac{\hbar}{2\pi} S''_{\text{out}}\,. \tag{27}$$

Here T_{ab} is the stress tensor at a point p. We have implicitly chosen a surface Σ that contains p and divides the world into two parts, such that the classical expansion and shear at Σ vanish.[h] The null vector k^a is orthogonal to Σ at p, and S_{out} is the von Neumann entropy of the quantum fields on one side of Σ. The double-prime

[g] Strictly, Θ is a functional derivative of the generalized entropy and so depends on Σ globally. It reduces to the ordinary derivative in Eq. (25) as $G\hbar \to 0$, which has no such dependence.

[h] An example of this is any infinite spatial plane in Minkowski space containing p. One can also consider more general settings.[62–64]

refers to a second functional derivative (with a Dirac delta-function stripped off), with respect to deformations of Σ in the direction of k^a at p.

The QNEC, Eq. (27), follows immediately from the QFC by combining Eqs. (1), (24), and (26), and taking the limit as $G \to 0$. By construction, the initial classical expansion and shear vanish at Σ. By Eq. (24), the expansion will be $O(G)$ and proportional to the stress tensor on the null hypersurface orthogonal to Σ. Thus the classical piece, in this special setting, is of the same order as the quantum correction arising from the entropy. After cancelling out G, this results in Eq. (27).

The QNEC is the first lower bound on the local stress tensor in quantum field theory. It implies the Averaged Null Energy Condition[65] on the integrated stress tensor, so the QNEC is stronger than the ANEC. Operationally, the QNEC can be thought of in terms of an observer who examines a signal or region of space up to a boundary. We can ask how the information gained by the observer would increase if they deformed the boundary to include a greater region. The QNEC tells us that the energy density of the signal limits the second derivative of the information content under this deformation.

We can integrate the QNEC twice, to obtain a precise version of Bekenstein's bound.[23] Thus the QNEC is a stronger, differential version of the bound. It captures in the most local way possible the relation between energy, spacetime, and information that underlies both the Bekenstein bound and the holographic entropy bounds.[i]

The QNEC is purely a statement about quantum field theory, and it can be proven in that setting. This was done for free theories in Ref. 62, for interacting theories with a holographic dual in Ref. 67, and for general interacting theories in Ref. 68. The field theory proofs are rather elaborate, totaling around 100 pages. By contrast, as noted earlier, it is trivial to obtain the QNEC from a conjecture about semiclassical gravity, the QFC.

This takes us back to the controversy surrounding the Bekenstein bound (Subsec. 4.2). Unruh and Wald[26] were right when they argued that quantum field theory already has properties sufficient to protect the GSL; in this sense, the Bekenstein bound was not "needed." But the properties of QFT that guarantee this are somewhat obscure and conspiratorial from the QFT viewpoint. They can be much more directly understood as being required by the compatibility of field theory with gravity. This contrast is far more apparent today, as quantum gravity considerations have yielded more precise, provable new results in QFT, such as Eq. (27) and Refs. 9, 30, 31 and 61.

This exposes a major new facet of the unity of Nature, one that Bekenstein first glimpsed. Quantum gravity not only governs extreme regimes such as the beginning of the universe or the singularity inside a black hole. It also governs the information

[i]For interacting theories there is a precise sense in which this relation is actually an equality, $T_{ab}k^a k^b = (\hbar/2\pi)S''_{\mathrm{out}}$, namely when Σ is deformed twice at the same point.[66] The inequality in Eq. (27) thus arises entirely from the off-diagonal contributions to the second functional derivative, when Σ is defomed at two different points.

content of relativistic quantum fields at low energies. It will be important to probe this regime directly, and we will not have to wait for a Planck collider to do it.

References

1. J. D. Bekenstein, "Black holes and the second law," *Nuovo Cim. Lett.* **4**, 737–740 (1972).
2. J. D. Bekenstein, "Black holes and entropy," *Phys. Rev. D* **7**, 2333 (1973).
3. J. D. Bekenstein, "Generalized second law of thermodynamics in black hole physics," *Phys. Rev. D* **9**, 3292 (1974).
4. W. Fischler and L. Susskind, "Holography and cosmology," `hep-th/9806039`.
5. R. Bousso, "A covariant entropy conjecture," *JHEP* **07**, 004 (1999) `hep-th/9905177`.
6. R. Bousso, "Holography in general space-times," *JHEP* **06**, 028 (1999) `hep-th/9906022`.
7. J. D. Bekenstein, "Black hole thermodynamics," *Physics Today* **33**, 24–31 (1980).
8. R. M. Wald, "Jacob Bekenstein and the development of black hole thermodynamics," `arXiv:1805.02302 [physics.hist-ph]`.
9. A. C. Wall, "A proof of the generalized second law for rapidly changing fields and arbitrary horizon slices," *Phys. Rev. D* **85**(6), 104049 (2012) `arXiv:1105.3445 [gr-qc]`.
10. J. Koeller, S. Leichenauer, A. Levine and A. Shahbazi-Moghaddam, "Local modular Hamiltonians from the quantum null energy condition," *Phys. Rev. D* **97**(6), 065011 (2018) `arXiv:1702.00412 [hep-th]`.
11. H. Casini, E. Teste and G. Torroba, "Modular Hamiltonians on the null plane and the Markov property of the vacuum state," *J. Phys. A* **50**(36), 364001 (2017), `arXiv:1703.10656 [hep-th]`.
12. R. Bousso, Z. Fisher, S. Leichenauer and A. C. Wall, "Quantum focusing conjecture," *Phys. Rev. D* **93**(6), 064044 (2016) `arXiv:1506.02669 [hep-th]`.
13. R. M. Wald, "Black hole entropy is the Noether charge," *Phys. Rev. D* **48**, 3427–3431 (1993) `gr-qc/9307038`.
14. V. Iyer and R. M. Wald, "A comparison of Noether charge and Euclidean methods for computing the entropy of stationary black holes," *Phys. Rev. D* **52**, 4430–4439 (1995) `gr-qc/9503052`.
15. T. Jacobson, G. Kang and R. C. Myers, "On black hole entropy," *Phys. Rev. D* **49**, 6587–6598 (1994) `gr-qc/9312023`.
16. X. Dong, "Holographic entanglement entropy for general higher derivative gravity," *JHEP* **01**, 044 (2014) `arXiv:1310.5713 [hep-th]`.
17. D. N. Page, "Particle emission rates from a black hole. II. Massless particles from a rotating hole," *Phys. Rev. D* **14**, 3260 (1976).
18. R. Penrose, "Gravitational collapse and space-time singularities," *Phys. Rev. Lett.* **14**, 57–59 (1965).
19. A. C. Wall, "A proof of the generalized second law for rapidly-evolving Rindler horizons," *Phys. Rev. D* **82**, 124019 (2010) `arXiv:1007.1493 [gr-qc]`.
20. R. Bousso and N. Engelhardt, "New area law in general relativity," *Phys. Rev. Lett.* **115**(8), 081301 (2015) `arXiv:1504.07627 [hep-th]`.
21. R. Bousso and N. Engelhardt, "Proof of a new area law in general relativity," *Phys. Rev. D* **92**(4), 044031 (2015) `arXiv:1504.07660 [gr-qc]`.
22. R. Bousso and N. Engelhardt, "Generalized second law for cosmology," *Phys. Rev. D* **93**(2), 024025 (2016), `arXiv:1510.02099 [hep-th]`.
23. J. D. Bekenstein, "A universal upper bound on the entropy to energy ratio for bounded systems," *Phys. Rev. D* **23**, 287 (1981).

24. R. Bousso, "The holographic principle," *Rev. Mod. Phys.* **74**, 825 (2002) hep-th/0203101.
25. J. D. Bekenstein, "Non-Archimedean character of quantum buoyancy and the generalized second law of thermodynamics," gr-qc/9906058.
26. W. G. Unruh and R. M. Wald, "Entropy bounds, acceleration radiation, and the generalized second law," *Phys. Rev. D* **27**, 2271–2276 (1983).
27. J. D. Bekenstein, "On Page's examples challenging the entropy bound," gr-qc/0006003.
28. R. Bousso, "Light-sheets and Bekenstein's bound," *Phys. Rev. Lett.* **90**, 121302 (2003), hep-th/0210295.
29. E. E. Flanagan, D. Marolf and R. M. Wald, "Proof of classical versions of the Bousso entropy bound and of the generalized second law," *Phys. Rev. D* **62**, 084035 (2000) hep-th/9908070.
30. H. Casini, "Relative entropy and the Bekenstein bound," *Class. Quant. Grav.* **25**, 205021 (2008) arXiv:0804.2182 [hep-th].
31. R. Bousso, H. Casini, Z. Fisher and J. Maldacena, "Proof of a quantum Bousso bound," *Phys. Rev. D* **90**(4), 044002 (2014) arXiv:1404.5635 [hep-th].
32. R. Bousso, "Bound states and the Bekenstein bound," *JHEP* **02**, 025 (2004) hep-th/0310148.
33. R. Bousso, "Harmonic resolution as a holographic quantum number," *JHEP* **03**, 054 (2004) hep-th/0310223.
34. D. Marolf, D. Minic and S. F. Ross, "Notes on space-time thermodynamics and the observer dependence of entropy," *Phys. Rev. D* **69**, 064006 (2004) arXiv:hep-th/0310022 [hep-th].
35. J. Bisognano and E. Wichmann, "On the duality condition for quantum fields," *J. Math. Phys.* **17**, 303–321 (1976).
36. R. Haag, *Local quantum physics: Fields, particles, algebras* (Springer Verlag, Berlin, 1992).
37. E. Witten, "Notes on some entanglement properties of quantum field theory," arXiv:1803.04993 [hep-th].
38. H. Casini, M. Huerta and J. A. Rosabal, "Remarks on entanglement entropy for gauge fields," *Phys. Rev. D* **89**(8), (2014) 085012 (2014) arXiv:1312.1183 [hep-th].
39. W. Donnelly and A. C. Wall, "Entanglement entropy of electromagnetic edge modes," *Phys. Rev. Lett.* **114**(11), 111603 (2015) arXiv:1412.1895 [hep-th].
40. W. Donnelly and A. C. Wall, "Geometric entropy and edge modes of the electromagnetic field," *Phys. Rev. D* **94**(10), 104053 (2016) arXiv:1506.05792 [hep-th].
41. J. Maldacena, "The large N limit of superconformal field theories and supergravity," *Adv. Theor. Math. Phys.* **2**, 231 (1998) hep-th/9711200.
42. S. Ryu and T. Takayanagi, "Holographic derivation of entanglement entropy from AdS/CFT," *Phys. Rev. Lett.* **96**, 181602 (2006) arXiv:hep-th/0603001 [hep-th].
43. V. E. Hubeny, M. Rangamani and T. Takayanagi, "A covariant holographic entanglement entropy proposal," *JHEP* **0707**, 062 (2007) arXiv:0705.0016 [hep-th].
44. A. C. Wall, "Maximin surfaces, and the strong subadditivity of the covariant holographic entanglement entropy," *Class. Quant. Grav.* **31**(22), 225007 (2014) arXiv:1211.3494 [hep-th].
45. T. Faulkner, A. Lewkowycz and J. Maldacena, "Quantum corrections to holographic entanglement entropy," *JHEP* **1311**, 074 (2013), arXiv:1307.2892.
46. N. Engelhardt and A. C. Wall, "Quantum extremal surfaces: Holographic entanglement entropy beyond the classical regime," arXiv:1408.3203 [hep-th].

47. R. Bousso, S. Leichenauer and V. Rosenhaus, "Light-sheets and AdS/CFT," *Phys. Rev. D* **86**, 046009 (2012) `arXiv:1203.6619 [hep-th]`.
48. R. Bousso, B. Freivogel, S. Leichenauer, V. Rosenhaus and C. Zukowski, "Null geodesics, local CFT operators and AdS/CFT for subregions," *Phys. Rev. D* **88**, 064057 (2013) `arXiv:1209.4641 [hep-th]`.
49. V. E. Hubeny and M. Rangamani, "Causal holographic information," *JHEP* **1206**, 114 (2012) `arXiv:1204.1698 [hep-th]`.
50. B. Czech, J. L. Karczmarek, F. Nogueira and M. Van Raamsdonk, "The gravity dual of a density matrix," *Class. Quant. Grav.* **29**, 155009 (2012) `arXiv:1204.1330 [hep-th]`.
51. D. L. Jafferis, A. Lewkowycz, J. Maldacena and S. J. Suh, "Relative entropy equals bulk relative entropy," *JHEP* **06**, 004 (2016) `arXiv:1512.06431 [hep-th]`.
52. X. Dong, D. Harlow and A. C. Wall, "Reconstruction of bulk operators within the entanglement wedge in gauge-gravity duality," *Phys. Rev. Lett.* **117**(2), 021601 (2016) `arXiv:1601.05416 [hep-th]`.
53. A. Almheiri, X. Dong and D. Harlow, "Bulk locality and quantum error correction in AdS/CFT," *JHEP* **04**, 163 (2015) `arXiv:1411.7041 [hep-th]`.
54. D. Harlow, "The Ryu-Takayanagi formula from quantum error correction," *Commun. Math. Phys.* **354**(3), 865–912 (2017) `arXiv:1607.03901 [hep-th]`.
55. G. 't Hooft, "Dimensional reduction in quantum gravity," `gr-qc/9310026`.
56. L. Susskind, "The world as a hologram," *J. Math. Phys.* **36**, 6377–6396 (1995) `hep-th/9409089`.
57. A. Ashtekar and B. Krishnan, "Isolated and dynamical horizons and their applications," *Living Rev. Rel.* **7**, 10 (2004) `arXiv:gr-qc/0407042 [gr-qc]`.
58. S. W. Hawking and G. F. R. Ellis, *The large scale stucture of space-time* (Cambridge University Press, Cambridge, England, 1973).
59. R. M. Wald, *General Relativity* (The University of Chicago Press, Chicago, 1984).
60. R. Bousso, "The holographic principle," *Rev. Mod. Phys.* **74**, 825 (2002) `hep-th/0203101`.
61. R. Bousso, H. Casini, Z. Fisher and J. Maldacena, "Entropy on a null surface for interacting quantum field theories and the Bousso bound," *Phys. Rev. D* **91**(8), 084030 (2015) `arXiv:1406.4545 [hep-th]`.
62. R. Bousso, Z. Fisher, J. Koeller, S. Leichenauer and A. C. Wall, "Proof of the quantum null energy condition," *Phys. Rev. D* **93**(2), 024017 (2016) `arXiv:1509.02542 [hep-th]`.
63. S. Leichenauer, "The quantum focusing conjecture has not been violated," `arXiv:1705.05469 [hep-th]`.
64. C. Akers, V. Chandrasekaran, S. Leichenauer, A. Levine and A. Shahbazi Moghaddam, "The quantum null energy condition, entanglement wedge nesting, and quantum focusing," `arXiv:1706.04183 [hep-th]`.
65. R. M. Wald and U. Yurtsever, "General proof of the averaged null energy condition for a massless scalar field in two-dimensional curved space-time," *Phys. Rev. D* **44**, 403–416 (1991).
66. S. Leichenauer, A. Levine and A. Shahbazi-Moghaddam, "Energy is entanglement," `arXiv:1802.02584 [hep-th]`.
67. J. Koeller and S. Leichenauer, "Holographic proof of the quantum null energy condition," `arXiv:1512.06109 [hep-th]`.
68. S. Balakrishnan, T. Faulkner, Z. U. Khandker and H. Wang, "A general proof of the quantum null energy condition," `arXiv:1706.09432 [hep-th]`.

The Bekenstein Bound

Don N. Page

Department of Physics, 4-183 CCIS,
University of Alberta, Edmonton, Alberta T6G 2E1, Canada
profdonpage@gmail.com

Bekenstein's conjectured entropy bound for a system of linear size R and energy E, namely $S \leq 2\pi ER$, has counterexamples for many of the ways in which the "system," R, E, and S may be defined. One consistent set of definitions for these quantities in flat Minkowski spacetime is that S is the total von Neumann entropy and E is the expectation value of the energy in a "vacuum-outside-R" quantum state that has the vacuum expectation values for all operators entirely outside a sphere of radius R. However, there are counterexamples to the Bekenstein bound for this set of definitions. Nevertheless, an alternative formulation ten years ago by Horacio Casini for the definitions of S and of $2\pi ER$ have finally enabled a proof for this particular formulation of the Bekenstein bound.

1. Introduction

Jacob Bekenstein was an outstanding scientist who made fundamental contributions to black hole thermodynamics and other related areas of theoretical physics. He is perhaps best known for proposing that a black hole has an entropy proportional to its event horizon surface area A and obeys a generalized second law, that the total black hole entropy plus the entropy of ordinary matter outside black holes never decreases.[1–3] This prediction was confirmed theoretically when Stephen Hawking discovered that black holes emit thermal radiation and have an entropy[4–6]

$$S_{\mathrm{BH}} = \frac{kc^3 A}{4\hbar G} = \frac{A}{4}, \tag{1}$$

where for the last quantity, and henceforth, I am using Planck units with $\hbar = c = G = k_{\mathrm{Boltzmann}} = 4\pi\epsilon_0 = 1$. Note that the subscript "BH" on the entropy can mean either "black hole" or "Bekenstein–Hawking," and this black hole entropy is indeed often called the Bekenstein–Hawking entropy.

In this paper I shall focus on another interesting proposal that Jacob Bekenstein made that was motivated by the generalized second law but which exists independently for nongravitational systems, a conjectured upper bound of the entropy S of a system of given energy E and radius R,[7] known as the Bekenstein bound,

$$S \leq 2\pi ER. \tag{2}$$

Bekenstein and his colleagues developed many arguments and examples to support this conjecture.[7–28] On the other hand, many counterarguments and counterexamples have also been put forward.[29–46] Whether the conjectured bound (1) holds or not depends on what systems are considered and how R, E, and S are defined.

Perhaps the simplest procedure[14,16] would be just to consider quantum fields inside some bounded region within a sphere of radius R and put boundary conditions on the fields at the boundary of the region. However, this procedure leads to a large number of counterexamples to Bekenstein's conjectured bound. For example,[30] the Casimir effect can make $E < 0$ for certain states of quantum fields confined within a certain region of radius $\leq R$, violating the bound. If states with $E < 0$ are excluded by definition, one can still consider a mixed state with arbitrarily small positive E that violates the bound. Even if E is redefined to be the nonnegative energy excess over that of the ground state,[14,16] one can violate the bound by a mixed state that is almost entirely the ground state and a tiny incoherent mixture of excited states, at least if the entropy is defined to be $S = -\mathrm{tr}\,\rho \ln \rho$.[32] If S is instead defined to be $S = \ln n$ for a mixture of n orthogonal pure states (which would agree with $S = -\mathrm{tr}\,\rho \ln \rho$ if the mixture had equal probabilities $1/n$ for each of those n pure states), then one can violate the bound by an equal mixture of the ground state and the first excited state of certain interacting fields with certain boundary conditions that have the two lowest states nearly degenerate in energy (separated by exponentially small tunnelling effects).[30,42,43] If interacting fields are excluded from the definition of allowable systems, one can get a violation by considering a sufficiently large number N of identical free fields, giving $n = N$ degenerate first excited states of finite energy but sufficiently large entropy $S = \ln n = \ln N$ to violate (1).[30,31] And even for a single free electromagnetic field, $S = \ln n$ can exceed $2\pi ER$ by an arbitrarily large factor by using boundary conditions corresponding to an arbitrarily large number of parallel perfectly conducting plates within the region of radius R,[42] or by using boundary conditions corresponding to an arbitrarily long coaxial cable loop coiled up within the region.[43]

However, other than in his papers with Schiffer,[14,16] Bekenstein generally advocated taking E to be the total energy of a complete system.[7,9,10,12,22,27] This would disallow using just the energy of fields within a bounded region with boundary conditions, since that would ignore the energy of whatever it is that is providing the boundary conditions. Therefore, all of the counterexamples mentioned above would be excluded by this restriction. However, then the problem is to define what one means by the radius R of the system. In the flat spacetime case (quantum fields in Minkowski spacetime) that we shall focus on here, Bekenstein takes R to mean the radius of a sphere which circumscribes the system, which leaves the problem of what it means for a sphere to circumscribe the complete system.

In quantum field theory in Minkowski spacetime, the complete system is the quantum state of the fields. Since the quantum fields extend all the way out to radial infinity, a sphere circumscribing the entire system would have to be at $R = \infty$, which makes the Bekenstein bound true (at least for states of positive energy and finite entropy) but trivial. To get a nontrivial bound, one needs to suppose that a sphere of finite R can circumscribe the system. For example, one might try to say that the sphere encloses all of the excitations of the fields from the vacuum. However, it is also hard to get this to occur for a finite R. For example, the wave function for any single particle state that is a superposition of energy eigenstates of bounded energy will not vanish outside any finite radius R, since a one-particle wave function that does vanish outside a finite region must be a superposition of arbitrarily large momentum components, which will have unbounded energy. Even if one looks at a composite system, such as a hydrogen atom, and ignores the fact that its center of mass will have amplitudes to be outside any finite sphere if it is made of purely bounded energy components, the wave function for the relative position of the electron and proton does not drop identically to zero outside any finite separation distance for states that are superpositions of energy eigenstates of bounded energy. In particular, even if one fixed the center of mass of a hydrogen atom in its ground state and ignored the infinite energy from the resulting infinite uncertainty of the center of mass momentum, the density matrix for the electron position would decay only exponentially with distance from the center of mass and never go to zero outside any sphere of finite radius R.

Therefore, it is problematic to define the radius R of a sphere circumscribing a complete system in any quantum field theory. This issue was not addressed by Bekenstein and his collaborators, but without such a definition, there is no nontrivial formulation of the conjectured bound (1) for complete systems, but only its trivial truth for any complete system with positive energy and finite entropy that can only be circumscribed by the sphere enclosing all of space, $R = \infty$.

Here I shall first review ways I have proposed[44] to define systems and their radii R, energies E, and entropies S, so that for each, there is a bound on S for a given system as a function of finite R and E. These bounds do not have the form of Bekenstein's conjectured inequality (1), though in some cases they may obey that inequality.

Then I shall outline a new approach by Horacio Casini[49] that gives new definitions for S and for $2\pi RE$, for which one can give a rigorous proof of the Bekenstein bound.

2. Vacuum-Outside-R States

The main new element in Ref. 44 was the proposal to define a system of radius R (in flat Minkowski spacetime), not by imposing boundary conditions on the field itself, but by imposing conditions on the quantum state of the field so that outside a closed ball of radius R the quantum state is indistinguishable from the vacuum at some time. Such a state will be called a vacuum-outside-R state. (For simplicity, set this time to be $t = 0$, and take the closed ball, say B, to be the region $r \leq R$ on the $t = 0$ hypersurface, where r is the standard radial polar coordinate giving the proper distance from the coordinate origin on that hypersurface.) In other words, a vacuum-outside-R state of the system, say as expressed by its density matrix ρ, is such that the expectation value of any operator O which is completely confined to the region $r > R$ when written in terms of field and conjugate operators at $t = 0$, is precisely the same as the expectation value of the same operator in the vacuum state $|0\rangle\langle 0|$,

$$\mathrm{tr}(O\rho) = \langle 0|O|0\rangle \,. \tag{3}$$

In particular, all the n-point functions for the field and for its conjugate momentum in the state ρ are the same as in the vacuum state, if all of the n points are outside the ball of radius R and on the hypersurface $t = 0$. Of course, the n-point functions need not be the same as their vacuum values if some or all of the points are inside the ball.

If operators confined to the three-dimensional region $r > R$ and $t = 0$ (say C, to give a name to this achronal spacelike surface, the $t = 0$ hypersurface with the central closed ball B, $r \leq R$, excluded) have the same expectation value in the vacuum-outside-R state as in the vacuum state, the same will be true in any quantum field theory that I shall call "strongly causal" for all operators confined to the Cauchy development or domain of dependence[47] of C, the larger four-dimensional region $r > R + |t|$ (say D) that is the set of all points in the Minkowski spacetime such that every inextendible (endless) causal, or non-spacelike (everywhere timelike or lightlike), curve through such a point intersects the partial Cauchy surface C. Just as solutions of hyperbolic wave equations in D are determined by the data on C, so the part of the quantum state of a strongly causal field in D, as represented by the expectation values of operators confined to D, is determined by the part of the quantum state in C, as represented by the expectation values of operators confined to C. (For some interacting quantum field theories, the expectation values of operators confined to the three-dimensional spacelike surface C may be too ill-defined for these theories to be "strongly causal" in my sense, but a wider class of these theories may be "weakly causal" in the sense that sufficiently many operators smeared over, but confined to, an arbitrarily thin-in-time four-dimensional slab, say E, containing C within D, have well-defined expectation values that determine the expectation values of all operators smeared over, but confined to, any part of D.)

Henceforth I shall restrict attention to strongly causal and weakly causal quantum field theories, calling them simply causal quantum field theories for short. I shall also assume that any quantum field theory under consideration is a nongravitational Lorentz-invariant quantum field theory in Minkowski spacetime, and that it has a unique pure state of lowest Minkowski energy $E = 0$ (the expectation value of the Hamiltonian H that generates translations in the time coordinate t in some

Lorentz frame, with the arbitrary constant in the Hamiltonian being adjusted to give the lowest energy state zero energy).

Therefore, for such a causal nongravitational quantum field theory in Minkowski spacetime, I proposed[44] that the radius R be defined so that all of the operators constructed from field and conjugate momentum operators smeared over regions confined to the region D, $r > R + |t|$ in some Lorentz frame, have in the particular quantum state being considered (a vacuum-outside-R state) the same expectation values that they have in the vacuum state for that quantum field theory. The energy E of the state ρ can then be simply defined to be the expectation value,

$$E \equiv \mathrm{tr}(H\rho)\,, \tag{4}$$

of the Hamiltonian H that generates time translations in the same Lorentz frame. Because the energy E has been defined to have the minimum value of zero for the unique pure vacuum state, there is no problem here with negative Casimir energies. In other words, the energy is that of the complete system over all of Minkowski spacetime.

Obviously we would also like a definition of the entropy S that has a minimum value of zero, which it should attain for the pure vacuum state. One simple definition is the von Neumann entropy,

$$S = S_{\mathrm{vN}} \equiv -\mathrm{tr}\,\rho \ln \rho\,, \tag{5}$$

using the density matrix ρ for the full state of the quantum field, over the entire Minkowski spacetime.

3. Entropy Bounds for Vacuum-Outside-R States

Now we may conjecture that for any vacuum-outside-R state of any particular causal nongravitational quantum field theory in Minkowski spacetime, one which has the vacuum expectation values in the region D, $r > R + |t|$ (the region causally disconnected from the ball $r \leq R$ at $t = 0$), the von Neumann entropy is bounded above by some function σ_{vN} (depending on the quantum field theory in question) of the radius R and energy expectation value E:

$$S_{\mathrm{vN}} \leq \sigma_{\mathrm{vN}}(R, E)\,. \tag{6}$$

Define this function $\sigma_{\mathrm{vN}}(R, E)$ to be the least upper bound on the von Neumann entropy of any state which is vacuum outside the radius R and which has energy expectation value E.

In the case of a scale-invariant quantum field, such as a free massless field, or say a massless scalar field ϕ with a $\lambda\phi^4$ self-coupling potential, the least upper bound function $\sigma_{\mathrm{vN}}(R, E)$ will actually be a function of the single dimensionless variable

$$x \equiv 2\pi RE\,, \tag{7}$$

say

$$\sigma_{\mathrm{vN}}(R, E) = \sigma_{\mathrm{N}}(x)\,. \tag{8}$$

Bekenstein's conjectured entropy bound (1), if R, E, and S were defined as done above, would be $\sigma_{\rm vN}(R,E) \leq x$, whether or not the quantum field theory is scale invariant, or

$$B_{\rm vN}(R,E) \equiv \frac{\sigma_{\rm vN}(R,E)}{x} \equiv \frac{\sigma_{\rm vN}(R,E)}{2\pi RE} \leq 1\,. \tag{9}$$

If the quantum field theory is scale invariant, we can define

$$B_{\rm N}(x) \equiv \frac{\sigma_{\rm N}(x)}{x}\,, \tag{10}$$

which should also be less than or equal to unity if Bekenstein's bound applies.

For a set of one or more free massless fields and vacuum-outside-R states with $x \gg 1$, one would expect that the highest entropy would be given by a mixed state that at $t = 0$ is approximately a high-temperature ($RT \gg 1$) thermal radiation state for $r < R$, surrounded by vacuum for $r > R$. A high-temperature thermal radiation state has an energy density for massless fields of approximately $a_r T^4$, and hence an entropy density $(4/3)a_r T^3$, where

$$a_r = \frac{\pi^2}{30}\left(n_b + \frac{7}{8}n_f\right) \tag{11}$$

is the radiation constant for n_b independent bosonic degrees of freedom for each momentum (e.g., n_b different spin or helicity states) and for n_f fermionic degrees of freedom. Therefore, in this case with $x \gg 1$,

$$B_{\rm vN}(R,E) = \frac{\sigma_{\rm N}(x)}{x} \approx \left(\frac{2^7 a_r}{3^5\pi^2 x}\right)^{\frac{1}{4}} = \left[\frac{2^6}{3^6 5 x}\left(n_b + \frac{7}{8}n_f\right)\right]^{\frac{1}{4}}, \tag{12}$$

which is indeed less than 1, thus obeying Bekenstein's conjectured bound, for

$$x \geq \frac{2^7 a_r}{3^5\pi^2} = \frac{2^6}{3^6 5}\left(n_b + \frac{7}{8}n_f\right) = \frac{64 n_b + 56 n_f}{3645}\,, \tag{13}$$

if x is also large enough that Eq. (12) is a good approximation. Thus one would expect that Bekenstein's conjectured bound, using the definitions above for R, E, and S, holds for a fixed set of free massless quantum fields at sufficiently large $x \equiv 2\pi RE$.

On the other hand, the definitions above for R, E, and S still permit Bekenstein's conjectured bound applied to them to be violated for sufficiently small x, as we can see by the following construction:

A way to construct vacuum-outside-R states, quantum states of a free quantum field theory that have vacuum expectation values in the region D, $r > R + |t|$, is to apply to the vacuum state unitary operators constructed from fields and/or conjugate momenta smeared within the region $r < R$ at $t = 0$. In particular, if h is an hermitian operator constructed from fields and/or conjugate momenta smeared within $r < R$ at $t = 0$, then $U = e^{ih}$ is such a unitary operator, and

$$|\psi\rangle = U|0\rangle = e^{ih}|0\rangle \tag{14}$$

is a pure quantum state that has precisely the vacuum expectation values in the region D. This result can be seen formally from the fact that any operator O confined to the region D (the four-dimensional region $r > R + |t|$) that is causally disconnected from the ball B (the three-dimensional region $r \leq R$ on the $t = 0$ hypersurface) commutes with the operators h and U that are confined to that hypersurface, $[O, h] = [O, U] = 0$, so

$$\mathrm{tr}(O\rho) = \langle\psi|O|\psi\rangle = \langle 0|U^{-1}OU|0\rangle = \langle 0|U^{-1}UO|0\rangle = \langle 0|O|0\rangle\,. \tag{15}$$

If $\{h_i\}$ is a set of hermitian operators that each are confined to the ball B (i.e., are constructed from fields and momenta that are smeared only over that region), and if $\{q_i\}$ is a set of positive numbers that sum to unity, then

$$\rho = \sum_i q_i e^{ih_i}|0\rangle\langle 0|e^{-ih_i} \tag{16}$$

is a more general vacuum-outside-R state, since this density matrix gives vacuum expectation values, $\mathrm{tr}(O\rho) = \langle 0|O|0\rangle$, for any operator O confined to the region D that is causally disconnected from B (i.e., having no causal curves, either timelike or lightlike, intersecting both the ball B of $r \leq R$ at $t = 0$ and the region D with $r > R + |t|$).

Here let us consider the simple example in which i takes only the two values 1 and 2, and $h_1 = 0$ and $h_2 = h$. Let $q_1 = 1 - q$ and $q_2 = q$, and let

$$e^{ih}|0\rangle = U|0\rangle = |\psi\rangle = c|0\rangle + s|1\rangle \tag{17}$$

in terms of a decomposition of $|\psi\rangle$ into the two orthonormal states $|0\rangle$ and $|1\rangle = (|\psi\rangle - \langle 0|\psi\rangle|0\rangle)/\sqrt{1 - |\langle 0|\psi\rangle|^2}$, so

$$c = \langle 0|\psi\rangle = \langle 0|U|0\rangle = \langle 0|e^{ih}|0\rangle\,, \tag{18}$$

$$s = \sqrt{1 - |\langle 0|U|0\rangle|^2} = \sqrt{1 - |c|^2}\,. \tag{19}$$

Note that $|1\rangle\langle 1|$ by itself is not generically a vacuum-outside-R state.

Now Eq. (16) gives the density matrix as

$$\begin{aligned}\rho &= (1 - q)|0\rangle\langle 0| + q|\psi\rangle\langle\psi| \\ &= (1 - qs^2)|0\rangle\langle 0| + qcs|0\rangle\langle 1| + q\bar{c}s|1\rangle\langle 0| + qs^2|1\rangle\langle 1|\,,\end{aligned} \tag{20}$$

a density matrix in the two-dimensional space of pure states spanned by the two orthonormal pure states $|0\rangle$ and $|1\rangle$. The two eigenvalues of this density matrix are, say, p and $1 - p$ (since their sum is $\mathrm{tr}\,\rho = 1$), with product

$$\begin{aligned}y \equiv p(1-p) &= \frac{1}{2}\{[p + (1-p)]^2 - [p^2 + (1-p)^2]\} = \frac{1}{2}\{[\mathrm{tr}(\rho)]^2 - [\mathrm{tr}(\rho^2)]\} \\ &= q(1-q)(1 - |\langle 0|U|0\rangle|^2) = q(1-q)s^2\,.\end{aligned} \tag{21}$$

The expectation value of the energy of this mixed state is, since I have assumed $H|0\rangle = 0$,

$$E = \mathrm{tr}(H\rho) = q\langle\psi|H|\psi\rangle = q\langle 0|U^{-1}HU|0\rangle\,. \tag{22}$$

Then

$$x \equiv 2\pi RE = 2\pi Rq\langle 0|U^{-1}HU|0\rangle\,. \tag{23}$$

The von Neumann entropy of this mixed state is

$$\begin{aligned}S_{\rm vN}(y) &= -{\rm tr}(\rho\ln\rho) = -p\ln p - (1-p)\ln(1-p)\\ &= \frac{y}{\frac{1}{2}(1+\sqrt{1-4y})}\ln\frac{1}{y} + \sqrt{1-4y}\ln\frac{1}{\frac{1}{2}(1+\sqrt{1-4y})}\\ &\approx y\left[(1+y+2y^2)\ln\frac{1}{y} + \left(1-\frac{1}{2}y-\frac{5}{3}y^2\right)\right],\end{aligned} \tag{24}$$

a monotonically increasing function of $y \equiv q(1-q)s^2 \leq 1/4$, where the last approximate equality of Eq. (24) applies for $y \ll 1$.

As q and/or h is reduced toward zero, x, y, and S also decrease toward zero, but whereas x and y asymptotically decrease linearly with q, the dominant term of S has an extra logarithmic factor that grows with the reduction of y, so the ratio,

$$B \equiv \frac{S_{\rm vN}}{x} \equiv \frac{S_{\rm vN}}{2\pi RE} \approx \frac{y}{x}\left(\ln\frac{1}{y}+1\right) \tag{25}$$

when $y \ll 1$, increases without limit as y is reduced toward zero. Therefore, when y is made sufficiently small (e.g., by making q sufficiently small), Bekenstein's conjectured bound is violated for the definitions of R, E, and S used here.

I then went on[44] to give some examples in the free quantum field theory of a single massless scalar field, finding for a particular set of vacuum-outside-R states that

$$B \sim \frac{1}{\gamma}\left(\ln\frac{\gamma}{x}+1\right), \tag{26}$$

with numerical calculations giving

$$\gamma \approx 12.6048\,, \tag{27}$$

thus violating the Bekenstein bound, $B \leq 1$, for

$$x \equiv 2\pi RE < \gamma e^{1-\gamma} \approx 0.000115\,. \tag{28}$$

I further conjectured[44] that the true upper bound has the same asymptotic form with the same value of γ,

$$B_{\rm N}(x) \equiv \frac{\sigma_{\rm N}(x)}{x} \sim \frac{1}{\gamma}\left(\ln\frac{\gamma}{x}+1\right). \tag{29}$$

4. A Provable Formulation of the Bekenstein Bound

Another approach developed by Raphael Bousso[48] is to try to bound the size of a system by including interactions. This goes some ways toward giving a proof of the Bekenstein bound, though I am a bit confused as to how interactions could possibly bind nonzero energy eigenstates into structures that are strictly zero outside some finite distance (even if this distance is just a relative distance from the center of mass in a zero-momentum eigenstate, as was done to avoid the problem of confining the center of mass to a finite region, which would, as discussed above, lead to components of the quantum state having arbitrarily large momenta and hence components with arbitrarily large energy). However, this paper does not seem to give precise upper bounds on the entropy as a function of the energy and size of the system.

An approach that finally does seem to allow one to get a proof of at least one form of the Bekenstein bound is one developed by Horacio Casini,[49] building upon previous work.[38,50,51]

This paper[49] showed some awareness of my paper[44] by citing it and mentioning its definition of the localization of a system (vacuum-outside-R states), but Ref. 49 incorrectly assumed Ref. 44 defined the entropy by using the microcanonical ensemble for the number of such states below energy E_0, which does not work because such states do not form a vector space. However, the erroneous assumption that I did not have a good way to calculate the entropy may have been fortunate, because it does seem rather hard to make progress toward finding a way to prove, rather than disprove, the Bekenstein bound using vacuum-outside-R states, so the erroneous assumption perhaps prevented Casini from getting bogged down in this difficult approach and may have helped motivate him to seek a more fruitful direction for finding at least one way in which the Bekenstein bound seems to be true.

Casini[49] takes quantum field theory in the right Rindler wedge of flat Minkowski spacetime, $z > |t|$, with Cauchy surface the spatial half-plane V defined by $z > 0$ at $t = 0$. A quantum field theory state over the entire spatial plane $t = 0$ then leads to a mixed state ρ_V on the half-plane and its Cauchy development. Because of the sharp boundary of the half-plane at $z = 0$, the von Neumann entropy of ρ_V, namely

$$S(\rho_V) = -\mathrm{tr}\, \rho_V \ln \rho_V \,, \tag{30}$$

diverges. However, if one subtracts the von Neumann entropy of the state ρ_V^0 that is the restriction of the global vacuum state to the spatial half-plane, one gets the finite result

$$S_V = S(\rho_V) - S(\rho_V^0)\,. \tag{31}$$

The ingenious proposal of Casini[49] is to replace $2\pi ER$ by

$$K_V = \mathrm{tr}(K\rho_V) - \mathrm{tr}(K\rho_V^0)\,, \tag{32}$$

where

$$K = 2\pi \int dx\, dy \int_0^\infty z\, \mathcal{H}(x, y, z, 0)\,, \tag{33}$$

is the boost operator, with $\mathcal{H}(x, y, z, t)$ being the Hamiltonian density operator.

Then with these clever identifications of what the quantities S and $2\pi ER$ should be, the Bekenstein bound becomes $S_V \leq K_V$, or

$$S(\rho_V) - S(\rho_V^0) \leq \mathrm{tr}(K\rho_V) - \mathrm{tr}(K\rho_V^0)\,. \tag{34}$$

However, the state that is the restriction of the global vacuum state to the spatial half-plane is

$$\rho_V^0 = \frac{e^{-K}}{\operatorname{tr} e^{-K}}, \tag{35}$$

so $K = -\ln \rho_V^0 - \ln \operatorname{tr} e^{-K}$, and one gets (using $\operatorname{tr} \rho_V = \operatorname{tr} \rho_V^0 = 1$) that the Bekenstein bound is equivalent to the known nonnegativity of the relative entropy,[52–54]

$$S(\rho_V|\rho_V^0) \equiv \operatorname{tr}(\rho_V \ln \rho_V) - \operatorname{tr}(\rho_V \ln \rho_V^0) \geq 0\,. \tag{36}$$

These arguments[49] were made in terms of a theory with a cutoff, but Robert Longo and Feng Xu[56] were able to recast them in full quantum field theory, by operator algebraic methods, into a rigorous proof of the Bekenstein bound.[7]

One disadvantage of Casini's form of the Bekenstein bound as the inequality (34), or $S_V \leq K_V$, is that both sides of the inequality can be negative,[55] unlike the entropy S and energy E in the vacuum-outside-R formulation discussed earlier above. However, Blanco and Casini[55] have also given an alternative form of the Bekenstein bound in which both the expressions for the entropy S and for $2\pi ER$ are nonnegative.

5. Conclusions and Acknowledgments

In conclusion, we have found that one can formulate precise definitions for entropy bounds of a complete quantum field system over all of Minkowski spacetime by giving precise definitions for the size R of the system on the hypersurface $t = 0$. In particular, R may be defined for vacuum-outside-R states as the largest round two-sphere on the $t = 0$ hypersurface, outside of which all of the operators have the same expectation values as in the vacuum state. However, this definition, plus the definition of E as the expectation value of the Hamiltonian and S as the von Neumann entropy of the entire quantum state, leads to violations of the Bekenstein bound. Nevertheless, an alternative set of definitions for S and for $2\pi RE$ by Horacio Casini have finally succeeded in leading to a rigorous proof of one form of the Bekenstein bound.

This work was supported by the Natural Sciences and Engineering Research Council of Canada.

References

1. J. D. Bekenstein, "Black Holes and the Second Law," *Lett. Nuovo Cim.* **4**, 737–740 (1972), doi:10.1007/BF02757029.
2. J. D. Bekenstein, "Black Holes and Entropy," *Phys. Rev. D* **7**, 2333–2348 (1973), doi:10.1103/PhysRevD.7.2333.
3. J. D. Bekenstein, "Generalized Second Law of Thermodynamics in Black Hole Physics," *Phys. Rev. D* **9**, 3292–3300 (1974), doi:10.1103/PhysRevD.9.3292.
4. S. W. Hawking, "Black Hole Explosions?" *Nature* **248**, 30–31 (1974), doi:10.1038/248030a0.

5. S. W. Hawking, "Particle Creation by Black Holes," *Commun. Math. Phys.* **43**, 199–220 (1975) [Erratum: *ibid.* **46**, 206 (1976)], doi:10.1007/BF02345020, 10.1007/BF01608497.
6. S. W. Hawking, "Black Holes and Thermodynamics," *Phys. Rev. D* **13**, 191–197 (1976), doi:10.1103/PhysRevD.13.191.
7. J. D. Bekenstein, "A Universal Upper Bound on the Entropy to Energy Ratio for Bounded Systems," *Phys. Rev. D* **23**, 287–298 (1981), doi:10.1103/PhysRevD.23.287.
8. J. D. Bekenstein, "Energy Cost of Information Transfer," *Phys. Rev. Lett.* **46**, 623–626 (1981), doi:10.1103/PhysRevLett.46.623.
9. J. D. Bekenstein, "Black Holes and Everyday Physics," *Gen. Rel. Grav.* **14**, 355–359 (1982), doi:10.1007/BF00756269.
10. J. D. Bekenstein, "Specific Entropy and the Sign of the Energy," *Phys. Rev. D* **26**, 950–953 (1982), doi:10.1103/PhysRevD.26.950.
11. J. D. Bekenstein, "Entropy Bounds and the Second Law for Black Holes," *Phys. Rev. D* **27**, 2262–2270 (1983), doi:10.1103/PhysRevD.27.2262.
12. J. D. Bekenstein, "Entropy Content and Information Flow in Systems with Limited Energy," *Phys. Rev. D* **30**, 1669–1679 (1984), doi:10.1103/PhysRevD.30.1669.
13. J. D. Bekenstein and E. I. Guendelman, "Symmetry Breaking Induced by Charge Density and the Entropy of Interacting Fields," *Phys. Rev. D* **35**, 716–731 (1987), doi:10.1103/PhysRevD.35.716.
14. M. Schiffer and J. D. Bekenstein, "Proof of the Quantum Bound on Specific Entropy for Free Fields," *Phys. Rev. D* **39**, 1109–1115 (1989), doi:10.1103/PhysRevD.39.1109.
15. M. Schiffer and J. D. Bekenstein, "Do Zero Frequency Modes Contribute to the Entropy?," *Phys. Rev. D* **42**, 3598–3599 (1990), doi:10.1103/PhysRevD.42.3598.
16. J. D. Bekenstein and M. Schiffer, "Quantum Limitations on the Storage and Transmission of Information," *Int. J. Mod. Phys. C* **1**, 355–422 (1990), doi:10.1142/S0129183190000207 [quant-ph/0311050].
17. J. D. Bekenstein, "Entropy Bounds and Black Hole Remnants," *Phys. Rev. D* **49**, 1912–1921 (1994), doi:10.1103/PhysRevD.49.1912 [gr-qc/9307035].
18. J. D. Bekenstein and A. E. Mayo, "Black Hole Polarization and New Entropy Bounds," *Phys. Rev. D* **61**, 024022 (2000), doi:10.1103/PhysRevD.61.024022.
19. A. E. Mayo, "The Optimal Entropy Bound and the Selfenergy of Test Objects in the Vicinity of a Black Hole," *Phys. Rev. D* **60**, 104044 (1999), doi:10.1103/PhysRevD.60.104044 [gr-qc/9905007].
20. J. D. Bekenstein, "NonArchimedean Character of Quantum Buoyancy and the Generalized Second Law of Thermodynamics," *Phys. Rev. D* **60**, 124010 (1999), doi:10.1103/PhysRevD.60.124010 [gr-qc/9906058].
21. J. D. Bekenstein, "Holographic Bound from Second Law of Thermodynamics," *Phys. Lett. B* **481**, 339–345 (2000), doi:10.1016/S0370-2693(00)00450-0 [hep-th/0003058].
22. J. D. Bekenstein, "On Page's Examples Challenging the Entropy Bound," gr-qc/0006003.
23. J. D. Bekenstein, "The Limits of Information," *Stud. Hist. Phil. Sci. B* **32**, 511–524 (2001), doi:10.1016/S1355-2198(01)00020-X [gr-qc/0009019].
24. J. D. Bekenstein, "Quantum Information and Quantum Black Holes," *NATO Sci. Ser. II* **60**, 1–26 (2002), doi:10.1007/978-94-010-0347-6-1 [gr-qc/0107049].
25. J. D. Bekenstein, "Limitations on Quantum Information from Black Hole Physics," *Acta Phys. Polon. B* **32**, 3555–3570 (2001), [quant-ph/0110005].
26. J. D. Bekenstein, "Black Holes and Information Theory," *Contemp. Phys.* **45**, 31–43 (2003), doi:10.1080/00107510310001632523 [quant-ph/0311049].

27. J. D. Bekenstein, "How Does the Entropy/Information Bound Work?," *Found. Phys.* **35**, 1805–1823 (2005), doi:10.1007/s10701-005-7350-7 [quant-ph/0404042].
28. J. D. Bekenstein, "Are There Hyperentropic Objects?," *Phys. Rev. D* **70**, 121502 (2004), doi:10.1103/PhysRevD.70.121502 [hep-th/0410106].
29. S. D. Unwin, "Possible Violations of the Entropy to Energy Ratio Bound," *Phys. Rev. D* **26**, 944–946 (1982), doi:10.1103/PhysRevD.26.944.
30. D. N. Page, "Comment on a Universal Upper Bound on the Entropy to Energy Ratio for Bounded Systems," *Phys. Rev. D* **26**, 947–949 (1982), doi:10.1103/PhysRevD.26.947.
31. W. G. Unruh and R. M. Wald, "Acceleration Radiation and Generalized Second Law of Thermodynamics," *Phys. Rev. D* **25**, 942–958 (1982), doi:10.1103/PhysRevD.25.942.
32. D. Deutsch, "Is There a Fundamental Bound on the Rate at Which Information Can Be Processed?," *Phys. Rev. Lett.* **48**, 286–288 (1982). doi:10.1103/PhysRevLett.48.286.
33. W. G. Unruh and R. M. Wald, "Entropy Bounds, Acceleration Radiation, and the Generalized Second Law," *Phys. Rev. D* **27**, 2271–2276 (1983). doi:10.1103/PhysRevD.27.2271.
34. J. Ambjorn and S. Wolfram, "Properties of the Vacuum. 1. Mechanical and Thermodynamic," *Annals Phys.* **147**, 1–32 (1983), doi:10.1016/0003-4916(83)90065-9.
35. M. J. Radzikowski and W. G. Unruh, "Correction to 'Acceleration Radiation and the Generalized Second Law of Thermodynamics.'," *Phys. Rev. D* **37**, 3059–3060 (1988), doi:10.1103/PhysRevD.37.3059.
36. W. G. Unruh, "Comment on 'Proof of the Quantum Bound on Specific Entropy for Free Fields'," *Phys. Rev. D* **42**, 3596–3597 (1990), doi:10.1103/PhysRevD.42.3596.
37. M. A. Pelath and R. M. Wald, "Comment on Entropy Bounds and the Generalized Second Law," *Phys. Rev. D* **60**, 104009 (1999), doi:10.1103/PhysRevD.60.104009 [gr-qc/9901032].
38. R. M. Wald, "Gravitation, Thermodynamics, and Quantum Theory," *Class. Quant. Grav.* **16**, A177–A190 (1999), doi:10.1088/0264-9381/16/12A/309 [gr-qc/9901033].
39. E. E. Flanagan, D. Marolf and R. M. Wald, "Proof of Classical Versions of the Bousso Entropy Bound and of the Generalized Second Law," *Phys. Rev. D* **62**, 084035 (2000), doi:10.1103/PhysRevD.62.084035.
40. W. G. Anderson, "Matters of Gravity," ed. J. Pullin, gr-qc/9909022 (1999).
41. R. M. Wald, "The Thermodynamics of Black Holes," *Living Rev. Rel.* **4**, 6 (2001), doi:10.12942/lrr-2001-6 [gr-qc/9912119].
42. D. N. Page, "Huge Violations of Bekenstein's Entropy Bound," gr-qc/0005111.
43. D. N. Page, "Subsystem Entropy Exceeding Bekenstein's Bound," hep-th/0007237.
44. D. N. Page, "Defining Entropy Bounds," *JHEP* **0810**, 007 (2008), doi:10.1088/1126-6708/2008/10/007 [hep-th/0007238].
45. D. Marolf and R. D. Sorkin, "On the Status of Highly Entropic Objects," *Phys. Rev. D* **69**, 024014 (2004), doi:10.1103/PhysRevD.69.024014 [hep-th/0309218].
46. S. N. Solodukhin, "Boundary Conditions and the Entropy Bound," *Phys. Rev. D* **63**, 044002 (2001), doi:10.1103/PhysRevD.63.044002 [gr-qc/0009103].
47. See, for example, S. W. Hawking and G. F. R. Ellis, *The Large Scale Structure of Space-Time* (Cambridge University Press, Cambridge, 1973), doi:10.1017/CBO9780511524646.
48. R. Bousso, "Bound States and the Bekenstein Bound," *JHEP* **0402**, 025 (2004), doi:10.1088/1126-6708/2004/02/025 [hep-th/0310148].
49. H. Casini, "Relative Entropy and the Bekenstein Bound," *Class. Quant. Grav.* **25**, 205021 (2008), doi:10.1088/0264-9381/25/20/205021 [arXiv:0804.2182 [hep-th]].

50. D. Marolf, D. Minic and S. F. Ross, "Notes on Space-Time Thermodynamics and the Observer Dependence of Entropy," *Phys. Rev. D* **69**, 064006 (2004), doi:10.1103/PhysRevD.69.064006 [hep-th/0310022].
51. D. Marolf, "A Few Words on Entropy, Thermodynamics, and Horizons," doi:10.1142/9789812701688-0008 [hep-th/0410168].
52. A. Wehrl, "General Properties of Entropy," *Rev. Mod. Phys.* **50**, 221–260 (1978), doi:10.1103/RevModPhys.50.221.
53. V. Vedral, "The Role of Relative Entropy in Quantum Information Theory," *Rev. Mod. Phys.* **74**, 197–234 (2002), doi:10.1103/RevModPhys.74.197.
54. M. A. Nielsen and I. L. Chuang, *Quantum Computation and Quantum Information* (Cambridge University Press, Cambridge, 2000).
55. D. D. Blanco and H. Casini, "Localization of Negative Energy and the Bekenstein Bound," *Phys. Rev. Lett.* **111**(22), 221601 (2013), doi:10.1103/PhysRevLett.111.221601 [arXiv:1309.1121 [hep-th]].
56. R. Longo and F. Xu, "Comment on the Bekenstein bound," arXiv:1802.07184 [math-ph].

Bekenstein's Entropy Bound and Variations Thereof

Gabriele Veneziano
Theory Department, CERN, CH-1211 Geneva 23, Switzerland
Collége de France, 11 place M. Berthelot, 75005 Paris, France
gabriele.veneziano@cern.ch

I will summarize some of the developments that followed from Jacob Bekenstein's original proposal of a general bound on entropy. More specifically, I will concentrate on developments that affected my own research on fundamental strings, on black holes, and on string-inspired cosmological models in which the Big Bang is replaced by a Big Bounce.

1. Introduction

Arguably, the most important contribution to physics by Jacob Bekenstein was the realization[1] that black holes must have entropy and that such entropy ought to be the area A of the horizon in Planck-length units[a] up to a numerical coefficient. After Hawking managed to compute the latter,[2] the formula

$$S_{\rm BH} = \frac{A}{4l_P^2}\,; \quad l_P^2 \equiv G\hbar \tag{1}$$

became universally known as the Bekenstein–Hawking entropy of a black hole. This result carried an enormous number of consequences, most notably that of Hawking's black hole evaporation[2] with a thermal spectrum of temperature $T = \hbar(8\pi GM)^{-1}$, the possibility of loss of quantum coherence[3] in processes involving black holes (so-called information paradox), etc.

This contribution to the memorial volume will deal with a different, yet related, aspect of Bekenstein's work, the one[4] about absolute upper bounds on entropy for any physical system. Interestingly enough, while both black hole entropy and Bekenstein's entropy bound involve quantum mechanics (through the explicit appearance of $\hbar$) only the former contains Newton's constant G. Thus Bekenstein's

[a] I will use units in which the speed of light c and Boltzmann's constant k_B are set to 1. I will keep G and $\hbar$ dependence explicit but neglect, most of the time, numerical factors of order 1. Although I will discuss the case of $D = 4$ uncompactified space–time dimensions, the extension to $D > 4$ dimensions is straightforward. When some of the latter are compactified, like in Kaluza–Klein or string theory, new interesting features emerge, but will not be addressed here.

bound appears, at least superficially, to be of completely general validity and independent of gravity.

In the following I will first formulate the bound and then, in successive sections, I will compare it with other entropy bounds and/or use it in connection with other theoretical frameworks with the aim of showing that it may have far reaching consequences.

2. The Bekenstein Bound

In 1981 Jacob Bekenstein proposed[4] what he called a "Universal Entropy Bound," referred to, hereafter, as the Bekenstein bound (BB). It states that, for any *isolated* system of mass M and size R, the (dimensionless) entropy[b] is bound by:

$$S \leq S_{\mathrm{BB}} = \frac{MR}{\hbar} \,. \tag{2}$$

I will not review here the arguments originally given by Bekenstein in favor of such a bound but just stress that, to my knowledge, no counter-examples to it have been given till now.

One realistic threat to its validity is the limit in which the number of particle species N_s is taken to be arbitrarily large. This does increase the l.h.s without any similar change on the r.h.s. An answer to this objection can be[5] that what enters in the l.h.s. is the difference between the entropy of the system and that of an empty region of the same size which also grows with N_s because of quantum fluctuations. The difference is also free from ultraviolet divergences.

Note that, since typically $M \sim R^3$, the bound, being more than extensive, appears, at least superficially, to be very generous for large systems and very tight for small ones. Another observation is that, although the bound does not contain Newton's constant G, it is saturated by the Bekenstein–Hawking entropy (1) of a black hole since, for a black hole of mass M, the role of R is played by the Schwarzschild radius $R_S \sim GM$ and $l_P^2 = G\hbar$[c].

The idea that black holes maximize entropy was also one of the motivations for another important concept in modern theoretical physics: holography. The basic assumption of holography[6,7] is that physics inside a compact region of space (the so-called bulk) bounded by a surface could be entirely described in terms of the degrees of freedom living on such a surface. For this to be possible the entropy in the bulk should not exceed the Bekenstein–Hawking entropy of a black hole having the same boundary as its own horizon. This imposes a new (holographic) bound on entropy that we will now introduce and compare with the BB.

[b]There are several concepts of entropy in the literature (thermodynamic, von Neumann, entanglement, ...). Here we shall think (perhaps too naively?) of it as the logarithm of the number of micro-states corresponding to a given macro-state. For more sophisticated discussions of the BB in the context of von Neumann/entanglement entropy, see e.g. Ref. 5.

[c]For a black hole to correspond to an isolated system one should consider the case of it being in thermal equilibrium with its surrounding Hawking radiation.

3. Holographic vs. Bekenstein Bound

The holographic bound on entropy, also referred to isolated bodies, is simply:

$$S \leq S_{\rm HOB} \sim \frac{R^2}{l_P^2}\,, \tag{3}$$

which, of course, is saturated again by $S_{\rm BH}$ of (1) since, for a black hole, $A \sim R^2$.

Since the BB and the HOB coincide (up to factors $O(1)$) for a black hole, one can ask which one is stronger for objects that are *not* black holes. Since for such objects $R > R_S \sim GM$ we find:

$$S_{\rm BB} = \frac{MR}{\hbar} = \frac{GMR}{l_P^2} \leq \frac{R^2}{l_P^2} = S_{\rm HOB}\,, \tag{4}$$

i.e. that the BB *implies* the HOB! This may seem paradoxical (since intuitively the HOB is sub-extensive while the BB is super-extensive) but, as we shall see later, there is no contradiction with this result.

4. Comparisons with String Entropy

A nice application of the BB can be found in the context of weakly coupled (or even free) string theory. In this case we know the entropy since the late sixties[8,9] and the question is whether the BB is satisfied. Again modulo numerical factors, we have:[d]

$$S_{\rm string} = \frac{Ml_s}{\hbar}\,; \quad l_s = \sqrt{\alpha'\hbar}\,, \tag{5}$$

where $l_s = \sqrt{\alpha'\hbar}$ is the fundamental length scale of (quantum) string theory.

Note that neither $S_{\rm BB}$ nor $S_{\rm string}$ contain Newton's constant or Planck's length. The latter is related to l_s by some (positive) power of the string coupling g_s. Obviously, compatibility between (2) and (5) demands:

$$R \geq l_s\,. \tag{6}$$

In other words, no (quantum) string can be smaller in size than l_s.[10] There are many indications that this is indeed the case (note that such a statement is clearly wrong for classical strings for which the size can be rescaled to arbitrarily small values), e.g. from the study of trans-Plankian-energy collisions of strings.[11]

Since we have already argued that the BB implies the HOB we expect to find no further constraints on weakly coupled strings by imposing $S_{\rm string} \leq S_{\rm HOB}$. Indeed:

$$S_{\rm string} = \frac{Ml_s}{\hbar} = \frac{GMl_s}{l_P^2} \leq \frac{GMR}{l_P^2} \leq \frac{{\rm Max}((GM)^2, R^2)}{l_P^2} = S_{\rm HOB}\,, \tag{7}$$

where in the last equality we have used the observation that, for a collapsed object, what matters for the HOB is the area of the horizon.

[d]This is true for a single string. Considering multi-string states of the same size and total mass would increase string entropy and thus make the consequences of the BB even stronger.

Let us finally compare string entropy and black-hole entropy. Again we expect no new constraint since black holes saturate the BB as well as the HOB. We compute:

$$\frac{S_{\text{string}}}{S_{\text{BH}}} = \frac{Ml_s}{\hbar}\frac{l_P^2}{R^2} = \frac{GMl_s}{R^2} = \frac{l_s}{GM}\,. \tag{8}$$

This ratio is larger than one if $GM < l_s$ which, in principle, is possible even for a fixed large mass M by taking the string coupling sufficiently small. The conclusion is again non-trivial:[10] in string theory there cannot be black holes whose Schwarzschild radius is smaller than l_s! But this is perfectly in line with what we have already concluded: if the size of a quantum string is always larger than l_s, a heavy string whose Schwarzschild radius is smaller than l_s is simply *not* a black hole.

Such a statement has deep implications on the fate of black-hole evaporation in string theory:[12] instead of terminating at a singularity the process terminates when the Schwarzschild radius becomes l_s and the Hawking temperature becomes the Hagedorn temperature of string theory. At that moment the black hole becomes simply a highly excited string decaying as any normal heavy string. This happens when the mass of the string reaches a specific value:[10,13,14]

$$M = \frac{l_s}{G} = g_s^{-2}M_s\,, \tag{9}$$

where $M_s = \hbar l_s^{-1}$ is the string mass scale and g_s is the so-called string coupling which controls the loop (genus) expansion in string theory as well as the ratio between the string and Planck lengths. I have dubbed such special strings "string holes" since they share properties of strings and black holes. There is a large literature (see e.g. Refs. 15, 16 and references therein) on the relation between black holes and fundamental strings satisfying (9) as well as other (in particular angular momentum and size) criteria. A closely related question is that of the meaning of the Hagedorn temperature: is it a true upper limit on temperature in string theory or just the signal of a phase transition, like in QCD? Such questions are nicely reviewed in Ref. 17.

5. Cosmological Entropy Bounds

So far we have dealt with isolated systems. The question then arises as to whether a suitable generalization of the various entropy bounds exists for generic non isolated systems such as the Universe itself as described by a cosmological solution to Einstein's equations.

Consider, in particular, an homogeneous-isotropic Universe described by the well-known (spatially flat) metric:

$$ds^2 = -dt^2 + a(t)^2 dx^i dx^i\,, \tag{10}$$

and sustained by a perfect fluid of energy and pressure density ρ and p. Einstein's equations take the well-known form:

$$H^2 \equiv \left(\frac{\dot{a}}{a}\right)^2 = \frac{8\pi G}{3}\rho\,; \quad \left(\frac{\ddot{a}}{a}\right) = -\frac{4\pi G}{3}(\rho - 3p)\,. \tag{11}$$

Consider now, at a given t, a sphere of radius R inside such a Universe. The energy inside that region scales like ρR^3 and therefore a naive extension of the BB would read:

$$S(R) \leq \frac{\rho R^4}{\hbar}, \tag{12}$$

while a (naive again) application of the HOB would give:

$$S(R) \leq \frac{R^2}{l_P^2}. \tag{13}$$

On the other hand, standard thermodynamic arguments give

$$S(R) \sim R^3, \tag{14}$$

the constant of proportionality being related to some intensive quantity like temperature.

We see clearly that, for sufficiently large R, the HOB will be violated while the BB will be comfortably satisfied. In other words, for large R the relation between BB and HOB is reversed with respect to that of an isolated body. Fischler and Susskind,[18] inspired by an earlier suggestion by Bekenstein,[19] were the first to try a generalization of the HOB to cosmology that would avoid the above mentioned violation by identifying a maximal R for which the HOB applies. But one could find counter-examples to their proposal, e.g. in inflationary cosmology after reheating, unless the bound itself excludes the entropy generated inside the region by non-adiabatic processes. On the other hand, the cosmological extension proposed in Ref. 19 appears to be too generous.

A hint about how a better bound can be found comes from comparing the total mass in a sphere of radius R to that of a black hole with the same radius:

$$\frac{M(R)}{M_{\mathrm{BH}}(R)} = \frac{\rho R^3}{R/G} = G\rho R^2 = (HR)^2, \tag{15}$$

where, in the last step, we have used Eq. (11). In other words, if $R > H^{-1}$ the mass inside the sphere exceeds that of a black hole having the same radius. This is just a hint that something may go wrong if we apply the naive bounds to regions exceeding the size of the cosmological horizon.

A more solid arguments goes as follows.[20] We have already stated that, for isolated objects of a given size, a black hole maximizes entropy (and saturated both the BB and the HOB). There are however strong arguments (see for instance Refs. 21 and 22) for the non-existence of black holes larger than the Hubble radius H^{-1}. Such putative black holes would be highly unstable towards fissioning into smaller ones. But then a natural bound on entropy in cosmology for an *arbitrarily large* spatial region would be to divide that region in many Hubble-size subregions each one containing a Hubble-size black hole. This gives[20,23,24] a bound, dubbed the Hubble-bound (HUB), of the form:

$$S(R) \leq S_{\mathrm{HUB}} = n_H S_{\mathrm{BH}}(R = H^{-1}) = (RH)^3 H^{-2} l_P^{-2} = R^3 H l_P^{-2};$$
$$\Rightarrow \sigma \equiv \frac{S}{V} \leq H l_P^{-2} \sim \frac{\sqrt{\rho/G}}{\hbar}. \tag{16}$$

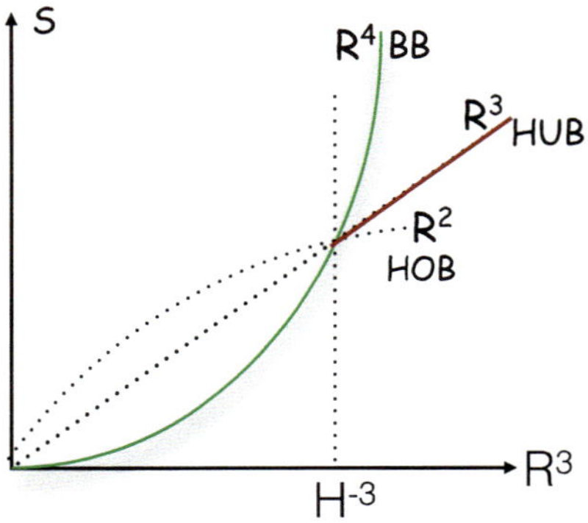

Fig. 1. Comparison of the three cosmological bounds as functions of the size R of the spatial region. For sub-Hubble regions the BB is the most stringent and presumably correct; for a Hubble-size region the three bounds coincide; for a super-Hubble region the naive HOB is the most stringent one, but incorrect, while the HUB (as well as Bousso's) should give the tightest reliable bound.

Note that the HUB, unlike the BB and the HOB, is extensive i.e. scales like the volume (if the region being considered is super-Hubble-sized). Indeed let us compare the three bounds. We easily find:

$$\frac{S_{\rm HUB}}{S_{\rm BB}} = (HR)^{-1}\,; \quad \frac{S_{\rm HUB}}{S_{\rm HOB}} = HR\,. \tag{17}$$

The three bounds therefore coincide for a Hubble-size region. For a super-horizon region ($HR > 1$) the HOB would be the strongest but is apparently incorrect. In that situation the HUB looks to be fine and implies the BB. For a sub-horizon region ($HR < 1$) the BB is the strongest one and is presumably correct. The HOB is less stringent, while the argument leading to the HUB does not apply. Comparison of the three entropy bounds is illustrated in Fig. 1.

6. More General Entropy Bounds

The HUB looks very reasonable but seems to apply only to very special (nearly homogeneous) cosmologies. Can one generalize the bound to generic space–times? One such generalization was given in an elegant paper by Raphael Bousso.[25] Unfortunately Bousso's bound only applies to light-like hyper surfaces.

The HUB would rather apply to space-like regions and its natural generalization would be in terms of the usual ADM foliation of space–time in space-like hyper surfaces and the concept of extrinsic curvature K^i_j and "expansion" ($\Theta = K^i_i$). For the special case of (10) $\Theta = 3H$. A problem with both the HUB and its naive Θ generalization is that the bound degenerates at the bounce where $\Theta = 0$. To overcome this difficulty, one can look at the theory of cosmological perturbations and identify a combination of quantities that define a scale of causal connection above which perturbations grow exponentially. The proposal of Ref. 26, based on the above criterion, is to replace the HUB by a covariant expression. For a space-like region defined by $\sigma < 0$, it takes the form:

$$S_{CEB} = l_P^{-2} \int_{\sigma<0} d^4x\sqrt{-g}\delta(\tau)\sqrt{\text{Max}_\pm[(G_{\mu\nu} \pm R_{\mu\nu})\partial^\mu\tau\partial^\nu\tau]}$$

$$= l_P^{-1}\hbar^{-1/2} \int_{\sigma<0} d^4x\sqrt{-g}\delta(\tau)\sqrt{\text{Max}_\pm\left[\left(T_{\mu\nu} \pm T_{\mu\nu} \mp \frac{1}{2}g_{\mu\nu}T\right)\partial^\mu\tau\partial^\nu\tau\right]}\,. \quad (18)$$

The above bound and Bousso's appear to be consistent with each other. Furthermore, by shrinking the region to a point they become bounds on a local entropy current s_μ such that $S = \int d^4x\sqrt{-g}\delta(\tau)s_\mu\partial^\mu\tau$. Then (18) becomes equivalent to

$$s_\mu\lambda^\mu \leq l_P^{-1}\hbar^{-1/2}\sqrt{\text{Max}_\pm\left[\left(T_{\mu\nu} \pm T_{\mu\nu} \mp \frac{1}{2}g_{\mu\nu}T\right)\lambda^\mu\lambda^\nu\right]}\,, \quad (19)$$

with λ^μ an arbitrary time-like vector. This is also in line with other claims in the literature.[27]

7. Some Applications of the Cosmological Entropy Bounds

As a first application of the cosmological entropy bounds discussed in the previous section, let us consider[18] entropy and its evolution during the usual adiabatic cosmological expansion where entropy is conserved in a fixed comoving volume. As a consequence entropy density has the following time dependence:

$$\sigma = S/V = S/V_{com}a^{-3} \sim a^{-3} = t^{-\frac{2}{1+w}}\,, \quad (20)$$

during an epoch dominated by an energy source with equation of state $p/\rho = w$. If we compare this with the HUB, where generically $H \sim t^{-1}$, we find

$$\frac{\sigma}{\sigma_{\text{HUB}}} \sim t^{-\frac{1-w}{1+w}}\,. \quad (21)$$

In words, the entropy bound becomes looser and looser as time goes on if $-1 < w < +1$, which is the "normal" range for w. Just for stiff matter ($w = 1$) the entropy varies exactly like the bound itself. The opposite happens for a contracting Universe in which, for any value of $w \neq 1$, the bound becomes tighter and tighter and would be violated at a big crunch singularity unless w goes to 1 or something else (a bounce?) happens.

Actually Bekenstein himself[19] argued along similar lines that the big bang singularity should be avoided at the quantum level. This latter conclusion is only reinforced if we abandon the hypothesis of adiabaticity. If irreversible phenomena occur this will increase entropy in a fixed comoving volume. Even if the naive bound of Ref. 18 needs to be amended, the more appropriate bounds discussed in the previous section become even more vulnerable. We will see below an application of this idea to a particular example of string cosmology.

For the moment let us remark that the well-known puzzle about what fixes the arrow of time in a theory which, microscopically, satisfies to a high degree of accuracy time-reversal invariance may find an explanation without assuming a particularly low initial entropy. Even if the entropy immediately after the big bang saturated the bound, the subsequent expansion makes the bound grow very fast in time relative to the actual entropy. There is thus plenty of room for further entropy increase and this establishes the arrow of time to coincide with that of an expanding Universe (whereas in the direction of contraction we would immediatly violate the bound).

In the old (pre-inflationary) cosmology the Universe magically emerges very nearly homogeneous, isotropic and spatially flat, from the big bang singularity (say at $t \sim t_P$) with a Planckian temperature and Planckian Hubble parameter. The HUB is thus automatically saturated at $t \sim t_P$. In the inflationary paradigm, instead, the very peculiar properties of the big bang are "explained" by a (sufficiently long) previous inflationary epoch which cools down the Universe, makes it spatially flat, and drastically dilutes the entropy density (see Eq. (21)). The role of the conventional Big Bang is now played by the reheating of the Universe after inflation[28] without which we cannot have the observed CMB or Big Bang Nucleosynthesis. The reheating temperature has to be sufficiently high, but certainly not of Planckian order. Consequently, the HUB is comfortably satisfied at the Big Bang and, because of Eq. (21), also thereafter.

Let us finally consider the case of string-inspired bouncing cosmologies, focussing for simplicity on the so-called pre-Big Bang scenario (see Refs. 29, 30 and references therein). In PBB cosmology the role of the old Big Bang is played by the exit from a previous phase of accelerated expansion (just like in ordinary inflation) except that the accelerated expansion is not caused by the potential energy of an "inflaton" (fundamental or effective) field. In the string frame metric (in which l_s is constant and l_P changes in time) the accelerated expansion is induced by the growth of the dilaton field (and consequently of the ratio l_P/l_s). Such a phase, called dilaton-driven inflation, is perturbative and fully under theoretical control.

In the Einstein frame (in which l_P is constant and l_s changes in time) dilaton-driven inflation corresponds to a contraction. The dilaton kinetic term is the dominate source of energy and the effective equation of state is $w = 1$. As already mentioned, in a contracting Universe the HUB tends to be violated unless $w = 1$ in which case, if the bound is initially satisfied, it remains so afterwards *provided the evolution is adiabatic.* This phase is supposed to end when either the curvature becomes $O(l_s^{-2})$ (making α' corrections important) or when the string coupling g_s becomes sufficiently large (to make look corrections important) whichever comes first. Which one of the two scenarios is realized depends on some "integration constants." Let us assume, for definiteness, that the first scenario is realized.

In a contracting-Universe smaller and smaller scales continuously go out of the horizon like they do in ordinary inflation. The resulting spectrum of adiabatic

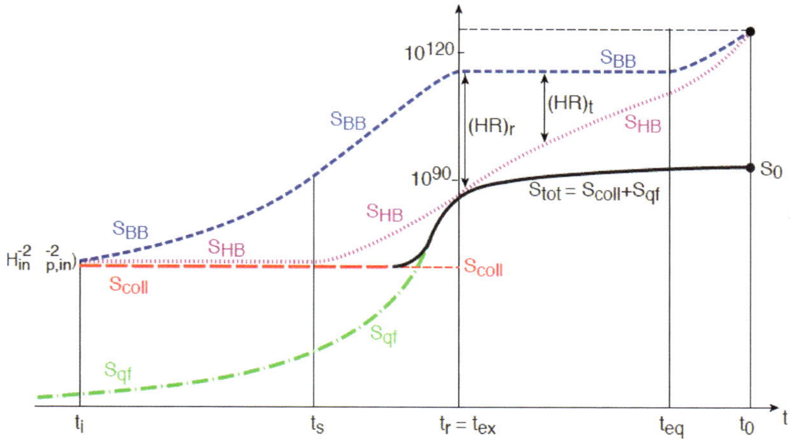

Fig. 2. Diagram illustrating the evolution of various entropy bounds and of actual entropy in the pre-big bang scenario. The tightest reliable bound is the Hubble bound (or its improved form in Ref. 26). It is dangerously threatened by the copious entropy production in amplified quantum fluctuations unless a curvature bounce occurs.

perturbations is strongly blue tilted ($n_s = 4$, $n_T = 3$)[e] and the entropy-density σ generated by this irreversible particle production is of $O(N_{eff}H^3)$ where N_{eff} is the effective number of light species that is produced (dilatons, gravitons, axions, gauge fields, moduli). This is negligibly small with respect to the HUB ($\sigma_{\rm HUB} = Hl_P^{-2} = Hl_s^{-2}g_s^{-2}$) as long as one is in the low-curvature and/or weak string coupling regimes. This phase is illustrated in the left-most part of Fig. 2.

However, dilaton driven inflation is an evolution towards both strong curvature and strong coupling. In our example one first reaches string-scale curvatures and one can argue[31] that the evolution gods towards a de Sitter attractor and constant dilaton-increase rate. Eventually one reaches a point at which

$$\frac{\sigma}{\sigma_{\rm HUB}} \sim N_{eff}g_s^2 \to O(1)\,. \tag{22}$$

At this point the only way to keep satisfying the bound appears to be an exit to an expanding Universe in the Einstein frame. At exit the temperature is expected to be of order M_s and the HUB is saturated with $g_s^2 \sim N_{eff}^{-1}$, which may be pretty close to estimates of $(M_s/M_P)^2$.

A picture of the evolution of the Universe while the curvature is frozen at the string scale and the coupling is still growing was attempted in Ref. 32 (see also Ref. 33) in terms of a closely packed gas of string holes which, eventually, would explode into radiation when the condition (22) is reached. From there on the evolution will be essentially standard with the HUB becoming increasingly loose leaving plenty of room for further entropy production. From a more microscopic point of view it is believed that this exit is induced by a combination of α' and

[e] The usual nearly scale-invariant adiabatic density perturbations have to be produced via the so-called curvaton mechanism, see Refs. 29 and 30.

string-loop corrections but these are hardly under control. We can thus appreciate here having entropy bounds as qualitative tools in spite of the fact that they are only conjectural. For another model of a big crunch based on entropy bounds see Ref. 34.

8. A Highly Speculative Application to Black-Hole Interiors

We want to finish this contribution with an even more speculative use of the BB and its descendants: an attempt to describe qualitatively the interior of a black hole assuming that is saturates locally some entropy bound. Since all the bounds we have described contain Planck's constant it is clear that we need to improve on a classical GR description of a black hole interior.

The starting point of this picture is the already mentioned string-black hole correspondence[15,16] for a particular class of strings of mass of mass $0(M_s g_s^{-2})$ and size of $O(l_s)$. As we already argued lighter strings are *not* black holes. How about heavier ones? The naive string entropy calculation would immediately imply that there are too few such strings to be able to match the entropy of black holes. One can try to argue, along the lines of Ref. 16, that self gravity effects reduce the physical mass of such heavy strings (provided they are sufficiently compact) thus increasing they number per unit mass and that such a mechanism may allow for the two entropies to match even for $M \gg M_s g_s^{-2}$. At a closer scrutiny, however, one realizes that this is very difficult to achieve (see, however, a recent attempt along these lines[35]). A better way appears to be to try and get the needed entropy not by a single heavy string but by a quantum system consisting of many gravitationally interacting light ones.

Let us then start increasing the mass of a single string-hole by adding to it a light string of energy E imposing it to be gravitationally bound to the SH and to be in thermal equilibrium with it by having $E \sim M_s$, the BH temperature of a string hole. The binding energy for the light string will be of order $GEM_s g_s^{-2}/d = El_s/d$ if d is the (average) distance between the light and heavy string. Thus, the two strings form a bound state if $d \sim l_s$. In other words the light string has to be closely packed w.r.t. the string hole.

Extending trivially the above procedure, one can then argue that by adding a distribution of light strings with a phase space distribution of the form:

$$\frac{dN}{d^3x d^3k} = \Phi(\omega r)\frac{r^2}{l_P^2}\,, \quad r = |\vec{x}|\,, \quad \omega = |\vec{k}|\,, \tag{23}$$

with $\Phi(\omega r)$ an appropriate distribution (e.g. a Planck distribution at temperature $\hbar/r$) endowed with an exponential cutoff above $\omega r = 1$, the number of light strings lying at a distance r equals the entropy of a BH of size r. Also each new "shell" of light strings at radius r is gravitationally bound to the inner shells and the local entropy bound (18) is saturated.

The resulting picture has therefore many appealing features. Two challenges remain to make it more than just a picture:

The first is whether one can construct an explicit metric corresponding to such a bound state. Probably the answer, based on many previous studies of similar problems in GR (see for instance Ref. 36) is that one cannot do so "classically." The object is a truly quantum state and any classical description of it is doomed. Indeed, recent attempts[35,37] try to take quantum effects into account in order to avoid some no-go theorems in GR.

The second question is how one can produce such an object from a gravitational collapse. Here the answer could be that the classical evolution of the interior of a collapsing object, after the horizon has formed, is like that of a big crunch with the singularity at $r = 0$ playing the rôle of a (space-like) big-crunch singularity. However, as we have already mentioned an accelerated contraction, like an accelerated expansion, leads to particle and entropy production. In order to maximize the latter the few initial high-energy, low-entropy quanta that go into forming (or accreating) the black hole should be converted, at very high curvatures, into many soft quanta (with a large increase of entropy) forming eventually the kind of quantum bound state I just described.[f]

We are now entering the domain of wild speculations about what the Bekenstein bound can eventually teach us. I should better, stop although I am sure this will not be the case for the many applications still to come of the Bekenstein entropy bound and its variations.

Acknowledgments

During the many visits I made to the Hebrew University of Jerusalem (the physics department as well as the Institute for Advanced Studies) I always looked for Jacob and tried to find out what he was up to. Each time I came out enriched by our conversation in which he was presenting his views in his typical modest — yet firmly convinced — way. I will deeply miss him each time I will visit again the Givat Ram Campus.

I would like to thank Eliezer Rabinovici for many useful remarks that greatly helped improve a first draft of this contribution.

References

1. J. D. Bekenstein, "Black holes and entropy," *Phys. Rev. D* **7**, 2333 (1973).
2. S. W. Hawking, "Particle creation by black holes," *Commun. Math. Phys.* **43**, 199 (1975) [Erratum *ibid.* **46**, 206 (1976)].
3. S. W. Hawking, "Breakdown of predictability in gravitational collapse," *Phys. Rev. D* **14**, 2460 (1976).

[f]This picture is not completely unrelated to the one advocated sometime ago by Dvali and Gomez.[38]

4. J. D. Bekenstein, "A universal upper bound on the entropy to energy ratio for bounded systems," *Phys. Rev. D* **23**, 287 (1981).
5. H. Casini, "Relative entropy and the Bekenstein bound," *Class. Quant. Grav.* **25**, 205021 (2008) [arXiv:0804.2182 [hep-th]].
6. G. 't Hooft, "Dimensional reduction in quantum gravity," *Conf. Proc. C* **930308**, 284 (1993) [gr-qc/9310026].
7. L. Susskind, "The World as a hologram," *J. Math. Phys.* **36**, 6377 (1995) [hep-th/9409089].
8. S. Fubini and G. Veneziano, "Level structure of dual-resonance models," *Nuovo Cim. A* **64**, 811 (1969).
9. K. Bardakci and S. Mandelstam, "Analytic solution of the linear-trajectory bootstrap," *Phys. Rev.* **184**, 1640 (1969).
10. G. Veneziano, "A stringy nature needs just two constants," *Europhys. Lett.* **2**, 199 (1986).
11. D. Amati, M. Ciafaloni and G. Veneziano, "Can space-time be probed below the string size?," *Phys. Lett. B* **216**, 41 (1989).
12. M. J. Bowick, L. Smolin and L. C. R. Wijewardhana, "Role of string excitations in the last stages of black hole evaporation," *Phys. Rev. Lett.* **56**, 424 (1986).
13. L. Susskind, "Some speculations about black hole entropy in string theory," in *The Black Hole*, ed. C. Teitelboim, 118–131 [hep-th/9309145].
14. G. Veneziano, "Hagedorn's temperature and the dual resonance model: A 25 year old love affair," in *Proceeding Nato Advanced Research Workshop "Hot Hadronic Matter: Theory and Experiment*, Divonne, France, June 1994.
15. G. T. Horowitz and J. Polchinski, "Selfgravitating fundamental strings," *Phys. Rev. D* **57**, 2557 (1998) [hep-th/9707170].
16. T. Damour and G. Veneziano, "Selfgravitating fundamental strings and black holes," *Nucl. Phys. B* **568**, 93 (2000) [hep-th/9907030].
17. J. L. F. Barbon and E. Rabinovici, "Touring the Hagedorn ridge," in *From Fields to Strings*, Vol. 3, Shifman, M. (ed.) *et al.* 1973–2008 [hep-th/0407236].
18. W. Fischler and L. Susskind, "Holography and cosmology," hep-th/9806039.
19. J. D. Bekenstein, "Is the cosmological singularity thermodynamically possible?," *Int. J. Theor. Phys.* **28**, 967 (1989).
20. G. Veneziano, "Pre-bangian origin of our entropy and time arrow," *Phys. Lett. B* **454**, 22 (1999) [hep-th/9902126].
21. B. J. Carr and S. W. Hawking, "Black holes in the early Universe," *Mon. Not. Roy. Astron. Soc.* **168**, 399 (1974).
22. B. J. Carr, "The Primordial black hole mass spectrum," *Astrophys. J.* **201**, 1 (1975).
23. R. Easther and D. A. Lowe, "Holography, cosmology and the second law of thermodynamics," *Phys. Rev. Lett.* **82**, 4967 (1999) [hep-th/9902088].
24. D. Bak and S. J. Rey, "Cosmic holography," *Class. Quant. Grav.* **17**, L83 (2000) [hep-th/9902173].
25. R. Bousso, "A Covariant entropy conjecture," *JHEP* **9907**, 004 (1999) [hep-th/9905177].
26. R. Brustein and G. Veneziano, "A Causal entropy bound," *Phys. Rev. Lett.* **84**, 5695 (2000) [hep-th/9912055].
27. E. E. Flanagan, D. Marolf and R. M. Wald, "Proof of classical versions of the Bousso entropy bound and of the generalized second law," *Phys. Rev. D* **62**, 084035 (2000) [hep-th/9908070].
28. G. Veneziano, "A quantum universe before the big bang(s)?," *J. Phys. Conf. Ser.* **880**, 012001 (2017).

29. M. Gasperini and G. Veneziano, "The Pre-big bang scenario in string cosmology," *Phys. Rept.* **373**, 1 (2003) [hep-th/0207130].
30. J. E. Lidsey, D. Wands and E. J. Copeland, "Superstring cosmology," *Phys. Rept.* **337**, 343 (2000) [hep-th/9909061].
31. M. Gasperini, M. Maggiore and G. Veneziano, "Towards a nonsingular pre-big bang cosmology," *Nucl. Phys. B* **494**, 315 (1997) [hep-th/9611039].
32. G. Veneziano, "A Model for the big bounce," *JCAP* **0403**, 004 (2004) [hep-th/0312182].
33. J. Quintin *et al.*, to appear.
34. T. Banks and W. Fischler, "Black crunch," hep-th/0212113.
35. R. Brustein and A. J. M. Medved, "Resisting collapse: How matter inside a black hole can withstand gravity," arXiv:1805.11667 [hep-th].
36. H. Bondi, "Massive spheres in general relativity," *Proc. Roy. Soc. Lond. A* **282**, 303 (1964).
37. P. O. Mazur and E. Mottola, "Surface tension and negative pressure interior of a non-singular black hole?," *Class. Quant. Grav.* **32**(21), 215024 (2015) [arXiv:1501.03806 [gr-qc]].
38. G. Dvali and C. Gomez, "Black holes as critical point of quantum phase transition," *Eur. Phys. J. C* **74**, 2752 (2014) [arXiv:1207.4059 [hep-th]].

Entropy, Gravity and Galactic Dynamics

Erik P. Verlinde
Institute of Physics, University of Amsterdam,
Science Park 904, 1098 XH Amsterdam
e.p.verlinde@uva.nl

Jacob Bekenstein's brilliant insights on black hole entropy and entropy bounds have greatly influenced our understanding of gravity and are central to the current development of emergent gravity. In this contribution I will explain the connections between holographic principle, the Bekenstein bound and another important topic of his research: the Milgromian dynamics of galaxies.

Bekenstein's Argument for the Black Hole Entropy formula

The name of Jacob Bekenstein will for ever be connected to the celebrated formula for the entropy of black holes, which he found together with Stephen Hawking.[1,2] It takes the well-known form

$$S = \frac{Ac^3}{4G\hbar} \tag{1}$$

where A stands for the area of the horizon. This beautiful formula combines all the constants of Nature and connects the geometry of space with the microscopic entropy of the underlying theory. It is taking physicists a lot of time to understand and appreciate the precise meaning of this formula: this scientific process that is still not fully completed.

The original arguments of Jacob Bekenstein that led him to conclude that black holes carry an entropy proportional to the horizon area are among the most insightful and conceptually deep ideas that I have encountered in my career. Bekenstein's reasoning have influenced my own thinking a lot and play a crucial role in developing the theory of emergent gravity, together with the results of Hawking and Unruh[2,3] for the temperature at the horizon. In particular, the fact that the horizon temperature is expressed in terms of the surface acceleration g as

$$T = \frac{\hbar g}{2\pi c} \tag{2}$$

gives an important clue towards the origin of the gravitational force.

In its simplest form Bekenstein's argument goes as follows. He postulated that when a particle with mass m is gradually lowered towards a black hole and is dropped at one Compton wavelength from the horizon, the entropy increases by precisely one unit. It follows from the Einstein equations that in that case also the horizon area increases precisely by one unit measured in Planck units. In this way one reaches the conclusion that the area and entropy are linearly related. By continuing this argument he could also already have predicted that the temperature of a black hole is proportional to the surface acceleration. Namely, the force that is applied to the mass when it is near the horizon is simply

$$F = mg\,. \tag{3}$$

The change in the mass, or energy, of the black hole, when the particle is dropped from one Compton wavelength is equal to the amount of work

$$F\Delta x = mg \times \frac{\hbar}{mc}\,. \tag{4}$$

Next one uses the first law of thermodynamics to relate the work to the change in entropy via

$$F\Delta x = T\Delta S\,. \tag{5}$$

It then follows from the postulate that ΔS is of order one that the temperature is proportional to $\hbar g/c$, which is indeed the Hawking–Unruh result. To get the correct normalization one must fix the change ΔS of the entropy to be equal to 2π.

The Bekenstein Bound and the Role of Inertia

Around 2009 I started thinking about how gravity and the inertial force or Newton might emerge from a microscopic perspective using entropic arguments. I realized quickly that one can reverse Bekenstein's reasoning to derive the inertial force from the known results for the horizon temperature.[4] His entropy postulate can be reformulated as saying that when we drop a particle from a small distance Δx to the horizon, that the entropy change ΔS is equal to the ratio of Δx and the Compton wavelength. In other words, one postulates the formula

$$\Delta S = 2\pi \frac{mc\Delta x}{\hbar}\,. \tag{6}$$

One easily verifies by multiplying both sides by the Hawking–Unruh temperature that one recovers the expression for the work done by the inertial force $F = mg$.

The above formula for the entropy change of the horizon is equivalent to the famous Bekenstein bound on the entropy of a system of size R and mass M. Following essentially the same reasoning he concludes that the maximal entropy equals

$$S_B = 2\pi \frac{McR}{\hbar}\,. \tag{7}$$

I always felt that this formula was under-appreciated by many physicists. This was partly due to the fact that the reasoning behind it was kind of intuitive, and also that the precise normalization constant is not easily fixed. Thirdly, the issue of the species problem made some people believe that it cannot hold in general.

A more modern version of the bound was found by Casini,[5] who related the Bekenstein bound to the notion of relative entropy. By the first law of entanglement entropy he proved a general bound on the change in entropy in terms of the modular Hamiltonian K. In particular cases K may be identified with the boost generator in the local flat Minkowski geometry along the horizon. His work also provided a resolution of the species problem.

It is striking that the Bekenstein bound is independent of Newton's constant, even though its derivation seems to use the laws of gravity. However, on closer inspection it only employs the law of inertia. Indeed, it follows again from equating the work done by the inertial force with the change in energy computed using the first law of thermodynamics. The fact that both the temperature as well as the inertial force are proportional to the surface gravity g ensures that the result for the bound is independent of g.

The idea to derive the Einstein equations from the holographic principle goes back to Jacobson.[6] The crucial insight that I learned from Bekenstein is that, before trying to derive the Einstein equations, one should first understand the origin of the inertial force. Inertia is a more fundamental concept than gravity. In fact, gravity can be seen as a modification of the laws of inertia caused by the fact that the Bekenstein-Hawking entropy is finite. In recent years it has become clear that this entropy formula also applies to Rindler horizons, and then describes the entanglement entropy of the microscopic theory. This interpretation is well understood in Anti-de Sitter space, where it is leads to the Ryu–Takayanagi formula.[7]

Emergent Gravity and Galactic Dynamics

More interesting are the implications for de Sitter space, since this case is closer to our own Universe. The cosmological horizon also carries an entropy that is given by same formula

$$S_{dS} = \frac{A(L)c^3}{4G\hbar} \qquad \text{where} \qquad L = \frac{c}{H_0} \tag{8}$$

equals the Hubble radius and $A(L) = 4\pi L^2$ is the horizon area. If we approximate our current universe as given by an exact de Sitter geometry, the horizon entropy would be enormous and would be of the order of 10^{123}. The horizon temperature of de Sitter space is is given by the Gibbons–Hawking result[8]

$$T = \frac{H_0}{2\pi}\,. \tag{9}$$

For our universe this temperature is immeasurably small. Yet, as I will argue, the presence of the entropy and the temperature have observable consequences.

The precise interpretation of this entropy is still a topic of scientific debate. In the black hole context it can be interpreted as the entanglement entropy across the horizon, as well as the logarithm the number of microscopic states. This distinction is related to the question whether one is dealing with an eternal two-sided black hole geometry, or a black hole that is formed by gravitational collapse: in the first case the entropy can be seen as measuring the amount of entanglement, while in the latter situation it counts the possible microscopic quantum states. In the de Sitter case either of these interpretations is possible.

Soon after having developed the idea of entropic gravity I became aware of the Milgrom-phenomenology of galaxy rotations curves, a research topic in which Bekenstein also played an important role. I will present it here in a way in which it connects to the previous topic. First of all, observationally the flattening of rotation curves in galaxies occurs when the gravitational acceleration drops below a value (up to small numerical constant) given by

$$\frac{GM}{R^2} < \frac{cH_0}{2}. \tag{10}$$

When the acceleration is much smaller, the observed acceleration does no longer obey Newton's law in terms of the observable baryonic matter M, but obeys a scaling relation of the form

$$g^2 = \text{const.} \times \frac{GM}{R^2}\, cH_0\,. \tag{11}$$

Observationally the constant is of the order of $1/6$, as was emphasized by Milgrom and Bekenstein.[9,10] These two equations play a central role in the Milgrom phenomenology and led to the proposal of modified Newtonian dynamics.

Both relations have a very natural interpretation in terms of emergent gravity. First, he criterion that sets the acceleration scale can be rewritten as[11]

$$2\pi\frac{McR}{\hbar} < \frac{A(R)c^3}{4G\hbar}\cdot\frac{R}{L}. \tag{12}$$

On the left-hand side we recognize the Bekenstein entropy associated with the mass M inside a region of size R. The right-hand side equals the Bekenstein–Hawking entropy of a black hole of size R times the fraction R/L. This total quantity therefor grows like the volume instead of the area. Furthermore, it equals the full entropy of de Sitter space when we take $R = L$. Hence, the natural interpretation of this formula is that it describes the contribution of an entropy density that when integrated gives the total de Sitter entropy. Indeed, as I argued in my paper on emergent gravity and the dark universe,[11] it describes the entropy that is carried by the dark energy in de Sitter space. After all, it is the addition of dark energy that led to the appearance of the cosmological horizon in the first place.

The scaling relation (11) for the observed acceleration in terms of the observable matter is also naturally rewritten as

$$\frac{1}{4\pi G}\int_{r\leq R} g^2 dV = \text{const.}\; 2\pi\frac{McR}{\hbar}\cdot\frac{H_0}{2\pi} \tag{13}$$

where we assumed that the mass M is concentrated near the origin. The left-hand side is equal to twice the gravitational self-energy of the galaxy contain in a sphere of radius R. On the right-hand side we again find the Bekenstein entropy, but now multiplied with the de Sitter temperature. It turns out that this relation can also be interpreted in the context of emergent gravity. For details see (11). I believe that Jacob Bekenstein indeed envisaged that Milgromian dynamics somehow must be related to entropic dynamics of de Sitter space. In particular he and Milgrom noted the relationship with the Unruh–Hawking formula for the de Sitter temperature and the typical acceleration scale in galactic dynamics. My work was an attempt to make this connection more explicit.

Conclusion

The last time I saw Jacob Bekenstein was during a conference in January 2014 at the Weizman institute. After a discussion with Moti Milgrom on galactic dynamics, Jacob and I walked to the conference-site and spoke our views on physics. I am happy that shortly after his profound contributions were recognized with the Einstein prize of the APS. The following paragraph is from my letter that supported his nomination.

The deep insights of Bekenstein in the nature of gravity, and in particular his work on black holes, have reshaped theoretical physics in a very profound way. We are currently witnessing a very important shift in our theoretical understanding of gravity. In my view these developments will turn out to be just as important as the development of quantum mechanics, and will eventually lead to a totally new paradigm for gravity, cosmology and the origin of space-time. Bekenstein's brilliant discovery of black hole entropy stands at the beginning of this development.

References

1. J. D. Bekenstein, Black holes and entropy, *Phys. Rev. D* **7**, 2333–2346 (1973).
2. S. W. Hawking, Particle creation by black holes, *Commun. Math. Phys.* **43**, 199–220 (1975).
3. W. G. Unruh, Notes on black hole evaporation, *Phys. Rev. D* **14**, 870 (1976).
4. E. P. Verlinde, On the origin of gravity and the laws of Newton, *JHEP* **1104**, 029 (2011).
5. H. Casini, Relative entropy and the Bekenstein bound, *Class. Quant. Grav.* **25**, 205021 (2008).
6. T. A. Jacobson, Thermodynamics of spacetime: The Einstein equation of state, *Phys. Rev. Lett.* **75**, 1260–1263 (1995).
7. S. Ryu and T. Takayanagi, Holographic derivation of entanglement entropy in AdS/CFT, *Phys. Rev. Lett.* **96**, 181602 (2006).
8. G. W. Gibbons and S. W. Hawking, Cosmological event horizons, thermodynamics, and particle creation, *Phys. Rev. D* **15**, 2738–2751 (1977).
9. M. Milgrom, A modification of the Newtonian dynamics — Implications for galaxies, *Astrophys. J.* **270**, 371–389 (1983).
10. J. D. Bekenstein and M. Milgrom, Does the missing mass problem signal the breakdown of Newtonian gravity? *Astrophys. J.* **286**, 7–14 (1984).
11. E. P. Verlinde, Emergent gravity and the dark universe, *SciPost Phys.* **2**(3), 016 (2017).

Other Contributions in Memory of Jacob Bekenstein

Jacob Bekenstein had really broad scientific interests. In this section we have collected some of the contributions that describe other interests of Jacob and also contributions that the authors find important to publish in a memorial volume like this one.

From the memories of his former student Ofer Lahav the reader will perhaps be somewhat surprized to find that the topic which Jacob formulated for his Master thesis was entitled "Cooling processes of formation of the first stars in the universe."

Yaron Oz in his paper derives a new analytical formula for the anomalous scaling in incompressible fluid turbulence. His interest in this problem began with a study of the classical dynamics of the horizon of black holes and was inspired by discussions with Jacob Bekenstein.

One of the Jacob's research fields is related with relativistic formulation of Modified Newtonian Dynamics theories (MOND). In his paper Mordehai Milgrom gives a comprehensive review of such theories and presents an interesting account of his discussions and collaboration with Bekenstein on this topic.

In the paper by Tsvi Piran (with Kenta Hotekezaka), where the origin of the merging black hole binaries detected by LIGO is discussed, the interest of Jacob Bekenstein in astrophysics and his not so well known papers on these subject are pointed out.

Finally, the papers by two close friends of Jacob, Eduardo Guendelman (with Emil Nissimov and Svetlana Pacheva) and Masud Chaichian (with Markku Oksanen and Anca Tureanu) are devoted to cosmology and induced gravity respectively.

Jacob Bekenstein's Universes: From Black Holes to Modified Theories of Gravity

Ofer Lahav
Department of Physics & Astronomy,
University College London, Gower Street, London WC1E 6BT, UK
o.lahav@ucl.ac.uk

Jacob Bekenstein's pioneering contributions to the foundations of Gravity cover a broad range of scales, from Black Holes to the entire Universe. He is well known as the father of Black Hole Thermodynamics and the inventor of the TeVeS modified gravity theory. I reflect here on my interaction with him, from the time he supervised me for my MSc project at Ben-Gurion University in the early 1980s to his sabbatical visit at UCL in 2014, about a year before he passed away (sadly just a month before the first LIGO discovery of Gravitational Waves from a Binary Black Hole). His quote "I feel more comfortable in the world if I understand why things are happening" reflects the way he viewed the world as a Physicist as well as his personal belief. He also allegedly said "Sometimes theories are smarter than their inventors."

1. Introduction

On the early morning of 17 August 2015, I was about to catch a Eurostar train to France. A sad email arrived that Jacob Bekenstein had passed away unexpectedly the previous day of a heart attack, while visiting Helsinki. He was only 68 years old. Bekenstein was the father of "Black Hole Thermodynamics". I had the privilege to be supervised by him for my MSc degree back in Ben-Gurion University in Beer-Sheva Israel in the early 1980s. He visited me at University College London (UCL) on his sabbatical term just a year before he passed away. Jacob was born in Mexico in 1947. After studying in New York and Princeton he was a Post-doc in Texas. He made an unusual career move to immigrate to Israel, to the Ben-Gurion University in Beer-Sheba (then the youngest university in the country). In recognition of his pioneering work Bekenstein received the Wolf Prize, the Israel Prize, the Einstein Prize and other awards.

2. The First Stars in the Universe

My MSc thesis with Bekenstein was not on Black Holes, but on the formation of the first stars in the Universe. In a memorial service for Jacob at the Israeli Academy of Sciences in Jerusalem (in July 2016) I recalled meeting him for the first time. After three years studying for a BSc in Physics at Tel Aviv University, I was in the 'Atuda' (a scheme which allows one to study at university first, before doing military service). I served for five years at a base of the Israeli Air force near Beer-Sheba. It made good sense, as a way of keeping my sanity in a five-year service, to enrol at nearby Ben-Gurion University for an MSc degree in Astrophysics. It was highly recommended to me to have Bekenstein as my project supervisor. On a first meeting with him, in a modest office he shared with David Owen (who later lectured to me), Bekenstein said he was not available for taking on a student, as he was busy working himself on a research project (later I understood it was about his ideas to modify Einstein's gravity, to overcome the dark matter problem, see below). I am not sure how I had the courage to approach him again some weeks later, and what made him change his mind. He was actually very generous with his time, and very kind to me. Taking into account that I was a part-time student, he 'cooked' for me a project on a 'quiet topic', which he correctly envisaged would become more relevant in years to come. The project was entitled "cooling processes of formation of the first stars in the Universe".

The question was, given the limited chemical reactions possible in the primeval Universe, how could a cloud of gas collapse? The fate of a collapsing cloud depends on the competition between gravity and pressure. For the gravity to win over pressure, the pressure has to be reduced by cooling of the gas. It is believed the cooling can be achieved by molecular hydrogen. This work resulted in a paper.[1] I was hoping to have Bekenstein as a co-author, but he declined saying he had not done 50% of the work. Later this became an important topic, with these reactions implemented in sophisticated hydrodynamical simulations. He always emphasized the importance of *thinking* about the science question. Straight after completing both the MSc and my service in the Air Force I moved to study for my PhD at the University of Cambridge, where I was mentored by George Efstathiou and by Donald Lynden-Bell (who sadly passed away in February 2018). I then combined Post-doctoral fellowships in Cambridge and in Princeton, before becoming a member of staff at the Institute of Astronomy in Cambridge. In 2004 I moved to UCL as Perren Chair of Astronomy, to start there the Cosmology area. Without the foundations I received by working with Bekenstein I doubt that my career in Astrophysics would have developed in such a fortunate way.

3. Black Hole Thermodynamics

Bekenstein is recognized as a pioneer of the idea of 'Black Hole Entropy'. This fairly abstract concept has led to one of the most important equations in the whole of Physics. This brings together two topics: The first, Entropy, is of great relevance to daily life. The second, Black Holes, is very remote from human experience. Entropy, usually labelled by the letter S, is the measure of 'disorder' (or randomness) of a system. A common example in explaining Entropy is the way a perfume sprayed in one corner of a room (in an 'ordered' state) will eventually be distributed in the entire room (in a 'disordered' state). The reverse process has never been observed. The Second Law of Thermodynamics states that the amount of Entropy (disorder) of an isolated system can never decrease over time. Ludwig Boltzmann (1844–1906) formulated Entropy as

$$S = k_B \ln(W)\,, \tag{1}$$

where W is the number of states of the system, and k_B is Boltzmann's constant. This equation appears on Boltzmann's gravestone in Vienna.

The other part of the story is that of Black Holes. In short, a Black Hole is a compact region with such strong gravity that nothing can escape from it, not even light. What is the connection between Entropy and Black Holes? Until the early 1970s the accepted picture was that a Black Hole hides everything. John Wheeler, one day in the early 1970s, allegedly asked his then PhD student Jacob Bekenstein, "What would happen if you poured a cup of hot tea into a Black Hole?" This question may sound naive, but is actually deep and inspiring. Bekenstein rescued the Second Law of Thermodynamics from an embarrassment, by allowing the Entropy of the cup of tea to be converted to the Black Hole Entropy (proportional to the Event Horizon area A). This way the Entropy of the Black Hole system never decreases.

In his influential 1973 paper[2] Bekenstein wrote, "There are a number of similarities between Black-Hole physics and thermodynamics. Most striking is the similarity in the behaviours of Black-Hole area and of Entropy: both quantities tend to increase irreversibly." The Black-Hole Entropy was eventually written as:

$$S_{BH} = \frac{k_B A}{4 l_P^2}\,, \tag{2}$$

where $l_p = \sqrt{G\hbar/c^3}$ is the Planck length. The history of the Bekenstein & Hawking iterations leading to final form is interesting by itself, but beyond the scope of this article. I will only add that I recall in the late 1980s a lecture by Denis Sciama in Cambridge. Sciama (1926–1999), the father of the modern British cosmology supervised Stephen Hawking and Martin Rees, among many others. In that lecture Sciama wrote on the blackboard the equation for S_{BH} and said that the subscripts 'BH' may stand for either 'Black Hole' or 'Bekenstein-Hawking'. The equation for the Black Hole temperature appears on Stephen Hawking's gravestone in Westminster Abbey. It is not obvious why the Black Hole Entropy depends on the area (or the boundary) rather that on the volume. This may remind us of Gauss' theorem, that the electric fields integrated over the surface tells us about the total electric charge within the boundary. It is a kind of topological property. Black Hole Entropy has led physicists to speculate that anything that happens in a volume of space-time can be described by the information on the boundary (surface) of that volume. This is called the 'Holographic Principle', and Jacob Bekenstein published several papers on this and on the connection to information theory. An overview is given in Bekenstein's book[3] and in his Scientific American article.[4]

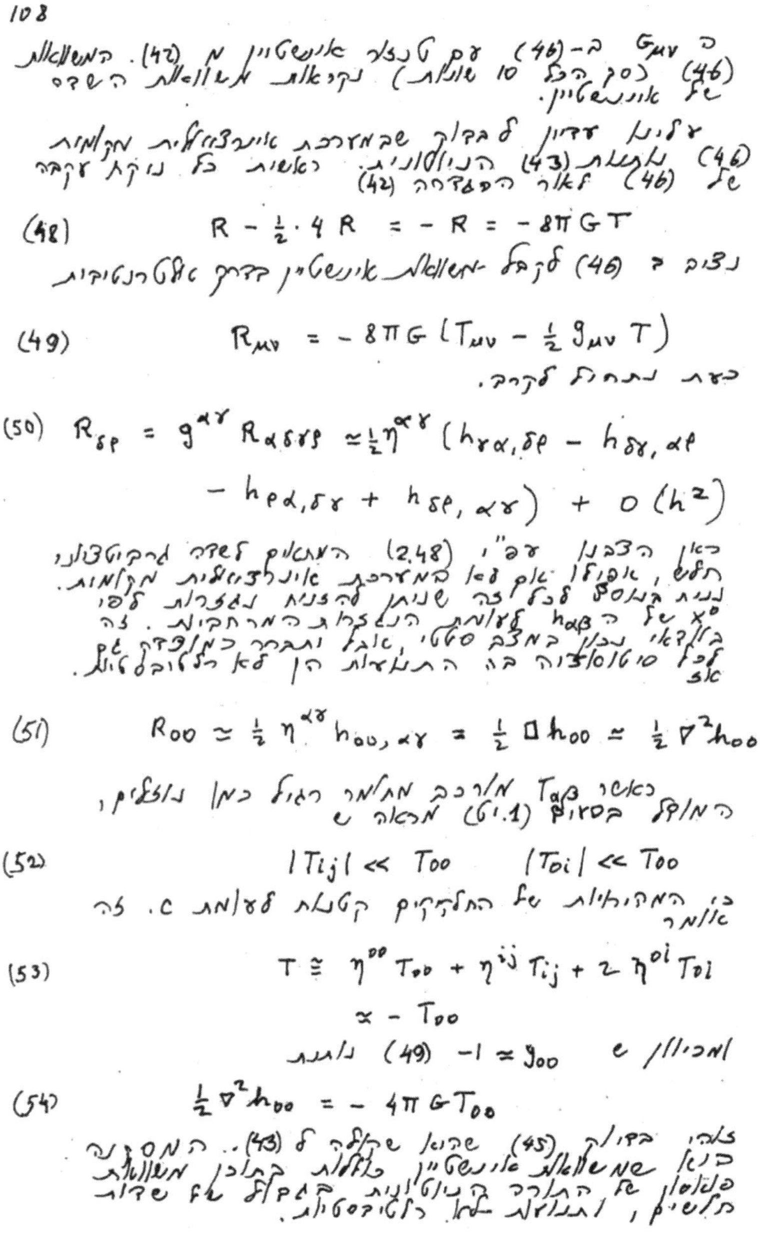

Fig. 1. Notes in Hebrew from Bekenstein's lectures on General Relativity given at the Hebrew University. Taken from his website http://www.phys.huji.ac.il/∼bekenste/.

4. The TeVeS Modified Gravity

In the early 1980s Bekenstein was inspired by Moti Milgrom's idea[6] to explain flat galaxy rotation curves by modifying Newtonian Dynamics (MOND), without the need to invoke Dark Matter. After various attempts to generalise MOND to a relativistic theory, Bekenstein devised TeVeS (Tensor, Vector, Scalar). It is derived from the action principle, so it obeys conservation laws. As a relativistic theory, it can predict effects of gravitational lensing (in both the weak and strong regimes). TeVeS has been praised as remarkable theoretical achievement, extending Einstein's General Relativity. However, in its simplest form TeVeS has been challenged in explaining for example the Cosmic Microwave Background peaks, and the Bullet Cluster.

Sadly, Bekenstein did not witness the LIGO discovery of Gravitational Waves. The first event GW150914, of a Binary Black Hole, was detected by LIGO[7] on 14 September 2015, just a month after Bekenstein passed away. The LIGO & Virgo detection[8] of Binary Neutron Star (BNS) GW170817 and electromagnetic followups turned out to be relevant for constraining cosmological modes. Several ground and space telescopes, including our Dark Energy Survey camera[9,10] in Chile captured this Gravitational Wave 'after glow'. In particular, the Fermi Gamma-ray space telescope detected a flash from BNS system only 1.7 seconds after the Gravitational Wave measurement. Given the distance of about 40 Mpc to the host galaxy NGC4993, this implies that the speed of Gravitational Waves is close to the speed of light to within one part in 10^{15}. In TeVeS the differences between the speed of gravity and light will be due to coupling of the curvature to other fields.[11] This may put severe constraints on TeVeS in its simplest form, although some other variants of TeVeS may survive that test. Indeed, it was shown recently[12] that in a previously unknown class of models, inspired by TeVeS, the speed of gravitational waves equals to the speed of light. More work is needed for a fair comparison of TeVeS and its extensions with current and future observational data.[13]

5. Reflections

Bekenstein was very methodological and very precise in both English and in Hebrew, orally as well as in writing. We told me once how he mastered the Hebrew language, which was not his native language. He read the Hebrew version of the Sears & Zemansky, that was the standard High-School Physics textbook in my generation. As he knew the Physics, he could very easily pick up the Hebrew words...

It was a pleasure to host Jacob and his wife Bilha in London for four months in 2014, a year before he passed away. We discussed various science topics, in what seems now as 'closing the loop' more than 30 years after working with him back at Ben-Gurion University.

In a popular article[14] we quoted different views of philosophers on how paradigm shifts happen. This included a a reference to the French physicist and philosopher of

Fig. 2. Jacob Bekenstein and Ofer Lahav at the main court of UCL, in July 2014.

science Duhem (1861–1916), who suggested that scientists follow their 'good sense' when facing dilemmas of this kind. Bekenstein commented by email to me on 5 May 2014:

Ofer

Thanks. Interesting paper. Some examples were not known to me. Neither some views. If you ask me, Duhem's is the most sensible point of view. He was a professional physicist and left a piece of useful physics, the Gibbs-Duhem equation. Personally I think the other views are stylized and bear little relation to how scientists actually work and reason. As for the central question what to do with Dark Energy and Dark Matter, as you say the evidence will eventually be overwhelming one way or another.

Jacob

This gives an insight to the way Bekenstein considered the way theories describe nature. In an interview in Hebrew he said: "I feel more comfortable in the world if I understand why things are happening." This reflects the way he viewed the world as a Physicist and his personal belief. Allegedly he also said "Sometimes theories are smarter than their inventors." Jacob Bekenstein is greatly missed in the new era of Gravitational Wave Physics and cosmological surveys of the deep Universe.

Acknowledgments

I thank Ed Copeland, Scott Dodelson, Pedro Ferreira, Moti Milgrom, Jonathan Oppenheim and Tom Zlosnik for helpful discussions on the interpretation of Bekenstein's theories in the light of recent observations. I also acknowledge support from a European Research Council Advanced Grant FP7/291329 and from the UK Science and Technology Research Council (STFC) Grant No. ST/M001334/1.

References

1. O. Lahav, *MNRAS* **220**, 259 (1986).
2. J. D. Bekenstein, *Phys. Rev. D* **7**, 2333 (1973).
3. J. D. Bekenstein, *Of Gravity, Black Holes and Information* (Di Renzo Editore, Italy, 2003).
4. J. D. Bekenstein, *Scientific American* **289**, 61 (2003).
5. J. D. Bekenstein, *Phys. Rev. D* **70**, 083509 (2004).
6. M. Milgrom, *ApJ* **270**, 365 (1983).
7. B. P. Abbott *et al.*, *Phys. Rev. Lett.* **116**, 061102 (2016).
8. B. P. Abbott *et al.*, *Phys. Rev. Lett.* **119**, 161101 (2017).
9. M. Soares-Santos *et al.*, *ApJ Lett.* **848**, L16 (2017).
10. A. Palmese *et al.*, *ApJ Lett.* **849**, L34 (2017).
11. E. Sagi, *Phys. Rev. D* **81**, 064031 (2010).
12. C. Skordis and T. Zlosnik, arXiv:1905.09465 [gr-qc].
13. J. M. Ezquiaga and M. Zumalacárregui, *Phys. Rev. Lett.* **119**, 251304 (2017).
14. O. Lahav and M. Massimi, *Astronomy & Geophysics* **55**, 3.13 (2014).

Turbulence and Random Geometry

Yaron Oz
Raymond and Beverly Sackler School of Physics and Astronomy, Tel Aviv University, Tel Aviv 69978, Israel
yaronoz@post.tau.ac.il

We outline our proposal for a field theory description of steady state incompressible fluid turbulence at the inertial range of scales in a general number of space dimensions. The theory consists of a Kolmogorov linear scaling mean field theory dressed by a Nambu–Goldstone dilaton mode that induces a random measure on the inertial range. We derive a KPZ-type formula for the anomalous scalings of the velocity structure functions, the velocity gradients and the local energy dissipation, and relate the dimensionless intermittency parameter to the boundary conformal anomaly.

Dedicated to the memory of Jacob Bekenstein

1. Introduction

Jacob Bekenstein was a deep thinker. When Jacob was a Ph.D. student thinking about black hole mechanics the mathematics of the black hole laws was known. Bardeen, Carter and Hawking have already formulated the four laws of black hole mechanics in analogy with the laws of thermodynamics, where the area A of the event horizon and the surface gravity of the black hole κ being analogous to the entropy and temperature, respectively. In particular, Hawking's theorem on the non-decrease with time of the event horizon area of a black hole $\delta A \geq 0$ and the fact that when two black holes coalesce, the area of the final event horizon is greater than the sum of the areas of the initial horizons $A_3 > A_1 + A_2$ established the analogy with the second law of thermodynamics. Jacob was there to uplift the veil and reveal that this is not just a mathematical result but rather the black hole entropy obeying the second law of thermodynamics. When I asked Jacob about the history of that period and how come all the experts at that time missed this observation, his answer was "Sociology". Jacob's scientific career has been centered around black holes. My interest in turbulence began when studying the classical dynamics of the black hole horizon. This paper is dedicated to Jacob, a deep thinker and a friend.

Most fluid motions in our universe are turbulent. However, despite centuries of research we still lack an analytical description and understanding of fluid flows in the non-linear regime. Turbulence is considered as the most important unsolved problem of classical physics. The turbulent incompressible fluid flows exhibit highly complex spatial and temporal intermittent structures. When considering their statistical average properties a universal structure is revealed in the inertial range of scales that exhibits homogeneity, isotropy and anomalous scalings. Our aim is to gain insight to the nature of the anomalous scalings and construct an analytical formula to calculate them.

In the following we will outline our proposal for a field theory description of steady state incompressible fluid turbulence at the inertial range of scales. We will consider a general number of space dimensions $d \geq 2$, with a particular attention being given to the physical two and three space dimensions. Physically, our theory consists of a Kolmogorov linear scaling mean field theory (K41)[1] dressed by a Nambu–Goldstone (NG) gapless dilaton mode. The NG mode arises from a spontaneous breaking due to the energy flux of the separate scale and time symmetries of the inviscid Navier–Stokes (NS) equations to a K41 scaling with a dynamical exponent $z = \frac{2}{3}$.[2] Mathematically, the dilaton mode induces a random measure on the inertial range and the emerging random geometry structure is reminiscent of two-dimensional quantum gravity,[3,4] with the dilaton being the analog of the Liouville mode.

We will derive a KPZ-type[3,4] analytical formula for the non-linear anomalous scalings of the velocity structure functions, the velocity gradients and the local energy dissipation:

$$\Delta - \Delta_{K41} = \mathcal{G}^2(d)\Delta(1-\Delta)\,, \tag{1}$$

where Δ_{K41} is the K41 linear scaling dimension of the fluid observable and Δ is the exact anomalous dimension. $\mathcal{G}(d)$ is a dimensionless parameter that depends on the number of space dimensions and quantifies intermittency and the deviation from Kolmogorov linear scaling. It can be measured numerically and experimentally and we will relate it to the boundary conformal anomaly of the K41 mean field theory.[2] Formula (1) can be viewed as a physically and mathematically consistent completion[5] of the random cascade Kolmogorov–Obukhov lognormal model[6,7] to the strong intermittency regime. It is in agreement with the experimental and numerical data and provides an infinite number of new predictions.[5]

The inspiration for the proposal comes from classical black hole dynamics. Black holes have a horizon: objects fall into the black hole through the horizon but cannot emerge. This breaks time reversal symmetry and allows Einstein equations to describe dissipative effects (see e.g. Refs. 8 and 9) that are crucial for turbulence. The evolution equations of the horizon are in the non-relativistic limit the NS equations that describe incompressible fluid flows.[10,11] The fluid pressure and velocity denote the deviation from equilibrium of the horizon hypersurface location and its normal, respectively. Since the horizon is characterized by a geometry, the gravity variables suggest that there should be a geometrical framework for studying the dynamics of fluids: *geometrization of turbulence* (for a brief review see Ref. 12).

We can map every solution of the incompressible NS equations to a geometrical configuration of the black hole horizon and its embedding. Hence, studying the statistical properties of fluid flows is mapped to studying the statistics of geometries, i.e. *random geometry*. By a random geometry we will mean changing the Euclidean measure dx on R^d to a random measure $d\mu(x) \sim e^{\gamma\phi(x)}dx$, where the Gaussian random field $\phi(x)$ is log-correlated (for a review see e.g. Ref. 13). We will identify the field ϕ with the dilaton and γ with the intermittency parameter $\mathcal{G}$.[2]

The paper is organized as follows. In Sec. 2 we will construct the field theory of turbulence, while Sec. 3 will be devoted to a discussion and outlook.

2. Field Theory of Turbulence

In this section we construct a field theory of turbulence in the inertial range of scales. The theory consists of a Kolmogorov linear scaling theory dressed by a NG dilaton that induces a random measure on the inertial range of scales.

2.1. *Flux states*

The incompressible NS equations provide a universal description of fluid flows at low Mach number, that is $v \ll v_s$ where v_s is the speed of sound. In flat space they read:

$$\partial_t v^i + v^j \partial_j v^i = -\partial_i p + \nu \partial_{jj} v^i + F^i \,, \qquad \partial_i v^i = 0, \quad i = 1, \ldots, d \,, \tag{2}$$

where $d \geq 2$ is the number of space dimensions, $v^i, i = 1, \ldots, d$ is the fluid velocity vector, p is the fluid pressure, ν is the kinematic viscosity and F^i is an external random force.

In a steady state we have an equality between the incoming energy flux due to the external force and the energy dissipation due to viscosity:

$$\int d^d x (F^i v^i) = \int d^d x \epsilon(x) \,, \tag{3}$$

where

$$\epsilon(x) = \frac{\nu}{2} (\partial_i v^j + \partial_j v^i)^2 \,, \tag{4}$$

is the local energy dissipation.

Fluid flows exhibit turbulence at large Reynolds numbers $\mathcal{R}_e = \frac{l_c v}{\nu} \simeq 10^3$, where l_c is a characteristic length scale and v is the velocity difference at that scale. In such a state the viscous term $\nu \partial_{jj} v^i$ is negligible compared to the nonlinear term $v^j \partial_j v^i$ in the NS equations (2), yet, $\epsilon(x)$ is non-vanishing and balances the incoming flux (3). This is a turbulent flux state. It is far from equilibrium and the Gibbs measure is inappropriate for quantifying its statistics. We will essentially work in the infinite Reynolds number limit.

One defines the inertial range to be the range of distance scales $l \ll r \ll L$, where the scales l and L are determined by the viscosity and forcing, respectively. Although the turbulent flow breaks all the symmetries of the NS equations, turbulence at the inertial range of scales exhibits statistical homogeneity, isotropy and anomalous scalings. We will be interested in the energy cascade steady state statistics and the equal time correlation functions.

2.2. *Spontaneous symmetry breaking of space and time dilatations*

In the absence of a viscosity term, the (inviscid) NS equations (2) exhibit two scale symmetries. These are independent scalings of space and time which we denote by $R_x \times R_t$:

$$R_x : x^i \to e^{\sigma} x^i \,, \quad R_t : t \to e^{z\sigma} t \,, \tag{5}$$

where σ and z are real parameters and an appropriate charge is being assigned to the random force. For any value of z, one can define the total dimension of an object that carries dimensions (r_x, r_t) under $R_x \times R_t$ as

$$\Delta_z = r_x + z r_t \,. \tag{6}$$

The dimension of v^i is $\Delta_z = 1 - z$ and the dimension of the different terms in the inviscid NS equations is $\Delta_z = 1 - 2z$.

The local energy dissipation (alternatively the flux) breaks spontaneously the symmetries of the inviscid NS equations to $R_{z=\frac{2}{3}}$, as follows from the requirement that

$$\Delta_z(\epsilon) = 2 - 3z = 0 \,. \tag{7}$$

This is the Kolmogorov linear scaling $\Delta_{z=\frac{2}{3}} \equiv \Delta_{K41}$:

$$\Delta_{K41}[v^i] = \frac{1}{3} \,. \tag{8}$$

Thus, we interpret the turbulent flux state as a state that spontaneously breaks the independent scalings of space and time to $z = \frac{2}{3}$.

When a scale symmetry is spontaneously broken one expects a NG gapless mode called the dilaton. We denote the expectation value (VEV) that breaks $R_x \times R_t \to R_{z=\frac{2}{3}}$ by:

$$\langle \epsilon \rangle = \bar{\epsilon} \,, \tag{9}$$

where $\langle ... \rangle$ means integration over both space and the probability distribution function of the force F. When considering turbulence in flat space $\bar{\epsilon}$ is a constant. It will be space dependent when studying turbulent flows in curved spaces.

The dilaton $\tau(x)$ is the fluctuation:

$$\epsilon(x) = \bar{\epsilon} e^{\delta \tau(x)} \,, \tag{10}$$

where $\delta(d)$ is a dimension dependent constant that will be related to intermittency and $\delta = 0$ means zero intermittency.

The dilaton is invariant under the unbroken $R_{z=\frac{2}{3}}$ symmetry:[14]

$$\tau(x) \to \tau(x) + \alpha\sigma + \frac{2}{3}\beta\sigma \,, \quad \alpha + \frac{2}{3}\beta = 0 \,, \tag{11}$$

and $\Delta_{z=\frac{2}{3}}[e^{\tau}] = 0$. We will denote its charges under $R_x \times R_t$ as $(-Q, \frac{3}{2}Q)$. We will relate Q to the intermittency.

2.3. The dilaton field theory

The steady state dilaton effective action in the inertial range of scales should be invariant under the symmetry (5) and under Galilean boosts which forbid time derivative terms in the Lagrangian. The dilaton action reads:[2]

$$S_D(\tau, \hat{g}) = \frac{d}{\Omega_d(d-1)!} \int_M d^d x \sqrt{\hat{g}} (\tau \mathcal{P}_{\hat{g}} \tau + 2Q \mathcal{Q}_{\hat{g}} \tau) , \tag{12}$$

where we use the notation of Ref. 15. When considering turbulence in flat space, $M = B_d(L)$ is the d-dimensional closed ball in Euclidean space R^d with radius L, whose boundary is the $(d-1)$-sphere S^{d-1} of radius L and $\hat{g}$ is the Euclidean metric. This will be the case most interesting to us.

Q is a dimensionless parameter called a background charge and $-Q$ is the dilaton charge under scaling of space as defined above. $\Omega_d = \frac{2\pi^{\frac{d+1}{2}}}{\Gamma[\frac{d+1}{2}]}$ is the surface volume of the d-dimensional sphere S^d.

$\mathcal{P}_{\hat{g}}$ are the conformally covariant GJMS operators:[16]

$$\mathcal{P}_{\hat{g}} = (-\Box)^{\frac{d}{2}} + \text{lower order} , \tag{13}$$

where $\Box = \hat{g}^{ab} \nabla_a \nabla_b$ with ∇_a being the covariant derivative on M and is the partial derivative on $M = B_d(L)$ whose metric is flat. They are local in even number of dimensions, for instance $\mathcal{P}_{d=2} = -\Box$, and non-local in odd number dimensions.[17] The dilaton is a log-correlated field reminiscent of the Liouville field in two-dimensional quantum gravity. $\mathcal{P}_{\hat{g}}$ gets a contribution from the boundary that depends on the boundary conditions.

$\mathcal{Q}_{\hat{g}}$ is the $\mathcal{Q}$-curvature scalar:[18]

$$\mathcal{Q}_{\hat{g}} = \frac{1}{2(d-1)} (-\Box)^{\frac{d}{2}-1} R + \cdots , \tag{14}$$

where R is the scalar curvature with respect to $\hat{g}$. It is local in even number of dimensions, for instance $\mathcal{Q}_{d=2} = \frac{1}{2}R$, and non-local in odd number of dimensions.[17] $\mathcal{Q}_{\hat{g}}$ gets a contribution from the boundary, for instance $\mathcal{Q}_{d=2} = \frac{1}{2}R + K$ where K is the extrinsic curvature. The bulk terms in $\mathcal{Q}_{\hat{g}}$ vanish on the ball $B_d(L)$ and it is determined by the boundary terms.

On conformally flat manifolds, such as the ball, $\mathcal{Q}_{\hat{g}}$ is related to the Euler density E_d:

$$\int_M d^d x \sqrt{\hat{g}} \mathcal{Q}_{\hat{g}} = \frac{1}{2} \Omega_d (d-1)! \chi(M) , \tag{15}$$

where $\chi(M)$ is the Euler characteristic of M, $\chi(B_d) = 1$, and we normalized the Euler density as:

$$\int_{S^d} \sqrt{g} d^d x E_d = \Omega_d d! . \tag{16}$$

Under a Weyl transformation $\hat{g}_{ab} \to e^{2\sigma} \hat{g}_{ab}$:

$$\mathcal{P}_{e^{2\sigma}\hat{g}} = e^{-d\sigma} \mathcal{P}_{\hat{g}} , \quad \mathcal{Q}_{e^{2\sigma}\hat{g}} = e^{-d\sigma} (\mathcal{Q}_{\hat{g}} + \mathcal{P}_{\hat{g}} \sigma) , \tag{17}$$

and this relation holds both for even and odd number of dimensions. We have:

$$S_D(\tau - Q\sigma, e^{2\sigma}\hat{g}) = S_D(\tau, \hat{g}) - S_D(Q\sigma, \hat{g}) = S_D(\tau, \hat{g}) - \frac{dQ^2}{\Omega_d(d-1)!} S_L(\sigma, \hat{g})\,, \quad (18)$$

where the Liouville action reads:

$$S_L(\phi, \hat{g}) = \int_M d^dx \sqrt{\hat{g}} (\phi \mathcal{P}_{\hat{g}} \phi + 2\mathcal{Q}_{\hat{g}} \phi)\,, \quad (19)$$

and

$$S_L(\phi - \phi_0, e^{2\phi_0}\hat{g}) = S_L(\phi, \hat{g}) - S_L(\phi_0, \hat{g})\,. \quad (20)$$

The linear shift in the dilaton under Weyl transformation (18) means that it can be viewed as a NG boson of a spontaneous breaking of the Weyl symmetry by the choice of the fixed metric $\hat{g}$.

The dilaton action is changed by a constant under a constant shift:

$$S_D(\tau + \tau_0, \hat{g}) = S_D(\tau, \hat{g}) + dQ\chi(M)\tau_0\,, \quad (21)$$

which imposes a selection rule in correlation functions.

The semiclassical limit of the dilaton field theory (12) is the large charge limit $Q \to \infty$ and the convenient parametrization is in terms of the rescaled dilaton field $\tau_c = \frac{\tau}{Q}$. This limit will correspond to the zero intermittency limit of the turbulence field theory.

Consider $M = B_d(L)$. The two-point function of the dilaton reads:

$$\langle \tau(x)\tau(y) \rangle = \frac{1}{d} \log\left(\frac{L}{x-y}\right) + \cdots\,, \quad (22)$$

where dots refer to a constant and terms that vanish in the limit $L \to \infty$. The dimension of the operator $V_\alpha(x) = e^{d\alpha\tau(x)}$ reads:

$$\Delta_\alpha = \frac{d}{2}\alpha(Q - \alpha)\,. \quad (23)$$

The dilaton theory (12) is a generalization of the two-dimensional Liouville conformal field theory to higher dimensions.[15]

2.4. *Dilaton dressing and random geometry*

Define the random measure on the inertial range of scales $d\mu(x)$:

$$d\mu(x) = \sqrt{g} d^dx\,, \quad \sqrt{g} = \sqrt{\hat{g}} e^{d\gamma\tau}\,, \quad (24)$$

where $\sqrt{\hat{g}} d^dx$ is the fixed measure, which when considering turbulence in flat space is the Euclidean measure as noted above. Physically we identify this random measure with the local energy dissipation (10), with $\sqrt{g} = \bar{\epsilon}$ and $\delta = d\gamma$ which vanishes when $\gamma \to 0$.

Requiring that $\Delta_\gamma = d$ (23):

$$\frac{d}{2}\gamma(Q - \gamma) = d\,, \quad (25)$$

we get:

$$Q = \gamma + \frac{2}{\gamma}, \tag{26}$$

and

$$\gamma_\pm = \frac{Q \pm \sqrt{Q^2 - 8}}{2}. \tag{27}$$

There are two branches $\gamma_\pm$. We require γ to be real, hence $Q^2 \geq 8$. Without loss of generality we take $Q \geq 0$. We have:

$$0 \leq \gamma_-^2 \leq 2, \quad \gamma_+^2 \geq 2, \quad \gamma_-\gamma_+ = 2. \tag{28}$$

The critical value $\gamma = \sqrt{2}$ corresponds in two-dimensional Liouville quantum gravity to the $c = 1$ barrier.[3,4]

The operators in the theory are K41 operators O_{K41} dressed by a dilaton factor $e^{d\alpha\tau}$:

$$O(x) = e^{d\alpha\tau} O_{K41}(x), \quad \alpha = \gamma(1 - \Delta), \tag{29}$$

where $d\Delta_{K41}$ is the undressed dimension of O_{K41}. Note, that both O_{K41} and its dilaton factor dressing are made of the fluid velocity and pressure. Requiring the conformal weight of O to be d we get:

$$\frac{d}{2}\alpha(Q - \alpha) + d\Delta_{K41} = d, \tag{30}$$

which using (26) and (29) reads:

$$\Delta - \Delta_{K41} = \frac{\gamma^2}{2}\Delta(1 - \Delta). \tag{31}$$

The partition of the theory in a fixed volume $V = \mathcal{L}^d$ with respect to the random metric scales as $\mathcal{L}^{-2d(1+\frac{1}{\gamma^2})}$. The expectation value of the one-point function of the integrated dressed operator (29) in a fixed volume scales as $\mathcal{L}^{d(1-\Delta)}$. $d(1-\Delta)$ is the dimension of a set in the random geometry of the inertial range and $d(1 - \Delta_{K41})$ is the dimension with respect to the fixed Euclidean metric on it. In general, the expectation value of the n-point function in a fixed volume scales as $\mathcal{L}^{d\sum_{i=1}^{n}(1-\Delta_i)}$.

Note, that in the field theory framework the natural units for the dimensions (conformal weights) of operators are inverse length, while when we will be applying (31) to turbulence we will associate dimensions of length to the fluid observables. In the fluid setup there is an IR scale L, which can be used in order to relate the two.

2.5. *Anomalous scaling*

Formula (31) reads as (1) with $\mathcal{G}^2(d) = \frac{\gamma^2}{2}$ that quantifies intermittency and the deviation from Kolmogorov linear scaling. The limit of zero intermittency $\mathcal{G}^2 \to 0$ is the semiclassical limit of the dilaton theory $Q \to \infty$ with $\mathcal{G}^2 \sim \frac{2}{Q^2}$. Δ_{K41} is the K41 dimension of the operator and Δ is the exact anomalous dimension. We have:

$$0 \leq \mathcal{G}_-^2 \leq 1, \quad \mathcal{G}_+^2 \geq 1, \quad \mathcal{G}_+\mathcal{G}_- = 1. \tag{32}$$

We will assume that the formula can be used when $\Delta_{K41} > 0$.

Consider the longitudinal structure functions S_n:

$$S_n(r) = \langle (\delta_r v)^n \rangle \sim r^{\xi_n}\,, \tag{33}$$

where $\delta_r v$ is the longitudinal velocity difference between points separated by a fixed distance $r = |\vec{r}|$:

$$\delta_r v = (\vec{v}(\vec{r},t) - \vec{v}(0,t)) \cdot \frac{\vec{r}}{r}\,. \tag{34}$$

The K41 scaling dimension of $(\delta_r v)^n$ is $\Delta_{K41} = \frac{n}{3}$, thus

$$\xi_n - \frac{n}{3} = \mathcal{G}^2(d)\xi_n(1-\xi_n)\,. \tag{35}$$

The numerical and experimental data in two (inverse cascade) and three space dimensions suggest:[5]

$$\mathcal{G}^2_{d=2} \sim 0\,, \quad \mathcal{G}^2_{d=3} \sim 0.16\,, \tag{36}$$

and implies that we should pick the $-$ branch (32). The corresponding charges are $Q^2_{d=3} \sim 17$, and if we take $\mathcal{G}^2_{d=2} \sim 0.01$ we get $Q^2_{d=2} \sim 205$. Both may be considered much larger than one, thus intermittency in these cases should be explained to a good accuracy by the semiclassical dynamics of the dilaton theory.

Formula (35) predicts a generic growth of the anomalous scalings at large n:

$$\xi_n \sim \sqrt{n}\,, \quad n \gg 1\,. \tag{37}$$

This is a rather different behavior compared to other existing proposals for the anomalous scalings. Unfortunately, at this point an accurate experimental or numerical data of high enough moments is not available and we cannot check this prediction.

The anomalous dimensions of the transverse structure functions of the velocity vector field as well as the mixed cases is the same as that of the longitudinal structure functions since the transverse and the longitudinal velocity increments have the same K41 scaling dimension.

The normalized local space average of the energy dissipation (4) over a d-dimensional ball of radius r, $B_d(r)$, centered at a point x reads:

$$\epsilon_r(x) = \frac{1}{Vol(B_d(r))}\int_{|x'-x|\leq r} d^d x' \epsilon(x')\,. \tag{38}$$

Consider the quantity $(r\epsilon_r)^{\frac{1}{3}}$. Its K41 scaling dimension is $\frac{1}{3}$. Formula (1) suggests that its statistics is the same as that of $\delta_r v$. Thus, if

$$\langle \epsilon_r^n \rangle \sim r^{\zeta_n}\,, \tag{39}$$

we get:

$$\zeta_{\frac{n}{3}} = \xi_n - \frac{n}{3}\,, \tag{40}$$

with ξ_n given in (35).

Next, we define the quantity $D_r^k \delta_r v$ where $D_r^k = (r\partial_r)^k$. It has the same K41 scaling dimension as $\delta_r v$, thus the anomalous dimensions of these operators are:

$$\langle (D_r^k \delta_r v)^n \rangle \sim r^{\xi_n}\,. \tag{41}$$

2.6. *Comparison to random cascade models*

Random cascade models for the redistribution of fluid energy via splitting into finer scale eddies allows a rich phenomenological framework for studying fluid turbulence and intermittency (for a review see e.g. Refs. 13 and 19). When the random variable is lognormally distributed one gets the Kolmogorov–Obukhov lognormal model.[6,7]

At leading order in the intermittency parameter $\mathcal{G}^2$, ξ_n of (35) coincides with the lognormal model:

$$\xi_n - \frac{n}{3} = \mathcal{G}^2 \frac{n}{3}\left(1 - \frac{n}{3}\right) . \tag{42}$$

They differ at higher orders in $\mathcal{G}^2$ and (35) may be viewed as a completion of the Kolmogorov–Obukhov formula (42) to the strong intermittency regime.

Unlike (42) which implies physically inconsistent supersonic velocities at large n and a violation of the convexity inequality, its completion (35) is mathematically and physically consistent.

In the lognormal model, the moments of the local energy dissipation ϵ_r scale as:

$$\langle \epsilon_r^n \rangle \sim r^{\mathcal{G}^2 n(1-n)} , \tag{43}$$

and

$$\langle (\delta_r v)^n \rangle \sim \langle (r\epsilon_r)^{\frac{n}{3}} \rangle \sim \langle \epsilon_r^{\frac{n}{3}} \rangle r^{\frac{n}{3}} \sim r^{\mathcal{G}^2 \frac{n}{3}(1-\frac{n}{3})+\frac{n}{3}} , \tag{44}$$

as in (42). In the random cascade models one separates the dynamics of the K41 theory from the that of the random measure and this is the reason for the factorization in (44). However, this separation of the dynamics is only an approximation that can be made at weak intermittency. In our language, the dilaton field interacts with the K41 theory and we cannot separate their dynamics.

More generally, in the random cascade models with a random variable W one has:

$$\langle \epsilon_r^n \rangle \sim r^{\zeta_n} , \quad \zeta_n = -\log_2(\langle W^n \rangle) , \tag{45}$$

which coincides with (43) when W is lognormally distributed, and

$$\xi_n = \frac{n}{3} - \log_2(\langle W^{\frac{n}{3}} \rangle) . \tag{46}$$

In the random geometry framework we get:

$$\xi_n = \frac{n}{3} - \log_2(\langle W^{\xi_n} \rangle) , \tag{47}$$

which coincides with (35) when W is lognormally distributed.

2.7. *Conformal anomaly and intermittency*

On the ball $B_d(L)$ the conformal anomaly coefficient a is obtained by varying the field theory partition function Z_{B_d}:

$$a \sim \frac{\partial}{\partial \log L} Z_{B_d}(L) . \tag{48}$$

Another way to see the anomaly is by the non-invariance of the partition function under a Weyl transformation that results in a term of the form e^{-aS_L}, where S_L is the Liouville action (19) and we neglect terms that are suppressed in the inertial range.[2] We will require that the total conformal anomaly vanishes:

$$a_{\text{total}} = a_{\text{dilaton}} + a_{K41} = 0 \,. \tag{49}$$

We may justify this requirement by asking that the inertial range universal structure and in particular the anomalous scalings should not depend on L. In an even number of dimensions we have:[2,15,21]

$$a_{\text{dilaton}} = \frac{2}{\Omega_d (d!)^2} \int_0^{\frac{d}{2}} dt \prod_{i=0}^{\frac{d}{2}-1} (i^2 - t^2) + \frac{(-1)^{\frac{d}{2}}}{\Omega_d (d-1)!} Q^2 \,. \tag{50}$$

Since $Q^2 \geq 8$ we see that the first term in (50) can be neglected and

$$a_{\text{dilaton}} \simeq \frac{(-1)^{\frac{d}{2}}}{\Omega_d (d-1)!} Q^2 \,, \tag{51}$$

that is its classical value (18). Using (49) and the large Q behavior of the intermittency parameter $\mathcal{G}^2 \sim \frac{2}{Q^2}$ we get:

$$\mathcal{G}^2(d) \simeq \frac{2}{\Omega_d (d-1)! |a_{K41}(d)|} \,. \tag{52}$$

This relation holds also for odd number of dimensions, where in both cases the contribution to the anomaly comes from the boundary.[2,22]

3. Discussion and Outlook

We outlined our proposal for a field theory description of the universal structure of incompressible fluid turbulence at the inertial range of scales, derived an analytical formula for anomalous scalings and related the intermittency parameter to the conformal anomaly coefficient. Note, that if the K41 theory is not conformally invariant[23] the same analysis as above holds but a_{K41} (52) gets contributions on the ball not only from the Euler density term. Mathematically, the structure of the field theory is that of a dynamical system (K41 theory) coupled to a random geometry. Physically, the random geometry structure is induced by a NG dilaton that is a consequence of a spontaneous symmetry breaking in the turbulent flux state and the dilaton field is responsible for the intermittency. There are infinite number of anomalous scaling dimensions predicted by the theory that may be checked numerically or experimentally.

There are numerous directions to follow. A calculation of the anomaly coefficient of the K41 mean field theory a_{K41} will enable via the relation (52) to calculate the intermittency parameter and compare with numerical and experimental data. For this we need to gain a better understanding of K41 theory, which is a non-local field theory. A calculation of equal time correlations of the local energy dissipation such as $\langle(\epsilon_r(x))^p(\epsilon_R(x))^q\rangle$, $\langle(\epsilon_r(x_1))^p(\epsilon_r(x_2))^q\rangle$ and velocity differences $\langle(\delta_r v)^p(\delta_R v)^q\rangle$ (see e.g. Ref. 24), where $r \neq R$ are in the inertial range, can shed more light on the theory and may be compared with numerical and experimental data.

The random cascade phenomenological framework for studying fluid turbulence and intermittency is not directly related to the NS equations that govern the motion of fluid flows. The spontaneous symmetry breaking mechanism that we introduced provides a direct link between the two via the dilaton that introduces the random structure. This topic deserves further study, in particular since the spontaneous symmetry breaking is in the flux state which is far from equilibrium.

It has been stressed by Polyakov[25] that the IR boundary conditions at the forcing scale L are of much importance for turbulence and that IR condensates break spontaneously conformal symmetry. Our dilaton theory is a conformal field theory[15] and it would be valuable to understand the IR effects on it as well as on the K41 mean field theory. In our framework the natural units for dimensions are inverse length and an explicit introduction of the IR scale L can explain the change to units of length.

A study of the non-local dilaton theories in odd number of dimensions may shed more light on odd-dimensional turbulence in general and on three-dimensional turbulence in particular. Also, studying turbulence in curved space may offer a new viewpoint on its dynamics. In our framework it amounts to introducing a random dressing of a fixed curved metric on the inertial range.

Higher-dimensional turbulence is mathematically intriguing and may also provide a $\frac{1}{d}$ expansion scheme. Denote by u the velocity one-form, then du is the vorticity two-form and the incompressibility condition reads $d^*u = 0$. In addition to the energy one has a conserved helicity made from the wedge products $\mathcal{H} = \int_M (du)^m$ when dim$M = 2m$ is even and $\mathcal{H} = \int_M u \wedge (du)^m$ when $dimM = 2m+1$ is odd.[26] Studying the turbulent cascade as a function of the number of space dimensions is likely to provide valuable insights.

One may try to generalize our framework to other systems. An interesting case is Burgers turbulence, which will correspond to a one-dimensional random geometry. One may expect that the shock wave solutions of Burges equations that determine the anomalous scaling correspond to singular one-dimensional manifolds.

Finally, the structure of our proposed field theory of steady state inertial range turbulence is reminiscent of the macroscopic structure of quantum gravity as a sum over spacetime geometries. It will be interesting to obtain a better understanding of this relationship.

Acknowledgments

I would like to thank L. Biferale, V. Mukhanov and A. Polyakov for discussions. This work is supported in part by the I-CORE program of Planning and Budgeting Committee (grant number 1937/12), the US-Israel Binational Science Foundation, GIF and the ISF Center of Excellence.

References

1. A. N. Kolmogorov, "The local structure of turbulence in incompressible viscous fluid for very large Reynolds number", *Dokl. Akad. Nauk. SSSR* **30**, 299 (1941); "Dissipation of energy in in locally isotroptic turbulence", *ibid.* **32**, 16 (1941) [Reproduced in *Proc. R. Soc. London, Ser. A* **434**, 9 and 15 (1991)].
2. Y. Oz, "Spontaneous symmetry breaking, conformal anomaly and incompressible fluid turbulence", *JHEP* **1711**, 040 (2017) [arXiv:1707.07855 [hep-th]].
3. V. G. Knizhnik, A. M. Polyakov and A. B. Zamolodchikov, "Fractal structure of 2D quantum gravity", *Mod. Phys. Lett. A* **3**, 819 (1988).
4. J. Distler and H. Kawai, "Conformal field theory and 2D quantum gravity", *Nucl. Phys. B* **321**, 509 (1989).
5. C. Eling and Y. Oz, "The anomalous scaling exponents of turbulence in general dimension from random geometry", *JHEP* **1509**, 150 (2015) [arXiv:1502.03069 [nlin.CD]].
6. A. N. Kolmogorov, "A refinement of previous hypotheses concerning the local structure of turbulence in a viscous incompressible fluid at high Reynolds number", *J. Fluid Mech.* **13**, 82 (1962).
7. A. M. Obukhov,"Some specific features of atmospheric turbulence", *J. Fluid Mech.* **13**, 77 (1962).
8. P. Kovtun, D. T. Son and A. O. Starinets, "Viscosity in strongly interacting quantum field theories from black hole physics", *Phys. Rev. Lett.* **94**, 111601 (2005) doi:10.1103/PhysRevLett.94.111601 [hep-th/0405231].
9. S. Bhattacharyya, V. E. Hubeny, S. Minwalla and M. Rangamani, "Nonlinear fluid dynamics from gravity", *JHEP* **0802**, 045 (2008) [arXiv:0712.2456 [hep-th]].
10. T. Damour, "Surface effects in black-hole physics", *Proceedings of the Second Marcel Grossmann Meeting on General Relativity*, 587 (1982).
11. C. Eling, I. Fouxon and Y. Oz, "The incompressible Navier–Stokes equations from membrane dynamics", *Phys. Lett. B* **680**, 496 (2009) [arXiv:0905.3638 [hep-th]].
12. C. Eling, I. Fouxon and Y. Oz, "Gravity and a geometrization of turbulence: An intriguing correspondence", *Contemporary Physics* **52**, 43 (2011), [arXiv:1004.2632 [hep-th]].
13. R. Rhodes and V. Vargas, "Gaussian multiplicative chaos and applications: A review", *Probab. Survey* **11**, 315 (2014), arXiv:1305.6221 [math.PR].
14. I. Hason, "Dual scale invariance spontaneous symmetry breaking and turbulence", arXiv:1708.08294 [hep-th].
15. T. Levy and Y. Oz, "Liouville conformal field theories in higher dimensions", *JHEP* **1806**, 119 (2018) [arXiv:1804.02283 [hep-th]].
16. C. R. Graham, R. Jenne, L. J. Mason and G. A. J. Sparling, "Conformally invariant powers of the Laplacian, I: Existence", *J. London Math. Soc.* **s2-46**, 557 (1992).
17. C. Fefferman and C. R. Graham, "Q-curvature and Poincare metrics", arXiv:math/0110271.
18. T. Branson, "Differential operators canonically associated to a conformal structure", *Math. Scand.* **57**, 293 (1985).
19. U. Frisch, *Turbulence: The Legacy of A. N. Kolmogorov* (Cambridge University Press, 1995).
20. S. Deser and A. Schwimmer, "Geometric classification of conformal anomalies in arbitrary dimensions", *Phys. Lett. B* **309**, 279 (1993) [hep-th/9302047].
21. J. S. Dowker, "Entanglement entropy for even spheres", arXiv:1009.3854 [hep-th].
22. C. R. Graham and E. Witten, "Conformal anomaly of submanifold observables in AdS/CFT correspondence", *Nucl. Phys. B* **546**, 52 (1999) [hep-th/9901021].

23. Y. Oz, "On scale versus conformal symmetry in turbulence", *Eur. Phys. J. C* **78**(8), 655 (2018) doi:10.1140/epjc/s10052-018-6147-8 [arXiv:1801.04388 [hep-th]].
24. R. Benzi, L. Biferale and F. Toschi, "Multiscale velocity correlations in turbulence", *Phys. Rev. Lett.* **80**, 3244 (1998).
25. A. M. Polyakov, "The theory of turbulence in two-dimensions", *Nucl. Phys. B* **396**, 367 (1993) doi:10.1016/0550-3213(93)90656-A [hep-th/9212145].
26. V. Arnold, Vladimir and B. Khesin, *Topological Methods in Hydrodynamics* (Springer, 1998).

MOND from a Brane-World Picture

Mordehai Milgrom

Department of Particle Physics and Astrophysics,
Weizmann Institute of Science, Rehovot 76100, Israel
moti.milgrom@weizmann.ac.il

Yacov (Jacob) Bekenstein contributed greatly to the development of MOND theory. His TeVeS was the bellwether to diverse relativistic-MOND formulations. However, none of the theories we have today is fully satisfactory, and, importantly, we still lack understanding of the roots of the MOND phenomenology. Groping for such understanding I describe a heuristic picture where (nonrelativistic) MOND dynamics emerge in a universe viewed as a nearly spherical brane embedded in a higher-dimensional flat space $\mathcal{E}$. The brane (centered at $\xi = 0$) is described by $\xi(\Omega)$; ξ and Ω are the radial and angular coordinates in $\mathcal{E}$. It couples to a radial potential $\varepsilon(\xi)$ in $\mathcal{E}$ as $\varepsilon\sigma$; σ is the brane density. The brane tension, T, balances the ε force at radius $\xi = \ell_0$, implying $\hat{a}_0 \equiv |\varepsilon'(\ell_0)| \approx T/\sigma\ell_0$, yielding for the velocity of small brane perturbations $c^2 \approx T/\sigma \approx \ell_0\hat{a}_0$. Local mass density $\rho(\Omega) \ll \sigma$ also couples to ε through $\varepsilon\rho$. The radial $\varepsilon'\rho$ force induces shallow brane indentations $\zeta(\Omega) = \xi(\Omega) - \ell_0$, $|\zeta| \ll \ell_0$, which, in turn, affect masses in a way that makes $\phi \equiv \varepsilon[\xi(\Omega)] \approx \phi_0 + \hat{a}_0\zeta$ play the role of gravitational potential. To get MOND dynamics, the brane Lagrangian is *assumed* a functional of the local brane orientation, whose normal makes an angle θ with the radial direction, $|\tan\theta| = |\vec{\nabla}\zeta| = |\vec{\nabla}\phi|/\hat{a}_0$. Thus, some $a_0 \sim \hat{a}_0 \approx c^2/\ell_0$ plays the role of the MOND acceleration constant in local gravitational dynamics. Aspects of MOND that may emerge naturally as geometrical properties are: a. The special role in MOND of an acceleration, a_0, that marks the transition from dynamics much above a_0 (large brane slopes) to scale-invariant dynamics much below a_0 (small slopes). b. The intriguing connection of a_0 with cosmology. c. The Newtonian limit corresponds to local departure $|\zeta| \ll \ell_0$; i.e., $\phi - \phi_0 \sim a_0\zeta \ll a_0\ell_0 \sim c^2$ – whereas relativity enters (but not treated here) when $|\zeta| \not\ll \ell_0$. The model also opens new vistas for extension, e.g., it points to possible dependence of a_0 on ϕ, and to a_0 losing its status and meaning altogether in the relativistic regime. It also reinforces the view that global (cosmological) and local (e.g. galactic) MOND dynamics have to be understood together as inseparable parts of the same construction. The 'old' cosmological-constant problem is addressed: the 'cosmological' effects of large brane energy density T are cancelled (balanced) by the action of the ε field. The balance is achieved at some radius ℓ_0 with space attaining a nearly spherical geometry. This geometry, when interpreted within general relativity, is characterized by a 'cosmological constant'

$\Lambda \approx \ell_0^{-2}$. With a view to generalizing the model to space-time branes I discuss possible connections with the nearly-de-Sitter nature of our Universe.

Dedicated to the memory of Yacov (Jacob) Bekenstein

1. Bekenstein and MOND

Yacov (Jacob) Bekenstein's contributions to black-hole thermodynamics have established him as a central figure in this field, and have turned out to have even further reaching consequences than initially imagined. These will surely be covered by other contributions to this volume. Here I concentrate on Jacob's influential involvement in the MOND program, most notably on his leading role in developing MOND theory.

Not only did Jacob play a major part in the development of MOND, but MOND seems to have played a major part in Jacob's scientific life and career, as we can learn from an interview he gave to "Scientific American" in March 2005 on the occasion of his receiving the Israel Prize (http://www.sciam.co.il/archive/archives/2453). Here are some excerpts translated from the Hebrew.

SA: The Israel Prize expresses appreciation of the totality of your scientific work; what do you think is your main achievement?

J: This has two answers. One is how people see me, which is expressed in the Prize citation, concentrating on my work on thermodynamics of black-holes. I started dealing with that in the early (19)70s, and you may say that by the end of the 70s my part in the subject ended... . The second answer is from my own point of view; today I work on other things... . I think that I am still doing important things, but it takes time until the public, and even important scientists, grasp what you have done.

SA: What are you investigating today?

J: Today, I am trying to change general relativity, still within the principles set by Einstein... in order to reconcile it with relatively new astronomical phenomena, such as that of the dark mass. ...(DM) is not an explanation but an adjustment to match the facts. In contradistinction, if one accepts... (MOND, which propounds)... that the true theory departs from Newton's theory for small accelerations... (the observations)... follow naturally, as direct consequences of the (MOND) equations... . There is a prescription that works well, MOND, but the reason it works so well is not known. You may say that MOND tells us how the real theory of gravity should look. And this is what I am trying to achieve. I have today a relativistic theory... (more on MOND and its achievements)... . I think that if this theory will work in the end, then this is more important than the thermodynamics of black holes.

Jacob was the first physicist to hear from me about MOND. It was 1982; Jacob was still at Ben-Gurion University in Beer Sheba. I went there from Rehovot with my initial MOND trilogy of preprints in my bag to tell Jacob and benefit from his advice. I had not prepared him for what was in these preprints. Jacob was immediately captured, but he also warned me – as I vividly remember – that this is going to encounter much opposition, but also that I should not heed such opposition. He was drawing on his own experience with black-hole entropy and on how his ideas had been received a decade earlier. Our meeting started a very fruitful and enjoyable collaboration that was to last for 30 years.

The development of MOND paradigm owes much to Jacob. Without doubt, his two peak contributions were the following: a. His very first contribution to MOND in work we collaborated on, describing the first full-fledged nonrelativistic MOND theory (Jacob dubbed it 'AQUAL').[1] This theory was derived from an action with the standard symmetries, and so it enjoyed the standard conservation laws, an important advance on my initial formulation. This work also clarified an important issue concerning the center-of-mass motion of composite bodies. b. After several attempts to construct a relativistic MOND theory – initially with me, but then mainly by himself, and notably with Bob Sanders – Jacob came up with his famous TeVeS theory,[2] building on ideas by Bob Sanders (TeVeS sprang from Sanders's 'stratified' MOND theory[3]).

The advent of TeVeS has greatly advanced the case for the whole MOND paradigm. Among other things, it attracted to MOND people who were interested in relativity, and as a result several other relativistic MOND formulations were proposed.

About a quarter of Jacob's publications since his first hearing about it were on MOND (about 20 odd out of 80 odd papers). Jacob also gave numerous talks (e.g., Refs. 4 and 5), and wrote reviews on MOND (e.g., Refs. 6 and 7).

Jacob's general attitude towards MOND had seen ups and downs, steered in part by his own momentary successes and failures in his attempts to advance it. Following are some examples of excerpts from his letters.

Letter from September 8, 1985: "I tried a new approach to Relativistic Mond which is guaranteed to be causal by definition, but the first indication is that it is not giving the right non-relativistic form. Right now I am pretty close to giving up on relativistic theories. The continual failures do not bode well for the whole idea, and I am less than ever interested in going back to empirical tests."

January 3, 1986 (translated from Hebrew); "Today when I read your letter anew your arguments seem more valid, and I understand less why I thought there are problems. By and large my feeling about MOND is now better... . I have a new idea for a relativistic theory... ."

August 21, 1998 (discussing a colleague he met in France): "...He knows about MOND and I explained to him the problems in turning it relativistic. He had a suggestion to use a theory where the physical metric is put together from the Einstein metric and a $B_{\mu\nu}$ antisymmetric potential. ...So far as I can see this will also fail to have MOND behavior... . Right now I am exploring an idea where the connection does not come from the metric but has independent dynamics. It is slow going."

September 7, 2000: "Yes, one feels the tide will turn; perhaps this is not the beginning of the end of DM but probably the end of the beginning...; just waiting to do some thinking about relat. MOND, which, I think, is becoming urgent."

November 16, 2003 while writing the TeVeS paper: "For a month or so I have been back at writing that paper on relativistic MOND. I find, as usually happens, that I did not do all the needed calculations before I started writing, so progress is slow. However, I am now at the stage having to do with MOND vs Newtonian limits, PPN tests and the like, and that is the last of the big issues in the paper."

Jacob's influence on me was twofold. I learned much science from him, especially that at the time I knew only little general relativity. I also greatly admired Jacob's knack for presentation and exposition, his clear and precise use of words always impressed me. So much so that in the early days of our collaboration I found myself, many times, trying to emulate his style.

2. Introduction

Vying with the dark matter paradigm, MOND[8] contends to account for the mass discrepancies observed in the Universe without dark components. It does so by modifying Newtonian dynamics (ND), and with it, perforce, general relativity (GR). References 9 and 10 are recent extensive reviews of MOND. MOND introduces a new acceleration constant, a_0, below which dynamics depart from ND and GR. For much higher accelerations; i.e., in the formal limit $a_0 \to 0$, a MOND theory should tend to standard dynamics. (In analogy with the 'correspondence principle' of quantum mechanics going to the classical for $\hbar \to 0$.) In the opposite, 'deep-MOND limit' (DML) $a_0 \to \infty$, keeping $a_0 G$ fixed, MOND dynamics become space-time scale invariant (SI).[11]

There are various 'galactic laws' that MOND predicts[8,9,12] which have been thoroughly tested and vindicated. For example, a tight relation between the asymptotic rotational speed in a galaxy and its total baryonic mass (tested, e.g., in Ref. 13); a general, 'mass-discrepancy-acceleration relation' (tested, e.g., in Refs. 14–17); and a tight correlation between the central surface density of a (disc) galaxy, and that attributed to a putative dark-matter halo[18,19] (tested, e.g., in Ref. 20). MOND predicts detailed rotation curves of individual disc galaxies, given only their baryon distribution (e.g., Ref. 9 and many others).

The MOND constant, a_0, has been consistently determined from various observed manifestations of these MOND laws (e.g., Refs. 15, 21–23). After some initial vacillation in the range $a_0 \sim (1-2) \times 10^{-8}$ cm s^{-2}, the value has converged on $a_0 \approx 1.2 \times 10^{-8}$ cm s^{-2} (beginning in the early 1990s with Ref. 22). As noted early on (e.g., Refs. 8, 24, 25), a_0 is near in value to some accelerations of cosmological significance:

$$\bar{a}_0 \equiv 2\pi a_0 \approx a_H(0) \equiv cH_0 \approx a_\Lambda \equiv c^2/\ell_\Lambda\,, \tag{1}$$

where $a_H \equiv cH$ is the acceleration associated with the cosmological expansion rate, H (the Hubble constant), and $a_H(0)$ is its present value, and $\ell_\Lambda = (\Lambda/3)^{-1/2}$ is the radius associated with Λ – the observed equivalent of a cosmological constant. Thus, the MOND length, $\ell_M \equiv c^2/a_0$, is of the order of the de Sitter radius of our Universe, ℓ_Λ, or of the present Hubble radius, $\ell_H \equiv c/H_0$. This numerical 'coincidence', if fundamental, may have far-reaching ramifications (e.g., Ref. 26).

While the salient predictions of MOND follow essentially from only the above basic tenets, we need to dress these tenets with a theory, and – no less desirable – to understand the origin of MOND, as all signs point to its being emergent from more fundamental concepts.

Indeed, theoretical endeavors (reviewed in Ref. 26) have proceeded roughly along two avenues. On one hand, full fledged MOND theories – both nonrelativistic (NR) and relativistic – that are derivable from an action principle have been put forth. Such theories are conceptually on a par with ND or GR. They are very useful; for example, for demonstrating various matter-of-principle issues in MOND, or for detailed numerical simulations in MOND, since they capture the essence of MOND, satisfy its basic tenets, share the salient predictions, and obey all the standard conservation laws. Yet, it is quite clear that none of the existing theories can be the final word on MOND.

On the other hand, it has been stressed repeatedly that MOND, as it is now understood, must emerge as an effective, approximate theory from a more basic world picture. Such an underlying theory might tell us why it is an acceleration that marks the boundary between the two regimes of dynamics – the DML, with its SI, and the high acceleration, ND/GR regime where SI is lost – and whether the value of a_0 has a special significance in the wider context, explaining perhaps relations (1).

Suggestions to derive dynamics as an effective description from a deeper, microscopic stratum go back far in time even for standard dynamics, such as those attempting to derive Newtonian inertia a-la Mach's principle, or the suggestions that the Einstein–Hilbert action for GR gravity emerges from effects of the quantum vacuum,[27] or from thermodynamical considerations, e.g., Refs. 28–30, to name but a few examples. The continued quest for a satisfactory theory of quantum gravity is also an impetus to dig deeper.

For MOND, the impetus to do so is even stronger. The added stimulus is twofold: 1. The appearance of a new 'constant of nature' in MOND – which may call for a more basic interpretation, and, in particular, its possible relation to the state of the Universe at large pointed to by Eq. (1). 2. The fact that all existing, full-fledged theories interpolate 'by hand' from the low- to the high-acceleration regime. Surely this interpolation should result from some more basic theory.

Such a realization has lead to diverse suggestions on how to obtain MOND phenomenology from microscopic considerations. Most of these are listed in Refs. 10 and 26. Such ideas, while promising to various degrees, have not materialized into full-fledged theories.

For example, in Ref. 25, I showed how, in the background of a de Sitter (hereafter dS) universe – such as ours is approximately – the length ℓ_Λ can appear in local dynamics in the guise of an acceleration $a_0 \sim a_\Lambda$ in just the way MOND requires. It was based on the observation that the quantum vacuum might serve as an absolute inertial frame. Following that, many more suggestions along similar lines have been discussed.[31–40]

Other microscopic approaches aim to reproduce – at least on galactic scales – a theory on the lines of the first full-fledged NR MOND theory proposed by Jacob and me in Ref. 1. There, the gravitational potential, ϕ, is governed by the nonlinear generalization of the Poisson equation

$$\vec{\nabla} \cdot [\mu(|\vec{\nabla}\phi|/a_0)\vec{\nabla}\phi] = 4\pi G\rho(\mathbf{x}) \,, \tag{2}$$

which is gotten from the Lagrangian density $\mathcal{L} \propto a_0^2\mathcal{F}[(\vec{\nabla}\phi)^2/a_0^2]$ $[\mu(x) \equiv \mathcal{F}'(x^2)]$ replacing the Poisson Lagrangian $\mathcal{L} \propto (\vec{\nabla}\phi)^2$. Correspondence with the Newtonian limit dictates $\mu(x \to \infty) \to 1$, while SI in the DML dictates $\mu(x \ll 1) \propto x$.

It is worth noting at this point that the DML of Eq. (2) ($a_0 \to \infty$, $\mathcal{A}_0 \equiv a_0 G$ fixed),

$$\vec{\nabla} \cdot (|\vec{\nabla}\phi|\vec{\nabla}\phi) = 4\pi\mathcal{A}_0\rho(\mathbf{x}) \,, \tag{3}$$

[the equality corresponds to the standard normalization of a_0 for which $\mu(x \ll 1) \approx x$] is not only invariant to space scaling $\mathbf{r} \to \lambda\mathbf{r}$ – which follows from the space-time SI of the DML in this time-independent equation – but has a higher symmetry: it is a conformally invariant in the 3-D Euclidean space.[41] With the other, usual, symmetries that this theory enjoys (rotations and translations), its full spatial symmetry group is isomorphic to the isometry group of a dS space-time. This might be another hint to a connection of MOND, as manifested locally, to the state of the Universe at large, perhaps in the framework of a conformal-field-theory-de-Sitter duality.[a]

Attempts to obtain effective (NR) gravitational dynamics described by Eq. (2) usually begin with known physical systems that are governed by a similar equations. Several such analog systems have been listed in Refs. 41, 43, 44. For example, nonlinear electrostatics with the dielectric coefficient depending on the electric-field strength (a situation that also occurs in Born-Infeld electrostatics). Another analog is the (continuity) equation for compressible, irrotational (potential) flows of barotropic fluids. An equation of state $P \propto \rho^3$ is needed to get the DML.

So, for example, the microscopic approach to MOND lead by Blanchet[45–47] builds on the nonlinear dielectric analogy, and involves a gravitationally polarizable, omnipresent medium. Another example is the program lead by Khoury[48,49] involving an omnipresent fluid, which reproduces DML phenomenology in its superfluid phase if one assumes that it has a $P \propto \rho^3$ equation of state.

The present contribution was inspired by yet another analog problem: The shape of a 'membrane', embedded in a (Euclidean) space with codimension one is

[a]QUMOND, another NR modified-gravity MOND version,[42] may also serve as a target for microscopic approaches. Its DML, $\Delta\phi = a_0^{1/2}\vec{\nabla} \cdot (|g_N|^{-1/2}g_N)$, where g_N is the Newtonian acceleration field, is SI but not conformally invariant.

determined by extremizing some 'energy functional'. In the simple case where the energy of the membrane is proportional to its volume, the shape is determined by an equation of the form (2), with $\mu(x) = (1 + x^2)^{-1/2}$, and vanishing right-hand side.

This model does not include the analog of sources, and the form of $\mu(x)$ is not what MOND dictates. But, as outlined in Ref. 44, it allows for generalizations that do. Naturally, this approach envisages our Universe as a 3-D spatial membrane (hereafter 'brane' as is the common term for higher-than-two-dimensional membrane) embedded in a higher dimensional space – which I will take to be Euclidean – where in the NR limit the extra dimension appears in dynamics within the brane as the MOND gravitational potential.

Brane analogs of gravity, and brane representations of our Universe – so called 'brane-world' models – have been long and amply discussed in the literature. An early brane imagery of gravity is the effective attraction between masses placed on a stretched, elastic membrane placed horizontally in the gravitational field near the Earth surface. The sort of gravity that emerges depends on the 'elastic' properties – the 'energy function' of the brane.

In 'brane-world' scenarios, the 3-space (plus 1-time) dimensional arena we shortsightedly perceive as 'our Universe' is embedded in a higher dimensional space-time ('the bulk'). While the full dynamics are taking place in the bulk cave, only their 'shadow' appears as effective, or emergent, dynamics on the our perceived cave-wall Universe, when we describe phenomena that do not probe directly the higher dimensions.

Interest in such brane-world pictures has greatly intensified in recent decades, largely due to the advent and developments of string theory, and the realization that it naturally gives rise to such diverse world-models, both because string theory entails higher dimensions and because branes play important roles there. References 50,51 are reviews of such theories, giving some history and extensive references. Following several predecessors, the brane-world models of Refs. 52, 53 have greatly boosted studies of such pictures.

The brane-world models studied to date come in great variety. They differ, in the first place, in the aspects of the observed Universe they are trying to reproduce. They also differ greatly in the underlying features invoked to such ends. For example, in the nature of the embedding space – e.g., its dimension, topology, and geometry (e.g., whether it is compactified or not); in the number of branes involved in addition to the one we live on, and their presumed mutual effects; in the (brane and bulk) degrees of freedom involved, and in the interactions and potentials that dictate the dynamics of these; in the initial conditions assumed when time evolution of the model is studied; etc.

By and large, these models, with their various invoked idiosyncrasies, do not follow – certainly not in detail – from more fundamental theories. Rather, they are largely constructed 'by hand' to address various issues or conundrums in particle physics, cosmology, or gravity in general (classical and quantum).

To give but a few examples, the model of Ref. 52 envisages an anti-de-Sitter embedding space and has ramified into one-brane and two-brane classes of models. Reference 53 suggested and studied 4-D Gravity on a Brane in 5-D Minkowski Space, and Ref. 54 considered cosmology in this latter picture. Some other models view our Universe as the lowest dimensional echelon in a cascade of several nesting branes.[55] The earlier version of the Ekpyrotic universe[56] aimed to explain aspects of cosmology, otherwise attributed to inflation. It invokes the event that lead to the big bang as a collision of our brane Universe with another, parallel 'bulk brane' that 'peeled off' yet another 'boundary brane'.

Membrane-induced classical gravity models have also been much discussed extensively by Carter and coworkers, e.g. in Refs. 57–61.

In my superficial perusal of some of the relevant literature I have not identified elements that might specifically serve in constructing MOND-inspired brane-world models (but see a somewhat related approach in Ref. 62).

I comment here, as an aside, that BIMOND – the bimetric formulation of relativistic MOND,[63] and other bimetric theories, which involve two metrics defined on a single space-time, may perhaps be obtained as near-contact (small-distance) approximations of a system of two branes in a bulk, each with its own Einstein–Hilbert action plus a mutual interaction.

Here I greatly expand on the MOND-from-brane idea introduced in Ref. 44. A crucial novel feature of the idea is the presence of an external potential $\varepsilon(\mathbf{X})$ in the embedding space that acts on brane and matter. The potential is highly symmetric (planar, spherical, etc.). It defines a local, preferred direction, $\mathbf{U}$ – a unit vector in the direction of $\vec{\nabla}_{\mathbf{X}}\varepsilon$ – that plays several roles: 1. $\mathbf{U}$ points to the direction that appears in the internal brane dynamics as the gravitational potential. 2. In order to obtain MOND dynamics the energy density of the brane has to depend on the local angle between the normal to the brane and $\mathbf{U}$ (akin to the energy of a dipole layer in an electric field). 3. The field $\mathbf{U}$ determines the global average geometry of the brane (e.g., a spherically symmetric ε induces a nearly spherical brane). 4. The MOND constant emerges as the value of $|\vec{\nabla}_{\mathbf{X}}\varepsilon|$ at the position of the brane. 5. The same $|\vec{\nabla}_{\mathbf{X}}\varepsilon|$ appears in the global balance of the brane at some average radius ℓ_0, which leads to $a_0 \sim c^2/\ell_0$.

These models, simplistic and heuristic as they are, do suggest possible explanations to – or at least geometrical interpretations of – some important aspects of MOND. For example, the local orientation of the brane with respect to $\mathbf{U}$ is gravitational acceleration (in units of a_0). A slope of 1 appears as a_0; the DML corresponds to near alignment of the local brane normal with $\mathbf{U}$, while the Newtonian regime corresponds to high inclinations. More can be understood if the brane is a sphere or pseudosphere of radius ℓ_Λ whose shape is locally distorted by masses that cause 'indentations' in the brane. In such a picture NR systems are described by shallow indentations of depth $\ll \ell_\Lambda$, while highly relativistic gravitational fields correspond to indentations whose depth is $\sim \ell_\Lambda$. It is also possible to demonstrate why a_0, ℓ_Λ, and the propagation speed of perturbations on the brane (standing for c), are related by a relation resembling (1).

This is only a laying out of the main principles. Practically every aspect of the model I describe requires further elaboration to make the model truly relevant to our Universe. In Sec. 3, I start to describe the elements of the model concentrating on local aspects. In Sec. 4, I consider global aspects, viewing the brane as a distorted sphere. In Sec. 5, I reconsider the model in the context of a dS universe. Section 6 is a discussion.

3. A Brane Model for MOND

One starts with a d-dimensional surface, $\mathcal{M}$, embedded in a $d+1$ dimensional Euclidean space, call it $\mathcal{E}$, with Euclidean coordinates $\mathbf{X} = (x^1, ..., x^d, x^{d+1})$ of the point $\mathbf{X}$ in $\mathcal{E}$. The brane can be described by specifying one of the coordinates, say $\xi \equiv x^{d+1}$, as a function, $\xi(\mathbf{x})$, of the d-dimensional vector $\mathbf{x} = (x^1, ..., x^d)$. [The interpretation of ξ and dynamics of the system, described below, ensure that ξ does not become multivalued, as this would lead to infinite gravitational accelerations $\mathbf{g} \propto \vec{\nabla}_x \xi$; see Sec. 4.1.3 below.] The metric induced on the brane from the Euclidean metric of $\mathcal{E}$ is $g_{ij} = \delta_{ij} + \xi_{,i}\xi_{,j}$ whose determinant is $g = 1 + (\vec{\nabla}\xi)^2$ ($\vec{\nabla}\xi \equiv \vec{\nabla}_{\mathbf{x}}\xi$), so the volume of the brane is

$$V = \int g^{1/2} d^d x = \int d^d x [1 + (\vec{\nabla}\xi)^2]^{1/2} . \tag{4}$$

Configurations $\xi(\mathbf{x})$ that extremize V satisfy $\vec{\nabla} \cdot [\mu(|\vec{\nabla}\xi|)\vec{\nabla}\xi] = 0$, with $\mu(x) = (1 + x^2)^{-1/2}$. I will want to identify ξ with the NR gravitational potential; so here V may be identified with the action for this potential.

To turn this into an analog for NR MOND gravity as formulated, e.g., by Eq. (2), we need to do at least the following: a. Justify the identification of ξ with the NR gravitational potential, ϕ. b. Introduce matter. c. Account for the appearance of a_0, with its different roles, in local dynamics. d. Account for the specific MOND limits: SI in the DML, $|\vec{\nabla}\phi| \ll a_0$, and the Newtonian limit for $|\vec{\nabla}\phi| \gg a_0$. e. Account for the breakdown of this NR picture beyond the Newtonian and into the relativistic regime. f. Account for the coincidence (1). Accounting fully for relativistic phenomena is even a further goal. I will not be able to achieve even all the goals a-f above, but for some I will indicate possible directions.

In the discussion below I shall refer to the following three standardly used scales: 1. 'Global' will refer to the whole brane; e.g., whether it is on average planar or spherical. 2. 'Local' refers to local variations in ξ due to the presence of matter. 3. 'Average local' which will refer to properties averaged over scales much smaller than the brane's extent, but large enough to homogenize over local roughness of the brane. Homogeneity and isotropy of the 'local average' brane structure are assumed here as an expression of the 'cosmological principle'. I shall refer to a local average piece of the brane as $\langle\mathcal{M}\rangle$.

The action V in Eq. (4) is invariant under rotations of the whole brane in $\mathcal{E}$. It thus does not single out a direction the displacement along which can be identified with the potential. We need to invoke a preferred direction in $\mathcal{E}$ which will determine this direction dynamically via interaction with the brane and with matter. Call the unit vector in this direction at a local position on the brane, $\mathbf{U}$. A minimalistic way do this is to assume some potential, $\varepsilon(\mathbf{X})$ in $\mathcal{E}$ which couples to the brane and to matter, the local direction of whose gradient is $\mathbf{U}$.

To get some kind of gravity it is enough for ε to exert forces on matter (in the $\mathbf{U}$ direction). This however is not enough to get MOND dynamics. As explained below, accounting for local MOND gravity *requires* that there is also energetic cost to misalignment between $\mathbf{U}$ and the normal to the brane, $\mathbf{N}(\mathbf{x})$. Namely, the brane energy increases with increasing angle $\psi(\mathbf{N}, \mathbf{U})$. I shall come to this later; here I note that this has the added benefit that it dynamically makes the brane perpendicular to $\mathbf{U}$ on local average.

Had this not happened dynamically, we would have had to invoke it by fiat. A component of $\mathbf{U}$ tangent to $\langle\mathcal{M}\rangle$ would introduce a preferred direction in the brane that would brake the isotropy of our Universe, breaking the cosmological principle, which we want to retain.

In Sec. 4, I consider the global structure of $\mathcal{M}$, in particular taking it to be globally a sphere of radius identified with ℓ_Λ. In this case $\mathbf{U}$ is radial from some center, and ε is a function of $|\mathbf{X}|$. But, for simplicity of presentation I take for the nonce $\mathbf{U}$ to be constant; this causes $\mathcal{M}$ to be planar globally and locally on average, which is adequate for considering local dynamics on scales much smaller than ℓ_Λ.

3.1. *Matter and its Lagrangian*

'Matter' will be represented my 'masses' confined to the brane: either isolated objects characterized by a mass parameter m, having the dimensions of inertial mass, or a continuous distribution of such masses of density $\rho(\mathbf{x})$.

Masses are assumed to have inertia tangent to $\langle\mathcal{M}\rangle$ – i.e., perpendicular to $\mathbf{U}$ – but not in the $\mathbf{U}$ direction. Thus, inertia is *not* a property in $\mathcal{E}$. The kinetic Lagrangian of a mass is of the form $(1/2)m\mathbf{v}^2$, where $\mathbf{v} = d\mathbf{x}/dt$ [with no contribution from $(d\xi/dt)^2$].

Had we assumed isotropic inertia in $\mathcal{E}$, the kinetic term for masses would have been $(1/2)m[\mathbf{v}^2 + (d\xi/dt)^2] = (1/2)mv_i(\delta_{ij} + \xi_{,i}\xi_{,j})v_j$. As will become clear below, to reproduce MOND gravity we need to consider and account for all values of $\vec{\nabla}\xi$, large as small. Without the assumption of no inertia in the $\mathbf{U}$ direction, the wrong inertia will result. I missed this important point in Ref. 44.

Such confinement of inertia to $\langle\mathcal{M}\rangle$ may seem natural in a model in which all matter degrees of freedom are confined to $\mathcal{M}$. But note that here inertia is not in the local tangent to $\mathcal{M}$ but in the tangent to $\langle\mathcal{M}\rangle$, which point to long-rage cause of inertia. I think that all this points to inertia itself being an emergent attribute of matter.

For example, this will come about if inertia is an acquired attribute of masses – a-la Mach's principle – due to their interaction with some physical agent that is confined to $\mathcal{M}$ (such as the quantum vacuum[25]).

Alternatively, absence of inertia in the $\mathbf{U}$ direction might result if inertia emerges directly from effects associated with the ε field, which produces a force field in the $\mathbf{U}$ direction (somewhat in a similar vein to a magnetic field not impeding motion of charged particles along it, only perpendicular to it). For example, a kinetic Lagrangian of the form $(m/2)(\dot{\mathbf{X}} \times \mathbf{U})^2$ gives in the equation of motion the inertial term $m[\mathbf{A} - (\mathbf{A} \cdot \mathbf{U})\mathbf{U}] \equiv m\mathbf{a}$, where $\mathbf{a}$ is the projection of the acceleration $\mathbf{A}$ perpendicular to $\mathbf{U}$.

Masses are also the couplings of matter to the ε field, which appears in the Lagrangian as $m\varepsilon(\xi)$. I take m to be both the coupling strength of masses to ε and the coefficient of inertia – an assumption that requires justification. But, a. one needs to assume only proportionality, i.e., that the ratio of the two couplings is the same for all masses, any proportionality constant can be absorbed into ε; b. If inertia too has something to do with the agent that produces ε, as suggested above, such a proportionality is natural.

In summary, the matter degrees of freedom appear in the action as

$$I_M = \int d^d x \; \rho(\mathbf{x}) \left\{ \frac{1}{2}\mathbf{v}^2 - \varepsilon[\xi(\mathbf{x})] \right\} , \tag{5}$$

where the density of masses is $\rho = dm/d^d x$.[b]

From the action (5) we get the equation of motion of masses:

$$\mathbf{a} = \frac{d\mathbf{v}}{dt} = -\vec{\nabla}_x \varepsilon \, . \tag{6}$$

So clearly, $\phi(\mathbf{x}) \equiv \varepsilon[\xi(\mathbf{x})]$ can be interpreted as the gravitational potential for masses on the brane.

According to this picture we sense and interpret the projection of motion tangent to the branes 'local average' (here, parallel to the $\mathbf{x}$ hyperplane) as a change in location. But the motion in the ξ direction is dictated by the $\mathbf{x}$ motion due to the confinement to the brane, and is perceived as a change in the potential, not as motion. This sensation occurs because our experience tells us that we can move along the $\mathbf{x}$ hyperspace at will, but we cannot change ϕ at will, without 'moving'; i.e., without changing $\mathbf{x}$.

3.2. *The brane action*

The brane action translates to the free (kinetic) action of gravity, analogue to the Poisson or the Einstein–Hilbert actions. Here, the discussion is restricted to NR gravity; so masses are assumed to move on the brane at speeds much smaller than the speed of small brane perturbations (see Secs. 4 and 4.1.3). We shall naturally

[b] ρ is not defined as the mass per unit volume of the brane $dm/dV = \rho[1 + (\vec{\nabla}_{\mathbf{x}}\xi)^2]^{-1/2} = \rho\cos\psi$, where $\cos\psi = \mathbf{N}(\mathbf{x}) \cdot \mathbf{U}$.

identify this latter speed with the speed of gravity (or of gravitational waves). Thus, I consider, at present, only quasi static situations where the brane adjusts its shape instantaneously to the position of the masses: the kinetic action of the brane will thus be neglected. The Lagrangian of a given configuration of the brane is thus (minus) the energy density of the brane configuration, $\xi(\mathbf{x})$, which encapsules its 'elasticity' and interaction with the ε field.

I assume a local action so it is $I = \int \mathcal{L} d^d x$, where $\mathcal{L}$ is some local functional of $\xi(\mathbf{x})$. Without further guidance from a more basic, microscopic, theory of the brane dynamics, the a priori gamut of possibilities for picking $\mathcal{L}$ is very wide. There could be terms that depend on the intrinsic geometry of $\mathcal{M}$, or its extrinsic geometry in $\mathcal{E}$, which would involve second derivatives of ξ. This, in itself is not deleterious – only higher time derivatives lead to Ostrogradsky instabilities – but I will consider Lagrangian with only first space derivatives of the potential appearing. So in general $\mathcal{L} = \mathcal{L}(\xi, \vec{\nabla}\xi)$. Requiring rotational symmetry in $\mathbf{x}$ restricts us to $\mathcal{L} = \mathcal{L}(\xi, |\vec{\nabla}\xi|)$. In this preliminary discussion we ignore the interaction of the ε field with the brane mass analogous to the $\rho\varepsilon$ Lagrangian for matter. We shall see in Sec. 4 that this interaction stabilizes the brane at some average position ξ_0. Given that, we can take here in the planar (local) case $\mathcal{L}$ to be invariant to ξ translations; i.e., ξ independent. We thus end up here with a brane Lagrangian density that is a function of the angle ψ between $\mathbf{N}$ and the ξ axis (or $\mathbf{U}$), since $|\vec{\nabla}\xi| = \tan\psi$. This is a cornerstone of the present model. It might be likened qualitatively to the effect of an electric field, which applies a moment on a dipole layer forcing it to align perpendicular to the field.

Adding this action to the matter action (5) (taking henceforth $d = 3$) we have for the total action

$$I = -\int F(|\vec{\nabla}\xi|) d^3x + \int \rho(\mathbf{x}) \left\{ \frac{1}{2}\mathbf{v}^2 - \varepsilon[\xi(\mathbf{x})] \right\} d^3x \,. \tag{7}$$

So far, ξ has been our brane degree of freedom. To reproduce an equation such as (2) we want ϕ to be our degree of freedom. A direct translation can be made under the following approximation: In Sec. 4 it will become clear that the NR limit, which I treat here, corresponds to small departures of ξ from the average ξ_0 of the brane. 'Small' here means much smaller than the extent over which $\varepsilon'(\xi)$ varies appreciably. We can thus take in this NR approximation $\varepsilon(\xi) \approx \varepsilon(\xi_0) + \varepsilon'(\xi_0)(\xi - \xi_0)$. Defining $\hat{a}_0 \equiv \varepsilon'(\xi_0)$ – a constant with the dimensions of acceleration – we can write $\vec{\nabla}\xi \approx \vec{\nabla}\phi/\hat{a}_0$ in the argument of F. Local shape distortions of $\mathcal{M}$ from the flat are thus perceived as gravitational fields.

The Newtonian limit occurs for steep indentations, $|\vec{\nabla}\xi| = \tan\psi \gg 1$. The DML corresponds to shallow gradients $|\vec{\nabla}\xi| = \tan\psi \ll 1$.

For the Newtonian limit, we need $F(Z \to \infty) \to KZ^2 + const.$, where K sets the general normalization of F and carries its dimensions. In this limit we have

$$I = -\frac{1}{8\pi G} \int (\vec{\nabla}\phi)^2 d^3x + \int \rho(\mathbf{x}) \left[\frac{1}{2}\mathbf{v}^2 - \phi(\mathbf{x}) \right] d^3x \,, \tag{8}$$

where we identified the Newton constant G as

$$G = \frac{\hat{a}_0^2}{8\pi K}\,. \tag{9}$$

Define the dimensionless $f(Z) \equiv F(Z)/K$, in terms of which the brane action is $-(\hat{a}_0^2/8\pi G)\int f(Z)d^3x$.

In the limit of small gradients, which corresponds to the DML, we take guidance from the requirement that the theory be SI. The action for the brane thus has to be invariant to $\mathbf{x} \to \lambda\mathbf{x}$ when $Z \equiv |\vec{\nabla}\xi| \ll 1$. A deeper theory, which I do not have, should tell us why this is so; but in default of such a theory we impose this requirement on the brane Lagrangian. The equation of motion of masses being SI tells us that the potential, or ξ is not to change under the scaling operation. The free action for masses is thus scale invariant, and so must then the brane action be in the limit $|\vec{\nabla}\xi| \ll 1$. This means that to lowest orders in Z we can write

$$f(Z \ll 1) = f(0) + (2/3)\kappa Z^3\,, \tag{10}$$

where κ is a dimensionless parameter, which I assume to be of order unity (see footnote c). The constant $f(0)$ does not play a role in local dynamics, but it is important in setting the global structure of the brane, since $Kf(0)$ represents the tension, or energy density of the brane unperturbed by masses. Also, note importantly that $\hat{a}_0^2 f(0)$ plays the role of a cosmological constant, since $Z \ll 1$ almost everywhere in our toy 'universe'. We already see that it is of order $\hat{a}_0^2$, in agreement with the 'coincidence' (1). This will be discussed in Sec. 4. The SI simply means that the contribution to the Lagrangian due to the indentations caused by matter is invariant to stretching the brane uniformly (without changing the depth of the indentations).

Defining the MOND constant as[c] $a_0 \equiv \hat{a}_0/\kappa$, this limit of the action is

$$I = -\frac{1}{12\pi\mathcal{A}_0}\int[(\vec{\nabla}\phi)^2]^{3/2}d^3x + \int\rho(\mathbf{x})\left[\frac{1}{2}\mathbf{v}^2 - \phi(\mathbf{x})\right]d^3x\,, \tag{11}$$

$(\mathcal{A}_0 \equiv Ga_0)$ and we have in Eq. (11) the required DML action leading to the field DML equation (3).

More generally, the 'interpolating function' appearing in Eq. (2) (whose argument is $|\vec{\nabla}\phi|/a_0$) is related to $f(Z)$ by

$$\mu(x) = [f'(Z)/2Z]_{Z=x/\kappa}\,. \tag{12}$$

$f(Z)$ can be obtained from $\mu(x)$ as $f(Z) = 2\int_0^Z Z\mu(\kappa Z)dZ + f(0)$. In terms of the equivalent $\nu(y)$ interpolating function, also commonly used, [defined by $x = \kappa Z = y\nu(y)$], we can write $f(Z) = (2/\kappa)\int ydZ = 2\kappa^{-2}\{y^2(Z)\nu[y(Z)] - \int^{y(Z)}\nu(y')y'dy'\}$, where $y(Z) = \kappa Z\mu(\kappa Z)$.

[c]This definition of a_0 corresponds to the standard normalization with which the mass-asymptotic-speed relation reads $MGa_0 = V_\infty^4$. It is a convenient normalization when we consider DML physics. a_0 has also the role of the boundary between the Newtonian and DML regimes at $|\vec{\nabla}\xi| \sim 1$, or $|\vec{\nabla}\phi| \sim \hat{a}_0$; here $\hat{a}_0$ is perhaps a more convenient normalization. An implicit assumption that accompanies MOND's basic tenets is that there are no small dimensionless constants involved, which in particular says that $a_0 \sim \hat{a}_0$, or $\kappa \sim 1$, as assumed.

4. Global Considerations

Noting the connection that MOND points to, between local dynamics and global properties of our observable Universe, as in Eq. (1), we need to extend the above planar-brane picture to one with a brane of some characteristic radius. I take it to have a topology of a 3-D sphere embedded in a Euclidean 4-D space, $\mathcal{E}$. The brane geometry is that of an approximate sphere of radius ℓ_0 centered at some preferred point in $\mathcal{E}$, distorted by local 'indentations' caused by masses. As before, the perturbations on the perfect sphere constitute the gravitational field sensed by us in the brane. The preferred direction, $\mathbf{U}$ in $\mathcal{E}$, is now radial from some center, and ξ is the radial coordinate measured from this center.

If $(\xi, \chi, \theta, \varphi)$ are spherical coordinates in $\mathcal{E}$ with respect to this center, the brane is described by

$$\xi(\Omega) = \xi(\chi, \theta, \varphi) = \ell_0 + \zeta(\Omega), \tag{13}$$

with ζ having the same sign everywhere.[d]

Already at the outset, this picture raises several important questions:

(1) Since we are seeking to account for a relation of the form $\hat{a}_0 \sim c^2/\ell_0$, what is the interpretation of c, and how is it brought into the picture?

I shall argue that in the model, c is naturally to be identified with the speed of small perturbations on the brane. It enters as the (square root of the) brane's tension-to-density ratio, which is determined – for a balanced brane – by the supporting force-per-unit-mass ($\sim \hat{a}_0$), and by the global radius, ℓ_0.

(2) In the planar case, the only 'constant of nature' that appears besides G is $\hat{a}_0$ and the only transition we had to consider is from small to large gradients, $\vec{\nabla}\xi = \vec{\nabla}\phi/\hat{a}_0$. Now there is an additional 'constant', ℓ_0. In most of the discussion I will assume $|\zeta(\Omega)| \ll \ell_0$, namely local gravity constitutes small perturbation on the global spherical geometry of the brane. But what is the translation of this condition into in-brane dynamics, and what is the meaning of $\zeta \not\ll \ell_0$, where one would clearly need to modify the treatment?

I will argue that $|\zeta(\Omega)| \ll \ell_0$ corresponds to the NR regime $\phi \ll c^2$, while $\zeta \not\ll \ell_0$ takes us into the regime of relativistic gravity. This follows naturally once we have made the connection $\hat{a}_0 \sim c^2/\ell_0$, since $\phi = \hat{a}_0\zeta$.

(3) In the planar case we could ignore the global brane dynamics. But now we have to account for the external forces on the brane itself. If the brane is assumed static (constant ℓ_0) on average, what balances it? And if ℓ_0 is time dependent, what is the dynamics of these changes.

(4) In the planar case we do not have a natural length scale for variation of $\varepsilon(\xi)$, $\varepsilon'(\xi)$, etc. Here ℓ_0 provides such a scale. This could open whole new avenues

[d]Here I shall confine myself to perturbations $|\zeta| \ll \ell_0$ – which we shall see correspond to the NR limit – so the sign of ζ need not be specified. It will become important when we will want to consider, in the future, $|\zeta| \not\ll \ell_0$.

for extending MOND as we now know it. For example, ξ dependence of ε' translates effectively (but not quite) into dependence of a_0 on the potential.

Jacob proposed at some point[64] that letting a_0 depend on the local gravitational potential might account for the galaxy-cluster conundrum of MOND; and this proposal was studied in more detail in Ref. 65. But there a_0 is required to change substantially within a potential range $\ll c^2$. Here, if ℓ_0 is the only scale length that appears, I speak of effective changes in a_0 from the NR to the relativistic regime, or, more accurately, of a_0 losing its role and meaning altogether in the relativistic regime. But then again, if $\varepsilon'(\xi)$ does change appreciably with ξ near $\xi = \ell_0$ – which introduces another length scale $\ell_1 \ll \ell_0$ into the model – then a_0 could depend strongly on the potential over potential changes of order $(\ell_1/\ell_0)c^2$.[e] Without the geometrical underlying picture, such a dependence would seem rather ad-hoc and unnatural. Here this is quite natural since even adding a constant to ϕ in some region of space, is tantamount to pushing that region of the brane to another value of ξ, where $a_0 \propto \varepsilon'(\xi)$ may be different.

(5) Equation (2) has no solution on a boundaryless brane – such as treated here – unless the total source vanishes. So, if Eq. (2) is to emerge, how can this condition be satisfied?

We shall see that the elements required for brane balance introduce terms to the source of Eq. (2) that automatically annul the total source. This shows that the global dynamics of the brane must be accounted for even when we speak of local dynamics. It thus supports the claimed made repeatedly in the context of MOND that cosmology and local MOND dynamics must be understood as two aspects of the same construct (as opposed to first constructing a local MOND theory from which cosmology will also emerge).

The arena where we describe dynamics is the sphere S_0 of radius ℓ_0, where a convenient choice of coordinates is $\Omega = \chi, \theta, \varphi$. But we will want to write the dynamics in covariant form (with respect to the S_0 geometry), so that they will automatically reduce to the planar case for local dynamics.

The Euclidean geometry of $\mathcal{E}$ induces the metric $g_{\mu\nu}$ on S_0, and the metric $\hat{g}_{\mu\nu}$ on $\mathcal{M}$. In the (χ, θ, φ) coordinates, the metric determinants are respectively

$$g^{1/2} = \ell_0^3 \sin^2\chi \sin\theta; \quad \hat{g}^{1/2} = \xi^3 \sin^2\chi \sin\theta [1 + \eta_{,\chi}^2 + \eta_{,\theta}^2/\sin^2\chi + \eta_{,\varphi}^2/\sin^2\chi \sin^2\theta]^{1/2}, \tag{14}$$

where $\eta \equiv \ln\xi$. The volume element on S_0 is $dV = g^{1/2} d\chi d\theta d\varphi = \ell_0^3 d\Omega$. We can write the ratio of the volume element on $\mathcal{M}$ to its projection on S_0 as

$$(\hat{g}/g)^{1/2} = (\xi/\ell_0)^3 [1 + (\ell_0/\xi)^2 \xi_{,\mu} g^{\mu\nu} \xi_{,\nu}]^{1/2}. \tag{15}$$

If we put $\ell_0 = \xi$, the local value of ξ at some point, this ratio is the local value of $1/\cos\psi$, where ψ is the angle between the normal to the brane and the radial direction. So,

$$\cos\psi = [1 + \eta_{,\chi}^2 + \eta_{,\theta}^2/\sin^2\chi + \eta_{,\varphi}^2/\sin^2\chi \sin^2\theta]^{-1/2}, \tag{16}$$

[e] Expand $\varepsilon'(\xi) \approx \varepsilon'(\ell_0) + \varepsilon''(\ell_0)\zeta \approx \varepsilon'(\ell_0) + [\varepsilon''(\ell_0)/\varepsilon'(\ell_0)]\phi$, and define $\ell_1 = [\varepsilon''(\ell_0)/\varepsilon'(\ell_0)]^{-1}$. Then $\varepsilon'(\xi) \approx \ell_0^{-1} c^2 (1 + \ell_0\phi/\ell_1 c^2)$.

Or,

$$\cos\psi = (1 + Z^2)^{-1/2}; \quad \tan\psi = Z; \quad Z \equiv (\ell_0/\xi)(\zeta_{,\mu}g^{\mu\nu}\zeta_{,\nu})^{1/2}. \tag{17}$$

Under the assumption $|\zeta(\Omega)| \ll \ell_0$, which we restrict ourselves to, we can write to lowest order in ζ/ℓ_0

$$Z = (\zeta_{,\mu}g^{\mu\nu}\zeta_{,\nu})^{1/2}. \tag{18}$$

Similar to the planar case, we want an equation from which we can determine the 'gravitational potential', $\xi(\Omega)$, and that locally will be of the form of Eq. (2). We want the effective action for ξ and its field equation to be covariant with regard to the geometry of S_0. In other words, we want the field equation to be of the form

$$[\mu(Z)\zeta_{;}{}^{\mu}]_{;\mu} = 4\pi GS, \tag{19}$$

where all the geometrical operations (covariant derivatives, raising of indices, etc.) are for $g_{\mu\nu}$.

An important issue immediately arises as alluded to in point 5 above: An equation such as Eq. (19) does not have solutions in a space with the topology of a sphere, unless its sources total to zero: Applying Gauss's theorem to Eq. (19), the volume integral of the left-hand side gives a boundary term, which vanishes because S_0 has no boundary. Thus it follows that the total source $\int_{S_0} S\, dV = 0$.

The source term of this equation, S, would have to be, for some reason, not proportional to the local mass density ρ, but to some $\rho + s$ with $\int_{S_0}(\rho + s)dV = 0$. But, since we want to retain ρ as the local source of gravity to a good approximation we will have s itself nearly homogeneous. This will automatically introduce global quantities into the source of our field equation: a crucial element of the model. This will come about when we include in the model the global balance of the brane.

4.1. *Bubble universe*

To the geometry as described above we add the following physics. The brane and matter on the brane are subject to a central potential field $\varepsilon(\xi)$, each couples to it with its own density.

4.1.1. *The matter action*

For the reasons listed above, we assume, as before, that only motion tangent to S_0 is subject to inertia, and enter the kinetic Lagrangian of matter; motion in the ξ direction is not associated with inertia.

The matter degrees of freedom, $\mathbf{x}$ appear in the action as

$$\int dV\rho\left\{\frac{1}{2}\left[\frac{\xi(\mathbf{x})}{\ell_0}\right]^2 \mathbf{v}^2 - \varepsilon[\xi(\mathbf{x})]\right\}, \tag{20}$$

ρ is the mass per unit volume on S_0. The velocity that enters inertia is taken as $(\xi/\ell_0)\mathbf{v}$, where $\mathbf{v}$ is the velocity of the projection of the body's position (at radius ξ) on S_0.

The matter equation of motion is thus

$$\left(\frac{\xi}{\ell_0}\right)^2 \dot{\mathbf{v}} = -\varepsilon'(\xi)\vec{\nabla}\xi - \frac{2\xi}{\ell_0^2}\mathbf{v}(\mathbf{v}\cdot\vec{\nabla}\xi) + \frac{\xi}{\ell_0^2}\mathbf{v}^2\vec{\nabla}\xi\,. \tag{21}$$

Restricting to systems with $\zeta \ll \ell_0$ – which is tantamount to the NR limit – we take the lowest order in ζ/ℓ_0. So, as before, we define $\hat{a}_0 \equiv \varepsilon'(\ell_0)$, $\phi \equiv \hat{a}_0\zeta$, and put $\xi/\ell_0 \approx 1$ to get

$$\dot{\mathbf{v}} = -\vec{\nabla}\phi - \frac{2}{\hat{a}_0\ell_0}\mathbf{v}(\mathbf{v}\cdot\vec{\nabla}\phi) + \frac{1}{\hat{a}_0\ell_0}\mathbf{v}^2\vec{\nabla}\phi\,. \tag{22}$$

$\hat{a}_0$ plays the role of the MOND constant, $\hat{a}_0 \sim a_0$. In the real world, and as we shall see in the model too, $\hat{a}_0 \sim c^2/\ell_0$, where c is the propagation speed of small perturbations on the brane, with the role of the speed of light. Thus, the last two terms on the right-hand side of Eq. (22) are suppressed by a factor $\sim v^2/c^2$ compared with the first and can be neglected in our approximation ($v^2/c^2 \sim \phi/c^2 \sim \zeta/\ell_0 \ll 1$) to give

$$\dot{\mathbf{v}} = -\vec{\nabla}\phi\,. \tag{23}$$

4.1.2. *The brane action*

The action terms that involve the brane's degree of freedom $\xi(\Omega)$ and determine its shape are

$$I = \int \{F(\xi, Z) + \varepsilon(\xi)[\rho(\Omega) + \sigma]\}dV\,. \tag{24}$$

Here, F depends on the shape of the brane and *on its local orientation with respect to the external, preferred direction* **U**. The angle, ψ, between the local brane normal and **U** being $\tan\psi = Z = (\ell_0/\xi)(\xi_{,\mu}g^{\mu\nu}\xi_{,\nu})^{1/2}$. For example, if this Lagrangian density is proportional the volume element of the brane $F(\xi, Z) = q(\xi/\ell_0)^3(1 + Z^2)^{1/2}$. The third term is the interaction of the brane with the external potential; σ is assumed constant and can be interpreted as the mass density of the brane (again, per unit volume on S_0).

Extremizing this action over ξ gives:

$$(\ell_0^2\xi^{-2}Z^{-1}F_{,z}\xi_{;}{}^{\mu})_{;\mu} = g^{-1/2}(g^{1/2}\ell_0^2\xi^{-2}Z^{-1}F_{,z}g^{\mu\nu}\xi_{,\mu})_{,\nu} = \varepsilon'(\rho+\sigma) + F_{,\xi} - Z\xi^{-1}F_{,z}\,. \tag{25}$$

Applying Gauss's theorem now requires not that the total mass vanish, but that

$$\int_{S_0} dV[\varepsilon'(\rho+\sigma) + F_{,\xi} - Z\xi^{-1}F_{,z}] = 0\,. \tag{26}$$

There is, in fact, a continuum of integral relations that solutions obey, namely the general higher-moments 'virial relations' for action-governed theories:[66] Multiply the field equation (25) by $g^{1/2}\xi^{\alpha}$ and integrate over S_0 to give

$$\int_{S_0} dV\xi^{\alpha}[\varepsilon'(\rho+\sigma) + F_{,\xi} + (\alpha-1)Z\xi^{-1}F_{,z}] = 0\,. \tag{27}$$

The zero-source condition (26) is the $\alpha = 0$ case.

Condition (26) can be used to eliminate the 'unobservable' σ and write the equation in terms of only the quantities that are measurable by us in $\mathcal{M}$, viz. ξ (or ϕ) and ρ. This is done at the cost of ending up with a nonlocal theory, since the right-hand side of Eq. (25) becomes

$$S = \varepsilon'\rho - \langle\varepsilon'\rho\rangle^* + F_{,\xi} - Z\xi^{-1}F_{,Z} - \langle F_{,\xi} - Z\xi^{-1}F_{,Z}\rangle^*, \tag{28}$$

where for a quantity Q

$$\langle Q\rangle^* \equiv \frac{\int Q dV}{\varepsilon'^{-1}\int \varepsilon' dV}. \tag{29}$$

While $\langle Q\rangle^*$ is not quite $\langle Q\rangle$, the volume-weighted average of Q, as it depends on ξ through $\varepsilon'(\xi)$, we do have $\int\langle Q\rangle^* dV = \int Q\, dV$.

For a constant $\rho(\Omega) = \rho_0$ the solution is $\xi = \xi_0 = $ const., and taking $\ell_0 = \xi_0$, Eq. (26) gives the equilibrium condition for the spherical bubble which determines ℓ_0:

$$\varepsilon'(\ell_0)(\rho_0 + \sigma) + F_{,\xi}(\ell_0, 0) = 0. \tag{30}$$

Restricting again to the 'nonrelativistic' case (see Sec. 4.1.3), with $\varepsilon' \approx \varepsilon'(\ell_0) \equiv \hat{a}_0$; also $\langle Q\rangle^* \approx \langle Q\rangle$ is a constant, and $Z \approx (\zeta_{,\mu}g^{\mu\nu}\zeta_{,\nu})^{1/2}$. Then write Eq. (25) as

$$[Z^{-1}F_{,Z}(\ell_0, Z)\zeta_{;}{}^{\mu}]_{;\mu} = \hat{a}_0(\rho - \langle\rho\rangle) + \ell_0^{-1}[F(\hat{F}_\xi - \hat{F}_Z) - \langle F(\hat{F}_\xi - \hat{F}_Z)\rangle]_{\xi=\ell_0}, \tag{31}$$

where $\hat{F}_\xi \equiv \partial \ln F/\partial \ln \xi$, and $\hat{F}_Z \equiv \partial \ln F/\partial \ln Z$.

To get the Newtonian limit for $Z \gg 1$ we have to have in this limit

$$F(\xi, Z) \to K(\xi)Z^2 + \bar{F}(\xi). \tag{32}$$

$\bar{F}(\xi)$ does not appear in the NR field equation. Define then, generalizing the planar case, the dimensionless

$$f(\xi, Z) \equiv F(\xi, Z)/K(\xi), \tag{33}$$

and $G \equiv \hat{a}_0^2/8\pi K(\ell_0)$, $\phi = \hat{a}_0\zeta$, and write Eq. (31) as

$$(4\pi G)^{-1}[\mu(Z)\phi_{;}{}^{\mu}]_{;\mu} = \rho - \langle\rho\rangle + \hat{a}_0^{-1}\ell_0^{-1}[F(\hat{F}_\xi - \hat{F}_Z) - \langle F(\hat{F}_\xi - \hat{F}_Z)\rangle]_{\xi=\ell_0}, \tag{34}$$

where $\mu(Z) \equiv (2Z)^{-1}f_{,Z}(\ell_0, Z)$, and $Z = (\phi_{,\mu}g^{\mu\nu}\phi_{,\nu})^{1/2}/\hat{a}_0$.

To get the required SI of the deep-MOND limit we want

$$f(\xi, Z \ll 1) \approx f_0(\xi) + (2/3)\kappa(\xi)Z^3, \tag{35}$$

and define the phenomenological MOND acceleration as $a_0 = \hat{a}_0/\kappa(\ell_0)$, so that in this limit $\mu = (\phi_{,\mu}g^{\mu\nu}\phi_{,\nu})^{1/2}/a_0$.

It will be a great advance if we will be able to understand from the microscopics of the brane why its energy function enjoys such SI in this limit, because this would essentially determine the deep-MOND limit from first principles.

It can be shown that in the nonrelativistic case I consider here, the terms other than ρ in source on the right-hand side of Eq. (34) are suppressed relative to ρ near local masses (such as a galaxy, a star, etc.), as it should if we are to reproduce a MOND equation, $[\mu(Z)\phi_{;}{}^{\mu}]_{;\mu} = 4\pi G\rho$. This can be seen as follows: First, near matter that is highly clumped, clearly $\langle\rho\rangle \ll \rho$, and we can neglect $\langle\rho\rangle$ in local dynamics. Second, the contribution to the source of the terms other than ρ is suppressed by a factor of order ζ/ℓ_0 relative to ρ. To see this, write $F(\hat{F}_\xi - \hat{F}_Z)$ as its mean $\langle F(\hat{F}_\xi - \hat{F}_Z)\rangle$, which can be very large (possibly $\gg \rho$), plus a position dependent contribution. The first cancels in the source term. The position-dependent part comes from regions where Z and ξ vary due to the presence of nearby masses. $\hat{F}_\xi$ and $\hat{F}_Z$ are of order unity or less [see, e.g. Eqs. (32), (35)]; so consider only the variation in F, which has contributions $\delta F \sim F_{,Z}Z$ and $\delta F \sim F_{,\xi}\zeta$. Take an isolated, nonrelativistic mass of density distribution ρ, which gives rise to a potential $\phi = \hat{a}_0\zeta$. Multiply Eq. (26) by ℓ_0 and subtract from Eq. (27) for $\alpha = 1$; then eliminate σ to get

$$\int dVZF_{,Z} = -\int dV\zeta\{\varepsilon'\rho - \langle\varepsilon\rho\rangle^* + \xi^{-1}[F(\hat{F}_\xi - \hat{F}_Z) - \langle F(\hat{F}_\xi - \hat{F}_Z)\rangle^*]\}. \quad (36)$$

Thus the contribution of the extra source terms to the 'mass' in Eq. (34), which is $\sim (\ell_0\hat{a}_0)^{-1}\int dVZF_{,Z}$ is $\sim (\zeta/\ell_0)\int dV\rho$, and can be neglected in our approximation. The same can similarly be shown for $F_{,\xi}\zeta$.

In summary, if the brane energy density has the appropriate dependence on the brane's shape and orientation – for microscopic reasons that I cannot pinpoint at present – NR MOND gravitational dynamics can emerge. The resulting equation is, strictly speaking, nonlocal, in that the source term involves global integrals over the whole brane. This is required by the structure of the field equation, but is also justified by the dynamics of the model requiring global near equilibrium between the brane 'elasticity' with the external force on the brane due to the ε field.

The appearance of global contributions to the source is such that they nearly annul any position-independent contribution to the source. This may offer a possibility for solving the 'old' cosmological-constant (CC) problem; viz., understanding why the vacuum energy acting as a CC does not seem to contribute to the energy content that enters cosmological evolution. It has been suggested before that by modifying GR to become nonlocal (e.g. by introducing into the action integrals over the whole of space-time) this problem can be solved (see, e.g., Refs. 67–72). Interestingly, some of these, such as Refs. 70, 72 propose to usher in such nonlocal modifications starting from local actions in some brane-world picture, as here. Here, such nonlocal modifications are, furthermore, tied with MOND in a way that reproduces naturally the MOND cosmological coincidence, Eq. (1).

4.1.3. *Meaning of the relativistic limit*

Along our treatment of the brane dynamics we have assumed in many places that we are dealing with the NR limit. This entered, e.g., when we assumed that masses move much slower than brane perturbations; when we took everywhere $\zeta \ll \ell_0$, or $\phi \ll \ell_0\hat{a}_0$, and claimed that this is tantamount to $\phi \ll c^2$; where we neglected terms suppressed by $v^2/\ell_0\hat{a}_0$ [e.g., going from Eq. (22) to Eq. (23)], claiming that this is the same as $v^2/c^2 \ll 1$; where we neglected all the source terms in Eq. (34) except ρ because they are suppressed by ζ/ℓ_0, etc. Our treatment clearly breaks down for $\zeta \not\ll \ell_0$. But why are all the above approximations valid in the NR limit? Clearly, if it is correct that $\ell_0\hat{a}_0 \sim c^2$, where c is the speed of small brane perturbations, this interpretation is valid. With the values that $\hat{a}_0 \sim a_0$ and $\ell_0 \sim \ell_\Lambda$ attain in the real Universe this is clearly so due to relation (1). But does the model, self consistently imply that $\ell_0\hat{a}_0 \sim c^2$?

So far we have not discussed the kinetics of the brane: we have had no kinetic term in the brane action, assuming that the brane adjusts its shape instantaneously to the configuration of matter. But, if σ, the brane coupling to the potential ε can also be identified with its inertial-mass density, as was the case for matter, then the equilibrium condition (26) says that $\hat{a}_0\ell_0$ is the ratio of the brane tension to its mass density.[f] This is indeed the square of the velocity of propagation of small perturbations of the brane's shape, which we can identify with c^2 of our toy model.

The intriguing coincidence (1) thus finds a natural origin in our model.

Finite brane inertia leading to a finite speed of propagation of perturbations also affects the matter free action. Space does not allow to treat this here, but it may well be that (relativistic) inertia of matter can emerge from the matter-brane interaction that we already have in the model, together with brane inertia: Energetic cost on motion of masses may come from the motion of the brane induced by matter motion. This is analogue to the mechanisms described in Ref. 73 where bodies move in a compressible fluid with which they interact. There are, in fact many instances of emergent-inertia scenarios of bodies moving in a medium with which they interact.[g]

Note in this connection that the role of c as a limiting speed of masses on the brane emerges automatically: Motion with $v > c$ leads to 'shock' formation of the brane with infinite values of $\vec{\nabla}\phi$, and so infinite brane energy densities.

There are no systems that are both relativistic $\zeta \sim \ell_0$ and in the DML $\vec{\nabla}\xi \ll 1$ because this would mean that the extent of the system is $\gg \ell_0$ which is not possible. This is the model geometrical meaning of the observation that in the real Universe we cannot have systems with $\phi \sim c^2$ and $|\vec{\nabla}\phi| \ll a_0$. For example we cannot have DML black holes because this would require their Schwarzschild radius to be $\gg \ell_\Lambda$.

Because relativistic gravity is described by many degrees of freedom (e.g., the metric components and possibly additional fields) extending the model to the full relativistic regime may require embedding in a space of higher codimension, with enough extra dimensions to account for the many gravitational degrees of freedom. For example, accommodating six gauge-fixed components of a 4-D metric on a 4-D brane would require a ten-dimensional embedding space.

---

[f] I am neglecting here the contribution of matter to the total force on the brane, making the plausible assumption that $\sigma \gg \langle\rho\rangle$. (This is certainly true if σ has a large contribution from vacuum energy density.)

[g] For example, the effective mass of electrons in solids, Euler–Heisenberg correction to the Maxwell action due to vacuum effects, etc.

We saw that $\hat{a}_0$ enters NR local dynamics as $\varepsilon'(\ell_0)$, applicable for $\zeta \ll \ell_0$. But clearly $\hat{a}_0$ (or a_0) will lose its status in the extreme-relativistic domain. As alluded to in point (4) above, to first order we can see that $\hat{a}_0$ has to be replaced by $\varepsilon'(\xi) \approx \hat{a}_0 + \varepsilon''(\ell_0)\zeta = \hat{a}_0[1 + p\phi/c^2]$, where $p \equiv (d \ln \varepsilon'/d \ln \xi)_{\ell_0}$ is presumably of order 1 (unless ε' varies quickly around ℓ_0). This introduces 'relativistic' corrections to our preceding results (together with other corrections of the same order). This cannot be described simply as MOND with a potential-dependent a_0 (for example because $\hat{a}_0$ also entered the definition of G).

But clearly, in the relativistic regime a_0 loses its significance altogether – a possibility discussed oftentimes (see e.g., Ref. 26). Various existing relativistic extension of MOND can be described as a covariant theory of gravity that hinges on an acceleration constant a_0, tends to GR in the limit $a_0 \to 0$, and satisfies the MOND

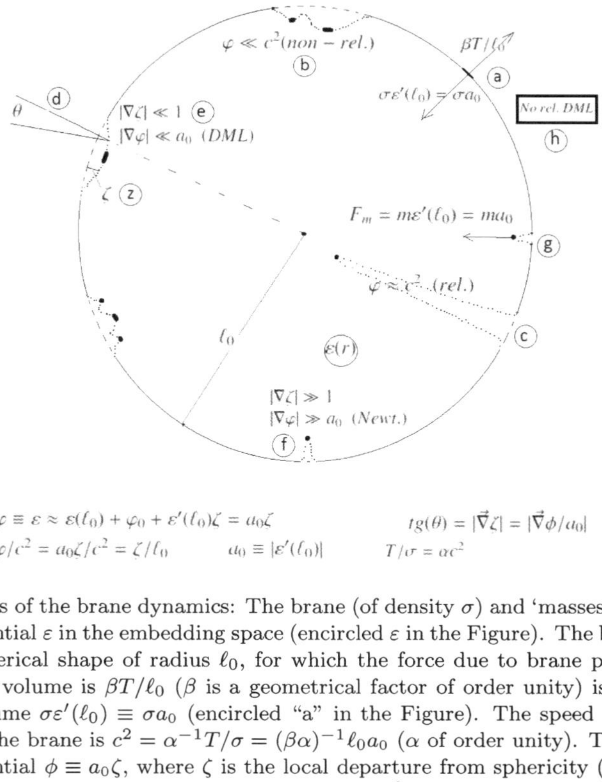

Fig. 1. Schematics of the brane dynamics: The brane (of density σ) and 'masses' (m) on it couple to a spherical potential ε in the embedding space (encircled ε in the Figure). The brane thus attains an on-average spherical shape of radius ℓ_0, for which the force due to brane pressure (tension), T, which per unit volume is $\beta T/\ell_0$ (β is a geometrical factor of order unity) is balanced by the force per unit volume $\sigma\varepsilon'(\ell_0) \equiv \sigma a_0$ (encircled "a" in the Figure). The speed of propagation of perturbations on the brane is $c^2 = \alpha^{-1}T/\sigma = (\beta\alpha)^{-1}\ell_0 a_0$ (α of order unity). The nonrelativistic gravitational potential $\phi \equiv a_0\zeta$, where ζ is the local departure from sphericity (encircled "z"). A shallow depression $\zeta \ll \ell_0$, which is equivalent to $\phi \ll c^2$, characterizes nonrelativistic gravity (encircled "b"), an approximation that is broken where $\zeta \not\ll \ell_0$ (encircled "c"). θ is the angle between the normal to the brane and the radial direction (encircled "d"). The limit $|\tan\theta| \ll 1$ – the same as $|\vec{\nabla}\zeta| \ll 1$, and the same as $|\vec{\nabla}\phi| \ll a_0$ – corresponds to the deep-MOND limit (encircled "e"), while $|\tan\theta| \gg 1$ is the Newtonian limit (encircled "f"). The ε force on a mass m is $m\varepsilon'$, which for 'nonrelativistic' depressions is $\approx m\varepsilon'(\ell_0) \approx ma_0$ (encircled "g"). This breaks down when $\phi \not\ll c^2$. A depression that is both relativistic ($\zeta \not\ll 1$), and in the deep-MOND limit ($|\vec{\nabla}\zeta| \ll 1$) is not possible, as it would have to have an extent $\gg \ell_0$ (encircled "h").

basic tenets in the NR regime. Our model points to the possibility that this is not the correct type of theory.

The general schematics and various relations that underlie the brane model are shown schematically in Fig. 1.

5. The de Sitter Context

Nonrelativistic dynamics addressed in the preceding sections can clearly not remain the last word. It is desirable to extend the picture to a model of our Universe as a 4-D space-time (ST) brane embedded, for example, in a higher-dimension Minkowski space-time. As mentioned in the introduction there are various such models in the literature proposed to various ends, but not, to my knowledge, ones that are directly relevant to MOND.

As also mentioned in the introduction, MOND seems to have some affinities with the nearly 4-D dS (hereafter dS^4) geometry of the Universe. Unlike a dS^4, our Universe was characterized by a 'big bang', and its dynamics in its cosmological past was dominated by matter. So our complete ST is certainly not dS^4. However, today it appears to be described nearly as a dS^4 (its energy density being dominated by a CC-like entity) and it will become even more so in the future. So as a first step it would be interesting to generalize the discussion in the preceding sections to a complete dS^4, as this might give relevant insights regarding our less symmetric Universe. I leave such idealized extensions to future work. Here I only point some relevant properties of dS^4 beyond those already discussed in the introduction.

A dS^4 is a maximally symmetric ST with Minkowskian signature and positive curvature. It can be described as a pseudo 4-D sphere, embedded in a 5-D Minkowski (M^5) ST, with coordinates X^A, $A = 0, 1, \ldots, 4$. The collection of all points satisfying (in this section I take $c = 1$)

$$X^2 \equiv X^A \eta_{AB} X^B = \ell_\Lambda^2 , \tag{37}$$

with the induced metric is a dS^4 space-time, of radius ℓ_Λ [$\eta_{AB} = \mathrm{diag}(-1, 1, 1, 1, 1)$].

The M^5 coordinates $X^A(\tau)$ of world line, $\mathcal{W}$, in the dS^4 satisfy Eq. (37) (τ is the proper time in both the M^5 and the dS^4). Taking the τ derivative, we get

$$X^A \eta_{AB} \dot{X}^B = 0 . \tag{38}$$

So $n^A = X^A/|X| = X^A/\ell_\Lambda$ – the unit radial vector in the M^5 at X^A – is perpendicular to all tangent vectors to the dS^4 at X^A; i.e., it is the local unit normal to the dS^4.

Differentiate Eq. (38) we get

$$\dot{X}^A \eta_{AB} \dot{X}^B + X^A \eta_{AB} \ddot{X}^B = 0 . \tag{39}$$

Since $\dot{X}^A \eta_{AB} \dot{X}^B = -1$ we have from Eq. (39)

$$n^A \eta_{AB} \ddot{X}^B = 1/\ell_\Lambda = a_\Lambda . \tag{40}$$

$a_5^A = \ddot{X}^A$ is the covariant acceleration on $\mathcal{W}$ viewed as a worldline in the M^5. We thus found that the component of a_5^A in the radial direction is always the constant a_Λ for any observer in the dS^4.

a_5^A is not the covariant acceleration on $\mathcal{W}$ viewed as a worldline in the dS^4, which is $a^\mu = D^2x^\mu/D\tau^2 = \ddot{x}^\mu + \Gamma^\mu_{\nu\alpha}\dot{x}^\nu\dot{x}^\alpha$, with $\Gamma^\mu_{\nu\alpha}$ the connection in the dS^4. So, for example, a freely falling (geodesic) observer in the dS^4 has $a^\mu = 0$, but $a_5^A = a_\Lambda n^A$. Writing $a^2 = g_{\mu\nu}a^\mu a^\nu$, where $g_{\mu\nu}$ is the metric in the dS^4, it is straightforward to show that, in accordance with Eq. (40), everywhere on any world line in the dS^4

$$a_5^2 = a^2 + a_\Lambda^2 \,. \tag{41}$$

In summary, dS^4 is a (pseudo) sphere of a fixed radius ℓ_Λ where every worldline has an acceleration component $a_\Lambda = 1/\ell_\Lambda$ in the radial direction in the M^5 embedding ST, resonating with the discussion of the previous sections.

This embedding and the accompanying facts are usually considered just a descriptive device. In the present context we should view this embedding as real with associated dynamics in the embedding ST which gives rise to the dynamics observed within the dS^4. For example, we would view the dS^4 as subject to a radial potential field $\varepsilon(\xi)$ where ξ is the radius in M^5. The action of ε on the dS^4 balances tension of the dS^4, part of which is the contribution of the possibly very large vacuum energy density. This balance occurs at some radius ℓ_Λ, thus making the effect of the large vacuum energy 'disappear' and so eliminating the 'old CC problem'. The value of ℓ_Λ at which balance occurs is very large compared with the 'natural' value for vacuum energy – e.g. the Planck length – and is observed as a small CC $\Lambda = 3\ell_\Lambda^{-2}$.

The constant radial acceleration a_Λ in $\mathcal{E}$ may seem as an artifact of no dynamical significance for matter in the dS^4. But in fact it can be said to have real observable effects within the dS^4 via the Gibbons–Hawking effect and via its modification of the Unruh temperature of accelerated observers in the dS^4 (see the discussion of these in Ref. 25 with references therein).

Extending the ideas in previous sections to dS^4 would require some serious generalizations: For example, dS^4 is not compact and would require treating contributions from the boundaries (at $t = \pm\infty$). These boundaries are Euclidean 3-spheres such as we have treated. Also, integrals over space which have appeared in our treatment will have to be generalized to ST integrals, introducing not only space nonlocality, but also time nonlocality, which, however, is not a stranger to MOND.[74] An alternative approach would be to treat the Euclidized dS^4 by transforming $t \to it$, which gives the compact 4-D sphere.

6. Discussion

Even the simplistic model discussed above already offers important insights on the interpretation of various aspects of gravity in general and of MOND in particular. This includes hints on how to extend MOND. For example, if the gravitational potential is but the distance from some center in $\mathcal{E}$ then it may give natural motivation for taking a_0 to depend on ϕ, or even more general extensions. In fact, this approach shows us how a_0 can altogether lose its role as a universal constant in the fully relativistic regime.

Some of the important elements that went into the model are as follows: (a) Our brane Universe and matter on it, couple to a potential $\varepsilon(\xi)$ that acts in the embedding space, $\mathcal{E}$. In the example model I discussed, the force acts in the radial, $\mathbf{U}$ direction (along the extra coordinate ξ) from some preferred center in $\mathcal{E}$. (b) The brane is also subject to a local moment that pulls its normal to local alignment with $\mathbf{U}$. This is required in order to get MOND dynamics. (c) It is required that matter, while it can move with the brane in the $\mathbf{U}$ direction, has no inertia in this direction: matter inertia is not isotropic in $\mathcal{E}$ but is confined to the hyperplane perpendicular to $\mathbf{U}$. Inertia of some degree of freedom is encapsuled in the free (kinetic) action; so the statement implies that only motion tangent to the local average brane enters the free action of matter. (d) The global brane configuration (e.g., a near sphere or pseudo-sphere of radius $\xi \approx \ell_0$) is determined by a balance between the brane tension and action of the external force on the brane. Local perturbations on the globally spherical brane are produced by a balance between the brane energy and the (radial) force on matter.

What we can get out from such a picture – worked out only in the 'NR' limit corresponding to small local departures of the brane radius from the constant ℓ_0 – is as follows: (a) Nonrelativistic, gravitational dynamics involving masses that may reproduce MOND behaviors. (b) The MOND constant a_0 is naturally introduced as the value of the ε force per unit mass at the brane radius: $a_0 \sim \varepsilon'(\ell_0)$. (b) The gravitational potential ϕ is proportional to the (small) local departure $\xi - \ell_0$ of the brane from a sphere. (c) The gravitational acceleration (gradient of the gravitational potential) measures the angle ψ between the local normal to the brane and $\mathbf{U}$. $|\vec{\nabla}\phi| = a_0$ corresponds to $\tan\psi \sim 1$. Thus the deep-MOND limit corresponds to very small departures of the brane normal from the radial, and the Newtonian regime corresponds to large departures. A brane Lagrangian that reproduces the exact MOND behavior that is required still has to be invoked, rather than derived from microscopics of the brane. (d) Balance of the brane, as in (d) above, implies that $a_0 \sim c^2/\ell_0$, where c is the velocity of perturbations; thus suggesting the origin of the observed MOND-cosmology 'coincidence' (1). (e) With this relation and the meaning of c, NR gravity $\phi - \phi_\infty \ll c^2$ corresponds to small departures of the brane from sphericity, $|\xi - \ell_0| \ll \ell_0$. (The requirement from matter velocities $|\mathbf{v}| \ll c$ also underlies the NR limit.) (f) The model points to a possible way out of the 'old' cosmological-constant problem, in that even a very large, but position independent contribution to the brane tension – such as a CC-like contribution – is automatically balanced out by the ε' field by having the brane adjust its radius ℓ_0. This occurs at a 'large' radius – large compared with the scales relevant to particle physics. The equilibrium radius is then perceived by us as a 'small' cosmological constant

$\ell_\Lambda^{-2} \sim \ell_0^{-2}$. To actually base a solution on this argument it will be necessary to generalize our discussion of the brane as a space-like sphere to a brane as a space-time pseudosphere. Initial look at dS^4 gives hope that this will be possible. Indeed, much remains to be elaborated on and understood perhaps from more basic principles. Almost every aspect of the model requires further elaboration. But even as it is, it seems to me that we already get out of the model more than we have put in.

References

1. J. Bekenstein and M. Milgrom, *Astrophys. J.* **286**, 7 (1984).
2. J. D. Bekenstein, *Phys. Rev. D* **70**, 083509 (2004).
3. R. H. Sanders, *Astrophys. J.* **480**, 492 (1997).
4. J. D. Bekenstein, arXiv:astro-ph/0412652 (2004).
5. J. D. Bekenstein, arXiv:0901.1524 (2009).
6. J. D. Bekenstein, *Contemp. Phys.* **47**, 387 (2006).
7. J. D. Bekenstein, *Phil. Trans. R. Soc. A* **369**, 5003 (2011).
8. M. Milgrom, *Astrophys. J.* **270**, 365 (1983).
9. B. Famaey and S. McGaugh, *Liv. Rev. Rel.* **15**, 10 (2012).
10. M. Milgrom, *Scholarpedia* **9**(6), 31410 (2014).
11. M. Milgrom, *Astrophys. J.* **698**, 1630 (2009).
12. M. Milgrom, *Mon. Not. R. Astron. Soc.* **437**, 2531 (2014).
13. S. S. McGaugh, *Phys. Rev. Lett.* **106**, 121303 (2011).
14. R. H. Sanders, *Astron. Astrophys. Rev.* **2**, 1 (1990).
15. S. S. McGaugh, *Astrophys. J.* **609**, 652 (2004).
16. X. Wu and P. Kroupa, *Mon. Not. R. Astron. Soc.* **446**, 330 (2015).
17. S. S. McGaugh *et al.*, *Phys. Rev. Lett.* **117**, 201101 (2016).
18. M. Milgrom, *Mon. Not. R. Astron. Soc.* **398**, 1023 (2009).
19. M. Milgrom, *Phys. Rev. Lett.* **117**, 141101 (2016).
20. F. Lelli *et al.*, *Astrophys. J. Lett.* **827**, 19L (2016).
21. M. Milgrom, *Astrophys. J.* **333**, 689 (1988).
22. K. G. Begeman *et al.*, *Mon. Not. R. Astron. Soc.* **249**, 523 (1991).
23. P. Li *et al.*, arXiv:1803.00022 (2018).
24. M. Milgrom, *Comm. Astrophys.* **13**(4), 215 (1989).
25. M. Milgrom, *Phys. Lett. A* **253**, 273 (1999).
26. M. Milgrom, *Canad. J. Phys.* **93**, 107 (2015).
27. A. D. Sakharov, *Sov. Phys. Doklady* **12**, 1040 (1968).
28. T. Jacobson, *Phys. Rev. Lett.* **75**, 1260 (1995).
29. T. Padmanabhan, *Rep. Prog. Phys.* **73** 046901 (2010).
30. E. P. Verlinde, *JHEP* **1104**, 29 (2011).
31. P. V. Pikhitsa, arXiv1010.0318 (2010).
32. V. V. Kiselev and S. A. Timofeev, arXiv:1009.1301 (2010).
33. C. M. Ho *et al.*, *Phys. Lett. B* **693**, 567 (2010).
34. X. Li and Z. Chang, *Commun. Theor. Phys.* **55**, 733 (2011).
35. F. R. Klinkhamer and M. Kopp, *Mod. Phys. Lett. A* **26**, 2783 (2011).
36. F. R. Klinkhamer, *Mod. Phys. Lett. A* **27**, 1250056 (2012).
37. E. Pazy and N. Argaman, *Phys. Rev. D* **85**, 104021 (2012).
38. E. Pazy, *Phys. Rev. D* **87**, 084063 (2013).

39. L. Smolin, *Phys. Rev. D* **96**, 083523 (2017).
40. E. P. Verlinde, *SciPost Phys.* **2**, 016 (2017).
41. M. Milgrom, *Phys. Rev. E* **56**, 1148 (1997).
42. M. Milgrom, *Mon. Not. R. Astron. Soc.* **403**, 886 (2010).
43. M. Milgrom, *J. Phys. A* **35**, 1437 (2002).
44. M. Milgrom, *New Astron. Rev.* **46**, 741 (2002).
45. L. Blanchet, *Class. Quant. Grav.* **24**, 3541 (2007).
46. L. Blanchet and A. Le Tiec, *Phys. Rev. D* **80**, 023524 (2009).
47. L. Blanchet and L. Heisenberg, *Phys. Rev. D* **96**, 083512 (2017).
48. L. Berezhiani and J. Khoury, *Phys. Lett. B* **753**, 639 (2016).
49. J. Khoury, *Phys. Rev. D* **93**, 103533 (2016).
50. P. Brax *et al.*, *Rept. Prog. Phys.* **67**, 2183 (2004).
51. R. Maartens and K. Koyama, *Liv. Rev. Rel.* **13**, 5 (2010).
52. L. Randall and R. Sundrum, *Phys. Rev. Lett.* **83**, 4690 (1999).
53. G. R. Dvali *et al.*, *Phys. Lett. B* **485**, 208 (2000).
54. C. Deffayet, *Phys. Lett. B* **502**, 199 (2001).
55. C. de Rham *et al.*, *Phys. Rev. Lett.* **100**, 251603 (2008).
56. J. Khoury *et al.*, *Phys. Rev. D* **64**, 123522 (2001).
57. B. Carter, arXiv:gr-qc/0012036 (2000).
58. B. Carter *et al.*, *Class. Quant. Grav.* **18**, 4871 (2001).
59. B. Carter, *Int. J. Theor. Phys.* **41**, 2201 (2002).
60. J. P. Uzan, *Int. J. Theor. Phys.* **17**, 2739 (2002).
61. B. Carter, arXiv:1112.2094 (2011).
62. R. G. Cai *et al.*, *JHEP* **10**, 9 (2018).
63. M. Milgrom, *Phys. Rev. D* **80**, 123536 (2002).
64. J. Bekenstein, Séminaires de l'IAP (2011).
65. H. Zhao and B. Famaey, *Phys. Rev. D* **86**, 067301 (2012).
66. M. Milgrom, *Phys. Lett. A* **190**, 17 (1994).
67. A. D. Linde, *Phys. Lett. B* **200**, 272 (1988).
68. N. Arkani-Hamed *et al.*, arXiv:hep-th/0209227 (2002).
69. A. A. Tseytlin, *Phys. Rev. Lett.* **66**, 545 (1991).
70. G. Dvali *et al.*, arXiv:hep-th/0208096 (2002).
71. A. Davidson and S. Rubin, *Class. Quant. Grav.* **26**, 235019 (2009).
72. G. Gabadadze, *Phy. Lett. B* **739**, 263 (2014).
73. M. Milgrom, *Phys. Rev. D* **73**, 084005 (2006).
74. M. Milgrom, *Ann. Phys.* **229**, 384 (1994).

Who Ordered That?
On the Origin of LIGO's Merging Binary Black Holes

Tsvi Piran
Racah Institute of Physics, The Hebrew University of Jerusalem,
Jerusalem 91904, Israel
https://scholars.huji.ac.il/tsvipiran/home
tsvi.piran@mail.huji.ac.il

Kenta Hotekezaka
Department of Astrophysical Sciences, Princeton University,
Princeton, NJ 08544, USA
kentah@astro.princeton.edu

LIGO's detection of gravitational waves from binary black hole mergers was an unexpected surprise that immediately raised the question — what is the origin of these black hole binaries? The "simplest" scenario is evolution of field massive stellar binaries. However, other possibilities involving capture have been proposed. We explore here one of the more interesting clues on this puzzle: the relatively modest spins of the resulting black holes that imply that the progenitor black holes were not spinning rapidly. More specifically we consider the implication of observed distribution of, $\chi_{\rm eff}$, the mass weighted projected (along the orbital axis) spins on the field evolution scenario. In all cases $\chi_{\rm eff}$ is small and in two of the cases the best fit value is negative. Only in one event the spin is positive at 90% credible. These observations are puzzling within the field binary scenario in which positive higher spins ($\chi_{\rm eff} \geq 0.5$) are expected. At first sight one may expect that this rules out the field evolutionary scenario. Indeed we show that with typical parameters a significant fraction ($\geq 25\%$) of the mergers should have high effective spin values. However, uncertainties in the outcome of the common envelope phase (the typical separation and whether the stars are rotating or not) and in the late stages of massive star evolution (the strength of the winds) make it impossible to rule out, at present, these scenarios. While observations of mergers with high effective spin will support this scenario, future observations of negative spin mergers would rule it out.

Keywords: Gravitational waves; black holes.

1. Prolog

Jacob Bekenstein is renowned for his deep insight and devotion to basic physics problems such as black hole's entropy, entropy bounds and TEVES. Less is known about his interest in astrophysics and his significant contributions to this field. Naturally, Jacob has worked mostly, but not only, on the astrophysics of black holes and gravitational collapse. Among this work Bekenstein 1973,[1] Bekenstein and Bowers 1974[2] and Bekenstein 1976[3] that deal with kicks give to black holes during gravitational collapse and with black holes in binary systems gained renewed significant forty years later with LIGO's discovery of merging black holes binaries, that we discuss here.

2. Introduction

Gravitational-wave astronomy begun on September 14, 2015 with LIGO's discovery[4] of GW150914, a binary black hole (BBH) merger.[a] Somewhat surprisingly this merger involved two massive black holes ($36m_\odot$ and $29m_\odot$). An additional BBH merger, GW151226, as well as a merger candidate, LVT151012 were discovered in LIGO's O1 run. Three other events, GW170104,[6] GW170608[7] and GW170814,[8] the latter one jointly with Virgo, were discovered in the O2 run. Most of these BBHs involved rather massive progenitor BHs that are larger than stellar mass black holes detected so far in X-ray binaries. The lightest black hole mass (observed in GW170608) is $7m_\odot$.

Among the most remarkable features of all six events are the relatively low values (ranging between -0.12 and $+0.21$) of the mass weighted projected (along the orbital axis) spins $\chi_{\rm eff}$ of the progenitors BHs. In two cases the best fit values are negative (but the errors don't exclude zero), in three cases it is practically zero and only in one case the best fit value is positive but even this value is small. These values are best fitted by a low-spin isotropic distribution[9,10]), suggesting that the progenitor black holes were not rotating rapidly and that their orientation was random.

This result is in some "tension" with the expectations from field binary evolution scenarios in which we expect that the individual spins should be both large and aligned with the orbital angular momentum axis. Namely, we expect that in this scenario a significant fraction of the mergers would have a large ($\gtrsim 0.5$) $\chi_{\rm eff}$.[11,12] The essence of the argument that suggests a large $\chi_{\rm eff}$ is the following[12,13]:

(i) To merge within a Hubble time, $t_{\rm H}$, the initial semi-major axis of the BBH at the moment of the formation of the second BH, a, should be small. Otherwise the gravitational waves merger time, that is proportional to a^4, would be too large.
(ii) With a relatively small separation the stars feel a significant tidal force and their spin tends to be synchronized with the orbital motion. This is particularly relevant for the spin of the second star that collapses.

[a]Indirect evidence for gravitational radiation was discovered in the binary Pulsar,[5] but LIGO's was the first direct detection.

(iii) If synchronized, the stellar spin, S, is large relative to the maximal angular momentum of a black hole with the same mass, Gm^2/c. that is $\chi_* \equiv Sc/Gm^2 > 1$.
(iv) If there is no kick during the collapse, the progenitor's spin, χ_*, will determine the black hole's spin, χ_{BH}.
(v) Hence we expect that χ_{BH} will be large and it will be oriented along the orbital axis. Therefore $\chi_*(t_{\rm H})/2 < \chi_{\rm eff} \lesssim 1/2$ if only the secondary has been synchronized and $\chi_*(t_{\rm H}) < \chi_{\rm eff} \lesssim 1$ if both progenitors have been synchronized. $\chi_*(t_{\rm H})$ is the spin ratio of a star in a binary system that merges within a Hubble time.[b]

The most favorable "standard"[c] evolutionary scenarios involves Wolfe–Rayet (WR) stars. These are massive stars that lost their envelopes. We compare here the observed distribution with those expected in this scenario. We begin in Sec. 3 with a discussion of the gravitational wave observations. We turn in Sec. 4 to observations of Galactic X-ray binaries containing BHs. In Sec. 5, we express the initial semi-major axis, a, in terms of the the merger time, t_c, and we estimate $\chi_*(t_c)$ in terms of the progenitor's parameters t_c. We include in these calculations also the effect of angular momentum loss via winds during the late stages of the stellar evolution. In Sec. 6 we discuss possible changes in the spin during the collapse to the black hole. In Sec. 7, we calculate, using these estimates, the expected spin distribution in different scenarios and compare it to the gravitational-wave observations. We conclude in Sec. 8 and summarize the results. We then address the question raised earlier whether the observations disfavor field binary evolution models[14–18] and support capture models,[19–28] in which the spins are expected to be randomly oriented.

3. Binary BH Mergers Observations

The relevant observed properties of the BBH merger events are summarized in Table 1. The most interesting ones for our purpose are the BHs' masses and their $\chi_{\rm eff}$ values. This latter quantity is defined as:

$$\chi_{\rm eff} \equiv \frac{m_1\chi_1 + m_2\chi_2}{m_{\rm tot}} \,, \tag{1}$$

where

$$\chi_{1,2} \equiv \frac{c\vec{S}_{1,2} \cdot \hat{L}}{Gm_{1,2}^2} \,, \tag{2}$$

and $m_{tot} = m_1 + m_2$ and $\hat{L}$ is a unit vector in the direction of the system's orbital angular momentum $\vec{L}$. The limits on $\chi_{\rm eff}$ are obtained from the observations of the gravitational-wave signals before and after the merger. In particular the lack

[b]For simplicity we assume in most, but not all, of the discussion that both black holes have similar masses.
[c]Mixed star scenario.

Table 1. Parameters of the BBH mergers detected during LIGO's O1 and O2 Run. The parameters are median values with 90% credible intervals. The values are taken from Refs. 6–8, 31 and 32.

Event	m_1 $[m_\odot]$	m_2 $[m_\odot]$	$m_{\rm tot}$ $[m_\odot]$	$\chi_{\rm eff}$	a_f
GW150914	$36.2^{+5.2}_{-3.8}$	$29.1^{+3.7}_{-4.4}$	$65.3^{+4.1}_{-3.4}$	$-0.06^{+0.14}_{-0.14}$	$0.68^{+0.05}_{-0.06}$
GW151226	$14.2^{+8.3}_{-3.7}$	$7.5^{+2.3}_{-2.3}$	$21.8^{+5.9}_{-1.7}$	$0.21^{+0.20}_{-0.10}$	$0.74^{+0.06}_{-0.06}$
LVT151012	23^{+18}_{-6}	13^{+4}_{-5}	37^{+13}_{-4}	$0.0^{+0.3}_{-0.2}$	$0.66^{+0.09}_{-0.10}$
GW170104	$31.2^{+8.4}_{-6.0}$	$19.4^{+5.3}_{-5.9}$	$50.7^{+5.9}_{-5.0}$	$-0.12^{+0.21}_{-0.30}$	$0.64^{+0.09}_{-0.20}$
GW170608	12^{+7}_{-2}	7^{+2}_{-2}	19^{+5}_{-1}	$0.07^{+0.23}_{-0.09}$	$0.69^{+0.04}_{-0.05}$
GW170814	$30.5^{+5.7}_{-3.0}$	$25.3^{+2.8}_{-4.2}$	$55.9^{+3.4}_{-2.7}$	$0.06^{+0.12}_{-0.12}$	$0.70^{+0.07}_{-0.05}$

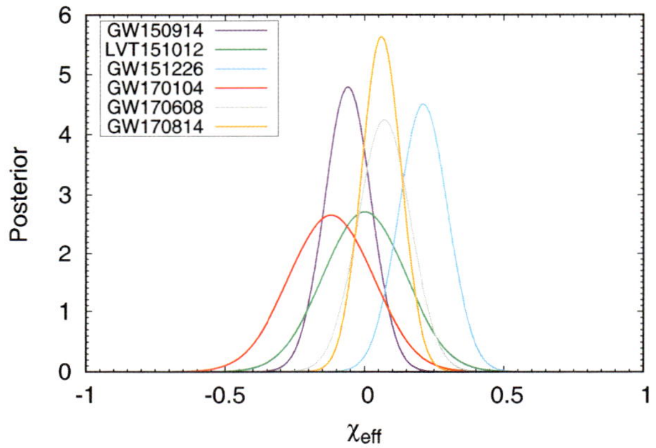

Fig. 1. The distribution of the observed spins. We have approximated each observed distribution as a Gaussian whose mean value and 90% credible interval are the values shown in Refs. 6–8, 31, 32 (see also Ref. 9).

of extended ringdown phases puts limits on the spins of the final BHs, a_f. The fact that $a_f \sim 0.6$–0.7 is an independent evidence that the initial aligned spins of the BHs were not close to unity.[29,30] Had the initial aligned spins been large, the final spin of the merged BHs would have been very close to unity and the GW signal would have had a long ringdown phase. Indeed, the final spin is slightly larger ($0.74^{+0.06}_{-0.06}$) for GW151226, for which the $\chi_{\rm eff}$ is largest. Figure 1 describes the observed $\chi_{\rm eff}$ distribution in terms of the corresponding six Gaussians corresponding to the posterior distributions of the observed events.

4. Galactic BHs in X-Ray Binaries

Observations of X-ray binaries involving BHs, albeit smaller mass ones, can also shed some light on the problem at hand. In particular observations of two such systems that include massive ($> 10m_\odot$) BHs, Cyg X-1 and GRS 1915+105, provide a good evidence that these massive BHs formed in situ, in a direct implosion and without a kick.[33] For example, Cyg X-1 moves at 9 ± 2 km/s relative to the stellar association Cygnus OB3, indicating that it could have lost at most $1 \pm 0.3 m_\odot$ at BH formation. Furthermore, the minuscule eccentricity of Cyg X-1, 0.018 ± 0.0003,[34] suggests that the orbit has been circularized during the binary evolution and the collapse didn't give the system a significant kick that disturbed the circular orbit. On the other hand, low mass BH binaries, GRO J1655-40 ($m_{\rm BH} = 5.3 \pm 0.7 m_\odot$) and V404 Cyg ($m_{\rm BH} = 9.0 \pm 0.6 m_\odot$), have larger peculiar velocities of 112 ± 13 km/s and 39.9 ± 5.5 km/s respectively.[33] These last two observations suggest a natal kick prescription of $v_{k,\rm BH} \approx m_{\rm ns} v_{k,\rm ns}/m_{\rm BH}$ for low mass BHs, where $m_{\rm ns} \approx 1.4 m_\odot$ and $v_{k,\rm ns} \approx 300$ km/s.[35] In addition, Ref. 36 shows that large (> 80 km/s) natal kicks are not required to explain the observed positions of low-mass X-ray binaries in the Galaxy.

A natal kick can change the orbital parameters of the binary. For instance, the misalignment angle between the final orbital angular momentum vector and the spin vectors, that are assumed to be aligned to the initial orbital angular momentum vector, is given by (e.g. Ref. 37):

$$\delta = \tan^{-1}\left(\frac{v_{k,\rm BH} \sin\theta \sin\phi}{\sqrt{(v_{\rm rel} + v_{k,\rm BH}\cos\theta)^2 + (v_{k,\rm BH}\sin\theta\cos\phi)^2}}\right), \tag{3}$$

where θ is the angle between the kick vector and the initial orbital plane, ϕ is measured in the plane perpendicular to the orbital velocity vector of the collapsing progenitor star, and $v_{\rm rel}$ is the relative velocity between the two stars:

$$v_{\rm rel} = 680\,\text{km/s}\left(\frac{1+q}{2q^{2/3}}\right)^{3/8}\left(\frac{t_c}{1\,\text{Gyr}}\right)^{-1/8}\left(\frac{m_2}{30 m_\odot}\right)^{1/8}, \tag{4}$$

where $q \equiv m_2/m_1$. This is larger than the magnitude of the natal kicks observed in Galactic X-ray binaries, thereby it is unlikely that BH natal kicks induce a large misalignment between the orbit and the spin vectors. In the case of $v_{k,\rm BH} \ll v_{\rm rel}$, the misalignment angle, inferred from the estimates of the natal kicks of the Galactic BHs, is $\delta \lesssim v_{k,\rm BH}/v_{\rm rel} \lesssim 0.15$. Therefore we conclude that BH natal kicks could affect the misalignment between the orbit and spins only if the natal kicks of BBH-merger progenitors are significantly larger than those observed in Galactic BH binaries.

Estimates of the spins of the BHs in Cyg X-1[38] and GRS 1915+105[39] suggest that in these two systems $a/m > 0.95$. Three other BHs, LMC X-1, M33 X-7, and 4U 1543-47, whose masses are larger than $9 m_\odot$, have $\chi > 0.8$. Only one BH with a mass $> 9 m_\odot$, XTE J1550-564 has a significantly lower value ($\chi = 0.34^{+.20}_{-.28}$). It is important to note that these large spins must be obtained at birth as accretion cannot spin up a massive BH to such a high spin value.[40,41]

To summarize, the massive Galctic BHs observed in X-ray binaries have large spins and were formed with no significant kicks. As these BHs are clearly a part of evolutionary systems this suggests that other BHs that form is such systems will also have large spins.

5. Merger Time, Orbital Separation and Synchronization.

The merger time, due to gravitational radiation driven orbital decay, is:

$$t_c \approx 10\ \mathrm{Gyr}\left(\frac{2q^2}{1+q}\right)\left(\frac{a}{44R_\odot}\right)^4\left(\frac{m_2}{30M_\odot}\right)^{-3}, \tag{5}$$

where $q = m_2/m_1$ is the mass ratio. Note that we assume circular orbits here and elsewhere. This simplifying assumption is based on the expectation that the orbit will be circularized during the binary evolution and that it won't be affected by the collapse of the secondary. It is supported by the observations of Galactic binaries containing massive BHs, discussed earlier (see Sec. 4).

Tidal forces exerted by the primary, denoted by the subscript 1, will tend to synchronize the secondary star, denoted by the subscript 2. If fully synchronized the final stellar dimensionless spin (normalized to the maximal spin of a BH with the same mass) would be:

$$\chi_2 \approx 0.5\ q^{1/4}\left(\frac{1+q}{2}\right)^{1/8}\left(\frac{\epsilon}{0.075}\right)\left(\frac{R_2}{2R_\odot}\right)^2\left(\frac{m_2}{30M_\odot}\right)^{-13/8}\left(\frac{t_c}{1\,\mathrm{Gyr}}\right)^{-3/8}, \tag{6}$$

where $\epsilon \equiv I_2/m_2R_2^2$ characterizes the star's moment of inertia. The progenitor's spin, χ_2, increases with the progenitors size and decreases when t_c increases.

The synchronization process takes place over $t_{\rm syn}$:

$$t_{\rm syn} \approx 20\ \mathrm{Myr}\ \frac{(1+q)^{31/24}}{q^{33/8}}\left(\frac{\epsilon}{0.075}\right)\left(\frac{E_2}{10^{-5}}\right)^{-1}\left(\frac{R}{2R_\odot}\right)^{-7}\left(\frac{m_2}{30M_\odot}\right)^{47/8}\left(\frac{t_c}{1\mathrm{Gyr}}\right)^{17/8}, \tag{7}$$

where E_2, is a dimensionless quantity[42] characterizing the inner structure of the star. E_2 is $\sim 10^{-7}$–10^{-4} for massive main sequence stars and Wolf-Rayet (WR) stars.[42,43] The characteristic values used in Eq. (7) correspond to a WR star. For WR stars, one can show[13] that $t_{\rm synWR}$ is almost independent of M_2 and it can be expressed as:

$$t_{\rm synWR} \approx 10\ \mathrm{Myr}\ q^{-1/8}\left(\frac{1+q}{2q}\right)^{31/24}\left(\frac{t_c}{1\ \mathrm{Gyr}}\right)^{17/8}. \tag{8}$$

Because of their short stellar lifetime, WR stars are not necessarily synchronized in binary systems even with t_c of a few hundreds Myr. In addition angular momentum can be lost due to winds during the last phases of the evolution of the star. Therefore the final stellar spin depends on:

(i) χ_i, the spins of the stars at the beginning of the WR phase;
(ii) the ratio of $t_{\rm syn}$ and the lifetime of the WR star, $t_{\rm WR}$;
(iii) the angular momentum loss timescale during the WR phase, $t_{\rm wind}$ (see Refs. 12 and 13).

We characterize the angular momentum loss due to winds by defining $t_{\rm wind} \equiv J_s/\dot{J}_s$, where J_s is the spin angular momentum of the star and $\dot{J}_s$ the angular momentum loss rate. Stronger winds corresponding to shorter $t_{\rm wind}$ values. To take account both of the stellar wind and the circularization process during the short stellar life time we solve the following equation to obtain the stellar spin parameter at the end of the WR phase[13]:

$$\dot{\chi}_* = \frac{\chi_{\rm syn}}{t_{\rm syn}} \left(1 - \frac{\chi_*}{\chi_{\rm syn}}\right)^{8/3} - \frac{\chi_*}{t_{\rm wind}}, \tag{9}$$

where $\chi_{\rm syn}$ is the stellar spin parameter in the synchronized state.

6. Collapse and the BH Spin

One can expect that, unless there is too much angular momentum (that is $\chi_* \leq 1$), the collapsing star implodes and the BH that forms swallows all the collapsing stellar mass.[d] If $\chi_* > 1$ a fraction of the matter will be ejected carrying the excess angular momentum and leading to a BH with $\chi \leq 1$.[44–46] Thus we expect that

$$\chi_{\rm BH} \approx \begin{cases} 1 & \text{if } \chi_* \geq 1, \\ \chi_* & \text{if } \chi_* < 1. \end{cases} \tag{10}$$

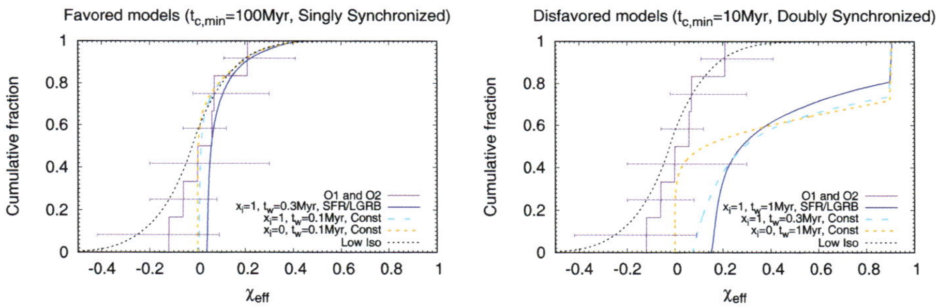

Fig. 2. The cumulative $\chi_{\rm eff}$ distribution for the O1 and O2 observing runs for the favored (*left*) and disfavored (*right*) models for WR binary scenario. We set the mass ratio, $q = 1$, and $m_{\rm tot} = 60M_\odot$ for all models. Also shown as a black dashed curve is the low-spin isotropic model of Ref. 9.

One may wonder if there are caveats to this conclusion. First, is it possible that matter is ejected during the collapse to a BH even if $\chi_* < 1$? This will, of course, change the relation between the progenitor's spin and the BH's spin. Second is mass ejected isotropically? If not the BH will receive a kick and the BBH will be put into an elliptical orbit (that will merge faster).[1] The kick may also change the resulting BH spin. Since the initial spin is in the direction of the orbital angular momentum the kick may reduce the spin component along this direction. Clearly these issues can only be addressed by a detailed numerical study of the collapse. However, as discussed in Sec. 3 observations of Cyg X-1 and GRS 1915+105. Galactic binaries containing massive ($> 10m_\odot$) BHs, provide a good evidence that massive BHs form

[d]The original stellar mass could be larger but this lost in an earlier phase due to winds.[33]

in situ in a direct implosion and without a kick.[33] Estimates of the spins of accreting massive BHs give an independent support to this conclusion.

7. A Comparison with Observations and Implications

The observed spin values are low (consistent with zero) even when compared with those expected for WR stars.[11,12] The "tension" appeared already in the first detection and it was intensified with the additional observations of low $\chi_{\rm eff}$ values and in particular with the observation of GW170104.

To quantify this issue we turn now to estimate the expected $\chi_{\rm eff}$ distribution of BBHs arising from WR binaries (see Refs. 11 and 12) and compare it with the observations. The distribution depends on the formation rate of these binaries. We consider BBH formation rates that follow the cosmic star formation rate (SFR)[47] and a constant BBH formation rate. Since WR stars are also considered to be the progenitors of type Ic Supernovae that accompany LGRBs we consider the possibility that the BBH formation rate follows the rate of Long GRBs (LGRBs). In fact, the rate of LGRBs is comparable to the rate of BBH mergers and an interesting possibility is that LGRBs accompany the formation of the massive BH that compose the BBH that eventually merge. Therefore we also consider a BBH formation history that follows the long GRB rate.[48] The resulting spin distribution arising from of the SFR model is not very different from the one that follows LGRBs. On the other hand, the results for a constant BBH formation rate are quite different as a constant rate produces a significant fraction of binaries formed at low redshifts. To merge sufficiently rapidly those systems must have small initial separations resulting in high spin mergers.

We expect that the rate of mergers follows the BBH formation rate with a time delay t_c whose probability is distributed as $\propto t_c^{-1}$. We consider a minimal time delay of 1 Myr (corresponding to an initial separation of $3 \cdot 10^{11}$ cm for a $\approx 30 m_\odot$ BBH), 10 Myr ($5.4 \cdot 10^{11}$ cm), or 100 Myr (10^{12} cm) between the formation of the BBH and its merger. These differences are important as the synchronization time depends strongly on the separation and hence on t_c. We also consider different wind timescales, $t_{\rm wind}$: 0.1, 0.3 and 1 Myr. Note that the shortest time scale considered, $t_{\rm wind} = 0.1$ Myr, roughly corresponds to a mass loss of $10^{-4.5} m_\odot$/yr which is at the level of the strongest winds observed in WR stars.[49,50] With these assumptions we obtain several probability distributions for the observed $\chi_{\rm eff}$ values. In general, the field binary scenario predicts a bimodal $\chi_{\rm eff}$ distribution with low and high spin peaks[e] (see Refs. 11, 12 for simple models and Refs. 15, 17 for population synthesis studies). Here the high spin peak corresponds to tidally synchronized binaries.

Figure 3 depicts the integrated observed distribution of $\chi_{\rm eff}$ compared with several WR models. One can see the variety of the resulting $\chi_{\rm eff}$ distribution: some models give 30–40% of high $\chi_{\rm eff}$ mergers, while $\chi_{\rm eff}$ is concentrated around zero

[e]The low and high spin peaks are around $\chi_{\rm eff} \approx 0$–0.1 and ≈ 1, respectively.

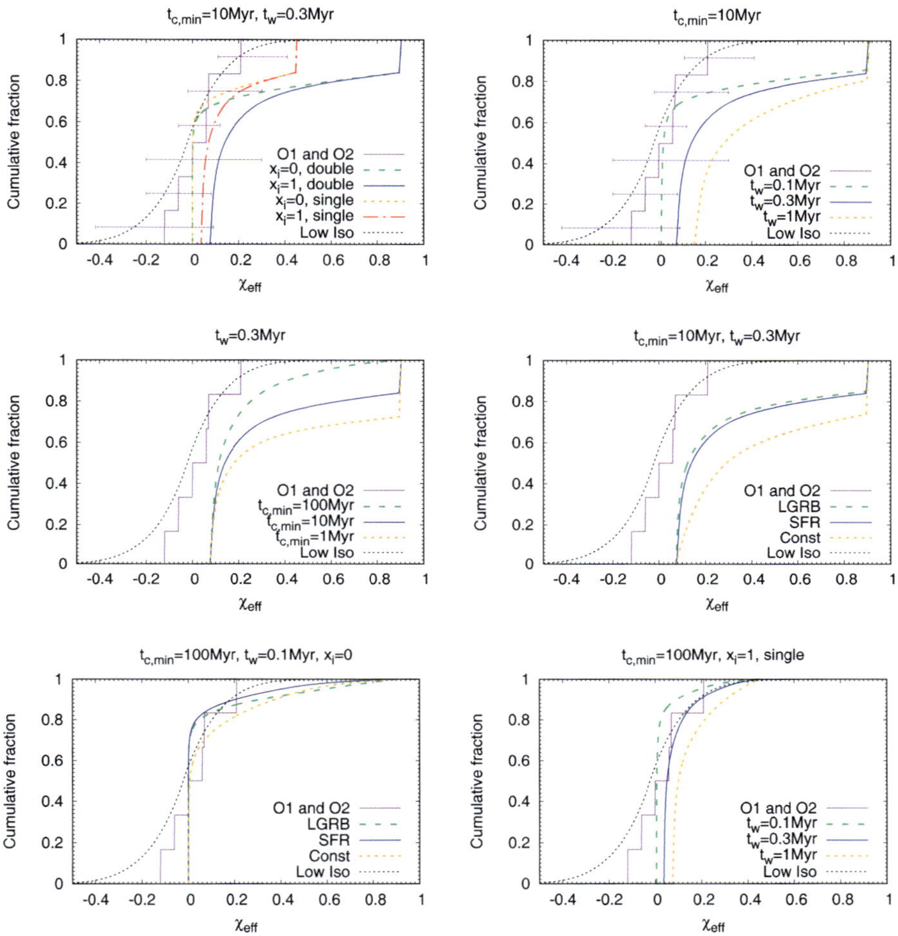

Fig. 3. The cumulative χ_{eff} distribution for the O1 and O2 observing runs for the WR binary scenario with different parameters. For the fiducial model (a blue solid line in the four upper panels), the BBH formation history follows the cosmic SFR, the two stars are synchronized at the beginning of the WR phase, the merger delay-time distribution is $\propto t^{-1}$ with a minimal time delay of $t_{c,\,\text{min}} = 10$ Myr and the wind timescale is $t_{\text{wind}} = 0.3\,\text{Myr}$. We set the mass ratio, $q = 1$, and $m_{\text{tot}} = 60M_{\odot}$ for all models. Also shown as a black dashed curve is the low-spin isotropic model in Ref. 9. The two bottom panels show models that deviate significantly from this fiducial choice (e.g. $t_{c,\,\text{min}} = 100$ Myr). In the bottom-left panel the WR stars are not synchronized initially ($\chi_i = 0$). With a strong wind and a long merger time delay distribution that follow the SFR or LGRB rates these systems produce very narrow low χ_{eff} distributions. A constant formation rate gives here a better fit to the data. In the bottom-right panel the stars are initially synchronized with a very strong wind and the long delay time the distribution is consistent with the observations.

for others. The models with the lowest χ_{eff} distributions are those in which the progenitors: (i) Are not synchronized at the beginning of the WR phase ($\chi_i = 0$); (ii) Have a strong ($t_{\text{wind}} = 0.1$ Myr) wind[f]; (iii) Have a long ($t_{c,\text{min}} = 100$ Myr)

[f]Note however that such winds might not be consistent with very massive remnants.

minimal time delay — corresponding to a large initial separation. The question whether one or two of the progenitors is influenced by the tidal interaction is secondary as it determines the largest $\chi_{\rm eff}$ values (> 0.4 or > 0.8) and those have not been observed so far.

Models with $\chi_i = 1$, a moderate wind ($t_{\rm wind} = 0.3$Myr) and a long delay (100 Myr) in which the BBH formation rate follows the SFR or LGRBs rates are consistent with the observations (apart from the nominal negative values, of course, but those could be due to the large measurement errors). The top four panels compare different models to a fiducial model in which both progenitors are spinning rapidly at the beginning of the WR phase $\chi_i = 1$, $t_{\rm wind} = 0.3$ Myr, $t_{c,\rm min} = 10$ Myr,

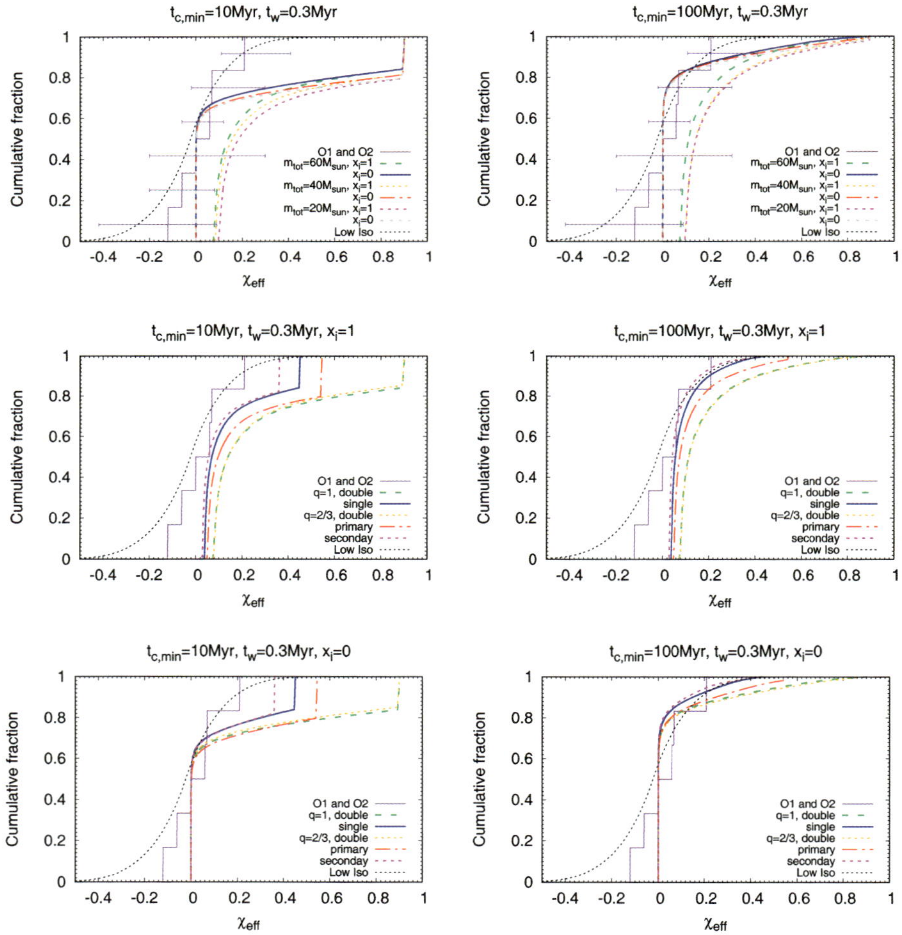

Fig. 4. The same as Fig. 3 but for different total masses and mass ratios. Here we use a BBH formation history proportional to the cosmic star formation history. In the middle and bottom panels, the curves labeled by "primary" ("secondary") show models in which one of the star i.e. the primary (secondary) is tidally synchronized.

and the BBH formation rate following the cosmic SFR. The lower two panels depict more extreme models. Here we find that if all the above conditions are satisfied then the resulting $\chi_{\rm eff}$ distribution (for SFR or LGRB rate) is too narrowly concentrated around zero. A better fit to the data is obtained under these conditions if the BBH formation rate is a constant (see bottom-left panel of Fig. 3). Even if the progenitor stars are synchronized at the beginning of the WR phase ($\chi_i = 1$) a strong enough wind ($t_{\rm wind} = 0.1$ Myr) can lead to sufficient angular momentum loss so that the final $\chi_{\rm eff}$ distributions would be concentrated close to zero (see bottom-right panel of Fig. 3).

Figure 4 shows the dependence of the $\chi_{\rm eff}$ distribution on the total mass and the mass ratio. Note that the dependence on these parameters is rather weak except for the cases in which one of the component stars is synchronized (singly synchronized). This is because of the dependence of Eq. (8) on the mass ratio. These differences do not change qualitatively our conclusions.

In order to take the relatively large measurement errors of $\chi_{\rm eff}$ into account when comparing different models with the observed data, we evaluate the odds ratios between the marginalized likelihoods of different field evolution models and the low-isotropic spin model of Ref. 9, the most favorable one among the synthetic models considered. We calculate the marginalized likelihood of each model for the six events, $p_i(d|M)$ and then combine them as $p(d|M) = \prod_i p_i(d|M)$. Tables 2 and 3 list the odds ratios of these different models of equal mass binaries with $m_{\rm tot} = 60M_\odot$ to the low-isotropic spin model, $p(d|M)/p(d|\text{Low Iso})$. The field binary evolution models with a long delay time (100 Myr) and a strong wind ($t_{\rm wind} = 0.1$ Myr) have an odds ratio of about unity. Furthermore, many of the models satisfying the conditions mentioned above have $p(d|M)/p(d|\text{Low Iso}) \gtrsim 0.1$. In spite of the two observed cases with negative mean $\chi_{\rm eff}$ values, these models cannot be ruled out with the current $\chi_{\rm eff}$ distribution of six observed events.

Table 2. Odds ratio of the models to the low-isotropic spin model for initially synchronized WR binaries and double (single) synchronization.

Model ($t_{\rm c,\,min}$)	$t_{\rm wind} = 0.1$ Myr	$t_{\rm wind} = 0.3$ Myr	$t_{\rm wind} = 1$ Myr
SFR (1Myr)	0.14 (0.24)	0.07 (0.30)	<0.001 (0.08)
LGRB (1Myr)	0.16 (0.26)	0.1 (0.36)	0.001 (0.12)
Const (1Myr)	0.04 (0.09)	0.003 (0.06)	<0.001 (0.004)
SFR (10Myr)	0.34 (0.51)	0.17 (0.70)	0.001 (0.21)
LGRB (10Myr)	0.36 (0.52)	0.24 (0.78)	0.003 (0.29)
Const (10Myr)	0.14 (0.30)	0.01 (0.20)	<0.001 (0.02)
SFR (100Myr)	1.05 (1.23)	0.52 (2.00)	0.004 (0.70)
LGRB (100Myr)	1.02 (1.17)	0.68 (2.06)	0.01 (0.89)
Const (100Myr)	0.97 (1.56)	0.08 (1.28)	<0.001 (0.14)

Table 3. Odds ratio of the models to the low-isotropic spin model for initially non-rotating WR binaries.

Model ($t_{c,\,\mathrm{min}}$)	$t_{\mathrm{wind}} = 0.1$ Myr	$t_{\mathrm{wind}} = 0.3$ Myr	$t_{\mathrm{wind}} = 1$ Myr
SFR (1Myr)	0.11 (0.22)	0.09 (0.19)	0.08 (0.22)
LGRB (1Myr)	0.12 (0.23)	0.10 (0.21)	0.09 (0.23)
Const (1Myr)	0.03 (0.08)	0.02 (0.06)	0.01 (0.08)
SFR (10Myr)	0.27 (0.47)	0.22 (0.42)	0.20 (0.38)
LGRB (10Myr)	0.28 (0.48)	0.24 (0.44)	0.22 (0.40)
Const (10Myr)	0.11 (0.28)	0.08 (0.22)	0.07 (0.18)
SFR (100Myr)	0.80 (1.10)	0.69 (1.13)	0.65 (1.07)
LGRB (100Myr)	0.79 (1.06)	0.68 (1.09)	0.64 (1.04)
Const (100Myr)	0.79 (1.43)	0.57 (1.31)	0.50 (1.19)

8. Conclusions

Before discussing the implications of these findings we turn, once more, to possible caveats. We have already argued that observations of Galactic X-ray binaries including massive BHs provide a good evidence for our model for the formation of massive BHs (no kick and no mass loss). The main open issues are all related to the late phases of the stellar binary evolution of very massive stars: (i) What are the spins of the BHs at the end of the common envelope phase? (ii) What is the separation at the end of the common envelope phase? (iii) How strong are the winds? The answers to these questions depend on better understanding these late phases.

Turning now to the results, the negative observed values are clearly inconsistent with the evolutionary model (unless there are significant kicks at the formation of the BHs[1]). However, while the observed low χ_{eff} values are at some "tension" with the expectations of the standard evolutionary scenario (see also Refs. 11–13), the large error bars of the χ_{eff} measurements don't allow us to rule out some of the field binary scenario models. These large errors combine with the small number statistics makes it impossible to make now clear conclusions.

Both the SFR and LGRB rates are favorable as proxies for the BBH formation rate. In both cases most of the formation takes place at early times, allowing for large initial separations. The resulting χ_{eff} distributions arising from these two scenarios are practically indistinguishable. A constant BBH formation rate implies more recent formation events and hence shorter merger times leading to larger χ_{eff} values. Still with extreme parameters even this distribution can be made consistent with the current data.

WR stars formed at high redshifts, $z \gtrsim 2$, are the best candidates for being progenitors with low χ_{eff}. They have a long merger time, allowing them to begin with a relatively large separation that implies much weaker synchronization. However, this is not enough and strong or moderate winds (for progenitors that are non-rotating at the end of the common envelope phase) are essential for consistency with the current distribution. A longer minimal time delay (corresponding to larger separations at the birth of BBHs) helps, but is insufficient to lead to consistency.

To conclude we note that a comparison of the currently observed O1 and O2 $\chi_{\rm eff}$ values with the models show some tension, however it does not rule out evolutionary models based on WR stars. In fact some models are almost as consistent as the best fitted low-isotropic spin model of Ref. 9. While many models predict a significant fraction ($> 25\%$) of large (> 0.4 for singly synchronized and > 0.8 for double synchronization) $\chi_{\rm eff}$ events, some produce distributions that are concentrated around positive very low $\chi_{\rm eff}$ values. The question which models are consistent depends on largely unexplored late stage evolution of very massive stars. Given these results it seems that while a significant fraction of high $\chi_{\rm eff}$ mergers will strongly support the field evolutionary scenario, lack of those will be hard to interpret. It may indicate another scenario or, for example, strong winds that remove the spin angular momentum. On the other hand a significant fraction of negative $\chi_{\rm eff}$ merger will be difficult to reconcile with this scenario, unless the BHs' angular momentum is dominated by very strong natal kicks.

Acknowledgments

KH is supported by the Flatiron Fellowship at the Simons Foundation and Lyman Spitzer Jr. Fellowship at Princeton University. TP is supported by an advanced ERC grant "TReX" and by the ISF-CHE I-Core center of excellence for research in Astrophysics.

References

1. J. D. Bekenstein, Gravitational-Radiation Recoil and Runaway Black Holes, *Ap. J.* **183**, 657 (July 1973).
2. J. D. Bekenstein and R. L. Bowers, Do OB runaways have collapsed companions?, *Ap. J.* **190**, 653 (June 1974).
3. J. D. Bekenstein, Supernovae in binaries, and the possible collapsed nature of the companion of HD 108, *Ap. J.* **210**, 544 (December 1976).
4. B. P. Abbott *et al.*, Observation of gravitational waves from a binary black hole merger, *Phys. Rev. Lett.* **116**, 061102 (February 2016).
5. R. A. Hulse and J. H. Taylor, Discovery of a pulsar in a binary system, *Ap. J. Lett.* **195**, L51 (January 1975).
6. B. P. Abbott *et al.*, GW170104: Observation of a 50-solar-mass binary black hole coalescence at redshift 0.2, *Phys. Rev. Lett.* **118**, 221101 (June 2017).
7. B. P. Abbott *et al.*, GW170608: Observation of a 19 solar-mass binary black hole coalescence, *Ap. J. Lett.* **851**, L35 (December 2017).
8. B. P. Abbott *et al.*, GW170814: A three-detector observation of gravitational waves from a binary black hole coalescence, *Phys. Rev. Lett.* **119**, 141101 (October 2017).

9. W. M. Farr *et al.*, Distinguishing spin-aligned and isotropic black hole populations with gravitational waves, *Nature* **548**, 426 (August 2017).
10. S. Vitale, D. Gerosa, C.-J. Haster, K. Chatziioannou and A. Zimmerman, Impact of Bayesian priors on the characterization of binary black hole coalescences, *Phys. Rev. Lett.* **119**, 251103 (December 2017).
11. M. Zaldarriaga, D. Kushnir and J. A. Kollmeier, The expected spins of gravitational wave sources with isolated field binary progenitors, *MNRAS* **473**, 4174 (January 2018).
12. K. Hotokezaka and T. Piran, Implications of the low binary black hole aligned spins observed by LIGO, *Ap. J.* **842**, 111 (June 2017).
13. D. Kushnir, M. Zaldarriaga, J. A. Kollmeier and R. Waldman, GW150914: spin-based constraints on the merger time of the progenitor system, *MNRAS* **462**, 844 (October 2016).
14. K. Belczynski, D. E. Holz, T. Bulik and R. O'Shaughnessy, The first gravitational-wave source from the isolated evolution of two stars in the 40-100 solar mass range, *Nature* **534**, 512 (June 2016).
15. K. Belczynski *et al.*, GW170104 and the origin of heavy, low-spin binary black holes via classical isolated binary evolution, arXiv e-prints (June 2017).
16. S. Stevenson *et al.*, Formation of the first three gravitational-wave observations through isolated binary evolution, *Nature Commun.* **8**, 14906 (April 2017).
17. K. Postnov and A. Kuranov, Black hole spins in coalescing binary black holes, arXiv e-prints (June 2017).
18. R. O'Shaughnessy, D. Gerosa and D. Wysocki, Inferences about supernova physics from gravitational-wave measurements: GW151226 spin misalignment as an indicator of strong black-hole natal kicks, *Phys. Rev. Lett.* **119**, 011101 (July 2017).
19. K. Ioka, T. Chiba, T. Tanaka and T. Nakamura, Black hole binary formation in the expanding universe: Three body problem approximation, *Phys. Rev. D* **58**, 063003 (September 1998).
20. C. L. Rodriguez, C.-J. Haster, S. Chatterjee, V. Kalogera and F. A. Rasio, Dynamical formation of the GW150914 binary black hole, *Ap. J. Lett.* **824**, L8 (June 2016).
21. R. M. O'Leary, Y. Meiron and B. Kocsis, Dynamical formation signatures of black hole binaries in the first detected mergers by LIGO, *Ap. J. Lett.* **824**, L12 (June 2016).
22. F. Antonini and F. A. Rasio, Merging black hole binaries in galactic nuclei: Implications for advanced-LIGO detections, *Ap. J.* **831**, 187 (November 2016).
23. I. Bartos, B. Kocsis, Z. Haiman and S. Márka, Rapid and bright stellar-mass binary black hole mergers in active galactic nuclei, *Ap. J.* **835**, 165 (February 2017).
24. M. Sasaki, T. Suyama, T. Tanaka and S. Yokoyama, Primordial black hole scenario for the gravitational-wave event GW150914, *Phys. Rev. Lett.* **117**, 061101 (August 2016).
25. S. Bird *et al.*, Did LIGO detect dark matter?, *Phys. Rev. Lett.* **116**, 201301 (May 2016).
26. S. Blinnikov, A. Dolgov, N. K. Porayko and K. Postnov, Solving puzzles of GW150914 by primordial black holes, *JCAP* **11**, 036 (November 2016).
27. A. Kashlinsky, LIGO gravitational wave detection, primordial black holes, and the near-IR cosmic infrared background anisotropies, *Ap. J. Lett.* **823**, L25 (June 2016).
28. C. L. Rodriguez *et al.*, Illuminating black hole binary formation channels with spins in advanced LIGO, *Ap. J. Lett.* **832**, L2 (November 2016).
29. L. Rezzolla *et al.*, The final spin from the coalescence of aligned-spin black hole binaries, *Ap. J. Lett.* **674**, L29 (February 2008).

30. L. Rezzolla *et al.*, Final spin from the coalescence of two black holes, *Phys. Rev. D* **78**, 044002 (August 2008).
31. B. P. Abbott *et al.*, Binary black hole mergers in the first advanced LIGO observing run, *Phys. Rev. X* **6**, 041015 (October 2016).
32. B. P. Abbott *et al.*, GW170104: Observation of a 50-solar-mass binary black hole coalescence at redshift 0.2, *Phys. Rev. Lett.* **118**, 221101 (June 2017).
33. I. F. Mirabel, Black holes formed by direct collapse: observational evidences, arXiv e-prints (November 2016).
34. J. A. Orosz *et al.*, The mass of the black hole in cygnus X-1, *Ap. J.* **742**, 84 (December 2011).
35. H.-T. Janka, Natal kicks of stellar mass black holes by asymmetric mass ejection in fallback supernovae, *MNRAS* **434**, 1355 (September 2013).
36. I. Mandel, Estimates of black hole natal kick velocities from observations of low-mass X-ray binaries, *MNRAS* **456**, 578 (February 2016).
37. T. M. Tauris *et al.*, Formation of double neutron star systems, *Ap. J.* **846**, 170 (September 2017).
38. J. E. McClintock, R. Narayan and J. F. Steiner, Black hole spin via continuum fitting and the role of spin in powering transient jets, *Space Sci. Rev.* **183**, 295 (September 2014).
39. T. Fragos and J. E. McClintock, The origin of black hole spin in galactic low-mass X-ray binaries, *Ap. J.* **800**, 17 (February 2015).
40. F. Valsecchi *et al.*, Formation of the black-hole binary M33 X-7 through mass exchange in a tight massive system, *Nature* **468**, 77 (November 2010).
41. T.-W. Wong *et al.*, Understanding compact object formation and natal kicks. IV. The case of IC 10 X-1, *Ap. J.* **790**, 119 (August 2014).
42. J.-P. Zahn, The dynamical tide in close binaries, *A&Ap* **41**, 329 (July 1975).
43. D. Kushnir, M. Zaldarriaga, J. A. Kollmeier and R. Waldman, Dynamical tides reexpressed, *MNRAS* **467**, 2146 (May 2017).
44. R. F. Stark and T. Piran, Gravitational-wave emission from rotating gravitational collapse, *Phys. Rev. Lett.* **55**, 891 (August 1985).
45. E. O'Connor and C. D. Ott, Black hole formation in failing core-collapse supernovae, *Ap. J.* **730**, p. 70 (April 2011).
46. Y. Sekiguchi and M. Shibata, Formation of black hole and accretion disk in a massive high-entropy stellar core collapse, *Ap. J.* **737**, 6 (August 2011).
47. P. Madau and M. Dickinson, Cosmic star-formation history, *Ann. Rev. Astron. & Astroph.* **52**, 415 (August 2014).
48. D. Wanderman and T. Piran, The luminosity function and the rate of Swift's gamma-ray bursts, *MNRAS* **406**, 1944 (August 2010).
49. P. A. Crowther, Physical properties of Wolf-Rayet stars, *Ann. Rev. Astron. & Astroph.* **45**, 177 (September 2007).
50. J. S. Vink and T. J. Harries, Wolf-Rayet spin at low metallicity and its implication for black hole formation channels, *A&Ap* **603**, A120 (July 2017).

Confinement/Deconfinement and Gravity-Assisted Emergent Higgs Mechanism in Quintessential Cosmological Model

Eduardo Guendelman
Physics Department, Ben-Gurion University of the Negev, Beer Sheva 8410501, Israel
guendel@bgu.ac.il

Emil Nissimov and Svetlana Pacheva
Institute for Nuclear Research and Nuclear Energy, Bulgarian Academy of Sciences, Sofia 1784, Bulgaria
nissimov@inrne.bas.bg, svetlana@inrne.bas.bg

Motivated by the ideas of Jacob Bekenstein concerning gravity-assisted symmetry breaking, we consider a non-canonical model of $f(R) = R + R^2$ extended gravity coupled to neutral scalar "inflaton", as well as to $SU(2) \times U(1)$ multiplet of fields matching the content of the bosonic sector of the electroweak particle model, however with the following significant difference — the $SU(2) \times U(1)$ iso-doublet Higgs-like scalar enters here with a standard positive mass squared and without quartic selfinteraction. Strong interaction dynamics and, in particular, QCD-like confinement effects are also considered by introducing an additional coupling to a strongly nonlinear gauge field whose Lagrangian contains a square-root of the standard Maxwell/Yang–Mills kinetic term. The latter is known to produce charge confinement in flat spacetime.

The principal new ingredient in the present approach is employing the formalism of non-Riemannian spacetime volume-forms — alternative generally covariant volume elements independent of the spacetime metric, constructed in terms of auxiliary antisymmetric tensor gauge fields of maximal rank. Although being almost pure-gauge, i.e. not introducing any additional propagating degrees of freedom, their dynamics triggers a series of physically important features when passing to the Einstein frame: (i) Appearance of two infinitely large flat regions of the effective "inflaton" scalar potential with vastly different energy scales corresponding to the "early" and "late" epochs of the Universe; (ii) Dynamical generation of Higgs-like spontaneous symmetry breaking effective potential for the $SU(2) \times U(1)$ iso-doublet scalar in the "late" Universe, and vanishing of the symmetry breaking in the "early" Universe; (iii) Dynamical appearance of charge confinement via the "square-root" nonlinear gauge field in the "late" Universe and deconfinement in the "early" Universe.

Keywords: Non-Riemannian volume-forms; quintessential evolution; confining gauge theories, dynamical generation of electroweak symmetry breaking.

1. Introduction

Jacob Bekenstein was a remarkable scientist and person. He had both the creativity and the courage to look at physics from a different perspective. That was evident from his very early idea that black holes must have entropy.[1–4] Bekenstein presented very strong arguments to support this idea but he was confronted with the opposition of S. Hawking, who strongly argued against his idea. The issue was later resolved in favor of Jacob when Hawking showed that the black holes emit radiation[5,6] in a way consistent with the entropy proposal.

Jacob had also his original approach to other fundamental physics problems, like the idea that gravity must be modified in order to reproduce effects of dark matter, without the dark matter being really present.[7]

Another topic Jacob was fascinated with, and on which we will focus here, was the search for new avenues for spontaneous symmetry breaking. One of us (E.G.) was for example involved in a joint research with Jacob on how symmetry breaking can be generated by density effects, even for a theory without spontaneous symmetry breaking when no background densities were present.[8] During this collaboration E.G. learned lots of physics, of course, but also he understood how it is possible for a great mind to be simultaneously a remarkable human being.

Bekenstein's search for new mechanisms for symmetry breaking lead him also to consider gravity-induced symmetry breaking effects instead of density effects. In an intriguing paper[9] from 1986 he proposed the remarkable idea of a gravity-assisted spontaneous symmetry breaking of electroweak (Higgs) type without invoking unnatural (according to his opinion) ingredients like negative mass squared and a quartic self-interaction for the Higgs field. By considering a model of gravity interacting with a standard Klein–Gordon scalar field (with small positive mass squared and without selfinteraction) coupled conformally to the scalar curvature he managed to obtain a prototype of dynamically induced Higgs-like spontaneous symmetry breaking scalar potential. A similar approach was further worked out in Ref. 10.

Motivated by Bekenstein's idea, we wrote an essay[11] to the gravity research foundation, where we considered a non-canonical model of gravity coupled to a neutral scalar "inflaton" as well as to a set of $SU(2) \times U(1)$ iso-doublet scalar and gauge fields corresponding to the bosonic sector of the electroweak particle model. Here the iso-doublet scalar field was introduced with a standard positive mass squared and without selfinteraction.

The essential non-standard feature of the model in Ref. 11 is its construction in terms of non-Riemannian spacetime volume-forms (alternative metric-independent generally covariant volume elements) defined in terms of auxiliary antisymmetric tensor gauge fields of maximal rank. The latter were shown to be almost pure-gauge — apart from few arbitrary integration constants they do not produce propagating field-theoretic degrees of freedom (see Appendices A of Refs. 11 and 13). Yet the non-Riemannian spacetime volume-forms trigger a series of important physical features unavailable in ordinary gravity-matter models with the standard Riemannian volume-form (given by the square-root of the determinant of the Riemannian metric):

(i) The "inflaton" φ develops a remarkable effective scalar potential in the Einstein frame possessing an infinitely large flat region for large negative φ describing the "early" universe evolution;

(ii) In the absence of the $SU(2) \times U(1)$ iso-doublet scalar field, the "inflaton" effective potential has another infinitely large flat region for large positive φ at much lower energy scale describing the "late" post-inflationary (dark energy dominated) universe;

(iii) Inclusion of the $SU(2) \times U(1)$ iso-doublet scalar field σ introduces a drastic change in the total effective scalar potential in the post-inflationary universe — the effective potential as a function of σ dynamically acquires exactly the electroweak Higgs-type spontaneous symmetry breaking form.

In the present paper we will extend the above model by introducing in the initial action a R^2-gravity term as well as coupling to an additional strongly nonlinear gauge field whose Lagrangian contains a square-root of the standard Maxwell/Yang–Mills kinetic term. The latter is known to describe charge confinement in flat spacetime[12] as well as in curved spacetime for static spherically symmetric field configurations (Appendix B in Ref. 13; see also Eq. (10) below). Thus, the addition of the "square-root" nonlinear gauge field will simulate the strong interactions QCD-like dynamics and, therefore, our extended model represents qualitatively a quintessential cosmological model incorporating the full bosonic content of the standard particle model. Now, in the physical Einstein frame alongside with the Bekenstein-inspired gravity-assisted dynamical generation of Higgs-type electroweak spontaneous symmetry breaking in the "late" universe, while there is no electroweak breaking in the "early" universe, we obtain gravity-assisted dynamical generation of charge confinement in the "late" universe as well as gravity-suppression of confinement, i.e., deconfinement in the "early" universe.

2. Non-Canonical Gravity Coupled to a Confining Nonlinear Gauge Field and the Bosonic Sector of the Electroweak Standard Model

2.1. *Non-standard $f(R)$-gravity model with non-Riemannian spacetime volume-forms*

We start with the following non-canonical $f(R) = R + R^2$ gravity-matter action constructed in terms of two different non-Riemannian volume-forms (generally covariant metric-independent volume elements) generalizing the actions in Refs. 11 and 13 (for simplicity we use units with the Newton constant $G_N = 1/16\pi$):

$$S = \int d^4x\, \Phi(A) \Big[R + L_1(\varphi, X) + L_2(\sigma, Y) - \frac{1}{2} f_0 \sqrt{-F^2}\Big] + \int d^4x\, \Phi(B) \Big[\epsilon R^2 - \frac{1}{4e^2} F^2 - \frac{1}{4g^2} \mathcal{F}^2(\mathcal{A}) - \frac{1}{4g'^2} \mathcal{F}^2(\mathcal{B}) + \frac{\Phi(H)}{\sqrt{-g}}\Big]. \quad (1)$$

Here the following notations are used:

- $\Phi(A)$ and $\Phi(B)$ are two independent non-Riemannian volume-forms given in terms of the dual field-strengths of rank-3 antisymmetric tensor gauge fields $A_{\nu\kappa\lambda}$ and $B_{\nu\kappa\lambda}$.

$$\Phi(A) = \frac{1}{3!}\varepsilon^{\mu\nu\kappa\lambda}\partial_\mu A_{\nu\kappa\lambda}\,, \quad \Phi(B) = \frac{1}{3!}\varepsilon^{\mu\nu\kappa\lambda}\partial_\mu B_{\nu\kappa\lambda}\,. \tag{2}$$

- $\Phi(H)$ is the dual field-strength of an additional auxiliary tensor gauge field $H_{\nu\kappa\lambda}$, whose presence is crucial for the consistency of the model (1):

$$\Phi(H) = \frac{1}{3!}\varepsilon^{\mu\nu\kappa\lambda}\partial_\mu H_{\nu\kappa\lambda}\,. \tag{3}$$

- We particularly emphasize that we start within the first-order *Palatini formalism* for the scalar curvature R and the Ricci tensor $R_{\mu\nu}$: $R = g^{\mu\nu}R_{\mu\nu}(\Gamma)$, where $g_{\mu\nu}$, $\Gamma^\lambda_{\mu\nu}$ – the metric and affine connection are *a priori* independent.
- $L_1(\varphi, X)$ is the "inflaton" Lagrangian:

$$L_1(\varphi, X) = X - f_1 e^{-\alpha\varphi}\,, \quad X \equiv -\frac{1}{2}g^{\mu\nu}\partial_\mu\varphi\partial_\nu\varphi\,, \tag{4}$$

  where α, f_1 are dimensionful positive parameters.
- $\sigma \equiv (\sigma_a)$ is a complex $SU(2)\times U(1)$ iso-doublet Higgs-like scalar field with Lagrangian:

$$L_2(\sigma, Y) = Y - m_0^2\sigma_a^*\sigma_a\,, \quad Y \equiv -g^{\mu\nu}(\nabla_\mu\sigma)_a^*\nabla_\nu\sigma_a\,, \tag{5}$$

  where gauge-covariant derivative acting on σ reads:

$$\nabla_\mu\sigma = \left(\partial_\mu - \frac{i}{2}\tau_A\mathcal{A}_\mu^A - \frac{i}{2}\mathcal{B}_\mu\right)\sigma\,, \tag{6}$$

  with $\frac{1}{2}\tau_A$ (τ_A – Pauli matrices, $A = 1,2,3$) indicating the $SU(2)$ generators and $\mathcal{A}_\mu^A$ ($A = 1,2,3$) and $\mathcal{B}_\mu$ denoting the corresponding electroweak $SU(2)$ and $U(1)$ gauge fields.
- The electroweak gauge field kinetic terms are of the standard Yang–Mills form (all $SU(2)$ indices $A, B, C = (1,2,3)$):

$$\mathcal{F}^2(\mathcal{A}) \equiv \mathcal{F}^A_{\mu\nu}(\mathcal{A})\mathcal{F}^A_{\kappa\lambda}(\mathcal{A})g^{\mu\kappa}g^{\nu\lambda}\,, \quad \mathcal{F}^2(\mathcal{B}) \equiv \mathcal{F}_{\mu\nu}(\mathcal{B})\mathcal{F}_{\kappa\lambda}(\mathcal{B})g^{\mu\kappa}g^{\nu\lambda}\,, \tag{7}$$

$$\mathcal{F}^A_{\mu\nu}(\mathcal{A}) = \partial_\mu\mathcal{A}^A_\nu - \partial_\nu\mathcal{A}^A_\mu + \epsilon^{ABC}\mathcal{A}^B_\mu\mathcal{A}^C_\nu\,, \quad \mathcal{F}_{\mu\nu}(\mathcal{B}) = \partial_\mu\mathcal{B}_\nu - \partial_\nu\mathcal{B}_\mu\,. \tag{8}$$

Finally, there is an additional coupling in the action (1) to another strongly nonlinear (Abelian) gauge field A_μ with the square-root Maxwell term $-\frac{1}{2}f_0\sqrt{-F^2}$ alongside the standard kinetic term $-\frac{1}{4e^2}F^2$:

$$F^2 \equiv F_{\mu\nu}F_{\kappa\lambda}g^{\mu\kappa}g^{\nu\lambda}\,, \quad F_{\mu\nu} = \partial_\mu A_\nu - \partial_\nu A_\mu\,. \tag{9}$$

As shown in Appendix B of Ref. 13, for static spherically symmetric fields in a static spherically symmetric spacetime metric the square-root term $-\frac{1}{2}f_0\sqrt{-F^2}$ produces an effective *"Cornell"-type confining potential*[14–16] $V_{\rm eff}(L)$ between charged quantized fermions, L being the distance between the latter:

$$V_{\rm eff}(L) = \sqrt{2}ef_0\, L - \frac{e^2}{2\pi\, L} + (L\text{-independent const})\,, \tag{10}$$

i.e., f_0 and e play the role of a confinement-strength coupling constant and of a "color" charge, respectively.

In fact, we could equally well take the "square-root" nonlinear gauge field A_μ to be non-Abelian — for static spherically symmetric solutions the non-Abelian model effectively reduces to the abelian one.[12] Thus, the "square-root" gauge field will simulate the QCD-like confining dynamics.

Let us note that the structure of action (1) is uniquely fixed by the requirement for invariance (with the exception of the regular mass term of the iso-doublet scalar σ_a) under the following global Weyl-scale transformations:

$$\begin{gathered} g_{\mu\nu} \to \lambda g_{\mu\nu}\,, \quad \varphi \to \varphi + \tfrac{1}{\alpha}\ln\lambda\,, \quad A_{\mu\nu\kappa} \to \lambda A_{\mu\nu\kappa}\,, \quad B_{\mu\nu\kappa} \to \lambda^2 B_{\mu\nu\kappa}\,, \\ \Gamma^\mu_{\nu\lambda}\,, \quad H_{\mu\nu\kappa}\,, \quad \sigma_a\,, \quad A_\mu\,, \quad \mathcal{A}^A_\mu\,, \quad \mathcal{B}_\mu \text{ — inert}\,. \end{gathered} \tag{11}$$

2.2. *Derivation of the Einstein-frame action*

Solutions of the equations of motion of the initial action (1) w.r.t. auxiliary tensor gauge fields $A_{\mu\nu\lambda}$, $B_{\mu\nu\lambda}$ and $H_{\mu\nu\lambda}$ yield the following algebraic constraints:

$$R + L_1(\varphi,X) + L_2(\sigma,Y) - \frac{1}{2}f_0\sqrt{-F^2} = -M_1 = \text{const}\,, \tag{12}$$

$$\epsilon R^2 - \frac{1}{4e^2}F^2 - \frac{1}{4g^2}\mathcal{F}^2(\mathcal{A}) - \frac{1}{4g'^2}\mathcal{F}^2(\mathcal{B}) + \frac{\Phi(H)}{\sqrt{-g}} = -M_2 = \text{const}\,, \tag{13}$$

$$\frac{\Phi(B)}{\sqrt{-g}} \equiv \chi_2 = \text{const}\,, \tag{14}$$

where M_1 and M_2 are arbitrary dimensionful and χ_2 arbitrary dimensionless *integration constants*. The algebraic constraint Eqs. (12)–(14) are the Lagrangian-formalism counterparts of the Dirac first-class Hamiltonian constraints on the auxiliary tensor gauge fields $A_{\mu\nu\lambda}$, $B_{\mu\nu\lambda}$, $H_{\mu\nu\lambda}$.[11,13]

The equations of motion of (1) w.r.t. affine connection $\Gamma^\mu_{\nu\lambda}$ (recall — we are using Palatini formalism):

$$\int d^4x\,\sqrt{-g}g^{\mu\nu}\Big(\frac{\Phi_1}{\sqrt{-g}} + 2\epsilon\,\frac{\Phi_2}{\sqrt{-g}}\,R\Big)\left(\nabla_\kappa\delta\Gamma^\kappa_{\mu\nu} - \nabla_\mu\delta\Gamma^\kappa_{\kappa\nu}\right) = 0 \tag{15}$$

yield a solution for $\Gamma^\mu_{\nu\lambda}$ as a Levi-Civita connection:

$$\Gamma^\mu_{\nu\lambda} = \Gamma^\mu_{\nu\lambda}(\bar{g}) = \frac{1}{2}\bar{g}^{\mu\kappa}\left(\partial_\nu\bar{g}_{\lambda\kappa} + \partial_\lambda\bar{g}_{\nu\kappa} - \partial_\kappa\bar{g}_{\nu\lambda}\right)\,, \tag{16}$$

w.r.t. to the following *Weyl-rescaled metric* $\bar{g}_{\mu\nu}$:

$$\bar{g}_{\mu\nu} = \big(\chi_1 + 2\epsilon\chi_2 R\big)g_{\mu\nu}\,, \quad \chi_1 \equiv \frac{\Phi_1(A)}{\sqrt{-g}}\,, \tag{17}$$

χ_2 as in (14). Upon using relation (12) and notation (14), Eq. (17) can be written as:

$$\bar{g}_{\mu\nu} = \left[\chi_1 - 2\epsilon\chi_2\left(L_1(\varphi, X) + L_2(\sigma, Y) - \frac{1}{2}f_0\sqrt{-F^2} + M_1\right)\right]g_{\mu\nu} \, . \tag{18}$$

Varying (1) w.r.t. the original metric $g_{\mu\nu}$ and using relations (12)–(14) we have:

$$\chi_1\left[R_{\mu\nu}+\frac{1}{2}\left(g_{\mu\nu}L^{(1)} - T^{(1)}_{\mu\nu}\right)\right]-\frac{1}{2}\chi_2\left[T^{(2)}_{\mu\nu}+g_{\mu\nu}\left(\epsilon R^2 + M_2\right)-4\epsilon R\, R_{\mu\nu}\right] = 0 \, , \tag{19}$$

with χ_1 and χ_2 as in (17) and (14), and $T^{(1,2)}_{\mu\nu}$ being the canonical energy-momentum tensors:

$$T^{(1,2)}_{\mu\nu} = g_{\mu\nu}L^{(1,2)} - 2\frac{\partial}{\partial g^{\mu\nu}}L^{(1,2)} \, . \tag{20}$$

of the scalar+gauge field Lagrangians in the original action (1):

$$\begin{aligned} L^{(1)} &\equiv L_1(\varphi, X) + L_2(\sigma, Y) - \frac{1}{2}f_0\sqrt{-F^2} \, , \\ L^{(2)} &\equiv -\frac{1}{4e^2}F^2 - \frac{1}{4g^2}\mathcal{F}^2(\mathcal{A}) - \frac{1}{4g'^2}\mathcal{F}^2(\mathcal{B}) \, . \end{aligned} \tag{21}$$

Taking the trace of Eq. (19) and using again relation (12) we solve for the ratio χ_1 (17):

$$\chi_1 = 2\chi_2\frac{T^{(2)}/4 + M_2}{L^{(1)} - \frac{1}{2}T^{(1)} - M_1} \, , \tag{22}$$

where $T^{(1,2)} = g^{\mu\nu}T^{(1,2)}_{\mu\nu}$. Explicitly we obtain from (22):

$$\chi_1 = \frac{1}{2\chi_2 M_2}(f_1 e^{-\alpha\varphi} + m_0\sigma^*\sigma - M_1) \, . \tag{23}$$

The Weyl-rescaled metric $\bar{g}_{\mu\nu}$ (18) can be written explicitly as:

$$\bar{g}_{\mu\nu} = \chi_1\Omega g_{\mu\nu} \, , \quad \Omega \equiv \frac{1 + \frac{\epsilon}{M_2}(f_1 e^{-\alpha\varphi} + m_0\sigma^*\sigma - M_1)^2}{1 + 2\epsilon\chi_2(\bar{X} + \bar{Y} - \frac{1}{2}f_0\sqrt{-\bar{F}^2})} \, , \tag{24}$$

$$\bar{X} \equiv -\frac{1}{2}\bar{g}^{\mu\nu}\partial_\mu\varphi\partial_\nu\varphi \, , \quad \bar{Y} \equiv -\bar{g}^{\mu\nu}(\nabla_\mu\sigma)^*_a\nabla_\nu\sigma_a \, , \quad \bar{F}^2 \equiv F_{\mu\nu}F_{\kappa\lambda}\bar{g}^{\mu\kappa}\bar{g}^{\nu\lambda} \, . \tag{25}$$

Now, we can bring Eq. (19) into the standard form of Einstein equations in the second-order formalism for the Weyl-rescaled metric $\bar{g}_{\mu\nu}$ (24), i.e., the *Einstein-frame* equations:

$$R_{\mu\nu}(\bar{g}) - \frac{1}{2}\bar{g}_{\mu\nu}R(\bar{g}) = \frac{1}{2}T^{\rm eff}_{\mu\nu} \tag{26}$$

with effective energy-momentum tensor corresponding according to the definition (20):

$$T^{\rm eff}_{\mu\nu} = \bar{g}_{\mu\nu}L_{\rm eff} - 2\frac{\partial}{\partial g^{\mu\nu}}L_{\rm eff} \tag{27}$$

to the following effective *Einstein-frame* matter Lagrangian (using short-hand notations (21)) and with χ_1 as in (23) and Ω as in (24)):

$$L_{\rm eff} = \frac{1}{\chi_1 \Omega}\left\{ L^{(1)} + M_1 + \frac{\chi_2}{\chi_1 \Omega}[L^{(2)} + M_2 + \epsilon(L^{(1)} + M_1)^2]\right\}. \qquad (28)$$

The full Einstein-frame action, where all quantities defined w.r.t. Einstein-frame metric (17) are indicated by an upper bar, explicitly reads:

$$S = \int d^4x \sqrt{-\bar{g}}[R(\bar{g}) + L_{\rm eff}(\varphi, \bar{X}; \sigma, \bar{Y}; \bar{F}^2, \bar{\mathcal{F}}(\mathcal{A})^2, \bar{\mathcal{F}}(\mathcal{B})^2)], \qquad (29)$$

where $\bar{X}$, $\bar{Y}$, $\bar{F}^2$ are as in (25) (and similarly for $\bar{\mathcal{F}}(\mathcal{A})^2$, $\bar{\mathcal{F}}(\mathcal{B})^2$), and where:

$$\begin{aligned} L_{\rm eff} &= (\bar{X} + \bar{Y})(1 - 4\epsilon\chi_2 \mathcal{U}(\varphi,\sigma)) + \epsilon\chi_2(\bar{X} + \bar{Y})^2(1 - 4\epsilon\chi_2\mathcal{U}(\varphi,\sigma)) \\ &\quad - (\bar{X} + \bar{Y})\sqrt{-\bar{F}^2}\epsilon\chi_2 f_{\rm eff}(\varphi,\sigma) - \frac{1}{2} f_{\rm eff}(\varphi,\sigma)\sqrt{-\bar{F}^2} \\ &\quad - \mathcal{U}(\varphi,\sigma) - \frac{1}{4e^2_{\rm eff}(\varphi,\sigma)}\bar{F}^2 - \frac{\chi_2}{4g^2}\bar{\mathcal{F}}^2(\mathcal{A}) - \frac{\chi_2}{4g'^2}\bar{\mathcal{F}}^2(\mathcal{B}). \end{aligned} \qquad (30)$$

In (30) the following notations are used:

- $\mathcal{U}(\varphi,\sigma)$ is the effective scalar field ("inflaton" + Higgs-like) potential:

$$\mathcal{U}(\varphi,\sigma) = \frac{\left(f_1 e^{-\alpha\varphi} + m_0 \sigma^* \sigma - M_1\right)^2}{4\chi_2\left[M_2 + \epsilon\left(f_1 e^{-\alpha\varphi} + m_0\sigma^*\sigma - M_1\right)^2\right]}. \qquad (31)$$

- $f_{\rm eff}(\varphi,\sigma)$ is the effective confinement-strength coupling constant:

$$f_{\rm eff}(\varphi,\sigma) = f_0(1 - 4\epsilon\chi_2\mathcal{U}(\varphi,\sigma)). \qquad (32)$$

- $e^2_{\rm eff}(\varphi,\sigma)$ is the effective "color" charge squared:

$$e^2_{\rm eff}(\varphi,\sigma) = \frac{e^2}{\chi_2}\left[1 + \epsilon e^2 f_0^2(1 - 4\epsilon\chi_2\mathcal{U}(\varphi,\sigma))\right]^{-1}. \qquad (33)$$

Note that (30) is of quadratic *"k-essence"* type[17–20] w.r.t. "inflaton" φ and the Higgs-like σ fields.

3. Quintessence, Confinement/Deconfinement and Gravity Assisted Emergent Higgs Mechanism

The nonlinear "confining" gauge field A_μ develops a nontrivial vacuum field-strength:

$$\left.\frac{\partial L_{\rm eff}}{\partial \bar{F}^2}\right|_{\bar{X},\bar{Y}=0} = 0 \qquad (34)$$

explicitly given by:

$$\sqrt{-\bar{F}^2}_{\rm vac} = f_{\rm eff}(\varphi,\sigma)\, e^2_{\rm eff}(\varphi,\sigma). \qquad (35)$$

Substituting (35) in (30) we obtain the following total effective scalar potential (with $\mathcal{U}(\varphi,\sigma)$ as in (31)):

$$\mathcal{U}_{\text{total}}(\varphi,\sigma) = \frac{\mathcal{U}(\varphi,\sigma)(1-\epsilon e^2 f_0^2) + e^2 f_0^2/4\chi_2}{1+\epsilon e^2 f_0^2\big(1-4\epsilon\chi_2\mathcal{U}(\varphi,\sigma)\big)}\,. \tag{36}$$

$\mathcal{U}_{\text{total}}(\varphi,\sigma)$ (36) has few remarkable properties.

First, $\mathcal{U}_{\text{total}}(\varphi,\sigma)$ possesses two infinitely large flat regions as function of φ when σ is fixed:

(a) $(-)$ flat "inflaton" region for large negative values of φ,
(b) $(+)$ flat "inflaton" region for large positive values of φ,

respectively, as depicted on Fig. 1 (for $m_0\sigma^*\sigma \leq M_1$) or Fig. 2 (for $m_0\sigma^*\sigma \geq M_1$).

(i) In the $(-)$ flat "inflaton" region:

- The effective scalar field potential reduces to:

$$\mathcal{U}(\varphi,\sigma=\text{fixed}) \simeq \frac{1}{4\epsilon\chi_2} \to \mathcal{U}_{\text{total}} \simeq \mathcal{U}^{(-)}_{\text{total}} = \frac{1}{4\epsilon\chi_2}\,, \tag{37}$$

  implying that all terms containing φ and σ disappear from the Einstein-frame Lagrangian (29), i.e., there is *no electroweak spontaneous breakdown in the $(-)$ flat "inflaton" region.*
- From (32) the first relation (37) implies $f_{\text{eff}} = 0$, i.e., there is *no confinement in the $(-)$ flat "inflaton" region.*

(ii) In the $(+)$ flat "inflaton" region:

- The effective scalar field potential becomes:

$$\mathcal{U}(\varphi,\sigma) \simeq \mathcal{U}_{(+)}(\sigma) = \frac{(m_0^2\sigma^*\sigma - M_1)^2}{4\chi_2[M_2+\epsilon(m_0^2\sigma^*\sigma - M_1)^2]} \tag{38}$$

$$\to \mathcal{U}_{\text{total}}(\varphi\sigma) \simeq \mathcal{U}^{(+)}_{\text{total}}(\sigma) = \frac{\mathcal{U}_{(+)}(\sigma)(1-\epsilon e^2 f_0^2) + e^2 f_0^2/4\chi_2}{1+\epsilon e^2 f_0^2(1-4\epsilon\chi_2\mathcal{U}_{(+)}(\sigma))} \tag{39}$$

  producing a dynamically generated *nontrivial vacuum for the Higgs-like field*:

$$|\sigma_{\text{vac}}| = \sqrt{M_1}/m_0\,, \tag{40}$$

  i.e., we obtain *"gravity-assisted" electroweak spontaneous breakdown in the $(+)$ flat "inflaton" region.*
- At the Higgs vacuum we have dynamically generated vacuum energy density (cosmological constant):

$$\mathcal{U}^{(+)}_{\text{total}}(\sigma_{\text{vac}}) \equiv 2\Lambda_{(+)} = \epsilon e^2 f_0^2\left[4\epsilon\chi_2(1+\epsilon e^2 f_0^2)\right]^{-1}. \tag{41}$$

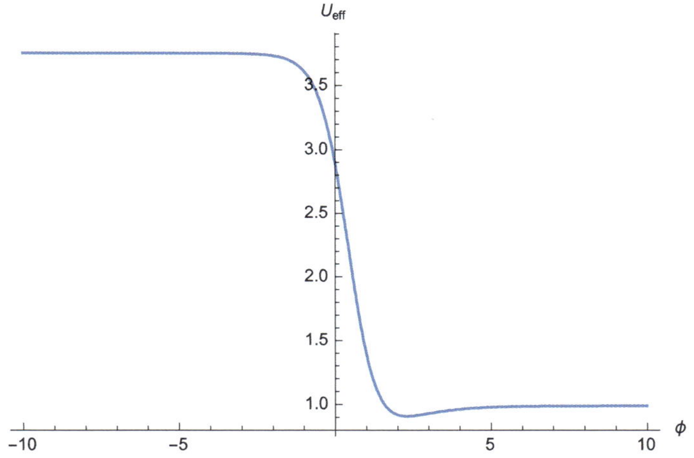

Fig. 1. Qualitative shape of the total effective scalar potential U_{total} (36) as function of the "inflaton" φ for fixed Higgs-like σ (when $m_0\sigma^*\sigma \leq M_1$).

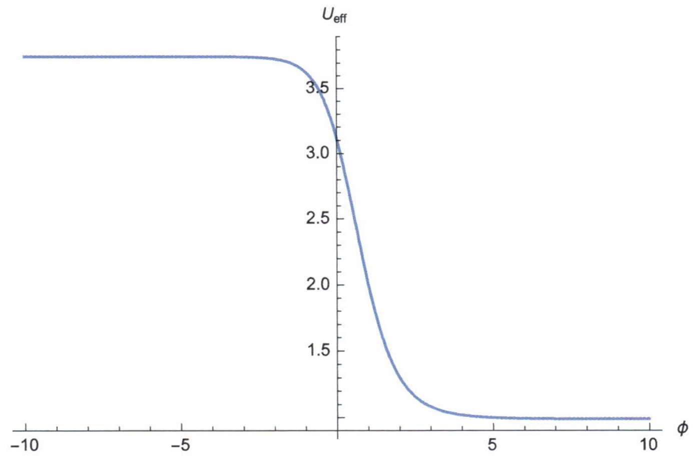

Fig. 2. Qualitative shape of the total effective scalar potential U_{total} (36) as function of the "inflaton" φ for fixed Higgs-like σ (when $m_0\sigma^*\sigma \geq M_1$).

- The effective confinement-strength coupling constant:

$$f_{\text{eff}} \simeq f_{(+)} = f_0(1 - 4\epsilon\chi_2\mathcal{U}_{(+)}(\sigma)) > 0\,, \tag{42}$$

therefore we obtain *"gravity-assisted" charge confinement in the (+) flat "inflaton" region.*

As seen from Fig. 1 or Fig. 2, the two flat "inflaton" regions of the total scalar potential given by $\mathcal{U}^{(-)}_{\text{total}} = \frac{1}{4\epsilon\chi_2}$ (37) and $\mathcal{U}^{(+)}_{\text{total}}(\sigma_{\text{vac}}) \equiv 2\Lambda_{(+)} = \epsilon e^2 f_0^2[4\epsilon\chi_2(1 + \epsilon e^2 f_0^2)]^{-1}$ (41), respectively, can be identified as describing the "early" ("inflationary") and "late" (today's dark energy dominated) epoch of the universe provided we take the following numerical values for the parameters in order to conform to the *PLANCK* data:[21,22]

$$\mathcal{U}^{(-)}_{\rm total} \sim 10^{-8} M_{\rm Pl}^4 \to \epsilon\chi_2 \sim 10^{8} M_{\rm Pl}^{-4}\,, \quad \Lambda_{(+)} \sim 10^{-122} M_{\rm Pl}^4 \to \frac{e^2 f_0^2}{\chi_2} \sim 10^{-122} M_{\rm Pl}^4\,, \tag{43}$$

where $M_{\rm Pl}$ is the Planck mass scale.

From the Higgs v.e.v. $|\sigma_{\rm vac}| = \sqrt{M_1}/m_0$ and the Higgs mass $\frac{M_1 m_0^2}{4\chi_2 M_2}$ resulting from the dynamically generated Higgs-like potential $\mathcal{U}^{(+)}_{\rm total}(\sigma)$ (39) we find:

$$m_0 \sim M_{\rm EW}\,, \quad M_{1,2} \sim M_{\rm EW}^4\,, \tag{44}$$

where $M_{\rm EW} \sim 10^{-16} M_{\rm Pl}$ is the electroweak mass scale.

4. Conclusions and Outlook

Here we have proposed a non-canonical model of $f(R) = R + R^2$ gravity coupled to non-standard matter incorporating two main building blocks — employing the formalism of non-Riemannian spacetime volume forms (generally covariant metric-independent volume elements) as well as introducing a special strongly non-linear gauge field with a square-root of the usual Maxwell/Yang–Mills kinetic term simulating QCD-like confinement dynamics. Due to the special interplay of the dynamics of the above principal ingredients our model is capable of producing in the Einstein frame:

- (i) Unified "quintessential" description of the evolution of the "early" and "late" universe due to a natural dynamical generation of vastly different vacuum energy densities thanks to the auxiliary non-Riemannian volume-form antisymmetric tensor gauge fields;
- (ii) Gravity-assisted dynamical generation of Higgs-like electroweak spontaneous symmetry breaking effective scalar potential in the "late" universe, as well as gravity-assisted charge confinement mechanism through the "square-root" nonlinear gauge field;
- (iii) Gravity-induced suppression of electroweak spontaneous symmetry breaking, as well as gravity-induced deconfinement in the "early" universe.

The non-Riemannian volume-form formalism has further physically relevant applications such as producing a novel mechanism for supersymmetric Brout–Englert–Higgs effect in supergravity through dynamical generation of a cosmological constant triggering spontaneous supersymmetry breaking and dynamical gravitino mass generation.[23,24]

Similarly, the QCD-simulating "square-root" nonlinear gauge field when interacting with gravity produces several other interesting effects:

- (a) black holes with an additional constant background electric field exercising confining force on charged test particles even when the black hole itself is electrically neutral;[25]

- (b) Coupling to a charged lightlike brane produces a charge-"hiding" lightlike thin-shell wormhole, where a genuinely charged matter source is detected as electrically neutral by an external observer.[26]
- (c) Coupling to two oppositely charged lightlike brane sources produces a two-"throat" lightlike thin-shell wormhole displaying a genuine QCD-like charge confinement, i.e., the whole electric flux is trapped within a tube-like spacetime region connected the two charged lightlike branes.[26]
- (d) Charge confining gravitational electrovacuum shock wave.[27]

The model here presented needs further amendments in order to avoid getting an unnaturally small value for the effective confinement strength coupling constant f_0 in the "late" universe resulting from the second relation (43) (condition for compatibility with the *PLANCK* data[21,22] for the value of today's cosmological constant).

Further obvious extension of the present model must be inclusion of the fermions in order to incorporate more faithfully the full standard particle model. To this end we can follow the steps outlined in several previous papers by some of us devoted to the study of modified gravity within the non-Riemannian volume element formalism coupled to fermionic matter fields, such as Ref. 28 (on the geometric origins of fermionic families), Ref. 29 (fermionic families and dark energy and dark matter), Ref. 30 (exotic low density fermionic states and neutrino dark energy).

Acknowledgments

We gratefully acknowledge support of our collaboration through the academic exchange agreement between the Ben-Gurion University in Beer-Sheva, Israel, and the Bulgarian Academy of Sciences. E.N. and E.G. have received partial support from European COST actions MP-1405 and CA-16104, and from CA-15117 and CA-16104, respectively. E.N. and S.P. are also thankful to Bulgarian National Science Fund for support via research grant DN-18/1.

References

1. J. Bekenstein, *Ph.D. thesis*, Princeton Univ. (1972).
2. J. Bekenstein, *Lett. Nuov. Cim.* **4**, 737 (1972).
3. J. Bekenstein, *Phys. Rev. D* **7**, 2333 (1973).
4. J. Bekenstein, *Phys. Rev. D* **9**, 3292 (1974).
5. S. Hawking, *Nature* **248**, 30 (1974).
6. S. Hawking, *Commun. Math. Phys.* **43**, 199 (1975).
7. J. Bekenstein, in *Particle Dark Matter: Observations, Models and Searches*, ed. G. Bertone (Cambridge Univ. Press, 2010) [arXiv:1001.3876].
8. J. Bekenstein and E. Guendelman, *Phys. Rev. D* **35**, 716–731 (1987).
9. J. Bekenstein, *Found. Phys.* **16**, 409 (1986).
10. P. Moniz, P. Crawford and A. Barroso, *Class. Quantum Grav.* **7**, L143 (1990).

11. E. Guendelman, E. Nissimov and S. Pacheva, *Int. J. Mod. Phys. D* **25**, 1644008 (2016) [arXiv:1603.06231].
12. P. Gaete and E. Guendelman, *Phys. Lett. B* **640**, 201–204 (2006) [arXiv:hep-th/0607113].
13. E. Guendelman, E. Nissimov and S. Pacheva, *Int. J. Mod. Phys. A* **30**, 1550133 (2015) [arXiv:1504.01031].
14. E. Eichten, K. Gottfried, T. Kinoshita, J. Kogut, K. Lane and T.-M. Yan, *Phys. Rev. Lett.* **34**, 369–372 (1975).
15. W. Buchmüller (ed.), *Quarkonia*, Current Physics Sources and Comments, Vol. 9 (North Holland, 1992).
16. M. Karliner, B. Keren-Zur, H. Lipkin and J. Rosner, *Ann. of Phys.* **324**, 2–15 (2009) [arXiv:0804.1575 [hep-ph]].
17. T. Chiba, T.Okabe and M. Yamaguchi, *Phys. Rev. D* **62**, 023511 (2000) [arXiv:astro-ph/9912463].
18. C. Armendariz-Picon, V. Mukhanov and P. Steinhardt, *Phys. Rev. Lett.* **85**, 4438 (2000) [arXiv:astro-ph/0004134].
19. C. Armendariz-Picon, V. Mukhanov and P. Steinhardt, *Phys. Rev. D* **63**, 103510 (2001) [arXiv:astro-ph/0006373].
20. T. Chiba, *Phys. Rev. D* **66**, 063514 (2002) [arXiv:astro-ph/0206298].
21. R. Adam *et al.* (Planck Collaboration), *Astron. Astrophys.* **571**, A22 (2014) [arXiv:1303.5082 [astro-ph.CO]].
22. R. Adam *et al.* (Planck Collaboration), *Astron. Astrophys.* **586**, A133 (2016) [arXiv:1409.5738 [astro-ph.CO]].
23. E. Guendelman, E. Nissimov and S. Pacheva, *Bulg. J. Phys.* **41**, 123 (2014) [arXiv:1404.4733].
24. E. Guendelman, E. Nissimov and S. Pacheva, in *Eight Mathematical Physics Meeting*, eds. by B. Dragovich and I. Salom (Belgrade Inst. Phys. Press, 2015), pp. 105–115 [arXiv:1501.05518].
25. E. Guendelman, E. Nissimov and S. Pacheva, *Phys. Lett. B* **704**, 230 (2011) [Erratum *ibid.* **705**, 545 (2011)] [arXiv:1108.0160].
26. E. Guendelman, E. Nissimov and S. Pacheva, *Int. J. Mod. Phys. A* **26**, 5211 (2011) [arXiv:1109.0453].
27. E. Guendelman, E. Nissimov and S. Pacheva, *Mod. Phys. Lett. A* **29**, 1450020 (2014) [arXiv:1310.1558].
28. E. Guendelman and A. Kaganovich, *Mod. Phys. Lett. A* **17**, 1227 (2002) [arXiv:hep-th/0110221].
29. E. Guendelman and A. Kaganovich, *Int. J. Mod. Phys. A* **19**, 5325 (2004) [arXiv:gr-qc/0408026].
30. E. Guendelman and A. Kaganovich, *Int. J. Mod. Phys. A* **21**, 4373 (2006) [arXiv:gr-qc/0603070].

Sakharov's Induced Gravity and the Poincaré Gauge Theory

Masud Chaichian, Markku Oksanen and Anca Tureanu

Department of Physics, University of Helsinki,
P.O. Box 64, FI-00014 Helsinki, Finland

We explore Sakharov's seminal idea that gravitational dynamics is induced by the quantum corrections from the matter sector. This was the starting point of the view that gravity has an emergent origin, which soon gained impetus due to the advent of black hole thermodynamics. In the generalized framework of Riemann–Cartan spacetime with both curvature and torsion, the induced gravitational action is obtained for free nonminimally coupled scalar and Dirac fields. For a realistic matter content, the induced Newton constant is obtained to be of the magnitude of the ultraviolet cutoff, which implies that the cutoff is of the order of the Planck mass. Finally, we conjecture that the action for any gauge theory of gravity at low energies can be induced by Sakharov's mechanism. This is explicitly shown by obtaining the Poincaré gauge theory of gravity.

1. Introduction

The standard approach for dealing with ultraviolet-divergent quantum corrections in quantum field theory is renormalization. Its techniques are well-developed mathematically, and the running of coupling constants are verified experimentally for the Standard Model. In curved spacetime, however, we encounter some problems. Particularly, the gravitational couplings are radiatively unstable in the sense that the cosmological constant and the Newton constant are extremely sensitive to any change of the parameters of the matter sector or a change of the Wilsonian cutoff scale of the matter action.

This problem could likely be solved by imposing constraints that keep certain matter parameters and quantities of the effective action of matter invariant under phase transitions and spontaneous symmetry breaking. The implementation of such constraints requires new physics beyond the Standard Model. Here we explore an alternative and historically significant approach to quantum corrections, which uses a cutoff scale that is related to the observed gravitational couplings and offers an illuminating view on the origin of gravity.

In Sakharov's approach to induced gravity,[1] classical gravity is considered to be induced by quantum corrections from the matter sector. At first matter fields are regarded to live on a spacetime that is curved but with a nondetermined geometry. The quantization of matter fields produces correction terms in the action, which involve the curvature of spacetime. In particular, one-loop corrections include the Einstein–Hilbert action of General Relativity (GR). Finally, elevating the nondetermined metric of spacetime to a dynamical variable turns the correction terms into a gravitational action. In other words, the regulated (but not renormalized) effective action of matter on a curved spacetime is identified as the gravitational action that determines the dynamics of spacetime geometry. In order to avoid curvature terms of arbitrarily high orders to be induced into the gravitational action, one usually assumes in Sakharov's approach that quantum corrections beyond one-loop order are somehow suppressed. Recall that already the squared Riemann curvature terms, which are present in the one-loop effective action of matter, generate new massive degrees of freedom that carry negative energy,[2,3] so-called ghosts, which generate severe problems.[4] When coupled to ordinary fields, ghosts cause the system to evolve to an infinitely excited state without a change in total energy. The inclusion of higher-order curvature terms only makes the situation worse. Such terms are typically present in the effective field theory of gravity, which is obtained as a low-energy limit of various quantum theories. In the induced gravitational action obtained in Sec. 4, the extra degrees of freedom are found to have masses around the Planck mass, since the dimensionless coupling constants of the squared curvature terms are smaller than one, and hence the characteristic length scale in their Yukawa potentials[2,3] is the Planck length. Thus, the effect of the extra degrees of freedom can be regarded to be negligible at low energies and long distances. We assume that a consistent description of quantum gravity and matter will eventually solve the ghost problem. Therefore, in this work, we adopt the view that higher-derivative contributions to the effective action can be ignored in the low-energy regime.

The most important lesson of Sakharov's vision is that any fundamental theory that includes or produces a curved spacetime manifold, on which a quantum field theory of matter can be set up, necessarily produces gravity as well. In this sense, gravity is an unavoidable and necessary companion of quantum matter. The idea to induce gravity from quantum effects and spontaneous symmetry breaking was further developed, particularly by Adler and Zee.[5–12] A recent perspective on Sakharov's induced gravity is given in Ref. 13, which includes a discussion on different interpretations of quantum corrections. The idea of induced gravity was naturally expanded to the realm of quantum gravity, where it has been used in particular to derive Einstein gravity as a low-energy effective theory of scale-invariant (and asymptotically-free) quantum field theory of gravity.[12,14–17]

When applied to matter fields, Sakharov's induced gravity is an early representative of the emergent approach to gravity, where gravity or at least gravitational dynamics is regarded to arise from a quantum theory that does involve a gravitational interaction in its initial definition. Sakharov's approach only addresses the induction of gravitational dynamics by producing the gravitational action, and consequently the field equations, but it does not produce spacetime, since the existence of a curved spacetime manifold is presumed. Since then many people have wondered whether spacetime too could be an emergent concept, and perhaps even gravity as a whole might have an emergent origin. An intriguing indication towards this view is the deep connection of gravity and thermodynamics. It all began from Bekenstein's discovery of the area law for black hole entropy,[18–21] which was predated by observations that the horizon surface area and the irreducible mass of black holes can never decrease in a classical process.[22–25] The identification of horizon surface area as entropy and surface gravity as temperature (both up to a constant factor) quickly led to a full analogy between black hole mechanics and thermodynamics,[26] and soon after to Hawking's discovery of thermal radiation of black holes.[27] The universal upper limit on the entropy that can be contained within a finite region of space which has a finite amount of energy, namely, the Bekenstein bound[28–31] means that a physical system with a finite energy in a finite space is described by at most a certain finite amount of information. A covariant generalization of the Bekenstein entropy bound has been achieved,[32] as well as a similar bound for asymptotically de Sitter spacetimes.[33] Black hole thermodynamics and the entropy bound were a major inspiration for 't Hooft's proposal of the holographic principle[34] and its subsequent string-theoretic interpretation by Susskind.[35] The gauge/gravity duality is the most rigorous realization of the holographic principle.[36] It has been argued that since the black hole entropy is, at least in part, an entanglement entropy,[37] it would be most satisfactory if the gravitational action is induced as Sakharov proposed, so that all black hole entropy would be entanglement entropy.[38,39] Black hole thermodynamics (and the holographic principle) has also had an influence on many other approaches to understand the relation between gravity and thermodynamics. The Einstein equation has been derived locally on Rindler causal horizons as a thermodynamic equation of state.[40,41] An extension of black hole thermodynamics to causal horizons has been considered.[42] The holographic relation of bulk and surface terms in gravitational actions[43] has been used in arguing that the field equations of any diffeomorphism invariant theory of gravity have a thermodynamic reinterpretation, and showing that the equipartition of energy on the microscopic degrees of freedom of a horizon can be used to derive gravity.[44–46] It has even been proposed that gravity is an entropic force caused by changes in the information associated with the positions of material bodies.[48] The existence of gravitationally bound quantum states[49] can be used to impose some conditions on the fundamental microscopic theory behind entropic gravity,[50,51] which is presently unknown. These results suggest a view of gravity and spacetime as emergent concepts, which may have a thermodynamic origin.

The gauge theory approach to gravity has been highly influential ever since the gauge invariance idea introduced by Weyl[52] for $U(1)$ was generalized to $SU(2)$ by Yang and Mills[53] and to all semisimple Lie groups by Utiyama,[54] who considered the gauging of the Lorentz group for the first time. The first consistent gauge theory of gravity was proposed by Kibble via gauging of the Poincaré group,[55] the symmetry group of the Minkowski spacetime, which was used to derive the Einstein–Cartan–Sciama–Kibble theory of gravity,[55,56] but more generally yields a Lagrangian that includes quadratic curvature and torsion terms.[57,58] From there on, gauge theories of every symmetry group related to gravity have been proposed and explored, including the group of translations,[59] the Weyl group (Poincaré group plus scale transformations),[60] the conformal and superconformal groups,[61] the affine group,[62] and so on. See Ref. 63 for a review of various gauge theories of gravity. The gauge theory approach has also been used in attempts to understand the relation of gravity and quantum mechanics. For example, several proposals for the gauge theory of gravity on noncommutative spacetime have been considered, e.g. Refs. 64–67.

In this work, we consider Sakharov's induced gravity on a Riemann–Cartan spacetime with both curvature and torsion. We shall derive the induced gravitational action at one-loop order for free scalar and Dirac fields. The mass scale that determines the induced gravitational constants, especially the induced Newton constant G_{ind}, is the ultraviolet cutoff Λ for the effective action of matter fields. For gravity to have the observed strength, $G_{\text{ind}}^{-1} = 8\pi M_{\text{P}}^2$, the ultraviolet cutoff Λ has to be comparable to the Planck mass M_{P}. The effect of torsion is generally weak except when the density of matter and spin is very high.[68,69] When the gravitational Lagrangian is the curvature scalar of the Riemann–Cartan spacetime, $\mathcal{L} = \frac{1}{2\kappa}\tilde{R}$, which gives the Einstein–Cartan–Sciama–Kibble theory of gravity, torsion does not propagate in vacuum. In more general theories, especially, in the generic Poincaré gauge theory of gravity (PG) with a Lagrangian that is quadratic in torsion and curvature, $\mathcal{L} = \tilde{R} + T^2 + \tilde{R}^2$, torsion does propagate in vacuum[70–72] (for recent development and further references, see Ref. 73), which may also improve the chances for its detection in the future. For a discussion of several physical implications of the torsion of spacetime, see Refs. 68, 69 and 74. The effect of the induced gravitational terms with dimensionless couplings, including the squared curvature terms, is found to be very small in the low-energy regime, since their induced coupling constants depend on the logarithm of the ultraviolet cutoff, which implies that those coupling constants are of the order of one or lower.

Including torsion into the gravitational theory is highly appealing, since then the intrinsic angular momentum of matter and gravity can be naturally incorporated into the theory. In general, both torsion and nonmetricity should be considered, in addition to curvature, in order to obtain a comprehensive understanding of the nature of induced gravity in the non-Riemannian setting. In this work, however, we confine the treatment to spacetimes with a vanishing nonmetricity. This limitation is motivated by our main goal, which is the emergence of PG as an induced theory of gravity via Sakharov's mechanism. PG is a viable alternative to GR, which has been studied extensively.[63] We show that the quantization of Dirac fields on Riemann–

Cartan spacetime induces the low-energy action of PG. The high-energy part of the induced action is found to contain additional terms compared to the action of PG. That is expectable, since the effective action is not limited to contain only terms quadratic in the field strengths, namely, in torsion and curvature. On a more general spacetime with a non-metric compatible connection, a general metric-affine gauge theory of gravity should be induced in the same way.

2. Sakharov's Approach

We generalize Sakharov's approach to Riemann–Cartan spacetime. As in Riemannian spacetime, the procedure can be considered to consist of five steps:

(1) Assume a Lorentzian spacetime manifold. The geometric notions of Riemann–Cartan spacetime can be derived by applying Einstein's Equivalence Principle to a Dirac spinor,[63] which can be considered to describe a neutron in a gravitational field, instead of applying it to a point mass as in GR.
(2) Leave the dynamics of the geometry undetermined, i.e., consider it as an arbitrary classical background and do not define an action for gravity.
(3) Quantize matter fields and determine their effective action. In the language of Feynman diagrams, the one-loop effective action represents the sum of all one-loop diagrams coupled to an arbitrary number of external gravitons. In this visualization, gravitons refer to small perturbations of the geometric background fields around a fixed background. Regularize the effective action, and obtain the contribution of the background geometry to the action.
(4) Elevate the geometric background fields to gravitational variables. We consider two possible choices for the variables: the vierbein and the spin connection (a first-order formalism), or the metric and the torsion (a second-order formalism).
(5) Identify the regulated one-loop effective action as the leading contribution to gravity. The gravitational action induced at one-loop order consists of contributions to vacuum energy, curvature terms up to second order and torsion terms up to fourth order.

Sakharov's approach, and many subsequent induced gravity approaches, involve several problems:

(1) The induced Newton constant is not guaranteed to be unique and positive. In general, the Feynman amplitudes for stress-energy tensor operators are complex, and should be continued analytically. This problem also appears in the approaches[8,10] where gravity is induced via symmetry breaking.[76] Some ways to address the problem have been proposed.[77] A further ambiguity is caused by the choice of a regularization method, since different methods imply different quantum corrections.

An ultraviolet-finite induced Newton constant can be obtained by including both scalar fields and spin-1/2 fields and by imposing fine-tuning constraints[39] on the numbers, masses and couplings of the scalar and spin-1/2 fields. Unfortunately, the masses of the constituent fields have to be comparable with the Planck mass, which causes problems in the presence of gravity, since quantum gravity effects should become significant at high energies.

(2) The introduction of dimensional parameters that determine the scale of gravitational couplings is not fully convincing. In our simplistic approach, cutoff regularization is used to set the mass parameter of the induced theory of gravity to be of the order of the Planck mass. Alternatively, one could use any other regularization method, for example, Pauli–Villars regulators with sufficiently high masses. If the scale of gravitational couplings is set by the masses of fundamental fields, one generally requires fields whose masses are comparable with the Planck, which is problematic.

(3) The elevation of the geometric notions to gravitational variables after the quantization of matter fields has no physical motivation (other than that it works). This problem is a consequence of the fact that Sakharov's approach does not address the emergence of spacetime geometry but rather the emergence of gravitational dynamics.

3. Geometric Definitions

The covariant derivatives associated with the connection involving the torsion $\tilde{\Gamma}$ and the torsionless connection Γ are denoted by $\tilde{\nabla}$ and ∇, respectively. The connections on Riemann–Cartan spacetime are defined to be metric compatible. Namely, the nonmetricity tensor is assumed to vanish throughout this work, $Q_{\mu\nu\rho} = \tilde{\nabla}_\mu g_{\nu\rho} = 0$. Relaxing the metric-compatibility would lead to more general spacetime geometries, e.g., Weyl–Cartan[60] or metric-affine,[62,78] which are not considered here. Greek indices $(\mu, \nu, \ldots)$ refer to a coordinate basis, and Latin indices $(a, b, \ldots)$ refer to an orthonormal noncoordinate basis.

The connection coefficient $\tilde{\Gamma}_{\mu\nu}{}^\rho$ can be written as the sum of the Christoffel symbol $\Gamma_{\mu\nu}{}^\rho$ and the contortion tensor $K_{\mu\nu}{}^\rho$,

$$\tilde{\Gamma}_{\mu\nu}{}^\rho = \Gamma_{\mu\nu}{}^\rho + K_{\mu\nu}{}^\rho\,, \tag{1}$$

where the contortion tensor $K_{\mu\nu}{}^\rho$ is defined by the torsion $T_{\mu\nu}{}^\rho$,

$$T_{\mu\nu}{}^\rho = 2\tilde{\Gamma}_{[\mu\nu]}{}^\rho = \tilde{\Gamma}_{\mu\nu}{}^\rho - \tilde{\Gamma}_{\nu\mu}{}^\rho\,, \tag{2}$$

as

$$K_{\mu\nu}{}^\rho = \frac{1}{2}(T_{\mu\nu}{}^\rho + T^\rho{}_{\mu\nu} + T^\rho{}_{\nu\mu})\,. \tag{3}$$

In the second-order formalism of gravity, the independent gravitational variables can be chosen as the metric and the torsion, which determine the connection (1). Torsion can be decomposed into three irreducible components: the trace vector $\mathcal{V}_\mu = T_{\mu\nu}{}^\nu$, the axial vector $\mathcal{A}^\mu = T_{\nu\rho\sigma}\epsilon^{\nu\rho\sigma\mu}$, and the tensor component $\mathcal{T}_{\mu\nu}{}^\rho$ with vanishing vector and axial vector parts, $\mathcal{T}_{\mu\nu}{}^{\nu} = 0$ and $\mathcal{T}_{\nu\rho\sigma}\epsilon^{\nu\rho\sigma\mu} = 0$. Then, in four-dimensional spacetime, we have

$$T_{\mu\nu\rho} = \frac{1}{3}(\mathcal{V}_\mu g_{\nu\rho} - \mathcal{V}_\nu g_{\mu\rho}) - \frac{1}{6}\epsilon_{\mu\nu\rho\sigma}\mathcal{A}^\sigma + \mathcal{T}_{\mu\nu\rho}\,. \tag{4}$$

The components of torsion couple to matter fields in different ways, which is discussed for scalar and Dirac fields in Sec. 4. The contortion tensor is expressed in terms of the components of torsion as

$$K_{\mu\nu}{}^{\rho} = \frac{1}{3}(g_{\mu\nu}\mathcal{V}^\rho - \delta_\mu{}^\rho\mathcal{V}_\nu) - \frac{1}{12}\epsilon_{\mu\nu\lambda\sigma}g^{\lambda\rho}\mathcal{A}^\sigma + \frac{1}{2}(\mathcal{T}_{\mu\nu}{}^{\rho} + \mathcal{T}^{\rho}{}_{\mu\nu} + \mathcal{T}^{\rho}{}_{\nu\mu})\,, \tag{5}$$

and we note that $K_{\mu\nu}{}^{\mu} = -K_{\mu}{}^{\mu}{}_{\nu} = -\mathcal{V}_\nu$ and $K_{\mu\nu\rho}\epsilon^{\mu\nu\rho\sigma} = \frac{1}{2}\mathcal{A}^\sigma$.

The curvature tensor of the connection (1) can be written in terms of the curvature tensor of the torsionless connection and the contortion tensor as[a]

$$\tilde{R}_{\mu\nu}{}^{\rho}{}_{\sigma} = R_{\mu\nu}{}^{\rho}{}_{\sigma} + 2\nabla_{[\mu}K_{\nu]\sigma}{}^{\rho} + 2K_{[\mu|\lambda}{}^{\rho}K_{|\nu]\sigma}{}^{\lambda}\,, \tag{6}$$

where

$$\tilde{R}_{\mu\nu}{}^{\rho}{}_{\sigma} = 2\partial_{[\mu}\tilde{\Gamma}_{\nu]\sigma}{}^{\rho} + 2\tilde{\Gamma}_{[\mu|\lambda}{}^{\rho}\tilde{\Gamma}_{|\nu]\sigma}{}^{\lambda}\,, \tag{7}$$

and $R_{\mu\nu}{}^{\rho}{}_{\sigma}$ is defined similarly in terms of $\Gamma_{\mu\nu}{}^{\rho}$. We may also express the curvature tensor of the torsionless connection as

$$R_{\mu\nu}{}^{\rho}{}_{\sigma} = \tilde{R}_{\mu\nu}{}^{\rho}{}_{\sigma} - 2\tilde{\nabla}_{[\mu}K_{\nu]\sigma}{}^{\rho} - T_{\mu\nu}{}^{\lambda}K_{\lambda\sigma}{}^{\rho} + 2K_{[\mu|\lambda}{}^{\rho}K_{|\nu]\sigma}{}^{\lambda}\,, \tag{8}$$

where everything on the right-hand side is defined in terms of the connection with torsion. Note that the order of tensor indices on curvature and torsion is such that the commutator of two covariant derivatives is written as

$$[\tilde{\nabla}_\mu, \tilde{\nabla}_\nu]V^\rho = \tilde{R}_{\mu\nu}{}^{\rho}{}_{\sigma}V^\sigma - T_{\mu\nu}{}^{\sigma}\tilde{\nabla}_\sigma V^\rho\,, \tag{9}$$

and the order will persist when the vierbein formalism and the spin representation are considered. The relations between the Ricci tensors, $\tilde{R}_{\mu\nu} = \tilde{R}_{\mu\rho}{}^{\rho}{}_{\nu}$ and $R_{\mu\nu} = R_{\mu\rho}{}^{\rho}{}_{\nu}$, and the scalar curvatures, $\tilde{R} = g^{\mu\nu}\tilde{R}_{\mu\nu}$ and $R = g^{\mu\nu}R_{\mu\nu}$, can be obtained from the relations (6) and (8). For the scalar curvatures we get

$$\tilde{R} = R - 2\nabla_\mu\mathcal{V}^\mu + \frac{2}{3}\mathcal{V}_\mu\mathcal{V}^\mu - \frac{1}{24}\mathcal{A}_\mu\mathcal{A}^\mu - \frac{1}{2}\mathcal{T}_{\mu\nu\rho}\mathcal{T}^{\mu\nu\rho}\,, \tag{10}$$

or the other way around,

$$R = \tilde{R} + 2\tilde{\nabla}_\mu\mathcal{V}^\mu + \frac{4}{3}\mathcal{V}_\mu\mathcal{V}^\mu + \frac{1}{24}\mathcal{A}_\mu\mathcal{A}^\mu + \frac{1}{2}\mathcal{T}_{\mu\nu\rho}\mathcal{T}^{\mu\nu\rho}\,. \tag{11}$$

In the first-order formalism, which is customarily used in the gauge theory approach to gravity, the independent variables are the vierbein $e^a{}_\mu$ and the spin connection $\tilde{\omega}_\mu{}^a{}_b$, or the corresponding one-forms, the coframe $\theta^a = e^a{}_\mu dx^\mu$ and the connection $\tilde{\omega}^a{}_b = \tilde{\omega}_\mu{}^a{}_b dx^\mu$. These variables are the gauge fields that are required to ensure the gauge invariance of the action. Additionally, in a metric-affine gauge

[a]Tensor indexes in brackets are antisymmetrized, $a_{[\mu}b_{\nu]} = \frac{1}{2}(a_\mu b_\nu - a_\nu b_\mu)$, and indexes between vertical lines are excluded from the antisymmetrization, $a_{[\mu|\rho}b_{|\nu]} = \frac{1}{2}(a_{\mu\rho}b_\nu - a_{\nu\rho}b_\mu)$.

theory like PG, we assume the presence of a metric, $g_{ab}\theta^a \otimes \theta^b$, which is here taken to be an orthonormal coframe, $g_{ab} = \eta_{ab}$ (the Minkowski metric), so that the metric in a coordinate basis is defined as

$$g_{\mu\nu} = \eta_{ab} e^a{}_\mu e^b{}_\nu \,. \tag{12}$$

We consider PG, which is based on the Lorentz connection, $\tilde{\omega}^{ab} = \tilde{\omega}^{[ab]}$, i.e., on an antisymmetric linear connection. More general gauge theories of gravity require a more general linear connection.[b] In the presence of a metric, it is possible to decompose the spin connection into a torsionless connection and a contortion component

$$\tilde{\omega}_\mu{}^{ab} = \omega_\mu{}^{ab} + K_\mu{}^{\nu\rho} e^a{}_\nu e^b{}_\rho \,, \tag{13}$$

which consists of the torsionless part $\omega_\mu{}^{ab}$ and the part proportional to contortion. The spin connection is related to the connection in a coordinate frame by the so-called tetrad postulate

$$\tilde{\nabla}_\mu e^a{}_\nu = \partial_\mu e^a{}_\nu + \tilde{\omega}_\mu{}^a{}_b e^b{}_\nu - \tilde{\Gamma}_{\mu\nu}{}^\rho e^a{}_\rho = 0 \,, \tag{14}$$

which can be equivalently written as

$$\tilde{\omega}_\mu{}^a{}_b = (\tilde{\Gamma}_{\mu\nu}{}^\rho e^a{}_\rho - \partial_\mu e^a{}_\nu) e_b{}^\nu \,. \tag{15}$$

The torsion two-form is defined in terms of $e^a{}_\mu$ and $\tilde{\omega}_\mu{}^a{}_b$ as the exterior covariant derivative of the orthonormal coframe,

$$T^a = d\theta^a + \tilde{\omega}^a{}_b \wedge \theta^b = \frac{1}{2} T_{\mu\nu}{}^a dx^\mu \wedge dx^\nu \,, \tag{16}$$

where the components are defined as

$$T_{\mu\nu}{}^a = 2\partial_{[\mu} e^a{}_{\nu]} + 2\tilde{\omega}_{[\mu|}{}^a{}_b e^b{}_{|\nu]} \,. \tag{17}$$

The curvature two-form is defined as

$$\tilde{R}^a{}_b = d\tilde{\omega}^a{}_b + \tilde{\omega}^a{}_c \wedge \tilde{\omega}^c{}_b = \frac{1}{2} \tilde{R}_{\mu\nu}{}^a{}_b dx^\mu \wedge dx^\nu \,, \tag{18}$$

where the components are written as

$$\tilde{R}_{\mu\nu}{}^a{}_b = 2\partial_{[\mu} \tilde{\omega}_{\nu]}{}^a{}_b + 2\tilde{\omega}_{[\mu|}{}^a{}_c \tilde{\omega}_{|\nu]}{}^c{}_b \,. \tag{19}$$

The curvature $R_{\mu\nu}{}^a{}_b$ for the torsionless connection $\omega_\mu{}^b{}_b$ is defined similarly as (19). In a coordinate basis, the components of the curvature are given as

$$\tilde{R}_{\mu\nu}{}^\rho{}_\sigma = \tilde{R}_{\mu\nu}{}^a{}_b e_a{}^\rho e^b{}_\sigma \,. \tag{20}$$

When expressed entirely in the orthonormal frame $\{e_a\}$ (and the coframe $\{\theta^a\}$), where $e_a = e_a{}^\mu \partial_\mu$, the components of the torsion and the curvature are given as

$$\tilde{R}_{ab}{}^c{}_d = 2e_{[a} \tilde{\omega}_{b]}{}^c{}_d + 2\tilde{\omega}_{[a|}{}^c{}_e \tilde{\omega}_{|b]}{}^e{}_d - c_{ab}{}^e \tilde{\omega}_e{}^c{}_d \,, \tag{21}$$

$$T_{ab}{}^c = 2\tilde{\omega}_{[a}{}^c{}_{b]} - c_{ab}{}^c \,, \tag{22}$$

---

[b]Extending the local Poincaré group with local scale transformations would require the connection to have a nonvanishing trace component, which leads to a gauge theory of gravity with Weyl–Cartan geometry.[60] A general metric-affine gauge theory of gravity is based on the local affine group, and it requires an unrestricted linear connection.[62,78]

where $\tilde{\omega}_a{}^c{}_b = e_a{}^\mu \tilde{\omega}_\mu{}^c{}_b$, and they involve the anholonomity of the basis:

$$[e_a, e_b] = c_{ab}{}^c e_c\,, \quad c_{ab}{}^c = 2e_{[a|}{}^\mu \partial_\mu e_{|b]}{}^\nu e^c{}_\nu\,. \tag{23}$$

4. Induced Gravitational Action from Quantized Matter Fields

One can consider matter fields that are coupled to curvature (and in the present case also to torsion) minimally, although there is no physical principle that would require such a restriction. In some cases, e.g. for scalar fields, nonminimal couplings are necessary in order to achieve renormalizability. Here our goal is not renormalization, but rather derivation of the one-loop effective action that is regulated but not renormalized. That way the effective action can be regarded as the origin of gravity. Therefore, the presence of nonminimal couplings is not necessary for the present approach, but we shall consider them for the sake of generality.

4.1. *Scalar fields*

First consider a free real scalar field ϕ on four-dimensional Riemann–Cartan spacetime. We can obtain the minimally-coupled Lagrangian from a free-field Lagrangian on Minkowski spacetime with the replacement $\partial_\mu \to \tilde{\nabla}_\mu$. The usual free-field Lagrangian on Minkowski spacetime with the kinetic part $\frac{1}{2}\eta^{\mu\nu}\partial_\mu\phi\partial_\nu\phi$ gives the same Lagrangian as in Riemannian spacetime, since $\tilde{\nabla}_\mu\phi = \nabla_\mu\phi = \partial_\mu\phi$, and hence the torsion does not appear in it. Expressing the kinetic part of the Lagrangian in Minkowski spacetime as $-\frac{1}{2}\phi\eta^{\mu\nu}\partial_\mu\partial_\nu\phi$ leads to a different Lagrangian with a coupling to the vector component of torsion as

$$-\frac{1}{2}\phi\tilde{\Box}\phi = -\frac{1}{2}\phi(\Box\phi + \mathcal{V}^\mu\nabla_\mu\phi)\,, \tag{24}$$

where

$$\tilde{\Box} = g^{\mu\nu}\tilde{\nabla}_\mu\tilde{\nabla}_\nu, \quad \Box = g^{\mu\nu}\nabla_\mu\nabla_\nu\,. \tag{25}$$

Thus, the minimal replacement rule does not lead to a unique scalar field Lagrangian in Riemann–Cartan spacetime. Unlike in a Riemannian spacetime, the form of the initial Lagrangian in flat spacetime matters. Since it would be a limited viewpoint to consider only the coupling to the vector component of torsion, we will include other torsion terms as well. Hence, we consider the parity-conserving free field Lagrangian with all nonminimal coupling terms

$$\mathcal{L} = \frac{1}{2}\left(g^{\mu\nu}\nabla_\mu\phi\nabla_\nu\phi - m^2\phi^2 - \sum_{i=1}^{5}\xi_i P_i\phi^2\right), \tag{26}$$

where ξ_i are dimensionless coupling constants and the corresponding even-parity curvature and torsion terms are

$$P_1 = R\,, \quad P_2 = \nabla_\mu\mathcal{V}^\mu\,, \quad P_3 = \mathcal{V}_\mu\mathcal{V}^\mu\,, \quad P_4 = \mathcal{A}_\mu\mathcal{A}^\mu\,, \quad P_5 = \mathcal{T}_{\mu\nu\rho}\mathcal{T}^{\mu\nu\rho}\,. \tag{27}$$

The Lagrangian (26) could alternatively be written in terms of $\tilde{\nabla}$, $\tilde{R}$ and the components of torsion.

Note that we choose to write the Lagrangian in terms of the torsionless covariant derivative and its curvature, since it makes the calculations more convenient by enabling the use of techniques developed for quantum fields on Riemannian spacetime (see the monographs Refs. 79 and 80). That is, the metric and the torsion are regarded as the background fields. Note, however, that the gravitational variables are not chosen yet. They will be chosen later, after the quantum effective action has been obtained. Similar to the case of spacetime without torsion, where only the first nonminimal coupling $\xi R\phi^2$ appears, the nonminimal couplings would be necessary for renormalization.[81] Naturally, including pseudoscalar fields and/or complex scalar fields would allow further nonminimal coupling terms, which would again be necessary for renormalization.[81] Recall, however, that our present goal is not renormalization.

The action is defined by the Lagrangian (26) as

$$S = \int d^4x\sqrt{-g}\,\mathcal{L} = -\frac{1}{2}\int d^4x\sqrt{-g}\,\phi D\phi\,, \tag{28}$$

where

$$D = \Box + m^2 + \sum_{i=1}^{5}\xi_i P_i\,. \tag{29}$$

The one-loop effective action is defined[80,82] in terms of D as

$$S_{\text{eff}}^{(1)} = \frac{i}{2}\ln\det(l^2 D) = \frac{i}{2}\operatorname{Tr}\ln(l^2 D)\,, \tag{30}$$

where the parameter l with dimension of length was introduced for dimensional reasons, so that $l^2 D$ is dimensionless. Since we consider only one-loop corrections, and hence there is no need to keep track of higher powers of $\hbar$, we have set $\hbar = 1$, along with $c = 1$. We can use the identity

$$\ln\left(\frac{D}{D_0}\right) = -\int_0^\infty \frac{d\tau}{\tau}[\exp(-i\tau D) - \exp(-i\tau D_0)]\,, \tag{31}$$

where D_0 is the operator for a suitable reference background $(g_0, \tilde{\Gamma}_0)$,[c] in order to obtain

$$S_{\text{eff}}^{(1)} = S_{\text{eff},0}^{(1)} - \frac{i}{2}\operatorname{Tr}\int_0^\infty \frac{d\tau}{\tau}[\exp(-i\tau D) - \exp(-i\tau D_0)]\,, \tag{32}$$

where $S_{\text{eff},0}^{(1)}$ is the one-loop effective action for the reference background. On a noncompact spacetime the reference background has to be chosen so that the physical action for induced gravity $S_{\text{phys}}^{(1)} = S_{\text{eff}}^{(1)} - S_{\text{eff},0}^{(1)}$ is well defined. We shall consider a compact spacetime for simplicity. This does not limit the generality of the derivation. The contribution of a reference background can be easily included into the

[c]Both operators D and D_0 are considered to have a small negative imaginary part in order to avoid divergence of the integration in (31).

induced action afterwards, in case one needs the result for a noncompact spacetime. Thence, instead of associating the operator D_0 with a reference background, we can freely choose it to be proportional to an identity operator as $D_0 = l^{-2}I$, so that we obtain

$$S^{(1)}_{\text{eff}} = -\frac{i}{2}\operatorname{Tr}\int_0^\infty \frac{d\tau}{\tau}\exp(-i\tau D) + \frac{i}{2}\lim_{\epsilon\to 0+}\int_0^\infty \frac{ds}{s}\exp(-is(1-i\epsilon))\operatorname{Tr} I\,. \quad (33)$$

In units of mass, the dimensions of the above constant l and the integration variables τ and $s = l^{-2}\tau$ are $[l] = -1$, $[\tau] = -2$ and $[s] = 0$. The second term in (33) is a constant, which is irrelevant dynamically, since it does not depend on D or on any geometric quantities. Therefore, in the following, we drop the constant term and write

$$S^{(1)}_{\text{eff}} = -\frac{i}{2}\operatorname{Tr}\int_0^\infty \frac{d\tau}{\tau}\exp(-i\tau D)\,. \quad (34)$$

We express the operator in the effective action in terms of a kernel function $K(\tau; x, y)$, which is defined as

$$\exp(-i\tau D)\phi(x) = \int d^4y\sqrt{-g}K(\tau; x, y)\phi(y)\,. \quad (35)$$

The kernel satisfies a Schrödinger-like equation

$$i\frac{d}{d\tau}K(\tau; x, y) = DK(\tau; x, y)\,, \quad (36)$$

with the boundary condition

$$K(\tau = 0; x, y) = I\delta(x, y)\,, \quad (37)$$

where I is the identity operator/matrix for the fields or field components. We are working with a spacetime of Lorentzian signature so that the kernel corresponds to a heat kernel with imaginary time.[82,83] Strictly speaking, the heat kernel technique is mathematically well defined only in Euclidean signature, when the squared distance $(x - y)^2$ of points is positive definite. The kernel discussed here should be regarded as an analytic continuation of the heat kernel, and it is only useful for finding the local contributions (in the limit $y \to x$) to the effective action. The trace of the operator in (34) is given by

$$\operatorname{Tr}\exp(-i\tau D) = \int d^4x\sqrt{-g}\,\operatorname{tr} K(\tau; x, x)\,, \quad (38)$$

where in the right-hand side the trace is taken over the field degrees of freedom.[d] The (divergent) gravitational terms that we are interested in come from the zero end of the integral over τ. We use the local series expansion of the kernel[79,80,82,83]

$$K(\tau; x, x) = \frac{i}{(4\pi i\tau)^2}\sum_{n=0}^{\infty}(i\tau)^n A_n(x)\,, \quad (39)$$

[d]When a field with several components or several fields are involved, the operator D and its kernel K are matrix-valued.

where $A_n(x)$ are constructed from the geometric quantities and parameters involved in D, i.e., from the covariant derivative, the curvature, the torsion, and the masses and couplings of the fields. We assume that the spacetime has no boundary. If the spacetime had a boundary, we would have to include boundary terms into the series expansion of the operator (38). In the presence of boundaries, each term in the series expansion except the zeroth term involves an additional boundary contribution,[83] and the series expansion (39) also involves half-integer terms, $n = \frac{1}{2}, \frac{3}{2}, \ldots$, which are purely boundary terms.

Since the operator (29) is readily in the Laplace form

$$D = \Box + B\,, \tag{40}$$

where the (endomorphism) term B does not involve derivatives,[e] one obtains the first three terms of the kernel (39) as[f]

$$A_0 = I\,, \tag{41}$$

$$A_1 = \frac{1}{6}RI - B\,, \tag{42}$$

$$\begin{aligned} A_2 = &\left(\frac{1}{180}R^{\mu\nu\rho\sigma}R_{\mu\nu\rho\sigma} - \frac{1}{180}R^{\mu\nu}R_{\mu\nu} + \frac{1}{72}R^2 - \frac{1}{30}\Box R\right) I \\ &+ \frac{1}{2}B^2 - \frac{1}{6}RB + \frac{1}{6}\Box B + \frac{1}{12}W^{\mu\nu}W_{\mu\nu}\,, \end{aligned} \tag{43}$$

where

$$W_{\mu\nu} = [\nabla_\mu, \nabla_\nu]\,. \tag{44}$$

For a single real scalar field, the identity is of course one-dimensional, $I = 1$, (44) vanishes, $W_{\mu\nu} = 0$, and

$$B = m^2 + \sum_{i=1}^{5} \xi_i P_i\,. \tag{45}$$

Which terms of the expansion of the kernel (39) appear in the regularized one-loop effective action depends on the chosen regularization method. For example, dimensional regularization involves only the term A_2, which contains quadratic curvature and torsion terms. In that sense, it is a too powerful regularization method for our purposes. We choose to use the cutoff regularization, since it involves all the given terms (41)–(43). The lower limit of the integral over τ is cut off at Λ^{-2} for the three divergent terms, where the ultraviolet cutoff parameter Λ has the dimension of mass, $[\Lambda] = 1$. For the third term with A_2 the upper limit of the

[e]We mean that B does not involve derivative operators acting on the field. Of course, B itself involves derivatives of the gravitational fields, namely, in the present case the curvature and torsion terms in (45).

[f]See Refs. 82, 85 and references therein for different techniques of derivation for the kernel coefficients, and e.g. Refs. 79, 80 and 83 for its use.

integral is cut off at $\tau_0 = \epsilon^{-2}$, where ϵ is an infrared cutoff. We obtain the kernel expansion of the cutoff-regularized one-loop effective action as

$$\begin{aligned} S^{(1)}_{\text{eff}} &= \frac{1}{32\pi^2}\sum_{n=0}^{\infty}\int_{0 \text{ or } \Lambda^{-2}}^{\infty \text{ or } \epsilon^{-2}} d(i\tau)(i\tau)^{n-3}\int d^4x\sqrt{-g}\,\mathrm{tr}\,A_n(x) \\ &= \frac{\Lambda^4}{64\pi^2}\int d^4x\sqrt{-g}\,\mathrm{tr}\,A_0 + \frac{\Lambda^2}{32\pi^2}\int d^4x\sqrt{-g}\,\mathrm{tr}\,A_1 \\ &\quad + \frac{\ln(\Lambda/\epsilon)}{16\pi^2}\int d^4x\sqrt{-g}\,\mathrm{tr}\,A_2 + \text{ultraviolet-finite terms}. \end{aligned} \tag{46}$$

The term of order Λ^4, i.e., the quartic divergence, is given as $\mathrm{tr}\,A_0 = 1$, which contributes only to the vacuum energy. The term of order Λ^2 in the one-loop effective action (46) is given as

$$\mathrm{tr}\,A_1 = -m^2 + \left(\frac{1}{6} - \xi_1\right)R - \sum_{i=2}^{5}\xi_i P_i\,. \tag{47}$$

The term proportional to the logarithm of the cutoff contains the second-order gravitational terms: quadratic curvature terms and torsion terms up to fourth power. The term $\frac{1}{2}B^2$ of the contribution (43) proportional to $\ln(\Lambda/\epsilon)$ also contains the first-order terms $m^2\sum_{i=1}^{5}\xi_i P_i$. In total, we have

$$\begin{aligned} \mathrm{tr}\,A_2 &= \frac{1}{180}R^{\mu\nu\rho\sigma}R_{\mu\nu\rho\sigma} - \frac{1}{180}R^{\mu\nu}R_{\mu\nu} + \frac{1}{72}R^2 - \frac{1}{30}\Box R \\ &\quad + \frac{1}{2}\left(m^2 + \sum_{i=1}^{5}\xi_i P_i\right)^2 - \frac{1}{6}R\left(m^2 + \sum_{i=1}^{5}\xi_i P_i\right) + \frac{1}{6}\sum_{i=1}^{5}\xi_i\Box P_i \\ &= \frac{1}{180}R^{\mu\nu\rho\sigma}R_{\mu\nu\rho\sigma} - \frac{1}{180}R^{\mu\nu}R_{\mu\nu} + \frac{1}{2}\left(\frac{1}{36} - \frac{\xi_1}{3} + \xi_1^2\right)R^2 \\ &\quad + \frac{1}{6}\left(\xi_1 - \frac{1}{5}\right)\Box R + \left(\xi_1 - \frac{1}{6}\right)R\sum_{i=2}^{5}\xi_i P_i + \frac{1}{2}\left(\sum_{i=2}^{5}\xi_i P_i\right)^2 \\ &\quad + \frac{1}{6}\sum_{i=2}^{5}\xi_i\Box P_i + \left(\xi_1 - \frac{1}{6}\right)m^2 R + m^2\sum_{i=2}^{5}\xi_i P_i + \frac{1}{2}m^4\,, \end{aligned} \tag{48}$$

where the sums in the latter expression are taken over the torsion terms, P_i with $i = 2, \ldots, 5$.

The Gauss–Bonnet–Chern term

$$G = R_{\alpha\beta\gamma\delta}R_{\mu\nu\rho\sigma}\epsilon^{\alpha\beta\mu\nu}\epsilon^{\gamma\delta\rho\sigma} = R^{\mu\nu\rho\sigma}R_{\mu\nu\rho\sigma} - 4R^{\mu\nu}R_{\mu\nu} + R^2\,, \tag{49}$$

whose integral is a topological invariant in four-dimensional spacetime, can be used to absorb the Riemann tensor squared term from the effective action.

Alternatively, we could write the effective action in terms of the connection $\tilde{\nabla}$ with torsion and its curvature (7) by using the relations between the two connections

(1) and their curvature tensors (8). That would be the appropriate way, if one chooses the first-order formalism, where $e^a{}_\mu$ and $\tilde{\omega}_\mu{}^a{}_b$ are the independent variables of the induced gravitational action. Such a case is considered in Sec. 5, where the induced action for PG is obtained.

4.2. *Spin-1/2 fields*

Next we consider a Dirac spinor field ψ. The Dirac field can couple to the vector and the axial vector components of torsion. Naturally, these couplings are of the same form as the couplings for any other vector and axial vector fields. In particular, the vector component $\mathcal{V}_\mu$ couples to the Dirac spinor in the same way as the electromagnetic field. In the Dirac action, spin-1/2 fields do not couple to curvature or to higher-order torsion invariants (27) due to dimensional reasons, since the mass dimensions are $[\psi] = \frac{3}{2}$, $[Riemann] = 2$ and $[Torsion] = 1$.[g]

The Hermitian Lagrangian for a free minimally coupled Dirac field is written as

$$\mathcal{L}_{\text{Dirac}}^{\text{min.}} = \frac{i}{2}(\bar{\psi}\underline{\gamma}^\mu\tilde{\nabla}_\mu\psi - \tilde{\nabla}_\mu\bar{\psi}\underline{\gamma}^\mu\psi) - m\bar{\psi}\psi\,, \tag{50}$$

where the spacetime-dependent γ matrices are defined as $\underline{\gamma}^\mu = e_a{}^\mu\gamma^a$, where $e_a{}^\mu$ is the inverse of the vierbein $e^a{}_\mu$. In four-dimensional spacetime, the constant γ matrices satisfy

$$\{\gamma^a,\gamma^b\} = 2\eta^{ab}I\,, \quad (\gamma^0)^2 = I, \quad (\gamma^i)^2 = -I\,, \quad (\gamma^a)^\dagger = \gamma^0\gamma^a\gamma^0\,, \tag{51}$$

where $a = 0,1,2,3$, $i = 1,2,3$, $(\eta^{ab}) = \text{diag}(1,-1,-1,-1)$ and I is the four-dimensional identity matrix. The fifth γ matrix is defined as

$$\gamma^5 = i\gamma^0\gamma^1\gamma^2\gamma^3\,, \tag{52}$$

and it satisfies

$$(\gamma^5)^2 = I\,, \quad (\gamma^5)^\dagger = \gamma^5\,, \quad \{\gamma^5,\gamma^a\} = 0\,. \tag{53}$$

The covariant derivative for the spinor and its conjugate $\bar{\psi} = \psi^\dagger\gamma^0$ is defined by

$$\begin{aligned}\tilde{\nabla}_\mu\psi &= \partial_\mu\psi + \frac{i}{2}\tilde{\omega}_\mu{}^{ab}\Sigma_{ab}\psi\,,\\ \tilde{\nabla}_\mu\bar{\psi} &= \partial_\mu\bar{\psi} - \frac{i}{2}\bar{\psi}\tilde{\omega}_\mu{}^{ab}\Sigma_{ab}\,,\end{aligned} \tag{54}$$

where $\Sigma_{ab} = -\frac{i}{4}[\gamma_a,\gamma_b]$ satisfies the Lie algebra of $SO(1,3)$. Likewise, the covariant derivative without torsion is

$$\begin{aligned}\nabla_\mu\psi &= \partial_\mu\psi + \frac{i}{2}\omega_\mu{}^{ab}\Sigma_{ab}\psi\,,\\ \nabla_\mu\bar{\psi} &= \partial_\mu\bar{\psi} - \frac{i}{2}\bar{\psi}\omega_\mu{}^{ab}\Sigma_{ab}\,.\end{aligned} \tag{55}$$

[g]We do not consider coupling constants with negative mass dimensions or couplings to fractional or negative powers of curvature and torsion.

The action for the minimally coupled Lagrangian (50) can be written in two forms, either in terms of $\tilde{\nabla}$ or ∇, as

$$S_{\rm Dirac}^{\rm min.} = \int d^4x\sqrt{-g}\bar{\psi}\left(i\underline{\gamma}^\mu\tilde{\nabla}_\mu - \frac{i}{2}\underline{\gamma}^\mu\mathcal{V}_\mu - m\right)\psi$$

$$= \int d^4x\sqrt{-g}\bar{\psi}\left(i\underline{\gamma}^\mu\nabla_\mu + \frac{1}{8}\underline{\gamma}^\mu\gamma^5\mathcal{A}_\mu - m\right)\psi\,, \tag{56}$$

where the boundary surface term coming from an integration by parts is assumed to vanish, $\int d^4x\partial_\mu(\sqrt{-g}\bar{\psi}\underline{\gamma}^\mu\psi) = 0$. According to the second expression of (56), a minimally coupled spinor couples only to the axial vector component of torsion. We shall consider a more general Dirac field that couples to both the vector and axial components of torsion.

Written in terms of the torsionless covariant derivative, a nonminimally coupled Dirac field has the action

$$\begin{aligned} S_{\rm Dirac}^{\rm non\text{-}min.} &= \int d^4x\sqrt{-g}\bar{\psi}D_{1/2}\psi\,, \\ D_{1/2} &= i\underline{\gamma}^\mu\mathcal{D}_\mu - m\,, \\ \mathcal{D}_\mu &= \nabla_\mu + i\alpha_1\mathcal{V}_\mu + i\alpha_2\gamma^5\mathcal{A}_\mu\,, \end{aligned} \tag{57}$$

where α_1 and α_2 are dimensionless coupling constants for the vector and the axial vector components of torsion, respectively. The minimally coupled case (50) corresponds to the couplings $\alpha_1 = 0$, $\alpha_2 = -\frac{1}{8}$.

The one-loop effective action is defined as

$$S_{\rm eff}^{(1)} = -i\ln\det(lD_{1/2})\,, \tag{58}$$

where the minus sign and the factor of two compared to the spin-0 case (34) come from the anticommuting and complex-valued nature of the Dirac field, respectively. Since the Dirac operator $D_{1/2}$ is a first-order differential operator, we cannot use the technique of Sec. 4.1 directly. However, we can square the operator $D_{1/2}$ as follows (for a more detailed proof, see Ref. 89). First we define a modification of the operator $D_{1/2}$ as

$$D_{1/2}^* = -i\underline{\gamma}^\mu\mathcal{D}_\mu - m\,. \tag{59}$$

In even-dimensional spacetime, the matrix γ^5, which is Hermitian and satisfies $(\gamma^5)^2 = I$, anticommutes with $\underline{\gamma}^\mu$. Hence, $\gamma^5 D_{1/2}^*\gamma^5 = D_{1/2}$,[h] and we obtain

$$\det(lD_{1/2}^*) = \det(l\gamma^5 D_{1/2}^*\gamma^5) = \det(lD_{1/2})\,. \tag{60}$$

Therefore, the operator $D_{1/2}$ in the effective action (58) can be squared as

$$S_{\rm eff}^{(1)} = -\frac{i}{2}\ln[\det(lD_{1/2})]^2 = -\frac{i}{2}\ln\det(l^2 D_{1/2}D_{1/2}^*)\,. \tag{61}$$

[h]In even-dimensional spacetime, the matrices $\{\underline{\gamma}^\mu\}$ and $\{-\underline{\gamma}^\mu\}$ form equivalent representations of the Clifford algebra (see e.g. Ref. 80).

Hence, the operator D in the effective action of a Dirac spinor is written as

$$D \equiv D_{1/2} D^*_{1/2} = (\underline{\gamma}^\mu \mathcal{D}_\mu)^2 + m^2 \,. \tag{62}$$

The operator D in (62) is not of the Laplace type for any of the derivatives $\mathcal{D}$, ∇ or $\tilde{\nabla}$. We can see this by expanding the first term in (62) as

$$(\underline{\gamma}^\mu \mathcal{D}_\mu)^2 = g^{\mu\nu} \mathcal{D}_\mu \mathcal{D}_\nu - 2i\alpha_2 \underline{\gamma}^\mu \underline{\gamma}^\nu \gamma^5 \mathcal{A}_\mu \mathcal{D}_\nu + \frac{1}{2} \underline{\gamma}^\mu \underline{\gamma}^\nu [\mathcal{D}_\mu, \mathcal{D}_\nu] \,, \tag{63}$$

where the last term can be written as

$$\frac{1}{2} \underline{\gamma}^\mu \underline{\gamma}^\nu [\mathcal{D}_\mu, \mathcal{D}_\nu] = \frac{1}{4} R + \frac{i}{4} \alpha_1 [\underline{\gamma}^\mu, \underline{\gamma}^\nu] V_{\mu\nu} + \frac{i}{4} \alpha_2 [\underline{\gamma}^\mu, \underline{\gamma}^\nu] \gamma^5 A_{\mu\nu} \,, \tag{64}$$

which involves the following second-rank tensors[i]

$$\begin{aligned} V_{\mu\nu} &= \nabla_\mu \mathcal{V}_\nu - \nabla_\nu \mathcal{V}_\mu \,, \\ A_{\mu\nu} &= \nabla_\mu \mathcal{A}_\nu - \nabla_\nu \mathcal{A}_\mu \,. \end{aligned} \tag{65}$$

Therefore, the operator (62) contains a first-order derivative term. However, for any partial differential operator D that contains first-order and second-order partial derivatives, with the second-order term taking the form $g^{\mu\nu}\partial_\mu \partial_\nu$, there exists[85] a unique connection $\hat{\nabla}$ and a unique endomorphism B for which the operator takes the Laplace form

$$D = \hat{\Box} + B \,, \quad \hat{\Box} = g^{\mu\nu} \hat{\nabla}_\mu \hat{\nabla}_\nu \,. \tag{66}$$

That connection consists of the Riemannian connection and a gauge (bundle) part. In the present case (62), the operator (66) is given by

$$\begin{aligned} \hat{\nabla}_\mu &= \mathcal{D}_\mu - i\alpha_2 \underline{\gamma}^\nu \underline{\gamma}_\mu \gamma^5 \mathcal{A}_\nu \\ &= \nabla_\mu + i\alpha_1 \mathcal{V}_\mu + \frac{i}{2} \alpha_2 [\underline{\gamma}_\mu, \underline{\gamma}_\nu] \gamma^5 \mathcal{A}^\nu \end{aligned} \tag{67}$$

and

$$B = \left(\frac{1}{4} R - 2\alpha_2^2 \mathcal{A}_\mu \mathcal{A}^\mu + m^2 \right) I + \frac{i}{4} \alpha_1 [\underline{\gamma}^\mu, \underline{\gamma}^\nu] V_{\mu\nu} + i\alpha_2 \gamma^5 \nabla_\mu \mathcal{A}^\mu \,. \tag{68}$$

The tensor $W_{\mu\nu}$ for the connection $\hat{\nabla}$ is given by

$$\begin{aligned} W_{\mu\nu} = &-\frac{1}{8} R_{\mu\nu\rho\sigma} [\underline{\gamma}^\rho, \underline{\gamma}^\sigma] + i\alpha_1 V_{\mu\nu} + i\alpha_2 \gamma^5 A_{\mu\nu} \\ &+ i\alpha_2 \gamma^5 \underline{\gamma}_\rho (\underline{\gamma}_\mu \nabla_\nu \mathcal{A}^\rho - \underline{\gamma}_\nu \nabla_\mu \mathcal{A}^\rho) \\ &- \alpha_2^2 (\underline{\gamma}_\mu \underline{\gamma}_\rho \underline{\gamma}_\nu \underline{\gamma}_\sigma - \underline{\gamma}_\nu \underline{\gamma}_\rho \underline{\gamma}_\mu \underline{\gamma}_\sigma) \mathcal{A}^\rho \mathcal{A}^\sigma \,. \end{aligned} \tag{69}$$

The kernel expansion and regularization of the effective action are performed in a way identical to that of the scalar field case. Only now the operator D, the kernel

[i] In a more general setting, the vector $\mathcal{V}_\mu$ and the axial vector $\mathcal{A}_\mu$ could contain not only the vector components of torsion but also gauge fields in some representation of a given gauge group. In that case, $V_{\mu\nu}$ and $A_{\mu\nu}$ would be extended to include the field strength tensors of the gauge fields.

K, the identity I and the tensor $W_{\mu\nu}$ are four-dimensional square matrices. The tensor $W_{\mu\nu}$ is given for the spinor in (69). The cutoff-regularized effective action has the same expression as for a scalar field (46) but with an opposite sign,

$$S^{(1)}_{\text{eff}} = -\frac{\Lambda^4}{64\pi^2}\int d^4x\sqrt{-g}\,\mathrm{tr}\,A_0 - \frac{\Lambda^2}{32\pi^2}\int d^4x\sqrt{-g}\,\mathrm{tr}\,A_1$$
$$- \frac{\ln(\Lambda/\epsilon)}{16\pi^2}\int d^4x\sqrt{-g}\,\mathrm{tr}\,A_2 + \text{ultraviolet-finite terms}\,. \tag{70}$$

Then we evaluate the traces of the terms (41)–(43) of the kernel expansion with (68) and (69). The first two terms of the effective action (70) are given by

$$\mathrm{tr}\,A_0 = 4\,, \tag{71}$$
$$\mathrm{tr}\,A_1 = -\frac{1}{3}R + 8\alpha_2^2\mathcal{A}_\mu\mathcal{A}^\mu - 4m^2\,, \tag{72}$$

using $\mathrm{tr}\,I = 4$ and $\mathrm{tr}\,B = R - 8\alpha_2^2\mathcal{A}_\mu\mathcal{A}^\mu + 4m^2$. Finally, after some lengthy algebra, we obtain the term proportional to $\ln(\Lambda/\epsilon)$ in the effective action (70) as

$$\mathrm{tr}\,A_2 = -\frac{7}{360}R^{\mu\nu\rho\sigma}R_{\mu\nu\rho\sigma} - \frac{1}{45}R^{\mu\nu}R_{\mu\nu} + \frac{1}{72}R^2 + \frac{1}{30}\Box R$$
$$+ \frac{2}{3}\alpha_1^2V^{\mu\nu}V_{\mu\nu} + \frac{2}{3}\alpha_2^2A^{\mu\nu}A_{\mu\nu} - \frac{4}{3}\alpha_2^2\Box(\mathcal{A}_\mu\mathcal{A}^\mu)$$
$$+ \frac{4}{3}\alpha_2^2\nabla_\mu(\mathcal{A}^\nu\nabla_\nu\mathcal{A}^\mu - \mathcal{A}^\mu\nabla_\nu\mathcal{A}^\nu)$$
$$+ \frac{1}{3}m^2R - 8\alpha_2^2m^2\mathcal{A}_\mu\mathcal{A}^\mu + 2m^4\,. \tag{73}$$

This term does not explicitly contain terms involving both curvature and torsion. Note, however, that the term $\nabla_\mu(\mathcal{A}^\nu\nabla_\nu\mathcal{A}^\mu - \mathcal{A}^\mu\nabla_\nu\mathcal{A}^\nu)$ contains such a cross term between the Ricci tensor and the axial vector component of torsion, $R_{\mu\nu}\mathcal{A}^\mu\mathcal{A}^\nu$.

It is interesting to compare the torsion terms of the one-loop effective actions for a scalar field and a Dirac field. The effective Lagrangian for a scalar field (46) contains the following torsion terms:

$$\begin{gathered} P_i\,, \quad \Box P_i\,, \quad (i = 2,\ldots,5) \\ P_iP_j\,, \quad (i = 1,\ldots,5,\ j = 2,\ldots 5) \end{gathered} \tag{74}$$

where P_i is defined in (27). The first term in (74) appears both as $\Lambda^2\xi_iP_i$ and $(\ln\Lambda/\epsilon)m^2\xi_iP_i$. The latter two terms in (74) appear in the term that is proportional to $\ln(\Lambda/\epsilon)$ in the effective action (46). On the other hand, the effective Lagrangian for a Dirac field (70) involves the following torsion terms:

$$\begin{gathered} \mathcal{A}_\mu\mathcal{A}^\mu\,, \quad \Box(\mathcal{A}_\mu\mathcal{A}^\mu)\,, \quad V^{\mu\nu}V_{\mu\nu}\,, \quad A^{\mu\nu}A_{\mu\nu}\,, \\ \nabla_\mu\mathcal{A}^\nu\nabla_\nu\mathcal{A}^\mu\,, \quad (\nabla_\mu\mathcal{A}^\mu)^2\,, \quad R_{\mu\nu}\mathcal{A}^\mu\mathcal{A}^\nu\,. \end{gathered} \tag{75}$$

There are no cross terms between the three components of torsion in the Dirac field case. It is noteworthy that the only common torsion terms in the effective

Lagrangians for a scalar field and a Dirac field are the following three terms involving the squared norm $\mathcal{A}_\mu \mathcal{A}^\mu$ of the axial component of torsion:

$$\Lambda^2 \mathcal{A}_\mu \mathcal{A}^\mu \,, \quad (\ln \Lambda/\epsilon) m^2 \mathcal{A}_\mu \mathcal{A}^\mu \,, \quad (\ln \Lambda/\epsilon) \Box (\mathcal{A}_\mu \mathcal{A}^\mu) \,. \tag{76}$$

Notice that the quartic term $(\ln \Lambda/\epsilon)(\mathcal{A}_\mu \mathcal{A}^\mu)^2$ does not appear in the Dirac field case (73). All the rest of the torsion terms in (74) and (75) appear only for either a scalar field or for a Dirac field.

4.3. *Induced gravitational couplings*

We compare the induced gravitational actions (46) and (70) with the gravitational action

$$S = \frac{1}{16\pi G} \int d^4x \sqrt{-g} (R - 2\lambda + \cdots) \,, \tag{77}$$

where the dots stand for all possible generally covariant terms constructed from the torsion and the curvature (except R). The induced cosmological constant is obtained from the induced actions (46) and (70) as

$$\frac{\lambda_{\text{ind}}}{G_{\text{ind}}} = \sum_f \frac{C_f^{(0)}}{4\pi} \left(\frac{\Lambda^4}{2} - \Lambda^2 m_f^2 + \ln\left(\frac{\Lambda}{\epsilon}\right) m_f^4 \right) \,, \tag{78}$$

where the sum is taken over all scalar and Dirac fields ($f = s, d$), and the constant $C_s^{(0)} = -1$ for a scalar field and $C_d^{(0)} = 4$ for a Dirac field, and the induced Newton constant G_{ind} will be discussed below. Since the masses m_f are taken to be much lighter than the ultraviolet cutoff, $m_f \ll \Lambda$, the dominant contribution to the cosmological constant is

$$\frac{\lambda_{\text{ind}}}{G_{\text{ind}}} \approx \frac{(4N_d - N_s)}{8\pi} \Lambda^4 \,, \tag{79}$$

where N_d and N_s are the number of Dirac fields and the number of scalar fields, respectively. Since the ultraviolet cutoff has to be at least above the electroweak scale, $\Lambda \gtrsim 1\,\text{TeV}$, up to where the Standard Model has been tested accurately, and setting G_{ind} to its observed order of magnitude, $G_{\text{ind}} \sim 10^{-1} M_\text{P}^{-2}$, we obtain $\lambda_{\text{ind}} \lesssim -10^{-64} M_\text{P}^2$ for a scalar field. A Dirac spinor produces a positive cosmological constant, $\lambda_{\text{ind}} \gtrsim 10^{-64} M_\text{P}^2$. As usual, the vacuum energy obtained from a quantum field theory is far too high compared to the minuscule observed value $\lambda \sim 10^{-122} M_\text{P}^2$. Actually, we will soon see that the ultraviolet cutoff is expected to be comparable to the Planck mass in this approach, so that the prediction for λ_{ind} with fermionic matter is much higher than the above estimate, around $\lambda_{\text{ind}} \sim M_\text{P}^2$. Sakharov's approach does not help us with the vacuum energy problem. Consequently, the vacuum energy term that is proportional to the volume of spacetime is usually ignored in Sakharov's approach. It might, however, be possible to resolve the problem with fine tuning and radiative instability of the cosmological constant. For this purpose, we would like to highlight the proposal of vacuum energy sequestering,[86] and in particular its recent local formulation,[87,88] where the perturbative instability of the vacuum energy contribution of quantized matter fields is tamed by letting the gravitational and cosmological constants become variables and introducing auxiliary volume four-forms for fixing the values of the given constants. Large contributions to the vacuum energy from the matter sector are cancelled by the sequestering mechanism. A potential problem with including the vacuum energy sequestering mechanism into Sakharov's induced gravity is that the gravitational terms (and consequently the gravitational and cosmological constants) are absent in the classical action prior to quantization of matter. Hence, the additional Lagrangian that is needed for the local sequestering mechanism has to be introduced by hand, unless one finds a way to induce those terms along with the gravitational action. A possible clue for such a direction might be that the gravitational action along with the vacuum energy sequestering mechanism can be considered to emerge from gauge fixing in an underlying theory.[88j]

Thus, the induced Newton constant is obtained as

$$\frac{1}{G_{\text{ind}}} = \sum_f \frac{C_f^{(1)}}{\pi}\left(\frac{\Lambda^2}{2} - \ln\left(\frac{\Lambda}{\epsilon}\right) m_f^2\right), \tag{80}$$

where $C_s^{(1)} = \frac{1}{6} - \xi_{1s}$ for a scalar field and $C_d^{(1)} = \frac{1}{3}$ for a Dirac spinor. Since $m_f \ll \Lambda$, the dominant contribution to the Newton constant is obtained as

$$\frac{1}{G_{\text{ind}}} \approx \frac{(2N_d + N_s - 6\sum_s \xi_{1s})}{12\pi}\Lambda^2. \tag{81}$$

This implies that for a single Dirac field $\Lambda \sim 10 M_{\text{P}}$. Including a scalar field or scalar fields with a large negative coupling $\xi_{1s} \ll -1$ would enable Λ to be set lower than M_{P}. Since such a strong nonminimal coupling of a scalar field to the curvature is not known, and the fundamental matter fields are fermions, we do not explore the case $\xi_{1s} \ll -1$ further here. Thus, for a realistic model of matter, where several fermionic fields are present, the ultraviolet cutoff Λ is comparable to the Planck mass.

We have obtained the one-loop effective actions for a scalar field and for a Dirac field with arbitrary nonminimal couplings ξ_i and α_j. In order to estimate the magnitude of the induced couplings of the torsion terms, we have to set the magnitudes of the constants ξ_i and α_j. For a scalar field (26) we assume that all the couplings satisfy $|\xi_i| \lesssim 1$, and the coupling to the scalar curvature also satisfies $\xi_1 \leq 1$, which ensures that the induced Newton constant (80) is not too negative, so that the positive contribution from fermionic matter can outweigh it. The given range of couplings also includes the special case $\frac{1}{6}R - \sum_{i=1}^{5} \xi_i P_i = \tilde{R}$, where all couplings are in the range $10^{-1} \lesssim |\xi_i| \lesssim 1$. In the leading Λ^2 order, that special

[j] In this extension of the theory,[88] the gravitational action with the vacuum energy sequestering mechanism is considered to be a gauge fixing action S_{gf} for an underlying theory. The BRST-invariant action is constructed as usual by adding the appropriate ghost action S_{gh}. The resulting gravitational action $S_{\text{g}} = S_{\text{gf}} + S_{\text{gh}}$ is found to be not only BRST invariant but also BRST exact, in the sense that $S_{\text{g}} \propto \delta_B F$, where $\delta_B F$ is the BRST transformation of a certain functional F.

case is Einstein–Cartan–Sciama–Kibble gravity. Since Einstein–Cartan–Sciama–Kibble gravity resides in the upper end of the chosen coupling range $|\xi_i| \lesssim 1$, it provides an estimate for the effect of the torsion terms in this range. The spin-spin contact interaction in Einstein–Cartan–Sciama–Kibble gravity is weak and becomes comparable to the effect of mass at very high mass densities,[69] around 10^{47} g/cm^3 for electrons and 10^{54} g/cm^3 for neutrons. These densities are so high that they are only encountered in black holes and in the early universe, but they are still much lower than the Planck density at which the quantum gravity effects are expected to dominate. Deviation from the special case of Einstein–Cartan–Sciama–Kibble theory does enable propagation of torsion, which is clearly possible in the generic induced action, but the magnitude of the couplings remains weak. Next we shall give a similar estimation for a Dirac field. We assume the couplings for a Dirac field (57) satisfy $|\alpha_j| \lesssim 1$. This range includes the minimally-coupled Dirac field, which corresponds to the couplings $\alpha_1 = 0$ and $\alpha_2 = -\frac{1}{8}$. Thus, in the given range of couplings, the leading-order torsion contributions in the induced action (70) are of a similar magnitude as in the case of the scalar field analyzed above.

In the low-energy realm of classical gravity, the curvature and torsion terms in the contributions (48) and (73), which are multiplied by $\ln(\Lambda/\epsilon)$ in the effective Lagrangian, are heavily suppressed compared to the leading Λ^2 contribution discussed above. The infrared cutoff ϵ can be chosen to be at most of the order of the mass of the lightest matter particles, which are the neutrinos, so that we can set $\epsilon \lesssim 10^{-3}\,$eV. Note that setting ϵ ten or twenty orders of magnitude lower than $10^{-3}\,$eV would still result in $\ln(\Lambda/\epsilon)$ being of the same order of magnitude, so that the present discussion does not depend much on the chosen infrared cutoff. Hence, in the coupling ranges chosen above, the dimensionless net couplings of the higher-order terms ($Riemann^2$, $\Box R$, $Torsion^4$, $\Box Torsion^2$, $(\nabla Torsion)^2$ and $Riemann \times Torsion^2$) in the effective Lagrangian are of the order one or below, and hence their effect on low-energy physics is marginal (apart from their possible impact on the propagation of torsion). This point of view can be justified by treating gravity as a low-energy effective field theory.

5. Induced Gauge Theory of Gravity

It is known that the two formulations of classical gravity on Riemann–Cartan spacetime are not generally equivalent. In the second-order formulation, the independent variables are the metric and the torsion. Induced gravity in the second-order formulation was considered in Sec. 4, and it was noted that the induced action can as well be written as a functional of the first-order variables. In the gauge theory formulation, which is a first-order formulation, the independent variables are the gauge fields required to achieve a desired local gauge symmetry. Here we consider Poincaré gauge symmetry, so that the gauge fields consist of the vierbein and the Lorentz connection. The two formulations of gravity with both curvature and torsion are equivalent classically for the degenerate case of Einstein–Cartan–Sciama–

Kibble gravity (see e.g. Ref. 69), but not for PG generally. The decomposition of a Lorentz connection to a Levi-Civita connection and contortion parts can always be inserted into the field equations of PG. Inserting the decomposition of the connection into the action, however, changes the theory significantly.[75]

Next we shall consider the induced gravitational action for PG. Recall that gravitational dynamics does not yet exist at this stage, since the gravitational fields play the role of (classical) background fields, while matter fields are quantized. Only after the one-loop effective action for matter has been obtained, we will choose the independent variables for gravity, which will be the vierbein and the Lorentz connection, after which the variational principle can be applied to derive the gravitational field equations. Since the elementary object of special relativity in the setting leading to gravity on a Riemann–Cartan spacetime is a Dirac spinor[63] rather than a mass point or a scalar field, we primarily consider quantization of Dirac fields in this section.

With the intention of obtaining an induced Poincaré gauge theory of gravity, we start from the Hermitian action for a minimally-coupled Dirac spinor,

$$S^{\text{min.}}_{\text{Dirac}} = \int d^4x(\det e^a{}_\mu)\left[\frac{i}{2}(\bar{\psi}\gamma^a\tilde{\nabla}_a\psi - \tilde{\nabla}_a\bar{\psi}\gamma^a\psi) - m\bar{\psi}\psi\right], \tag{82}$$

where the Poincaré gauge covariant derivative is defined as $\tilde{\nabla}_a\psi = e_a{}^\mu\tilde{\nabla}_\mu\psi$ with the definition of $\tilde{\nabla}_\mu\psi$ given in (54).[k] The action for a Dirac spinor has a similar form as in the second-order formulation of Sec. 4.2 with (57) or without (56) nonminimal couplings, except that now the geometry of the background is determined by the independent variables $e^a{}_\mu$ and $\tilde{\omega}_\mu{}^{ab}$. The minimally coupled action (82) is rewritten as

$$S^{\text{min.}}_{\text{Dirac}} = \int d^4x(\det e^a{}_\mu)\bar{\psi}\left(i\gamma^a\tilde{\nabla}_a - \frac{i}{2}\gamma^a\mathcal{V}_a - m\right)\psi, \tag{83}$$

where the vector component of torsion is defined as $\mathcal{V}_a = e_a{}^\mu e_b{}^\nu T_{\nu\mu}{}^b$. Nonminimal couplings to torsion could be included in a similar way as in Sec. 4. However, we should note that the principle of gauge invariance does not require such nonminimal terms. Therefore, we will consider the minimally coupled case for simplicity.

The one-loop effective action is defined (as in Sec. 4.2) as

$$S^{(1)}_{\text{eff}} = -\frac{i}{2}\ln\det(l^2D) - \frac{i}{2}\operatorname{Tr}\ln(l^2D), \tag{84}$$

where the squared differential operator D for the action (83) is written as

$$D = \left[\gamma^a\left(\tilde{\nabla}_a - \frac{1}{2}\mathcal{V}_a\right)\right]^2 + m^2. \tag{85}$$

The kernel expansion of the operator D for a spinor on Riemann–Cartan spacetime has been studied before, particularly in Refs. 90 and 91. Such calculations

[k]Note that in the volume element, $(\det e^a{}_\mu) = (\det e_a{}^\mu)^{-1}$, if one prefers to use the inverse vierbein that appears in the covariant derivative.

are based on the decomposition of the connection into torsion-free and torsion components in one way or another. We will adopt a similar approach but with one crucial difference: the relation of the two connections (13) shall be used both ways. First the decomposition of the Lorentz connection $\tilde{\omega}^{ab}$ is used for the derivation of the one-loop effective action. Next the Lorentz connection $\tilde{\omega}^{ab}$ is composed back together, so that the effective action is expressed in terms of the background fields $e^a{}_\mu$ and $\tilde{\omega}_\mu{}^{ab}$, namely, in terms of the gauge fields of PG. After that we can elevate the said background fields into independent gravitational variables, and thereafter determine their dynamics by setting up the variational principle and deriving the field equations. Varying the action before the full Lorentz connection is composed would lead to inequivalent field equations,[75] which would ruin our chances to obtain parity with PG. We avoid this problem with the approach described above.

The operator (85) is written with (13) as

$$D = \left[\gamma^\mu \left(\tilde{\nabla}_\mu - \frac{i}{8}\gamma_5 \mathcal{A}_\mu\right)\right]^2 + m^2 . \tag{86}$$

The one-loop effective action is derived in the same way as in Sec. 4.2. Then we write it in terms of the variables $e^a{}_\mu$ and $\tilde{\omega}_\mu{}^{ab}$. We obtain it as

$$\begin{aligned} S^{(1)}_{\text{eff}} = &-\frac{\Lambda^4}{64\pi^2}\int d^4x(\det e^a{}_\mu)\,\mathrm{tr}\,A_0 - \frac{\Lambda^2}{32\pi^2}\int d^4x(\det e^a{}_\mu)\,\mathrm{tr}\,A_1 \\ &- \frac{\ln(\Lambda/\epsilon)}{16\pi^2}\int d^4x(\det e^a{}_\mu)\,\mathrm{tr}\,A_2 + \text{ultraviolet-finite terms}. \end{aligned} \tag{87}$$

The induced gravitational action at low energies is defined by the first two terms in the effective action (87), which are obtained as

$$\mathrm{tr}\,A_0 = 4\,, \tag{88}$$

$$\mathrm{tr}\,A_1 = -\frac{1}{3}\tilde{R} - \frac{2}{3}\tilde{\nabla}_a \mathcal{V}^a - \frac{4}{9}\mathcal{V}_a\mathcal{V}^a + \frac{1}{9}\mathcal{A}_a\mathcal{A}^a - \frac{1}{6}\mathcal{T}_{abc}\mathcal{T}^{abc} - 4m^2 . \tag{89}$$

Thus, at low energies, we obtain the induced gravitational action as

$$\begin{aligned} S^{\text{low-energy}}_{\text{induced PG}} = &\frac{1}{2\kappa_{\text{ind}}}\int d^4x(\det e^a{}_\mu) \\ &\times\left(\tilde{R} + \frac{4}{3}\mathcal{V}_a\mathcal{V}^a - \frac{1}{3}\mathcal{A}_a\mathcal{A}^a + \frac{1}{2}\mathcal{T}_{abc}\mathcal{T}^{abc} - 2\lambda_{\text{ind}}\right). \end{aligned} \tag{90}$$

This is the low-energy part of the PG action[58] with the relative coupling constants of the terms set to certain values. Thus, we have shown that the low-energy regime of PG is induced by quantized Dirac fields in Riemann–Cartan spacetime.[1] Note

[1]For the nonminimally coupled Dirac action,

$$S^{\text{non-min.}}_{\text{Dirac}} = \int d^4x(\det e^a{}_\mu)\bar{\psi}(i\gamma^a\tilde{\mathcal{D}}_a - m)\psi\,,$$

$$\tilde{\mathcal{D}}_a = \tilde{\nabla}_a + i\beta_1\mathcal{V}_a + i\beta_2\gamma_5\mathcal{A}_a\,,$$

that we have dropped the total derivative term $\tilde{\nabla}_a \mathcal{V}^a$ from the action (90). The induced gravitational couplings in the action (90) are given as

$$\frac{\lambda_{\text{ind}}}{\kappa_{\text{ind}}} = \frac{1}{8\pi^2}\left(\frac{N_d \Lambda^4}{2} - \sum_d \Lambda^2 m_d^2\right), \tag{91}$$

$$\frac{1}{\kappa_{\text{ind}}} = \frac{N_d}{48\pi^2}\Lambda^2, \tag{92}$$

where N_d is the number of free Dirac fields, m_d is the mass of each Dirac field, and we have omitted the terms of order $\ln(\Lambda/\epsilon)$. What was said about the induced couplings in Sec. 4.3 still applies, including the assessment of the orders of magnitudes for the cutoffs.

The third term in the one-loop effective action (87) provides the high-energy or strong-gravity regime of the induced gravitational action. The dimensionless coupling constants of those higher-order terms are of the order of one or lower, and hence their effect is weak at energies well below the ultraviolet cutoff $\Lambda \sim M_{\text{P}}$. The high-energy part of the induced gravitational action (87) does not exactly match the corresponding regime of PG. In PG, the Lagrangian is defined to be quadratic in the field strengths, namely, in curvature and torsion. Therefore, the high-energy part of the PG action consists of squared curvature terms. On the other hand, the induced action typically contains all terms which are gauge invariant and dimensionally permitted. In the present case, it means that the high-energy part of the induced action can also include quartic torsion terms, as well as terms which involve covariant derivatives, for example, $\tilde{\Box}\tilde{R}$. This is in general the case in the effective field theory approach to gravity. One should also note that further contributions to the high-energy part are induced at higher loops. Thus, it is practically impossible to achieve perfect parity with PG at high energies in the induced gravity approach.

6. Conclusions and Outlook

We have obtained the induced gravitational action on an Einstein–Cartan spacetime by identifying it as the cutoff-regularized one-loop effective action of quantized matter fields. This is the generalization of Sakharov's induced gravity[1] to a spacetime with both curvature and torsion. When the ultraviolet cutoff Λ is chosen to

where the coupling constants β_i are related to the couplings α_i of the action (57) as

$$\beta_1 = \alpha_1 + \frac{i}{2}, \quad \beta_2 = \alpha_2 + \frac{1}{8},$$

the induced low-energy action is the same as in Eq. (90) except that the contribution $-\frac{1}{3}\mathcal{A}_a\mathcal{A}^a$ of the axial component of torsion is replaced by

$$-\frac{1}{3}(1 - 2\beta_2 + 8\beta_2^2)\mathcal{A}_a\mathcal{A}^a.$$

be comparable to the Planck mass, $\Lambda \sim M_{\mathrm{P}}$, the induced Newton constant (80) has the observed magnitude. As usual, the induced cosmological constant (78) is much too large compared to the observed value, since the vacuum energy contribution is comparable to the square of the ultraviolet cutoff, $\lambda_{\mathrm{ind}} \sim \Lambda^2 \sim M_{\mathrm{P}}^2$. We speculated that it might be possible to use the local vacuum energy sequestering mechanism[87] in induced gravity for setting the correct value for λ_{ind} and avoiding its radiative instability. In a reasonable range of nonminimal couplings for the free matter fields, the contribution of torsion was found to be comparable to that in Einstein–Cartan–Sciama–Kibble gravity. Hence, the effect of torsion is quite weak except in very high matter densities. In general, however, the induced gravitational action is more general than the Einstein–Cartan–Sciama–Kibble theory, which implies that propagation of torsion is possible. In the part of the induced action that dominates at high energies, the dimensionless coupling constants were found to be of the order one, which implies that their effect on low-energy physics is marginal.

Then we have set out to show that the Poincaré gauge theory of gravity (PG) can be obtained by using the Sakharov induced gravity mechanism. We have shown that the quantization of free Dirac fields induces the low-energy part of the PG action (90) with certain relative couplings between the curvature and torsion terms. We conjecture that the result can be generalized to any gauge theory of gravity, in particular to the more general metric-affine gauge theories of gravity. The high-energy part of the induced action was observed to differ from the Poincaré gauge theory of gravity, since it does not contain only squared curvature terms but also terms that involve covariant derivatives and further contributions from torsion. This is to be expected in an approach based on effective field theory, since any term that is both invariant under the given symmetry and dimensionally allowed can, and often will, appear in the effective action. In conclusion, based on our derivation of the Poincaré gauge theory of gravity and the general structure of the effective action, we conjecture that the Sakharov mechanism can be used to induce the action for any gauge theory of gravity.

If we regard that gravity is induced via quantization of matter, what should we do with gravity itself? This is something that Sakharov's approach cannot address. Most of us believe that gravity should be quantized in one way or another. On the other hand, it is conceivable that space, time and gravity could emerge at a length scale above the Planck length, so that quantization in the conventional sense would be unnecessary. The fundamental theory behind all that, which would not involve a gravitational interaction, might of course be a quantum theory of some kind. Perhaps the strongest hint towards an emergent nature of gravity is the well known and deep connection of gravity and thermodynamics.

The merit of Sakharov's idea in this perspective is that, having a theory that includes a curved spacetime and also the quantized matter and gauge fields at the sub-Planckian energies, the gravitational interaction is necessarily produced as well. In other words, gravity emerges as an unavoidable companion of quantum matter.

Acknowledgments

We are much grateful to Stephen Adler, Stanley Deser, Friedrich W. Hehl and Yuri Obukhov for several illuminating correspondences. We also deeply thank Milutin Blagojević, Amir Ghalee, Yaghoub Heydarzade, Paddy Padmanabhan, Ilya Shapiro, Andrei Smilga, Dimitri V. Vassilevich and Anthony Zee for many valuable remarks. M.O. gratefully acknowledges support from the Emil Aaltonen Foundation.

References

1. A. D. Sakharov, "Vacuum quantum fluctuations in curved space and the theory of gravitation," *Sov. Phys. Dokl.* **12**, 1040 (1968); reprinted as *Gen. Rel. Grav.* **32**, 365 (2000) doi:10.1023/A:1001947813563.
2. K. S. Stelle, "Classical gravity with higher derivatives," *Gen. Rel. Grav.* **9**, 353 (1978) doi:10.1007/BF00760427.
3. K. S. Stelle, "Renormalization of higher derivative quantum gravity," *Phys. Rev. D* **16**, 953 (1977) doi:10.1103/PhysRevD.16.953.
4. J. Klusoň, M. Oksanen and A. Tureanu, "Hamiltonian analysis of curvature-squared gravity with or without conformal invariance," *Phys. Rev. D* **89**, 064043 (2014) doi:10.1103/PhysRevD.89.064043 [arXiv:1311.4141 [hep-th]].
5. S. L. Adler, "Order-R vacuum action functional in scalar-free unified theories with spontaneous scale breaking," *Phys. Rev. Lett.* **44**, 1567 (1980) doi:10.1103/PhysRevLett.44.1567.
6. S. L. Adler, "A formula for the induced gravitational constant," *Phys. Lett. B* **95**, 241 (1980) doi:10.1016/0370-2693(80)90478-5.
7. S. L. Adler, "Induced gravitation," *AIP Conf. Proc.* **68**, 915 (1980) doi:10.1063/1.2948651.
8. S. L. Adler, "Einstein gravity as a symmetry breaking effect in quantum field theory," *Rev. Mod. Phys.* **54**, 729 (1982) doi:10.1103/RevModPhys.54.729.
9. S. L. Adler, "Einstein gravitation as a long wavelength effective field theory," *Phil. Trans. R. Soc. Lond. A* **310**, 273 (1983) doi:10.1098/rsta.1983.0089.
10. A. Zee, "A broken symmetric theory of gravity," *Phys. Rev. Lett.* **42**, 417 (1979) doi:10.1103/PhysRevLett.42.417.
11. A. Zee, "Spontaneously generated gravity," *Phys. Rev. D* **23**, 858 (1981) doi:10.1103/PhysRevD.23.858.
12. A. Zee, "Einstein gravity emerging from quantum Weyl gravity," *Annals Phys.* **151**, 431 (1983) doi:10.1016/0003-4916(83)90286-5.
13. M. Visser, "Sakharov's induced gravity: A modern perspective," *Mod. Phys. Lett. A* **17**, 977 (2002) doi:10.1142/S0217732302006886 [arXiv:gr-qc/0204062].
14. A. V. Smilga, "Spontaneous generation of the Newton constant in the renormalizable gravity theory," in *Zvenigorod 1982, Proceedings, Group Theoretical Methods In Physics*, vol. **2**, 73–77. [arXiv:1406.5613 [hep-th]].
15. R. I. Nepomechie, "Einstein gravity as the low-energy effective theory of Weyl gravity," *Phys. Lett. B* **136**, 33 (1984) doi:10.1016/0370-2693(84)92050-1.
16. M. B. Einhorn and D. R. T. Jones, "Induced Gravity I: Real scalar field," *JHEP* **1601**, 019 (2016) doi:10.1007/JHEP01(2016)019 [arXiv:1511.01481 [hep-th]].
17. M. B. Einhorn and D. R. T. Jones, "Induced Gravity II: Grand unification," *JHEP* **1605**, 185 (2016) doi:10.1007/JHEP05(2016)185 [arXiv:1602.06290 [hep-th]].

18. J. D. Bekenstein, "Black holes and the second law," *Lett. Nuovo Cim.* **4**, 737 (1972) doi:10.1007/BF02757029.
19. J. D. Bekenstein, "Black holes and entropy," *Phys. Rev. D* **7**, 2333 (1973) doi:10.1103/PhysRevD.7.2333.
20. J. D. Bekenstein, "Generalized second law of thermodynamics in black hole physics," *Phys. Rev. D* **9**, 3292 (1974) doi:10.1103/PhysRevD.9.3292.
21. J. D. Bekenstein, "Statistical black hole thermodynamics," *Phys. Rev. D* **12**, 3077 (1975) doi:10.1103/PhysRevD.12.3077.
22. R. Penrose and R. M. Floyd, "Extraction of rotational energy from a black hole," *Nature* **229**, 177 (1971) doi:10.1038/physci229177a0.
23. D. Christodoulou, "Reversible and irreversible transformations in black hole physics," *Phys. Rev. Lett.* **25**, 1596 (1970) doi:10.1103/PhysRevLett.25.1596.
24. D. Christodoulou and R. Ruffini, "Reversible transformations of a charged black hole," *Phys. Rev. D* **4**, 3552 (1971) doi:10.1103/PhysRevD.4.3552.
25. S. W. Hawking, "Gravitational radiation from colliding black holes," *Phys. Rev. Lett.* **26**, 1344 (1971) doi:10.1103/PhysRevLett.26.1344.
26. J. M. Bardeen, B. Carter and S. W. Hawking, "The four laws of black hole mechanics," *Commun. Math. Phys.* **31**, 161 (1973) doi:10.1007/BF01645742.
27. S. W. Hawking, "Particle creation by black holes," *Commun. Math. Phys.* **43**, 199 (1975) doi:10.1007/BF02345020.
28. J. D. Bekenstein, "A universal upper bound on the entropy to energy ratio for bounded systems," *Phys. Rev. D* **23**, 287 (1981) doi:10.1103/PhysRevD.23.287.
29. J. D. Bekenstein, "Entropy content and information flow in systems with limited energy," *Phys. Rev. D* **30**, 1669 (1984) doi:10.1103/PhysRevD.30.1669.
30. M. Schiffer and J. D. Bekenstein, "Proof of the quantum bound on specific entropy for free fields," *Phys. Rev. D* **39**, 1109 (1989) doi:10.1103/PhysRevD.39.1109.
31. J. D. Bekenstein, "How does the entropy/information bound work?," *Found. Phys.* **35**, 1805 (2005) doi:10.1007/s10701-005-7350-7 [arXiv:quant-ph/0404042].
32. R. Bousso, "A covariant entropy conjecture," *JHEP* **9907**, 004 (1999) doi:10.1088/1126-6708/1999/07/004 [arXiv:hep-th/9905177].
33. R. Bousso, "Bekenstein bounds in de Sitter and flat space," *JHEP* **0104**, 035 (2001) doi:10.1088/1126-6708/2001/04/035 doi:hep-th/0012052.
34. G. 't Hooft, "Dimensional reduction in quantum gravity," *Conf. Proc. C* **930308**, 284 (1993) [arXiv:gr-qc/9310026].
35. L. Susskind, "The world as a hologram," *J. Math. Phys.* **36**, 6377 (1995) doi:10.1063/1.531249 [arXiv:hep-th/9409089].
36. J. M. Maldacena, "The large N limit of superconformal field theories and supergravity," *Int. J. Theor. Phys.* **38**, 1113 (1999) doi:10.1023/A:1026654312961 [*Adv. Theor. Math. Phys.* **2**, 231 (1998) doi:10.4310/ATMP.1998.v2.n2.a1] [arXiv:hep-th/9711200].
37. L. Bombelli, R. K. Koul, J. Lee and R. D. Sorkin, "A quantum source of entropy for black holes," *Phys. Rev. D* **34**, 373 (1986) doi:10.1103/PhysRevD.34.373.
38. T. Jacobson, "Black hole entropy and induced gravity," [arXiv:gr-qc/9404039].
39. V. P. Frolov and D. V. Fursaev, "Mechanism of generation of black hole entropy in Sakharov's induced gravity," *Phys. Rev. D* **56**, 2212 (1997) doi:10.1103/PhysRevD.56.2212 [arXiv:hep-th/9703178].
40. T. Jacobson, "Thermodynamics of space-time: The Einstein equation of state," *Phys. Rev. Lett.* **75**, 1260 (1995) doi:10.1103/PhysRevLett.75.1260 [arXiv:gr-qc/9504004].
41. C. Eling, R. Guedens and T. Jacobson, "Non-equilibrium thermodynamics of spacetime," *Phys. Rev. Lett.* **96**, 121301 (2006) doi:10.1103/PhysRevLett.96.121301 [arXiv:gr-qc/0602001].

42. T. Jacobson and R. Parentani, "Horizon entropy," *Found. Phys.* **33**, 323 (2003) doi:10.1023/A:1023785123428 [arXiv:gr-qc/0302099].
43. A. Mukhopadhyay and T. Padmanabhan, "Holography of gravitational action functionals," *Phys. Rev. D* **74**, 124023 (2006) doi:10.1103/PhysRevD.74.124023 [arXiv:hep-th/0608120].
44. D. Kothawala, S. Sarkar and T. Padmanabhan, "Einstein's equations as a thermodynamic identity: The cases of stationary axisymmetric horizons and evolving spherically symmetric horizons," *Phys. Lett. B* **652**, 338 (2007) doi:10.1016/j.physletb.2007.07.021 [arXiv:gr-qc/0701002].
45. T. Padmanabhan, "Equipartition of energy in the horizon degrees of freedom and the emergence of gravity," *Mod. Phys. Lett. A* **25**, 1129 (2010) doi:10.1142/S021773231003313X [arXiv:0912.3165 [gr-qc]].
46. T. Padmanabhan, "Surface density of spacetime degrees of freedom from equipartition law in theories of gravity," *Phys. Rev. D* **81**, 124040 (2010) doi:10.1103/PhysRevD.81.124040 [arXiv:1003.5665 [gr-qc]].
47. T. Padmanabhan, "Emergent gravity paradigm: recent progress," *Mod. Phys. Lett. A* **30**, 1540007 (2015) doi:10.1142/S0217732315400076 [arXiv:1410.6285 [gr-qc]].
48. E. P. Verlinde, "On the origin of gravity and the laws of Newton," *JHEP* **1104**, 029 (2011) doi:10.1007/JHEP04(2011)029 [arXiv:1001.0785 [hep-th]].
49. V. V. Nesvizhevsky *et al.*, "Quantum states of neutrons in the earth's gravitational field," *Nature* **415**, 297 (2002) doi:10.1038/415297a.
50. M. Chaichian, M. Oksanen and A. Tureanu, "On entropic gravity: The entropy postulate, entropy content of screens and relation to quantum mechanics," *Phys. Lett. B* **712**, 272 (2012) doi:10.1016/j.physletb.2012.04.065 [arXiv:1109.2794 [hep-th]].
51. M. Chaichian, M. Oksanen and A. Tureanu, "On gravity as an entropic force," *Phys. Lett. B* **702**, 419 (2011) doi:10.1016/j.physletb.2011.07.019 [arXiv:1104.4650 [hep-th]].
52. H. Weyl, "Elektron und Gravitation. I," *Z. Phys.* **56**, 330 (1929) doi:10.1007/BF01339504.
53. C. N. Yang and R. L. Mills, "Conservation of isotopic spin and isotopic gauge invariance," *Phys. Rev.* **96**, 191 (1954) doi:10.1103/PhysRev.96.191.
54. R. Utiyama, "Invariant theoretical interpretation of interaction," *Phys. Rev.* **101**, 1597 (1956) doi:10.1103/PhysRev.101.1597.
55. T. W. B. Kibble, "Lorentz invariance and the gravitational field," *J. Math. Phys.* **2**, 212 (1961) doi:10.1063/1.1703702.
56. D. W. Sciama, "On the analogy between charge and spin in general relativity," in *Recent Developments in General Relativity* (PWN, Warsaw; Pergamon Press, Oxford, 1962), p. 415.
57. Y. N. Obukhov, V. N. Ponomarev and V. V. Zhytnikov, "Quadratic Poincaré gauge theory of gravity: A comparison with the general relativity theory," *Gen. Rel. Grav.* **21**, 1107 (1989) doi:10.1007/BF00763457.
58. P. Baekler, F. W. Hehl and J. M. Nester, "Poincaré gauge theory of gravity: Friedman cosmology with even and odd parity modes. Analytic part," *Phys. Rev. D* **83**, 024001 (2011) doi:10.1103/PhysRevD.83.024001 [arXiv:1009.5112 [gr-qc]].
59. Y. Itin, "Energy momentum current for coframe gravity," *Class. Quant. Grav.* **19**, 173 (2002) doi:10.1088/0264-9381/19/1/311 [arXiv:gr-qc/0111036].
60. J. M. Charap and W. Tait, "A gauge theory of the Weyl group," *Proc. Roy. Soc. Lond. A* **340**, 249 (1974) doi:10.1098/rspa.1974.0151.
61. M. Kaku, P. K. Townsend and P. van Nieuwenhuizen, "Gauge theory of the conformal and superconformal group," *Phys. Lett. B* **69**, 304 (1977) doi:10.1016/0370-2693(77)90552-4.

62. E. A. Lord, "The metric affine gravitational theory as the gauge theory of the affine group," *Phys. Lett. A* **65**, 1 (1978) doi:10.1016/0375-9601(78)90113-5.
63. M. Blagojević and F. W. Hehl, *Gauge Theories of Gravitation* (Imperial College Press, London, 2013).
64. A. H. Chamseddine, "Complexified gravity in noncommutative spaces," *Commun. Math. Phys.* **218**, 283 (2001) doi:10.1007/s002200100393 [arXiv:hep-th/0005222].
65. A. H. Chamseddine, "Deforming Einstein's gravity," *Phys. Lett. B* **504**, 33 (2001) doi:10.1016/S0370-2693(01)00272-6 [arXiv:hep-th/0009153].
66. M. A. Cardella and D. Zanon, "Noncommutative deformation of four-dimensional Einstein gravity," *Class. Quant. Grav.* **20**, L95 (2003) doi:10.1088/0264-9381/20/8/101 [arXiv:hep-th/0212071].
67. M. Chaichian, M. Oksanen, A. Tureanu and G. Zet, "Gauging the twisted Poincaré symmetry as noncommutative theory of gravitation," *Phys. Rev. D* **79**, 044016 (2009) doi:10.1103/PhysRevD.79.044016 [arXiv:0807.0733 [hep-th]].
68. F. W. Hehl, "Spin and torsion in general relativity: I, Foundations," *Gen. Relat. Grav.* **4**, 333 (1973) doi:10.1007/BF00759853; "Spin and torsion in general relativity: II, Geometry and field equations," *ibid* **5**, 491 (1974) doi:10.1007/BF02451393.
69. F. W. Hehl, P. Von Der Heyde, G. D. Kerlick and J. M. Nester, "General relativity with spin and torsion: Foundations and prospects," *Rev. Mod. Phys.* **48**, 393 (1976) doi:10.1103/RevModPhys.48.393.
70. W. Adamowicz, "Plane waves in gauge theories of gravitation," *Gen. Rel. Grav.* **12**, 677 (1980) doi:10.1007/BF00771860.
71. M. Q. Chin, D. C. Chern, R. R. Hsu and W. B. Yeung, "Plane fronted torsion waves in a gravitational gauge theory with a quadratic Lagrangian," *Phys. Rev. D* **28**, 2094 (1983) doi:10.1103/PhysRevD.28.2094.
72. P. Singh and J. B. Griffiths, "A New class of exact solutions of the vacuum quadratic Poincare gauge field theory," *Gen. Rel. Grav.* **22**, 947 (1990) doi:10.1007/BF00763233.
73. M. Blagojević and B. Cvetković, "Generalized pp waves in Poincaré gauge theory," *Phys. Rev. D* **95**, 104018 (2017) doi:10.1103/PhysRevD.95.104018 [arXiv:1702.04367 [gr-qc]].
74. I. L. Shapiro, "Physical aspects of the space-time torsion," *Phys. Rept.* **357**, 113 (2002) doi:10.1016/S0370-1573(01)00030-8, [arXiv:hep-th/0103093].
75. J. M. Nester and C. H. Wang, "Can torsion be treated as just another tensor field?," *Int. J. Mod. Phys. Conf. Ser.* **07**, 158 (2012) doi:10.1142/S2010194512004229.
76. F. David, "A Comment on induced gravity," *Phys. Lett. B* **138**, 383 (1984) doi:10.1016/0370-2693(84)91922-1.
77. Y. V. Novozhilov and D. V. Vassilevich, "Induced classical gravity," *Lett. Math. Phys.* **21**, 253 (1991) doi:10.1007/BF00420376.
78. F. W. Hehl, J. D. McCrea, E. W. Mielke and Y. Ne'eman, "Metric affine gauge theory of gravity: Field equations, Noether identities, world spinors, and breaking of dilation invariance," *Phys. Rept.* **258**, 1 (1995) doi:10.1016/0370-1573(94)00111-F [arXiv:gr-qc/9402012].
79. N. D. Birrell and P. C. W. Davies, *Quantum Fields in Curved Space* (Cambridge University Press, 1982) doi:10.1017/CBO9780511622632.
80. L. E. Parker and D. Toms, *Quantum Field Theory in Curved Spacetime: Quantized Field and Gravity* (Cambridge University Press, 2009) doi:10.1017/CBO9780511813924.
81. I. L. Buchbinder and I. L. Shapiro, "On the renormalization of models of quantum field theory in an external gravitational field with torsion," *Phys. Lett. B* **151**, 263 (1985) doi:10.1016/0370-2693(85)90848-2.

82. I. G. Avramidi, "Heat kernel and quantum gravity," *Lect. Notes Phys. Monogr.* **64**, 1 (2000) doi:10.1007/3-540-46523-5.
83. D. V. Vassilevich, "Heat kernel expansion: User's manual," *Phys. Rept.* **388**, 279 (2003) doi:10.1016/j.physrep.2003.09.002 [arXiv:hep-th/0306138].
84. R. M. Wald, *General Relativity* (University of Chicago Press, 1984) doi:10.7208/chicago/9780226870373.001.0001.
85. P. B. Gilkey, "The spectral geometry of a Riemannian manifold," *J. Diff. Geom.* **10**, 601 (1975) URL:`http://projecteuclid.org/euclid.jdg/1214433164`.
86. N. Kaloper and A. Padilla, "Sequestering the standard model vacuum energy," *Phys. Rev. Lett.* **112**, 091304 (2014) doi:10.1103/PhysRevLett.112.091304 [arXiv:1309.6562 [hep-th]].
87. N. Kaloper, A. Padilla, D. Stefanyszyn and G. Zahariade, "A manifestly local theory of vacuum energy sequestering," *Phys. Rev. Lett.* **116**, 051302 (2016) doi:10.1103/PhysRevLett.116.051302 [arXiv:1505.01492 [hep-th]].
88. R. Bufalo, J. Klusoň and M. Oksanen, "Canonical formulation and path integral for local vacuum energy sequestering," *Phys. Rev. D* **94**, 044005 (2016) doi:10.1103/PhysRevD.94.044005 [arXiv:1606.01282 [hep-th]].
89. G. De Berredo-Peixoto, "A Note on the heat kernel method applied to fermions," *Mod. Phys. Lett. A* **16**, 2463 (2001) doi:10.1142/S0217732301005965 [arXiv:hep-th/0108223].
90. Y. N. Obukhov, "Spectral geometry of the Riemann–Cartan space-time and the axial anomaly," *Phys. Lett. B* **108**, 308 (1982) doi:10.1016/0370-2693(82)91199-6.
91. Y. N. Obukhov, "Spectral geometry of the Riemann–Cartan space-time," *Nucl. Phys. B* **212**, 237 (1983) doi:10.1016/0550-3213(83)90303-6.

Selected Works of Jacob Bekenstein

LETTERE AL NUOVO CIMENTO VOL. 4, N. 15 12 Agosto 1972

Black Holes and the Second Law (*).

J. D. BEKENSTEIN (**)

Joseph Henry Laboratories, Princeton University - Princeton, N. J.

(ricevuto il 22 Maggio 1972)

Black-hole physics seems to provide at least two ways in which the second law of thermodynamics may be transcended or violated:

a) Let an observer drop or lower a package of entropy into a black hole; the entropy of the exterior world decreases. Furthermore, from an exterior observer's point of view a black hole in equilibrium has only three degrees of freedom: mass, charge and angular momentum (1). Thus, once the black hole has settled down to equilibrium, there is no way for the observer to determine its interior entropy. Therefore, he cannot exclude the possibility that the total entropy of the universe may have decreased in the process. It is in this sense that the second law appears to be transcended (2).

b) A method for violating the second law has been proposed by GEROCH (3): By means of a string one slowly lowers a body of rest mass m and nonzero temperature toward a Schwarzschild black hole of mass M. By the time the body nears the horizon, its energy as measured from infinity, $E = m(1-2M/r)^{\frac{1}{2}}$, is nearly zero; the body has already done work m on the agent which lowers the string. At this point the body is allowed to radiate into the black hole until its rest mass is $m - \Delta m$. Finally, by expending work $m - \Delta m$, one hauls the body back up.
The net result: a quantity of heat Δm has been completely converted into work. Furthermore, since the addition of the radiation to the black hole takes place at a point where $(1-2M/r)^{\frac{1}{2}} \approx 0$, the mass of the black hole is unchanged. Thus the black hole appears to be unchanged after the process. This implies a violation of the second law: « One may not transform heat entirely into work without compensating changes taking place in the surroundings. »

In this note we formulate the second law in a form suitable for black-hole physics which resolves the transcendence problem. We also indicate why Geroch's gedanken experiment does not, in fact, violate the second law.

(*) Work supported by National Science Foundation Grant No. GP 30799 X.
(**) National Science Foundation Predoctoral Fellow.
(1) C. W. MISNER, K. S. THORNE and J. A. WHEELER: *Gravitation* (San Francisco, 1972).
(2) For transcendence of the law of baryon conservation see J. D. BEKENSTEIN: *Phys. Rev. Lett.*, **28**, 452 (1972); *Phys. Rev. D*, **5**, 1239 (1972) and to be published.
(3) R. GEROCH: *Colloquium at Princeton University* (Dec. 1971).

We state the second law as follows: « Common entropy plus black-hole entropy never decreases. » By common entropy we mean entropy in the black-hole's exterior. By black-hole entropy $S_{\text{b.h.}}$ we mean

$$S_{\text{b.h.}} = \eta k L_{\text{P}}^{-2} A \,, \tag{1}$$

where A is the area of the black hole (4) in question, L_{P} is the Planck length $(\hbar G/c^3)^{\frac{1}{2}} = 1.6\cdot 10^{-33}$ cm, k is Boltzmann's constant $1.38\cdot 10^{-16}$ erg/°K and η is a constant number of order unity.

The introduction of a black-hole entropy is necessitated by our first example. Without it the second law is definitely transcended. With black-hole entropy the second law becomes a well-defined statement susceptible to verification by an exterior observer. The choice of the area of a black hole as a measure of its entropy is motivated by the results of CHRISTODOULOU (5) and of HAWKING (6) that the area of a black hole never decreases, and that it increases for all but a very special class of black-hole transformations (5). The area of a black hole appears to be the only one of its properties having this entropylike behavior which is so essential if the second law as we have stated it is to hold when entropy goes down a black hole.

By choosing $S_{\text{b.h.}}$ proportional to A rather than to some monotonically increasing function of A, we ensure that the total black-hole entropy of a system of black holes (the sum of individual $S_{\text{b.h.}}$) depends only on the total horizon area—also a nondecreasing quantity which is of fundamental importance in Hawking's work (6).

The introduction of k and of a length squared in (1) is necessitated by dimensional considerations. We choose Planck's length because it is the only truly universal constant with units of length. (We shall show later on that no larger scale of length will do.) Although a black hole is a thoroughly classical entity, black-hole entropy contains $\hbar$ because it relates to the interaction of a black hole with material systems. It is well known that the expression for entropy of any material system always contains $\hbar$; thus, the appearance of $\hbar$ in $S_{\text{b.h.}}$ is understandable. (In what follows we use units with $G = c = 1$; thus $L_{\text{P}}^2 = \hbar$.)

Let us illustrate the operation of the second law with a simple example. Consider the case of a narrow beam of black-body radiation of temperature T which is directed into a black hole of mass M. We are clearly assuming that geometrical optics is applicable. Thus, the characteristic wavelengths of the radiation must be much smaller than M. This implies that

$$T \gg \hbar/kM \,. \tag{2}$$

We note also that, to a given energy E in the beam, there corresponds entropy (7) S, where

$$S = E/T \,. \tag{3}$$

The increase in area of the black hole (4) caused by the infall of radiation with energy E may be calculated from the conservation of energy and angular momentum.

(4) The area of a (Kerr-Newman) black hole of mass M, charge Q and angular momentum L is $A = 4\pi\{[M + (M^2 - Q^2 - L^2/M^2)^{\frac{1}{2}}]^2 + L^2/M^2\}$. An extreme Kerr black hole has $L = M^2$ and $Q = 0$.

(5) D. CHRISTODOULOU: *Phys. Rev. Lett.*, **25**, 1596 (1970); D. CHRISTODOULOU and R. RUFFINI: *Phys. Rev. D*, **4**, 3552 (1971).

(6) S. W. HAWKING: *Phys. Rev. Lett.*, **26**, 1344 (1971); *Comm. Math. Phys.*, **25**, 152 (1972).

(7) A black-body cavity emits entropy E/T as it emits energy E; the propagation of the beam is a reversible process.

(One neglects gravitational radiation losses since they are of $O(E^2/M)$.) This increase is minimized when the black hole is extreme Kerr, and when the beam of radiation travels in the equatorial plane, and adds to the black hole as much angular momentum as is consistent with the requirement that it be captured by the black hole (8), The increase is (8)

$$(\Delta A)_{\min} = 8\pi(2 - 3^{\frac{1}{2}})ME \,. \tag{4}$$

The corresponding minimum possible increase in $S_{\text{b.h.}}$ is

$$(\Delta S_{\text{b.h.}})_{\min} = 8\pi(2 - 3^{\frac{1}{2}})\eta kME/\hbar \,. \tag{5}$$

On the other hand, according to (2) and (3) the corresponding decrease of common entropy is

$$|\Delta S| \ll kME/\hbar \,. \tag{6}$$

Thus, provided η is of order unity, the overall entropy (common entropy plus $S_{\text{b.h.}}$) always increases for all M. It should be clear that the length L_{P} appearing in (1) cannot be chosen much larger than $\hbar^{\frac{1}{2}}$ if the second law is to hold in all cases.

At first sight it would appear that the second law will break down if $T \lessapprox \hbar/kM$. However, in this regime one cannot continue treating the radiation by geometrical optics, but must treat it by wave theory. If this is done, one finds (8) that statistical fluctuations in the radiation are all important, and thus the second law is irrelevant in the first place. Thus, for this gedanken experiment the second law as here formulated holds whenever fluctuations are unimportant.

CHRISTOULOU (5) has described a process by means of which a particle can be added to a black hole without there resulting an increase in the black hole's area. One could imagine that the particle in question carries entropy. It would then appear that this entropy goes down the black hole without there being a compensating increase in $S_{\text{b.h.}}$ as demanded by the second law.

In Cristodoulou's process the particle in question must be created at the black hole's horizon by the fission of a primary particle (the other fission product escaping to infinity). It follows that the primary particle is required to fall from far away to the horizon. Recent results (9) indicate that in such a fall the particle will radiate into the black hole gravitational waves with energy of the order of its rest mass. One can show (8,10) that this radiation will always cause a nonnegligible increase in the black hole's area corresponding to the increase in $S_{\text{b.h.}}$ demanded by the second law. In this manner one reconciles Christodoulou's process with the second law.

Let us now try to resolve the difficulty raised by Geroch's gedanken experiment. The key assumption made which leads to conflict with the second law is that one can calculate the work done by (on) the body by taking differences of its « potential energy » $m(1 - 2M/r)^{\frac{1}{2}}$ at two different points. Such a procedure assumes the process of lowering (raising) the body to be of a conservative or reversible character. However, the tendency of a black hole's area to increase irreversibly at the slightest provocation (6) warns us that there may be an element of irreversibility in the problem.

---

(8) J. D. BEKENSTEIN: to be published.
(9) M. DAVIS, R. RUFFINI and J. TIOMNO: *Phys. Rev. D*, to be published (1972).
(10) J. D. BEKENSTEIN: Ph. D. Thesis, Princeton University (1972) (unpublished).

Indeed, some calculations based on the metric for a Schwarzschild black hole placed in an exterior quadrupole gravitational field ([11]), coupled with simple physical arguments, suggest that the black hole in question is deformed, and its area increases all along as the body is lowered by the string ([8,10]). This increase in area is irreversible. Thus, when the body is hauled back up, the black hole returns to an undeformed Schwarzschild configuration of larger area (and hence larger mass) than the initial one. Therefore, the work expended to raise the body must be larger than the one calculated from the potential method; the work done by the body is less than Δm. This means that the heat Δm is, at best, only partially converted into work. No violation of the second law is entailed here. One should also notice that the « surroundings », *i.e.* the black hole, are changed in the overall process.

Further details on the above examples, other examples and implications of the second law for black-hole physics will be reported elsewhere.

We conclude on a speculative note. One sees from (1) that the natural unit of area of a black hole is the Planck length squared. One cannot help wondering about the possible connection between this feature and the expected quantum structure of space-time at a scale of the order of the Planck length ([12]).

* * *

The author thanks Prof. J. A. WHEELER for his deep interest in, and for his many comments on, the work reported here.

([11]) YA. B. ZEL'DOVICH and I. D. NOVIKOV: *Stars and Relativity* (Chicago, 1972), p. 138.

([12]) J. A. WHEELER: a chapter in *Relativity, Groups and Topology*, edited by B. DEWITT and C. DEWITT (New York, 1965).

PHYSICAL REVIEW D VOLUME 7, NUMBER 8 15 APRIL 1973

Black Holes and Entropy*

Jacob D. Bekenstein†
Joseph Henry Laboratories, Princeton University, Princeton, New Jersey 08540
and Center for Relativity Theory, The University of Texas at Austin, Austin, Texas 78712‡
(Received 2 November 1972)

There are a number of similarities between black-hole physics and thermodynamics. Most striking is the similarity in the behaviors of black-hole area and of entropy: Both quantities tend to increase irreversibly. In this paper we make this similarity the basis of a thermodynamic approach to black-hole physics. After a brief review of the elements of the theory of information, we discuss black-hole physics from the point of view of information theory. We show that it is natural to introduce the concept of black-hole entropy as the measure of information about a black-hole interior which is inaccessible to an exterior observer. Considerations of simplicity and consistency, and dimensional arguments indicate that the black-hole entropy is equal to the ratio of the black-hole area to the square of the Planck length times a dimensionless constant of order unity. A different approach making use of the specific properties of Kerr black holes and of concepts from information theory leads to the same conclusion, and suggests a definite value for the constant. The physical content of the concept of black-hole entropy derives from the following generalized version of the second law: When common entropy goes down a black hole, the common entropy in the black-hole exterior plus the black-hole entropy never decreases. The validity of this version of the second law is supported by an argument from information theory as well as by several examples.

I. INTRODUCTION

A black hole[1] exhibits a remarkable tendency to increase its horizon surface area when undergoing any transformation. This was first noticed by Floyd and Penrose[2] in an example of the extraction of energy from a Kerr black hole by means of what has come to be known as a Penrose process.[3] They suggested that an increase in area might be a general feature of black-hole transformations. Independently, Christodoulou[4,5] had shown that no process whose ultimate outcome is the capture of a particle by a Kerr black hole can result in the decrease of a certain quantity which he named the irreducible mass of the black hole, M_{ir}. In fact, most processes result in an increase in M_{ir} with the exception of a very special class of limiting processes, called reversible processes, which leave M_{ir} unchanged. It turns out that M_{ir} is proportional to the square root of the black hole's area[5,6] [see (1)]. Thus Christodoulou's result implies that the area increases in most processes, and thus it supports the conjecture of Floyd and Penrose. Christodoulou's conclusion is also valid for charged Kerr black holes.[4,6]

By an approach radically different from Christodoulou's, Hawking[7] has given a general proof that the black-hole surface area cannot decrease in *any* process. For a system of several black holes Hawking's theorem implies that the area of each individual black hole cannot decrease, and moreover that when two black holes merge, the area of the resulting black hole (provided, of course, that one forms) cannot be smaller than the sum of initial areas.

It is clear that changes of a black hole generally take place in the direction of increasing area. This is reminiscent of the second law of thermodynamics which states that changes of a closed thermodynamic system take place in the direction of increasing entropy. The above comparison suggests that it might be useful to consider black-hole physics from a thermodynamic viewpoint: We already have the concept of energy in black-hole physics, and the above observation suggests that something like entropy may also play a role in it. Thus, one can hope to develop a thermodynamics for black holes – at least a rudimentary one. In this paper we show that it is possible to give a precise definition of black-hole entropy. Based on it we construct some elements of a thermodynamics for black holes.

There are some precedents to our considerations. The idea of making use of thermodynamic methods in black-hole physics appears to have been first considered by Greif.[8] He examined the possibility of defining the entropy of a black hole, but lacking many of the recent results in black-hole physics, he did not make a concrete proposal. More recently, Carter[9] has rederived the result of Christodoulou[4,5] that the irreducible mass of a Kerr black hole is unchanged in a reversible transformation by applying to the black hole the criterion for a thermodynamically reversible transformation of a rigidly rotating star.[10] Carter's example shows the possibilities inherent in the use of thermodynamic arguments in black-hole physics.

In this paper we attempt a unification of black-hole physics with thermodynamics. In Sec. II we point out a number of analogies between black-hole physics and thermodynamics, all of which bring out the parallelism between black-hole area and entropy. In Sec. III, after a short review of elements of the theory of information, we discuss some features of black-hole physics from the point of view of information theory. We take the area of a black hole as a measure of its entropy—entropy in the sense of inaccessibility of information about its internal configuration. We go further in Sec. IV and propose a specific expression for black-hole entropy in terms of black-hole area. Earlier[11,12] we had proposed this same expression on different grounds; here we find the value of a previously unknown constant by means of an argument based on information theory. In Sec. V we propose a generalization of the second law of thermodynamics applicable to black-hole physics: When some common entropy goes down a black hole, *the black-hole entropy plus the common entropy in the black-hole exterior never decreases*.[11,12]

In Secs. VI and VII we construct several examples which provide support for the generalized second law. In addition, we analyze in Sec. VII a thought experiment proposed by Geroch[13] in which, with the help of a black hole, heat is apparently converted entirely into work in violation of the second law. We show that, in fact, due to fundamental physical limitations the conversion efficiency is somewhat smaller than unity. Moreover, the efficiency is no greater than the maximum efficiency allowed by thermodynamics for the heat engine which is equivalent to the Geroch process, so that this process cannot be regarded as violating the second law.

II. ANALOGIES BETWEEN BLACK-HOLE PHYSICS AND THERMODYNAMICS

We have already mentioned the resemblance between the tendency of black-hole area to increase, and the tendency of entropy to increase. Changes of a black hole or of a system of black holes select a preferred direction in time: that in which the black-hole area increases. Likewise, changes of a closed thermodynamic system select a preferred direction in time: that in which the entropy increases. This parallelism between black-hole area and entropy goes even deeper.

Black-hole area turns out to be as intimately related to the degradation of energy as is entropy. In thermodynamics the statement "the entropy has increased" implies that a certain quantity of energy has been degraded, i.e., that it can no longer be transformed into work. Now, as Christodoulou has emphasized,[4,5] the irreducible mass M_{ir} of a Kerr black hole, which is related to the surface area A of the black hole by[14]

$$M_{ir} = (A/16\pi)^{1/2}, \qquad (1)$$

represents energy which cannot be extracted by means of Penrose processes.[3] In this sense it is inert energy which cannot be transformed into work. Thus, an increase in A, and hence in M_{ir}, corresponds to the degradation (in the thermodynamic sense) of some of the energy of the black hole.

The irreducible mass of a Schwarzschild black hole is just equal to its total mass. Thus, no energy can be extracted from such a black hole by means of Penrose processes. However, the merger of two Schwarzschild black holes can yield energy in the form of gravitational waves.[7] The only restriction on the process is that the total black-hole area must not decrease as a result of the merger.[7] However, the sum of the irreducible masses of individual black holes may (in fact, does) decrease. We see that for a system of several black holes the degraded energy E_d is more appropriately given by

$$E_d = (\textstyle\sum A/16\pi)^{1/2} = (\sum M_{ir}^2)^{1/2} \qquad (2)$$

than by $\sum M_{ir}$. According to this formula the degraded energy of a system of black holes is smaller than the sum of degraded energies of the black holes considered separately. Thus by combining Schwarzschild black holes which are already "dead," one can still obtain energy.[7] Analogously, by allowing two thermodynamic systems which are separately in equilibrium to interact, one can obtain work, whereas each system by itself could have done no work. From the above observations the parallelism between black-hole area and entropy is again evident.

We shall now construct the black-hole analog of the thermodynamic expression

$$dE = TdS - PdV\ . \qquad (3)$$

For convenience we shall from now on write all our equations in terms of the "rationalized area" of a black hole α defined by

$$\alpha = A/4\pi\ . \qquad (4)$$

Consider a Kerr black hole of mass M, charge Q, and angular momentum $\vec{L}$. (3-vectors here refer to components with respect to the Euclidean frame at infinity.) Its rationalized area is given by[5,7]

$$\alpha = r_+{}^2 + a^2$$

$$= 2Mr_+ - Q^2, \quad (5)$$

where

$$\vec{a} = \vec{L}/M, \quad (6)$$

$$r_\pm = M \pm (M^2 - Q^2 - a^2)^{1/2}. \quad (7)$$

Differentiating (5) and solving for dM we obtain

$$dM = \Theta d\alpha + \vec{\Omega} \cdot d\vec{L} + \Phi dQ \,, \quad (8)$$

where

$$\Theta \equiv \tfrac{1}{4}(r_+ - r_-)/\alpha \,, \quad (9a)$$

$$\vec{\Omega} \equiv \vec{a}/\alpha \,, \quad (9b)$$

$$\Phi \equiv Qr_+/\alpha \,. \quad (9c)$$

In (8) we have the black-hole analog of the thermodynamic expression (3): The terms $\vec{\Omega}\cdot d\vec{L}$ and ΦdQ clearly represent the work done *on* the black hole by an external agent who increases the black hole's angular momentum and charge by $d\vec{L}$ and dQ, respectively. Thus $\vec{\Omega}\cdot d\vec{L} + \Phi dQ$ is the analog of $-PdV$, the work done *on* a thermodynamic system. Comparing our expression for work with the expressions for work done on rotating[15] and charged[16] bodies, we see that $\vec{\Omega}$ and Φ play the roles of rotational angular frequency and electric potential of the black hole, respectively.[5,9] The α in (8) resembles the entropy S in (3) as we have noted before: For any change of the black hole $d\alpha \geq 0$,[5-7] while for any change of a closed thermodynamic system $dS \geq 0$. Moreover, it is clear from (7) and (9a) that Θ, the black-hole analog of temperature T, is non-negative just as T is. From the above observations the formal correspondence between (3) and (8) is evident.

All the analogies we have mentioned are suggestive of a connection between thermodynamics and black-hole physics in general, and between entropy and black-hole area in particular. But so far the analogies have been of a purely formal nature, primarily because entropy and area have different dimensions. We shall remedy this deficiency in Sec. IV by constructing out of black-hole area an expression for black-hole entropy with the correct dimensions. Preparatory to this we shall now look at black-hole physics from the point of view of the theory of information.

III. INFORMATION AND BLACK-HOLE ENTROPY

The connection between entropy and information is well known.[17,18] The entropy of a system measures one's uncertainty or lack of information about the actual internal configuration of the system. Suppose that all that is known about the internal configuration of a system is that it may be found in any of a number of states with probability p_n for the nth state. Then the entropy associated with the system is given by Shannon's formula[17,18]

$$S = -\sum_n p_n \ln p_n \,. \quad (10)$$

This formula is uniquely determined by a few very general requirements which are imposed in order that S have the properties expected of a measure of uncertainty.[17]

It should be noticed that the above entropy is dimensionless. This simply means that we choose to measure temperature in units of energy. Boltzmann's constant is then dimensionless.

Whenever new information about the system becomes available, it may be regarded as imposing some constraints on the probabilities p_n. For example, the information may be that several of the p_n are, in fact, zero. Such constraints on the p_n always result in a decrease in the entropy function.[18] This property is formalized by the relation[17,18]

$$\Delta I = -\Delta S, \quad (11)$$

where ΔI is the new information which corresponds to a decrease ΔS in one's uncertainty about the internal state of the system. Equation (11) is the basis for Brillouin's identification of information with negative entropy.[18] All the above comments apply to such divers systems as a quantity of gas in a box or a telegram. A familiar example of the relation between a gain of information and a decrease in entropy is the following. Some ideal gas in a container is compressed isothermally. It is well known that its entropy decreases. On the other hand, one's information about the internal configuration of the gas increases: After the compression the molecules of the gas are more localized, so that their positions are known with more accuracy than before the compression.

The second law of thermodynamics is easily understood in the context of information theory. The entropy of a thermodynamic system which is not in equilibrium increases because information about the internal configuration of the system is being lost during its evolution as a result of the washing out of the effects of the initial conditions. It is possible for an exterior agent to cause a decrease in the entropy of a system by first acquiring information about the internal configuration of the system. The classic example of this is that of Maxwell's demon.[18] But information is never free. In acquiring information ΔI about the system, the agent inevitably causes an increase in the entropy of the rest of the universe which exceeds ΔI.[18] Thus, even though the entropy of the system decreases in accordance with (11), the over-all entropy of the universe increases in the process.

The conventional unit of information is the "bit" which may be defined as the information available when the answer to a yes-or-no question is precisely known (zero entropy). According to the scheme (11) a bit is also numerically equal to the maximum entropy that can be associated with a yes-or-no question, i.e., the entropy when no information whatsoever is available about the answer. One easily finds that the entropy function (10) is maximized when $p_{yes} = p_{no} = \frac{1}{2}$. Thus, in our units, one bit is equal to ln2 of information.

Let us now return to our original subject, black holes. In the context of information a black hole is very much like a thermodynamic system. The entropy of a thermodynamic system in equilibrium measures the uncertainty as to which of all its internal configurations compatible with its macroscopic thermodynamic parameters (temperature, pressure, etc.) is actually realized. Now, just as a thermodynamic system in equilibrium can be completely described macroscopically by a few thermodynamic parameters, so a bare black hole in equilibrium (Kerr black hole) can be completely described (insofar as an exterior observer is concerned) by just three parameters: mass, charge, and angular momentum.[1] Black holes in equilibrium having the same set of three parameters may still have different "internal configurations." For example, one black hole may have been formed by the collapse of a normal star, a second by the collapse of a neutron star, a third by the collapse of a geon. These various alternatives may be regarded as different possible internal configurations of one and the same black hole characterized by their (common) mass, charge, and angular momentum. It is then natural to introduce the concept of black-hole entropy as the measure of the *inaccessibility* of information (to an exterior observer) as to which particular internal configuration of the black hole is actually realized in a given case.

At the outset it should be clear that the black-hole entropy we are speaking of is *not* the thermal entropy inside the black hole. In fact, our black-hole entropy refers to the equivalence class of all black holes which have the same mass, charge, and angular momentum, not to one particular black hole. What are we to take as a measure of this black-hole entropy? The discussion of Sec. II predisposes us to single out black-hole area. But to be more general we shall only assume that the entropy of a black hole, S_{bh}, is some *monotonically increasing* function of its rationalized area:

$$S_{bh} = f(\alpha). \tag{12}$$

Although our motivating discussion for the introduction of the concept of the black-hole entropy made use of the specific properties of stationary black holes, we shall take (12) to be valid for any black hole, including a dynamically evolving one, since the surface area is well defined for any black hole. This choice is supported by the following observations.

As mentioned earlier, the entropy of an evolving thermodynamic system increases due to the gradual loss of information which is a consequence of the washing out of the effects of the initial conditions. Now, as a black hole approaches equilibrium, the effects of the initial conditions are also washed out (the black hole loses its hair)[1]; only mass, charge, and angular momentum are left as determinants of the black hole at late times. We would thus expect that the loss of information about initial peculiarities of the hole will be reflected in a gradual increase in S_{bh}. And indeed Eq. (12) predicts just this; by Hawking's theorem S_{bh} increases monotonically as the black hole evolves. This agreement is evidence in favor of the choice (12).

We mentioned earlier that the possibility of causing a decrease in the entropy of a thermodynamic system goes hand in hand with the possibility of obtaining information about its internal configuration. By contrast, an exterior agent cannot acquire any information about the interior configuration of a black hole. The one-way membrane nature of the event horizon prevents him from doing so.[1] Therefore, we do not expect an exterior agent to be able to cause a decrease in the black hole's entropy. Equation (12) is in agreement with this expectation; by Hawking's theorem S_{bh} never decreases. Here we have a new piece of evidence in favor of the choice (12).

One possible choice for f in (12), $f(\alpha) \propto \alpha^{1/2}$, is untenable on various grounds. Consider two black holes which start off very distant from each other. Since they interact weakly we can take the total black-hole entropy to be the sum of the S_{bh} of each black hole. The black holes now fall together, merge, and form a black hole which settles down to equilibrium. In the process no information about the black-hole interior can become available; on the contrary, much information is lost as the final black hole "loses its hair." Thus, we expect the final black-hole entropy to exceed the initial one. By our assumption that $f(\alpha)\propto\alpha^{1/2}$, this implies that the irreducible mass [see (1)] of the final black hole exceeds the sum of irreducible masses of the initial black holes. Now suppose that all three black holes are Schwarzschild ($M=M_{ir}$). We are then confronted with the prediction that the final black-hole mass exceeds the initial one. But this is nonsense since the total black-hole mass can only decrease due to gravitational radiation losses. We thus see that the choice $f(\alpha)\propto\alpha^{1/2}$ is untenable.

The next simplest choice for f is

$$f(\alpha)=\gamma\alpha\,, \tag{13}$$

where γ is a constant. Repetition of the above argument for this new f leads to the conclusion that the final black-hole area must exceed the total initial black-hole area. But we know this to be true from Hawking's theorem.[7] Thus the choice (13) leads to no contradiction. Therefore, we shall adopt (13) for the moment; later on we shall exhibit some more positive evidence in its favor.

Comparison of (12) and (13) shows that γ must have the units of (length)$^{-2}$. But there is no constant with such units in classical general relativity. If in desperation we appeal to quantum physics we find only one truly universal constant with the correct units[14]: $\hbar^{-1}$, that is, the reciprocal of the Planck length squared. (Compton wavelengths are not universal, but peculiar to this or that particle; they clearly have no bearing on the problem.) We are thus compelled to write (12) as

$$S_{bh}=\eta\hbar^{-1}\alpha\,, \tag{14}$$

where η is a dimensionless constant which we may expect to be of order unity. This expression was first proposed by us earlier[11,12] from a different point of view.

We need not be alarmed at the appearance of $\hbar$ in the expression for black-hole entropy. It is well known[15] that $\hbar$ also appears in the formulas for entropy of many thermodynamic systems that are conventionally regarded as classical, for example, the Boltzmann ideal gas. This is a reflection of the fact that entropy is, in a sense, a count of states of the system, and the underlying states of any system are always quantum in nature. It is thus not totally unexpected that $\hbar$ appears in (14). These observations also suggest that it would be somewhat pretentious to attempt to calculate the precise value of the constant $\eta\hbar^{-1}$ without a full understanding of the quantum reality which underlies a "classical" black hole. Since there is no hope at present of obtaining such an understanding, we bypass the issue, and in the next section we use a semiclassical argument to arrive at a value for $\eta\hbar^{-1}$ which should be quite close to the correct one.

IV. EXPRESSION FOR BLACK-HOLE ENTROPY

In our attempt to obtain a value for $\eta\hbar^{-1}$ we shall employ the following argument. We imagine that a particle goes down a Kerr black hole. As it disappears some information is lost with it. According to the discussion of Sec. III we expect the black-hole entropy, as the measure of inaccessible information, to reflect the loss of the information associated with the particle by increasing by an appropriate amount. How much information is lost together with the particle? The amount clearly depends on how much is known about the internal state of the particle, on the precise way in which the particle falls in, etc. But we can be sure that the absolute minimum of information lost is that contained in the answer to the question "does the particle exist or not?" To start with, the answer is known to be yes. But after the particle falls in, one has no information whatever about the answer. This is because from the point of view of this paper, one knows nothing about the physical conditions inside the black hole, and thus one cannot assess the likelihood of the particle continuing to exist or being destroyed. One must, therefore, admit to the loss of one bit of information (see Sec. III) at the very least.

Our plan, therefore, is to compute the minimum possible increase in the black hole's area which results from the disappearance of a particle down the black hole, then to compute the corresponding minimum possible increase of black-hole entropy by means of our original formula (12), and finally to identify this increase in entropy with the loss of one bit of information in accordance with the scheme (11). If our procedure is reasonable we should then recover the functional form of f given by (13), together with a definite value for γ.

There are many ways in which a particle can go down a black hole, all leading to varying increases in black-hole area. We are interested in that method for inserting the particle which results in the smallest increase. This method has already been discussed by Christodoulou[4-6] in connection with his introduction of the concept of irreducible mass. The essence of Christodoulou's method is that if a freely falling point particle is captured by a Kerr black hole from a turning point in its orbit, then the irreducible mass and, consequently, the area of the hole are left unchanged. For reasons that will become clear presently we wish to allow the particle to have a nonzero radius. As shown in Appendix A, Christodoulou's method can be generalized easily so as to allow for this, as well as for the possibility that the particle is brought to the horizon by some method other than by free fall. We find in Appendix A that the increase in area for the generalized Christodoulou process is no longer precisely zero. But interestingly enough, the minimum increase in rationalized area, $(\Delta\alpha)_{min}$, turns out to be independent of the parameters of the black hole. For a spherical particle of rest mass μ, and proper radius b,

$$(\Delta\alpha)_{min} = 2\mu b\,. \tag{15}$$

For a point particle $(\Delta\alpha)_{min}=0$; this is Christodoulou's result.

Expression (15) gives the minimum possible increase in black-hole area that results if a given particle is added to a Kerr black hole. We can try to make $(\Delta\alpha)_{min}$ smaller by making b smaller. However, we must remember that b can be no smaller than the particle's Compton wavelength $\hbar\mu^{-1}$, or than its gravitational radius 2μ, whichever is the larger. The Compton wavelength is the larger for $\mu \leq 2^{-1/2}\hbar^{1/2}$, and the gravitational radius is the larger for $\mu > 2^{-1/2}\hbar^{1/2}$ $(2^{-1/2}\hbar^{1/2} \approx 10^{-5}\,g)$. Thus, if $\mu \leq 2^{-1/2}\hbar^{1/2}$, then $2\mu b$ can be as small as $2\mu\hbar\mu^{-1}=2\hbar$. But if $\mu > 2^{-1/2}\hbar^{1/2}$, then $2\mu b$ can be no smaller than $4\mu^2 > 2\hbar$. We conclude that quantum effects set a lower bound of $2\hbar$ on the increase of the rationalized area of a Kerr black hole when it captures a particle. Moreover, this limit can be reached only for a particle whose dimension is given by its Compton wavelength. Of course, only such an "elementary particle" can be regarded as having no internal structure. Therefore, the loss of information associated with the loss of such a particle should be minimum. And indeed we find that the increase in black-hole entropy is smallest for just such a particle. This supports our view that $2\hbar$ is the increase in rationalized area associated with the loss of one bit of information.

Following our program we shall equate the minimum increase in black-hole entropy, $(\Delta S_{bh})_{min} = 2\hbar\, df/d\alpha$, with ln2, the entropy increase associated with the loss of one bit of information. Integration of the resulting equation gives $f(\alpha) = (\frac{1}{2}\ln 2)\hbar^{-1}\alpha$. Thus, we have arrived again at (13) by an alternate route, and have obtained the value of γ into the bargain. We now have

$$S_{bh} = (\tfrac{1}{2}\ln 2)\hbar^{-1}\alpha\,, \tag{16}$$

which is of the same form as (14). Our argument has determined the dependence of S_{bh} on α in a straightforward manner. However, our value $\eta = \frac{1}{2}\ln 2$ might presumably be challenged on the grounds that it follows from the assumption that the smallest possible radius of a particle is precisely equal to its Compton wavelength whereas the actual radius is not so sharply defined. Nevertheless, it should be clear that if η is not exactly $\frac{1}{2}\ln 2$, then it must be very close to this, probably within a factor of two. This slight uncertainty in the value of η is the price we pay for not giving our problem a full quantum treatment. However, in what follows we shall suppose that $\eta = \frac{1}{2}\ln 2$. Examples to be given later will show that this value leads to no contradictions.

How is the entropy of a system of several black holes defined? It is natural to define it as the sum of individual black-hole entropies. Then Hawking's theorem tells us that the total black-hole entropy of the system cannot decrease. But this is just what we would expect since the information lost down the black holes is unrecoverable. This observation lends support to our choice.

In conventional units (16) takes the form

$$\begin{aligned} S_{bh} &= (\tfrac{1}{2}\ln 2/4\pi)kc^3\hbar^{-1}G^{-1}A \\ &= (1.46\times 10^{48}\ \mathrm{erg}\ ^\circ\mathrm{K}^{-1}\,\mathrm{cm}^{-2})A\,, \end{aligned} \tag{17}$$

where k is Boltzmann's constant. We see that the entropy of a black hole is enormous. For example, a black hole of one solar mass would have $S_{bh} \approx 10^{60}$ erg $^\circ$K^{-1}. By comparison the entropy of the sun is $S \approx 10^{42}$ erg $^\circ$K^{-1}; those of a white dwarf or a neutron star of one solar mass even smaller. The large numerical value of black-hole entropy serves to dramatize the highly irreversible character of the process of black-hole formation. We may define a characteristic temperature for a Kerr black hole by the relation $T_{bh}^{-1} = (\partial S_{bh}/\partial M)_{L,Q}$ which is the analog of the thermodynamic relation $T^{-1} = (\partial S/\partial E)_V$. By using (8) and (16) we find

$$\begin{aligned} T_{bh} &= 2\hbar(\ln 2)^{-1}\Theta \\ &= (0.165\ ^\circ\mathrm{K\,cm})\,(r_+ - r_-)\,(r_+^2 + a^2)^{-1}\,, \end{aligned} \tag{18}$$

where $r_\pm$ and a are to be given in centimeters. We introduce this T_{bh} in anticipation of our discussion of an example in Sec. VII. But we emphasize that one should not regard T_{bh} as *the* temperature of the black hole; such an identification can easily lead to all sorts of paradoxes, and is thus not useful.

V. THE GENERALIZED SECOND LAW

Suppose that a body containing some common entropy goes down a black hole. The entropy of the visible universe decreases in the process. It would seem that the second law of thermodynamics is transcended here in the sense that an exterior observer can never verify by direct measurement that the total entropy of the whole universe does not decrease in the process.[19] However, we know that the black-hole area "compensates" for the disappearance of the body by increasing irreversibly. It is thus natural to conjecture that the second law is not really transcended provided that it is expressed in a generalized form: *The common entropy in the black-hole exterior plus the black-hole entropy never decreases.* This statement means that we must regard black-hole entropy as a genuine contribution to the entropy content of the universe.

Support for the above version of the second law comes from the following argument. Suppose that a body carrying entropy S goes down a black hole (which may have existed previously or may be formed by the collapse of the body). The S is the uncertainty in one's knowledge of the internal configuration of the body. So long as the body was still outside the black hole, one had the option of removing this uncertainty by carrying out measurements and obtaining information up to the amount S. But once the body has fallen in, this option is lost; the information about the internal configuration of the body becomes truly inaccessible. We thus expect the black-hole entropy, as the measure of inaccessible information, to increase by an amount S. Actually, the increase in S_{bh} may be even larger because any information that was available about the body to start with will also be lost down the black hole. Therefore, if we denote by ΔS_c the change in common entropy in the black-hole exterior ($\Delta S_c \equiv -S$), then we expect that

$$\Delta S_{bh} + \Delta S_c = \Delta(S_{bh} + S_c) > 0 . \qquad (19)$$

This is just the generalized second law which we proposed above: The generalized entropy $S_{bh} + S_c$ never decreases. Examples supporting this law will be given in Sec. VI-VII.

This is a good place to mention the question of fluctuations. We know that the common entropy of a closed thermodynamic system can decrease spontaneously as a result of statistical fluctuations, i.e., the second law, being a statistical law, is meaningful only if statistical fluctuations are small. Is black-hole entropy also subject to decreases of a statistical nature? Not classically – Hawking's theorem guarantees that. Quantum mechanically there are two ways by which the black-hole entropy can undergo statistical decreases. One of them depends on the quantum fluctuations of the metric of the black hole which one has reasons to expect.[1] Such fluctuations would be reflected in small random fluctuations in the area, and thus in the entropy of the black hole, and some of these fluctuations would be expected to be decreases in entropy. However, even if one regards a black hole as a purely classical object, it is still possible for its area and entropy to undergo small decreases when the black hole absorbs a single quantum under certain conditions.[20] However, the probability of such an event occurring in any given trial is very small. Therefore, the decrease in area and entropy is of a statistical nature, and is quite analogous to the decrease in entropy of a thermodynamic system due to statistical fluctuations. This discussion serves us warning that the law (19) is expected to hold only insofar as statistical fluctuations are negligible.

We noticed earlier (Sec. IV) the very large magnitude of black-hole entropy. In fact, one can say that the black-hole state is the maximum entropy state of a given amount of matter. The point is that in the gravitational collapse of a body into a black hole, the loss of information down the black hole is the maximum allowed by the laws of physics. Thus if the body collapses to form a Kerr black hole, all information about it is lost with the exception of mass, charge, and angular momentum.[1] These quantities are given in terms of Gaussian integrals,[1] and so information about them cannot be lost. But all other information about the body is eventually lost. Therefore, the resulting black hole must correspond to the maximum (generalized) entropy which can be associated with the given body.

VI. EXAMPLES OF THE GENERALIZED SECOND LAW AT WORK

In the examples which follow we endeavor to subject the generalized second law to the most stringent test possible in each case by maximizing the entropy going down the black hole with a given body while minimizing the associated increase in black-hole entropy.

A. Harmonic Oscillator

As a first example we take an harmonic oscillator composed of two particles of rest mass $\frac{1}{2}m$ each connected by a nearly massless spring of spring constant K. We imagine the oscillator to be enclosed in a spherical box and to be maintained at temperature T. We assume for simplicity that conditions are such that the oscillator is nonrelativistic ($T \ll m$). Let ω be the vibrational frequency of the oscillator. Then the (normalized) probability that the oscillator is in its nth quantum state is given by the canonical distribution

$$p_n = (1 - e^{-x})^{-1} e^{-nx}, \quad x = \hbar\omega/T. \tag{20}$$

The entropy of the oscillator as computed from (10) is

$$S = x(e^x - 1)^{-1} - \ln(1 - e^{-x}), \tag{21}$$

and the mean vibrational energy,

$$\langle E \rangle \equiv \sum p_n (n + \tfrac{1}{2})\hbar\omega,$$

is

$$\langle E \rangle = [(e^x - 1)^{-1} + \tfrac{1}{2}]\, \hbar\omega\,. \tag{22}$$

We remark that the thermal distribution (20) maximizes the entropy of the oscillator for given $\langle E \rangle$, and is thus ideally suited to our plan for subjecting the generalized second law to the most stringent test possible.

Suppose that the box goes down a Kerr black hole. The corresponding increase in black-hole entropy cannot be smaller than the lowest limit derived by the method of Appendix A. From (15) and (16) we have $\Delta S_{\rm bh} \geq \mu b \hbar^{-1} \ln 2$, where b is the outer radius of the box and μ is its total rest mass. Clearly b must be at least as large as half of the mean value $\langle y \rangle$ of the separation of the two masses y. And $\langle y \rangle$ in turn must clearly be larger than Δy, the root mean square of the thermal oscillation of y $[(\Delta y)^2 \equiv \langle (y - \langle y \rangle) \rangle]$, so that y will always be positive. Now according to the (quantum) virial theorem $\frac{1}{2}\langle E \rangle$ is equal to the mean potential energy of the oscillator $\frac{1}{2}K(\Delta y)^2$. Since the reduced mass of the oscillator is $\frac{1}{4}m$, we have $K = \frac{1}{4}m\omega^2$. We thus find from all the above that

$$b > \langle E \rangle^{1/2} m^{-1/2} \omega^{-1}.$$

Remembering that $\mu > m + \langle E \rangle$ (because the box itself must have some mass) we obtain

$$\Delta S_{\rm bh} > \langle E \rangle^{1/2} m^{-1/2} (\hbar\omega)^{-1} (m + \langle E \rangle) \ln 2\,. \tag{23}$$

We assume that the entropy given by (21) is the only contribution to the entropy in the box. This amounts to neglecting the contribution of the black body radiation in the box, etc., a sensible procedure if T is not very high. Then $\Delta S_c = -S$ and we have

$$\Delta(S_{\rm bh} + S_c) > \xi^{-1/2}(1 + \xi)[\tfrac{1}{2} + (e^x - 1)^{-1}] \ln 2 - x(e^x - 1)^{-1} + \ln(1 - e^{-x}), \tag{24}$$

where we have introduced the notation $\xi \equiv m\langle E \rangle^{-1}$ and used Eqs. (21) and (22) for S and $\langle E \rangle$. We now show that $\Delta(S_{\rm bh} + S_c) > 0$ as required by the generalized second law. The expression in (24) regarded as a function of x for given ξ has a single minimum at

$$x = x_m \equiv \xi^{-1/2}(1 + \xi) \ln 2$$

which has the value

$$\tfrac{1}{2} x_m + \ln[1 - \exp(-x_m)]\,.$$

Our assumption that the oscillator is nonrelativistic means that $\xi \gg 1$, and hence that $x_m \gg 2\ln 2$. Under these conditions the minimum is positive (in fact, it is positive for $\xi \gtrsim 1$). It follows immediately that $\Delta(S_{\rm bh} + S_c)$ is positive for all x and all ξ which are compatible with the requirement of a nonrelativistic oscillator. The generalized second law is obeyed over the entire regime for which our treatment is valid.

B. Beam of Light

As a second example we consider a beam of light which is aimed at a Kerr black hole. This example is particularly interesting because it shall bring us face to face with the issue of fluctuations as a limitation on the applicability of the second law. We shall restrict our attention only to those cases for which geometrical optics is a valid approximation. We shall thus represent the path of the beam by a null geodesic in the Kerr background.

We shall take it that the beam is thermalized at a certain temperature T. This implies that its entropy is a maximum for given energy. The entropy is easily calculated; in fact, the entropy and energy for each mode in the beam are given by the same expressions (21) and (22) which apply to a harmonic oscillator, except that one must omit the zero-point energy term $\frac{1}{2}\hbar\omega$. The total entropy S and mean energy $\langle E \rangle$ of the beam are obtained by integrating these expressions weighed by the conventional density of states

$$\rho = 2\omega^2 (2\pi)^{-3} V d\Omega \tag{25}$$

over all ω. In (25) V is the volume of the beam and $d\Omega$ is the solid angle it subtends. Integrating by parts the expression for S, one easily obtains the relation

$$S=\tfrac{4}{3}\langle E\rangle T^{-1}\,, \tag{26}$$

which, not surprisingly, is identical to that for radiation inside a black-body cavity of temperature T.[15] [In Ref. 12 (26) was given with an incorrect numerical factor.]

As the beam nears the black hole, it is deflected by the gravitational field. Insofar as its effects on electromagnetic radiation are concerned, a stationary gravitational field can be mocked up by an appropriate nonabsorbing refractive medium in flat spacetime.[21] But the propagation of a beam of light through such a medium is a reversible process.[22] We infer from this that the entropy of the beam will remain unchanged as the beam nears the black hole. Thus the entropy change of the visible universe when the beam goes down the hole is just

$$\Delta S_c = -\tfrac{4}{3}\langle E\rangle\, T^{-1}\,. \tag{27}$$

What is the increase in black-hole entropy associated with the process? From (8) we see that the increase in α is minimized when the angular momentum that the hole gains from the beam is maximized for given $\langle E\rangle$. Now, the gain in angular momentum is limited because the beam will not be captured if it carries too much angular momentum. In Appendix B we take this into account in calculating (in the geometrical optics limit) the minimum possible increase in α for given $\langle E\rangle$ of the beam. We find that $\Delta\alpha \geq \beta M\langle E\rangle$ where β ranges from 8 for the case of a Schwarzschild hole to $4(1-\sqrt{3}/2)$ for the case of an extreme Kerr hole, this last value being the smallest possible β. From (16) it follows that

$$\Delta S_{bh} \geq (\tfrac{1}{2}\beta \ln 2) M\hbar^{-1}\langle E\rangle\,. \tag{28}$$

Our assumption (Appendix B) that geometrical optics is always applicable means that the bulk of wavelengths in the beam are much shorter than the characteristic dimension of the hole $\approx M$. Thus, if ω_c is some characteristic frequency in the beam, then we require that $\omega_c \gg M^{-1}$. From the form of the Planck spectrum (22) we see that $\hbar\omega_c \approx T$; therefore (27) tells us that

$$|\Delta S_c| \ll \tfrac{4}{3} M\hbar^{-1}\langle E\rangle\,. \tag{29}$$

Comparison of (28) and (29) shows that a violation of the generalized second law (19) cannot arise in the regime under consideration.

In the above discussion the condition that geometrical optics be applicable prevented us from taking T to be arbitrarily small. As a result it turned out to be impossible for $|\Delta S_c|$ to exceed ΔS_{bh}, and so a violation of the second law was ruled out. But there is a way to circumvent the restriction on T. One simply selects the temperature T (arbitrarily) to be as small as one pleases, and arranges for all frequencies $\omega<\omega_c'$ to be filtered out of the beam. Here $\omega_c' \gg M^{-1}$ is a definite frequency unrelated to T. It should be clear that geometrical optics will be a valid approximation for this case also, so that we may take over the result (28). But the result (27) must be modified since we are here dealing with a truncated frequency spectrum. We are mostly interested in the regime $T \ll \hbar\omega_c'$. Then for all frequencies in the beam $x=\hbar\omega/T \gg 1$. It follows from (21) and (22) that for each mode the entropy to energy ratio is T^{-1} ($S\approx xe^{-x}$, $\langle E\rangle \approx \hbar\omega e^{-x}$). Therefore instead of (27) we have

$$\Delta S_c = -\langle E\rangle\, T^{-1}. \tag{30}$$

It now appears that if

$$T < T_c \equiv \hbar(\tfrac{1}{2}\beta M \ln 2)^{-1},$$

then $\Delta S_{bh} + \Delta S_c$ will be negative in contradiction with the generalized second law.

The resolution of the above paradox is that in the regime $T<T_c$ statistical fluctuations are already dominant so that our entire picture of the process is invalid. To verify the importance of fluctuations we calculate the mean number of quanta N in the beam by integrating the mean number of quanta per mode, $(e^x-1)^{-1}$, weighed by the density of states (25) over all $\omega>\omega_c'$. For $T=T_c$ we get (recall that $\hbar\omega_c'/T_c \gg 1$ by our assumptions)

$$N \approx \frac{V}{M^3}\,\frac{d\Omega}{4\pi}\,\delta^{-3}(\delta M\omega_c')^2 \exp(-\delta M\omega_c')\,, \tag{31}$$

where $\delta \equiv \frac{1}{2}\beta \ln 2$ ($0.2 \leq \delta \leq 2.8$). It is clear that for any beam aimed at the black hole $d\Omega/4\pi \ll 1$. Recalling that $M\omega_c' \gg 1$ by assumption, we see from (31) that each quantum occupies a mean volume much larger than M^3. But the cross section of the beam must be smaller than $\sim M^2$ if the beam is to go down the black hole. Thus the mean separation between quanta is much larger than M, the characteristic dimension of the black hole. In case $T<T_c$ the above effect is even more accentuated.

We conclude that in the regime $T \lesssim T_c$ for which the second law (19) appears to break down, our description of the process as a continuous beam going down the black hole is invalidated by the large fluctuations in the concentration of energy in the beam (or equivalently, the large fluctuations in the energy of each section of the beam). In this regime $\langle E\rangle$ is no longer a good measure of the actual energy. It appears, therefore, that statistical fluctuations are responsible for the breakdown of the second law in the context in which we have applied it here. But we can demonstrate that the law has not lost all its meaning by adopting a point of view more suitable to the circumstances at hand than the one used above.

We take the point of view that quanta are going down the black hole one at a time, rather than in a continuous stream. Thus we must check the validity of the generalized second law for the infall of each quantum. The analysis of Appendix B still leads to formula (28) for the increase in black-hole entropy except that $\langle E\rangle$ is replaced by $\hbar\omega$, the energy of the quantum. To compute the common entropy going down the black hole we reason as follows. From our point of view a quantum of definite frequency is going down the black hole. Thus we are no longer dealing with the probability distribution (20); instead we shall ascribe probability $\frac{1}{2}$ to each of the two possible polarizations of the quantum. Then according to (10) the entropy associated with the quantum is ln2. Therefore,

$$\Delta S_{\rm bh}+\Delta S_c \geq (\tfrac{1}{2}\beta\ \ln 2)M\omega - \ln 2. \qquad (32)$$

Since $\frac{1}{2}\beta \geq 0.268$, and since we are assuming that $M\omega \gg 1$, we see that $\Delta S_{\rm bh}+\Delta S_c$ is in fact positive: The generalized second law is upheld for the infall of each quantum.

VII. A PERPETUAL MOTION MACHINE USING A BLACK HOLE?

Geroch[13] has described a procedure using a black hole which appears to violate the second law of thermodynamics by converting heat into work with unit efficiency. He envisages a box filled with black-body radiation which is slowly lowered by means of a string from far away down to the horizon of a black hole, at which point its energy as measured from infinity vanishes. Therefore, if the box's rest mass is μ, then the agent lowering the string obtains work equal to μ out of the process. The box is then allowed to emit into the black-hole radiation of (proper) energy $\Delta\mu$. Finally, the agent retrieves the box; since its rest mass is now $\mu-\Delta\mu$, he must do work $\mu-\Delta\mu$ to accomplish this. Therefore, in the whole process the agent obtains net work $\Delta\mu$ at the expense of heat $\Delta\mu$ - conversion with unit efficiency. We shall now show that, in fact, due to fundamental physical limitations, the efficiency of the Geroch process is slightly smaller than unity, so that no violation of the second law is entailed here.

The box under consideration must have a nonzero radius (see below). Because of this its energy as measured from infinity is never quite zero when it is as close to the horizon as it can possibly be. We shall assume that the box is in the shape of a sphere of radius b. Then according to the analysis of Appendix C the minimum value of the energy is

$$E = 2\mu b\Theta\ , \qquad (33)$$

where Θ is defined by (9a). It follows that in lowering the box from infinity to the horizon, the agent obtains only work $\mu(1-2b\Theta)$ rather than μ. After the box has radiated into the black hole, its rest mass becomes $\mu-\Delta\mu$ and according to (33) its energy at the horizon is just $2(\mu-\Delta\mu)b\Theta$. Thus the agent must do work $(\mu-\Delta\mu)(1-2b\Theta)$ to retrieve the box to infinity where its energy is $\mu-\Delta\mu$. Therefore, in the over-all process the agent obtains net work $\Delta\mu(1-2b\Theta)$ in exchange for the expenditure of heat $\Delta\mu$. The efficiency of conversion is

$$\epsilon = 1-2b\Theta\ , \qquad (34)$$

which is smaller than unity. In practical situations $b \ll r_+$ so that $b\Theta \ll 1$ and the efficiency can be quite near to unity. But the departure of ϵ from unity, albeit small, serves to resolve the problem raised by Geroch's example: There is no violation of the Kelvin statement of the second law.[23]

We must now explain why b cannot be arbitrarily small. Physically the reason is that the box must be large enough for the wavelengths characteristic of radiation of some temperature T to fit into it. More formally we can argue as follows. The frequency of the photon ground state associated with the box, ω_0, cannot exceed that frequency ω_p at which the Planck photon-number spectrum

$$\propto \omega^2[\exp(\hbar\omega/T)-1]^{-1}$$

peaks. Otherwise the frequencies of all photon states would lie in the exponential tail of the spectrum, the occupation number of each state would be small, and the resulting large fluctuations would make the concept of temperature meaningless. We have the conventional relation $\omega_0 b' = \pi$, where b' is the interior radius of the box ($b' < b$), and we easily find that $\hbar\omega_p < 2T$. Therefore, $\omega_0 < \omega_p$ implies that $b > \pi\hbar/2T$. It is thus clear that there is a lower limit for b.

We may write the efficiency (34) in a more transparent form by recalling that $\Theta = \frac{1}{2}T_{\rm bh}\ \ln 2/\hbar$ [see (18)], where $T_{\rm bh}$ is the characteristic temperature associated with the black hole. Since $b > \pi\hbar/2T$ we find that

$$\epsilon < 1 - T_{bh}/T . \quad (35)$$

We now recall that the efficiency of a heat engine operating between two reservoirs, one at temperature T and the second at temperature $T_{bh} < T$, is restricted by $\epsilon \leq 1 - T_{bh}/T$. We thus see that the Geroch process is no more efficient than its "equivalent reversible heat engine." This observation makes it evident again that Geroch's process is not in violation of the second law of thermodynamics. Finally, we wish to remark that since our primary formula (33) is valid only when the box is small compared to the black hole ($b\Theta \ll 1$), we can vouch for the validity of (35) only when $T_{bh} \ll T$. However, due to the smallness of T_{bh} this condition will be satisfied in all cases of practical interest.

We now verify that the Geroch process is in accord with the generalized second law (19). We mentioned earlier that the agent obtains work $\Delta\mu(1-2b\Theta)$ for a decrease $\Delta\mu$ in the rest mass of the box. This means that the black hole's mass must increase by $2\Delta\mu b\Theta$ in the complete process. According to (8) and (16) the corresponding increase in black hole entropy is $\Delta S_{bh} = \Delta\mu b \hbar^{-1} \ln 2$ (angular momentum is not added to the hole; see Appendix C). But since $b > \pi\hbar/2T$ we have that

$$\Delta S_{bh} > (\tfrac{1}{2}\pi \ln 2)\Delta\mu/T . \quad (36)$$

On the other hand, the decrease in entropy of the box is clearly $\Delta\mu/T$ (heat/temperature). Thus

$$\Delta S_c = -\Delta\mu/T . \quad (37)$$

From (36) and (37) it follows that $\Delta(S_{bh} + S_c) > 0$ as required by the generalized second law.

ACKNOWLEDGMENTS

The author is grateful to Professor J. A. Wheeler for his interest in this work, his encouragement and his penetrating criticism. He also thanks Professor Karel Kuchař and Professor Brandon Carter, and Dr. Bahram Mashhoon for valuable suggestions and comments.

APPENDIX A

Here we shall calculate the minimum possible increase in black-hole area which must result when a spherical particle of rest mass μ and proper radius b is captured by a Kerr black hole. We are interested in the increase in area ascribable to the particle itself, as contrasted with any increase incidental to the process of bringing the particle to black-hole horizon. For example, there is some *circumstantial* evidence for believing that when the particle is lowered into the black hole by a string, there occurs an increase in black-hole area even as the particle is being lowered.[11] Furthermore, the area will experience an additional increase due to the gravitational waves radiated into the black hole by the string as it relaxes when the particle is dropped.[11] Similarly, if the particle falls freely to the horizon it emits gravitational waves into the hole even before it falls in; the amount of radiation may even be significant.[24] This radiation will also result in an increase in area. Here we shall ignore all these incidental effects and concentrate on the increase in area caused by the particle all by itself.

We assume that the particle is neutral so that it follows a geodesic of the Kerr geometry when falling freely. We shall employ Boyer-Lindquist coordinates for the charged Kerr metric[25]

$$ds^2 = g_{tt}\,dt^2 + 2g_{t\phi}\,dt\,d\phi + g_{\phi\phi}\,d\phi^2 + g_{rr}\,dr^2 + g_{\theta\theta}\,d\theta^2 . \quad (A1)$$

For later reference we give g_{rr}:

$$g_{rr} = (r^2 + a^2\cos^2\theta)\Delta^{-1} , \quad (A2)$$

where

$$\Delta \equiv r^2 - 2Mr + a^2 + Q^2 . \quad (A3)$$

The event horizon is located at $r = r_+$ where $r_\pm$ are defined by (7). We have

$$\Delta = (r - r_-)(r - r_+) . \quad (A4)$$

First integrals for geodesic motion in the Kerr background have been given by Carter.[25] Christodoulou[5] uses the first integral

$$E^2[r^4 + a^2(r^2 + 2Mr - Q^2)] - 2E(2Mr - Q^2)ap_\phi - (r^2 - 2Mr + Q^2)p_\phi^{\,2} - (\mu^2 r^2 + q)\Delta = (p_r\Delta)^2 \quad (A5)$$

as a starting point of his analysis. In (A5) $E = -p_t$ is the conserved energy, p_ϕ is the conserved component of angular momentum in the direction of the axis of symmetry, q is Carter's fourth constant of the motion,[25] μ is the rest mass of the particle, and p_r is its covariant radial momentum.

Following Christodoulou we solve (A5) for E:

$$E = Bap_\phi + \{[B^2 a^2 + A^{-1}(r^2 - 2Mr + Q^2)]p_\phi^2 + A^{-1}[(\mu^2 r^2 + q)\Delta + (p_r \Delta)^2]\}^{1/2}, \quad \text{(A6)}$$

where

$$A \equiv r^4 + a^2(r^2 + 2Mr - Q^2), \quad \text{(A7)}$$

$$B \equiv (2Mr - Q^2)A^{-1}. \quad \text{(A8)}$$

At the event horizon $\Delta = 0$ [see (A4)] so that there

$$A = A_+ = (r_+^2 + a^2)^2, \quad B = B_+ = (r_+^2 + a^2)^{-1}. \quad \text{(A9)}$$

Furthermore, at the horizon $Ba = \Omega$ [see (9b)], and the coefficients of p_ϕ^2 and $\mu^2 r^2 + q$ in (A6) vanish. However,

$$p_r \Delta = (r^2 + a^2 \cos^2\theta)p^r$$

does not vanish at the horizon in general. If the particle's orbit intersects the horizon, then we have from (A6) that

$$E = \Omega p_\phi + A_+^{-1/2}|p_r \Delta|_+.$$

As a result of the capture, the black hole's mass increases by E and its component of angular momentum in the direction of the symmetry axis increases by p_ϕ. Therefore, according to (8) the black hole's rationalized area will increase by $\Theta^{-1}A_+^{-1/2}|p_r\Delta|_+$. As pointed out by Christodoulou this increase vanishes *only if* the particle is captured from a turning point in its orbit in which case $|p_r\Delta|_+ = 0$. In this case we have

$$E = \Omega p_\phi. \quad \text{(A10)}$$

The above analysis shows that it is possible for a black hole to capture a *point* particle without increasing its area. How is this conclusion changed if the particle has a nonzero proper radius b? First we note that regardless of the manner in which the particle arrives at the horizon (being lowered by a string, splitting off from a second particle which then escapes, etc.), it must clearly acquire its parameters E, p_ϕ, and q while every part of it is still outside the horizon, i.e., while it is not yet part of the black hole. Moreover, as the particle is captured, it must already be detached from whatever system brought it to the horizon, so that it may be regarded as falling freely. Therefore, Eq. (A6) should always describe the motion of the particle's center of mass at the moment of capture.

It should be clear that to generalize Christodoulou's result to the present case one should evaluate (A6) not at $r = r_+$, but at $r = r_+ + \delta$, where δ is determined by

$$\int_{r_+}^{r_+ + \delta} (g_{rr})^{1/2}\, dr = b$$

($r = r_+ + \delta$ is a point a proper distance b outside the horizon). Using (A2) we find

$$b = 2\delta^{1/2}(r_+^2 + a^2\cos^2\theta)^{1/2}(r_+ - r_-)^{-1/2}. \quad \text{(A11)}$$

To obtain this we have assumed that $r_+ - r_- \gg \delta$ (black hole not nearly extreme). Expanding the argument of the square root in (A6) in powers of δ, replacing δ by its value given by (A11), and keeping only terms to $O(b)$ we get

$$E = \Omega p_\phi + [(r_+^2 - a^2)(r_+^2 + a^2)^{-1}p_\phi^2 + \mu^2 r_+^2 + q]^{1/2} \times \tfrac{1}{2}b(r_+ - r_-)(r_+^2 + a^2)^{-1}(r_+^2 + a^2\cos^2\theta)^{-1/2} \quad \text{(A12)}$$

Here we have already assumed that the particle reaches a turning point as it is captured since we know that this minimizes the increase in black-hole area. Equation (A12) is the generalization to $O(b)$ of the Christodoulou condition (A10).

What is q in (A12)? We can obtain a lower bound for it as follows. From the requirement that the θ momentum p_θ be real it follows that [25]

$$q \geq \cos^2\theta[a^2(\mu^2 - E^2) + p_\phi^2/\sin^2\theta]; \quad \text{(A13)}$$

the equality holds when $p_\theta = 0$ at the point in question. If we replace E in (A13) by Ωp_ϕ [see (A12)] we obtain

$$q \geq \cos^2\theta[a^2\mu^2 + p_\phi^2(1/\sin^2\theta - a^2\Omega^2)],$$

which is correct to zeroth order in b. We know that $1/\sin^2\theta \geq 1$; it is easily shown that $a^2\Omega^2 \leq \frac{1}{4}$ for a charged Kerr black hole. Therefore $q \geq a^2\mu^2\cos^2\theta$. Substituting this into (A12) we find

$$E \geq \Omega p_\phi + \tfrac{1}{2}\mu b(r_+ - r_-)(r_+^2 + a^2)^{-1} \quad \text{(A14)}$$

which is correct to $O(b)$. By retracing our steps we see that the equality sign in (A14) corresponds to the case $p_\phi = p_\theta = p^r = 0$ at the point of capture. The increase in black-hole area, computed by means of (8), (9a), and (A14), is

$$\Delta\alpha \geqslant 2\mu b\ . \tag{A15}$$

This gives the fundamental lower bound on the increase in black-hole area. We note that it is independent of M, Q, and L.

APPENDIX B

Here we shall calculate the minimum possible increase in black-hole area which must result when a light beam of energy $E>0$ coming from infinity is captured by a Kerr black hole. If the black hole is nonrotating the increase is simply obtained by setting $dM=E$ in (8):

$$\Delta\alpha = 8ME \quad \text{for} \quad a=0. \tag{B1}$$

If the black hole is rotating, $\Delta\alpha$ can be minimized by maximizing the angular momentum p_ϕ which is brought in by the beam [see (8)]. To accomplish this we consider the effective potential V for the motion of a massless particle in a Kerr background.

This V is just the value of E given by (A6) regarded as a function of r for $\mu=0$ and $p_r=0$ (E equals V at a turning point). This potential starts off at a value Ωp_ϕ at $r=r_+$ (see Appendix A), increases with r, reaches a maximum, and then falls off to zero as $r\to\infty$. For the beam to be captured by the hole it is necessary that p_ϕ be small enough for the peak of the potential barrier to be smaller than E of the beam. The optimum case we seek corresponds to the peak being just equal to E so that p_ϕ has its largest possible value.

It is clear that we must take q in (A6) as small as possible in order to have the lowest possible potential peak for given p_ϕ. Let us first take $q<0$. Then according to Carter[25] there are solutions to the geodesic equation only if $|p_\phi|< aE$. From (8) it follows that

$$\Delta\alpha = \Theta^{-1}(E-\Omega p_\phi)$$

$$>\Theta^{-1}E(1-\Omega a).$$

But since $\Omega a \leqslant \frac{1}{2}$ and $\Theta^{-1}\geqslant 8M$ it follows that

$$\Delta\alpha > 4ME \quad \text{for} \quad q<0. \tag{B2}$$

Next we take $q=0$. Two cases are possible[25]: Either $|p_\phi|<aE$ as above so that (B2) is again applicable, or else the orbit is purely equatorial. In the second case one may calculate the peak of the barrier numerically and then find the optimum increase in α. It turns out that $(\Delta\alpha)_{\min}$ decreases monotonically with a for fixed M. The limit of $(\Delta\alpha)_{\min}$ as $a\to M$ may be computed analytically because in this limit one can find the height of the potential analytically with sufficient accuracy. One finds

$$(\Delta\alpha)_{\min}\to 4\left(1-\tfrac{1}{2}\sqrt{3}\right)ME \quad \text{as} \quad a\to M. \tag{B3}$$

It is clear that for $q>0$ the potential peak will be higher and the increase in area will be larger than the one given by (B3). Thus we find that the minimum increase in area results when the beam is captured by an extreme Kerr black hole from a purely equatorial orbit.

APPENDIX C

Here we compute the value of the energy (as measured from infinity) of a particle of rest mass μ and proper radius b which is hanging from a string just outside the horizon of a Kerr black hole. It is clear that the particle will not be moving in the r or θ directions; hence $p^r=p_\theta=0$ for it. We cannot claim that the particle does not move in the ϕ direction. In fact, since it will be within the ergosphere in general, it cannot avoid moving in the ϕ direction.[1] Our intuitive notion that the particle is "not moving" must be applied only in a locally nonrotating (Bardeen) frame.[26] The particle is at rest in such a frame if for it

$$\frac{d\phi}{dt} = -\frac{g_{t\phi}}{g_{\phi\phi}}.$$

It follows that

$$p_\phi = \mu\left(g_{t\phi}\frac{dt}{d\tau} + g_{\phi\phi}\frac{d\phi}{d\tau}\right)=0.$$

If the particle were to be dropped, it would clearly keep its energy E and it would still have $p_\phi=p^r=p_\theta=0$, at least momentarily. We may thus compute E for the particle hanging in the string at a proper distance b from the horizon by setting $p_\phi=0$ in (A12). For q we take the value given by (A13) with the equality sign ($p_\theta=0$) $p_\phi=0$, and $E=0$ [since for $p_\phi=0$, E is of $O(b)$]. Thus

$$q=\mu^2a^2\cos^2\theta$$

and

$$E = \tfrac{1}{2}\mu b(r_+-r_-)(r_+^2+a^2)^{-1}$$

$$=2\mu b\Theta\ . \tag{C1}$$

*Based in part on the author's Ph.D. thesis, Princeton University, 1972; work supported in part by the National Science Foundation Grants No. GP 30799X to Princeton University and No. GP 32039 to the University of Texas at Austin.

†National Science Foundation Predoctoral Fellow when this work was initiated.

‡Present address.

[1]C. W. Misner, K. S. Thorne, and J. A. Wheeler, *Gravitation* (Freeman, San Francisco, 1973).

[2]R. Penrose and R. M. Floyd, Nature 229, 177 (1971).

[3]R. Penrose, Riv. Nuovo Cimento 1, 252 (1969).

[4]D. Christodoulou, Phys. Rev. Letters 25, 1596 (1970).

[5]D.Christodoulou, Ph.D. thesis, Princeton University, 1971 (unpublished).

[6]D. Christodoulou and R. Ruffini, Phys. Rev. D 4, 3552 (1971).

[7]S. W. Hawking, Phys. Rev. Letters 26, 1344 (1971); contribution to *Black Holes*, edited by B. DeWitt and C. DeWitt (Gordon and Breach, New York, 1973).

[8]J. M. Greif, Junior thesis, Princeton University, 1969 (unpublished).

[9]B. Carter, Nature 238, 71 (1972).

[10]Ya. B. Zel'dovich and I. D. Novikov, *Stars and Relativity* (University of Chicago Press, Chicago, 1971), p. 268.

[11]J. D. Bekenstein, Ph.D. thesis, Princeton University, 1972 (unpublished).

[12]J. D. Bekenstein, Lett. Nuovo Cimento 4, 737 (1972).

[13]R. Geroch, Colloquium at Princeton University, December 1971.

[14]We use units with $G = c = 1$ unless otherwise specified.

[15]L. D. Landau and E. M. Lifshitz, *Statistical Physics* (Addison-Wesley, Reading, Mass., 1969).

[16]L. D. Landau and E. M. Lifshitz, *Electrodynamics of Continuous Media* (Addison-Wesley, Reading, Mass., 1960), p. 5.

[17]The mathematical definition of information was first given by C. E. Shannon; the relevant papers are reprinted in C. E. Shannon and W. Weaver, *The Mathematical Theory of Communications* (University of Illinois Press, Urbana, 1949). An elementary introduction to information theory is given by M. Tribus and E. C. McIrvine, Sci. Amer. 225 (No. 3), 179 (1971). The derivation of statistical mechanics from information theory was first carried out by E. T. Jaynes, Phys. Rev. 106, 620 (1957); 108, 171 (1957).

[18]The relation between information theory and thermodynamics is discussed in detail by L. Brillouin, *Science and Information Theory* (Academic, New York, 1956), especially Chaps. 1, 9, 12–14.

[19]This was first pointed out to the author by J. A. Wheeler (private communication).

[20]J. D. Bekenstein, Phys. Rev. D 7, 949 (1973).

[21]See for example, A. M. Volkov, A. A. Izmest'ev, and G. V. Skrotskii, Zh. Eksp. Teor. Fiz. 59, 1254 (1970) [Sov. Phys. JETP 32, 686 (1971)].

[22]See E. Hisdal, Phys. Norv. 5, 1 (1971), and references cited therein.

[23]In Ref. 12 we gave an alternate resolution of the paradox posed by Geroch based on the apparent tendency of the black-hole area to increase as the box is being lowered.[11] The present approach is preferable in that it is independent of the validity of the interpretation given in Ref. 11, and in that it fits very well into the thermodynamic approach of this paper as will be evident presently.

[24]See M. Davis, R. Ruffini, and J. Tiomno, Phys. Rev. D 5, 2932 (1972) for the radiation of a particle falling radially into a Schwarzschild black hole. It is not clear whether the large amount of radiation found to go down the black hole is a device of the linearized approximation, or whether the effect will persist for other types of orbits or for Kerr black holes. Therefore we do not base any of our arguments here on this effect as we did in Refs. 11 and 12.

[25]B. Carter, Phys. Rev. 174, 1559 (1968).

[26]J. M. Bardeen, Astrophys. J. 161, 103 (1970).

PHYSICAL REVIEW D VOLUME 9, NUMBER 12 15 JUNE 1974

Generalized second law of thermodynamics in black-hole physics*

Jacob D. Bekenstein
Center for Relativity Theory, The University of Texas at Austin, Austin, Texas 78712
(Received 17 September 1973)

In previous work we introduced the concept of black-hole entropy, which we identified with the surface area of the black hole in question expressed in units of the Planck length squared. We suggested that the appropriate generalization of the second law for a region containing a black hole is that the black-hole entropy plus the common entropy in the black-hole exterior never decreases. Here we establish the validity of this law for the infall of an entropy-bearing system into a much larger and more massive generic stationary black hole. To do this we determine a general lower bound for the increase in black-hole entropy, and an upper bound for the entropy of the system, while allowing for quantum effects at each stage. In passing we show that the generalized second law is a statistical law which becomes overwhelmingly probable in the limit of a macroscopic system. We also consider briefly more general situations. Finally, we give two simple examples of predictions made by the generalized second law for black-hole formation processes.

I. INTRODUCTION

Black-hole physics mirrors thermodynamics in many respects. For example, Christodoulou[1] has shown that the efficiencies of processes for extracting energy from a black hole are limited by their irreversibility; the most efficient processes are the "reversible" ones. This result has a clear thermodynamic ring to it which has been well brought out by Carter.[2] Further analogies between black-hole physics and thermodynamics have been noted by the present author,[3,4] by Carter,[5] and by Bardeen, Carter, and Hawking.[6] The formal black-hole analog of the ordinary second law of thermodynamics (OSL), "the entropy of a closed system never decreases," is Hawking's theorem,[7] "the surface area of a black hole never decreases."

We have conjectured[3,8,4] that this last analogy is more than formal, namely, that the area of a black hole expressed in suitable units may be regarded as the black-hole entropy—entropy in the sense of information about the black-hole interior inaccessible to exterior observers. Logically related is a second conjecture,[3,8,4] the generalized second law of thermodynamics (GSL): *The sum of the black-hole entropy and the common (ordinary) entropy in the black-hole exterior never decreases*. Arguments and examples supporting these conjectures are set forth in detail in Ref. 4 (henceforth called paper I) together with reasons for taking the black-hole entropy as

$$S_{bh} = (\tfrac{1}{2}\ln 2)\hbar^{-1}\alpha\,. \tag{1}$$

Here α is the rationalized black-hole area (event-horizon surface area divided by 4π), and $\hbar$ is Planck's constant. Units with $G=c=$ Boltzmann's constant $k=1$ are used so that $\hbar = 2.6\times10^{-66}$ cm^2.

The GSL is the only guise in which the second law has a chance of retaining both its validity *and* its usefulness for regions of the universe near black holes. The OSL is inapplicable just to the exterior of the holes since they are sinks of entropy. And if applied to both interiors and exterior, the OSL is not useful since there is no way for exterior observers to measure the interior ordinary entropy and so to apply the law. We say that the OSL is transcended.[3,4] Not so the GSL, since black-hole entropy can be determined without delving into black-hole interiors. Thus by generalizing the concept "entropy" to include black-hole entropy, one can hope to make the second law valid *and* useful in the presence of black holes. One also hopes that the GSL will provide the sort of rapid and effortless solutions to certain problems that the OSL is noted for in more ordinary contexts. But first it is clearly imperative to establish the general validity of the GSL beyond reasonable doubt; some progress in this direction is reported here.

Considering first the case of a system entering a far larger and more massive generic stationary black hole, we show in several steps that the associated increase in black-hole entropy, ΔS_{bh}, is at least as large as the maximum entropy S_{max} that the system can carry down the hole, as required by the GSL. In Sec. II we find a general lower bound for ΔS_{bh} by treating the system as a small perturbation on the hole, while making allowance for the system's nonzero dimensions which, in the final analysis, are dictated by quantum effects. The result agrees with a less general one obtained in paper I. In Sec. III we obtain an upper bound on S_{max} for a macroscopic system by thermodynamic and quantum considerations, and show that it is in general smaller than the corresponding ΔS_{bh}. In Sec. IV we compute S_{max} for a simple microscopic system by quantum and statistical arguments, and show that in the mean $\Delta S_{bh} \geqslant S_{max}$. Statistical violations of the GSL are possible, but they become highly improbable as the system grows large. In Sec. V we consider situations more complicated than those studied earlier, and we elucidate the physical factors relevant to the operation of the GSL by means of an example. We also consider the prospects for a general proof of the GSL. Finally, in Sec. VI we show with two examples how one can make predictions with the GSL.

II. INCREASE IN BLACK-HOLE ENTROPY

We require a lower bound on the ΔS_{bh} associated with the absorption of a generic system by a much larger and more massive stationary black hole. Before we begin, it is worthwhile pointing out the inadequacy of treating the system as a point particle in our context. If we did this, we could argue, with Christodoulou,[1] Israel,[9] and Bardeen *et al.*,[6] that the increase in black-hole area can be made arbitrarily small (Christodoulou's "reversible" process). The argument is simplest for a Schwarzschild hole: Imagine the system deposited at rest (in stationary coordinates) arbitrarily near the horizon. Its energy (as measured from infinity) is red-shifted away so that in absorbing it the hole experiences no change in mass or area. Indeed, Israel, and Bardeen *et al.*, have inferred from such arguments that ΔS_{bh} can be made negligible compared to the entropy of the system, so that the GSL cannot work. However, as already recognized by Christodoulou, a reversible process is an ideal whose perfect attainment is prevented by the atomicity of matter. This may be unimportant in other contexts, but when the subject is entropy, whose very meaning springs from the atomicity, it cannot be ignored. Specifically, if one allows for the nonzero dimensions of the system required by the quantum nature of matter, then the center of mass of the system cannot be deposited at rest nearer to the horizon than the typical "radius" of the system, and the increase in area will no longer vanish. We shall now obtain a very general lower bound for this increase and show that it suffices to make the GSL work. An alternative but less general discussion leading to the same results has appeared in paper I.

We first recall some facts about black holes.[7,10] A generic black hole is a region of spacetime enclosed in an event horizon (one-way membrane), which is a null hypersurface generated by a congruence of null geodesics characterized locally by a (real) convergence ρ and shear σ. The proper 2-area δA of an element of horizon changes according to

$$\frac{d\delta A}{dv} = -2\rho\delta A\,, \tag{2}$$

where v is the affine parameter of a typical local generator. In turn ρ satisfies

$$\frac{d\rho}{dv} = \rho^2 + |\sigma|^2 + 4\pi T_{\beta\gamma} l^\beta l^\gamma\,, \tag{3}$$

where $T_{\beta\gamma}$ is the stress-energy tensor of the matter at the horizon, and $l^\beta = dx^\beta/dv$ is the (null) tangent vector to the local generator (as well as the outgoing normal to the horizon). We assume the weak energy condition[7]: $T_{\beta\gamma} l^\beta l^\gamma \geqslant 0$.

If we calculate $d^2\delta A/dv^2$ from (2), eliminate first derivatives with (2) and (3), integrate (over area) for given v, and then over v from v to $v=\infty$, we get

$$\frac{dA}{dv} = 2\int_v^\infty dv' \int_H (4\pi T_{\beta\gamma} l^\beta l^\gamma + |\sigma|^2 - \rho^2)\delta A(v')\,. \tag{4}$$

Here H is a spacelike section of the horizon for fixed v and $A(v) = \int_H \delta A(v)$ is the total area of this section; we have assumed that $dA/dv \to 0$ as $v \to \infty$. Thus far the formalism can describe both a system collapsing to form a black hole and one falling into a pre-existing hole. To make the problem tractable we specialize to the latter case and consider only a stationary black hole (possibly more general than Kerr) and a system which makes only a small perturbation on it, i.e., one of small mass and moderate density. We may then regard $T_{\beta\gamma}$ as a first-order quantity. Now both σ and ρ vanish for a perfectly stationary hole.[7,10] Thus for our perturbed hole both σ and ρ will be first-order quantities, so that $|\sigma|^2 - \rho^2$ may be neglected in (4) to first order.

The system first makes contact with the horizon at some v, say $v=0$. We obtain a *lower bound* for the total increase in area by integrating (4) from $v=0$ to $v=\infty$ (Ref. 11):

$$\Delta\alpha \geqslant 2\int_0^\infty dv \int_v^\infty dv' \int_H T_{\beta\gamma} l^\beta l^\gamma \delta A(v')\,. \tag{5}$$

Here $\alpha = A/4\pi$ is the rationalized area. Let us now reexpress (5) in terms of the rest mass and dimensions of the system.

We assume that the system is small in size compared to the hole. Then we may work in a locally inertial frame (not necessarily the rest frame) whose origin lies within the system as the latter crosses the horizon. Let the frame be equipped with an orthonormal tetrad λ_a^ν such that $\lambda_a^\nu \lambda_{b\nu}=\eta_{ab}$ (Minkowski metric).[12] The λ_0^ν, the time axis, is the 4-velocity of the frame; λ_i^ν for $i=1,2,3$ is the space triad. For a generator which intersects the system we may write (for the interval encompassed in the frame)

$$l^\nu = K(\lambda_0^\nu + n^i \lambda_i^\nu), \tag{6}$$

where K and n^i are four appropriate parameters. Since $l^\nu l_\nu=0$, $n^i n^i=1$. Now in our particular frame $l^0=K$, $l^i=Kn^i$, and the Christoffel symbols vanish. Then the geodesic equation for a generator shows that K and n^i are constants along each generator. Due to the smallness of the system, they will also be constant from generator to generator inside the system to within fractional errors of the order of the size of the system divided by that of the hole.

We shall choose λ_0^ν to be orthogonal to one of the (constant v) H surfaces at an event within the system. To the same approximation used above, λ_0^ν will be orthogonal to the H surface at all points within the system (if H is not pathological) so that H will be a constant-time surface. Now on a given generator two H surfaces differing by dv are separated by the interval $dx^\nu = l^\nu dv$. By the constancy of K, the time difference between them, $dt=dx^0=Kdv$, is constant within the system. Hence, all relevant H surfaces are constant-time surfaces within the system (and λ_0^ν is orthogonal to all) in our approximation. Since l^ν is the null outgoing normal to the H's, it is clear from (6) that the n^i are the components of the unit spacelike (outgoing) normal to these surfaces. Thus the normal distance dx between H surfaces differing by dv is also Kdv, and the surfaces are parallel within the system.

We now introduce $T_{ab}=T_{\beta\gamma}\lambda_a^\beta \lambda_b^\gamma$, the ordinary components of the stress-energy tensor in our frame. In terms of these

$$T_{\beta\gamma} l^\beta l^\gamma = K^2(T_{00} + 2T_{0i}n^i + T_{ij}n^i n^j), \tag{7}$$

which we substitute in (5). Consider first $\int_H T_{0i} n^i \delta A$. Since $T_{ab}=0$ outside the system, the integral is unchanged if extended to a closed constant-time 2-surface enclosing that part of the system *outside* the horizon, and coinciding with H inside the system. On H, n^i is the *inward* normal to this surface; we redefine n^i to mean this same normal for the rest of the surface, and δA to mean the corresponding area element. Then by Gauss's theorem the integral is just $-\int_V T_{0i,i}dV$, where V denotes the volume of the part of the system outside the horizon at the given time. Conservation of energy as expressed in our frame implies that $T_{0i,i}=\partial T_{00}/\partial t$. Thus

$$\int_H T_{0i} n^i \delta A = -\frac{\partial}{\partial t}\int_V T_{00} dV - \int_H T_{00}\delta A, \tag{8}$$

where we allow for the facts that V decreases in time, while $dt=dx$ for the interval between H surfaces.

Now consider $\int_H T_{ij} n^i n^j \delta A$. It is consistent with our earlier remarks to regard the n^i as constants everywhere inside the system at a given time. Using $T_{ij,j}=\partial T_{i0}/\partial t$ and repeating the above procedure we get

$$\int_H T_{ij} n^i n^j \delta A = -\frac{\partial}{\partial t}\int_V T_{i0} n^i dV - \int_H T_{i0} n^i \delta A. \tag{9}$$

But

$$\int_V T_{i0} n^i dV = \int_x^\infty dx' \int_\Sigma T_{i0} n^i dA, \tag{10}$$

where dA is the area element on a 2-surface $\Sigma(x')$ parallel and simultaneous to H, x' measures distance normal to such surfaces starting from $x'=0$ at the innermost point of the system, and the integral over x' goes from the H surface (at $x'=x$) to the outermost point of the system. As before, we transform the surface integral in (10) to one over the volume V' of the system lying outward of $\Sigma(x')$. Since $dx=dt$ we have

$$\int_V T_{i0} n^i dV = -\frac{\partial}{\partial t}\int_x^\infty dx' \int_{V'} T_{00} dV' - \int_V T_{00} dV. \tag{11}$$

Substituting (8) and (9) into the integral of (7) and eliminating all reference to T_{0i} by means of (8) and (11) we get

$$\int_H T_{\beta\gamma} l^\beta l^\gamma \delta A = K^2 \frac{\partial^2}{\partial t^2}\int_x^\infty dx' \int_{V'} T_{00} dV'. \tag{12}$$

To integrate over v as in (5) we replace Kdv by dt in accordance with our earlier remark, and the integrals become trivial. After an integration by parts we get

$$\Delta\alpha \geq 2\int_V x T_{00} dV, \tag{13}$$

where the integral is evaluated at $v=0$, i.e., at the moment when the system first makes contact with the horizon; $x=0$ labels points of the system right at the horizon. In making use of an inertial frame in the calculations, we have assumed implicitly that the self-gravity of the system is negligible. Then T_{00} represents the energy density of the system, and according to (13) $\Delta\alpha$ is at least twice the product of the system's relativistic energy $\mathcal{E}$ and the normal distance $\bar{x}$ from its center of mass (c.m.) to the horizon at $v=0$ ($t=0$), all quantities measured in the inertial frame.

The $\mathcal{E}$ exceeds the system's *rest* mass μ by a Lorentz factor $\gamma \geq 1$ corresponding to the relative motion between the c.m. and the frame. We can make this motion have no component in the x direction (normal to the H surfaces) by applying to our frame an appropriate Lorentz boost in that direction. This does not affect the constant-time character of the surfaces, or the argument leading to (13). Because there is no Lorentz contraction of the system in the x direction in the *new* inertial frame, $\bar{x}=b$ where b is the normal distance from the c.m. to the horizon at the instant of contact as measured in the *rest* frame of the system. Thus we get our final result,

$$\Delta\alpha \geq 2\mu b. \tag{14}$$

Equation (14) agrees with the less general result (A15) obtained in paper I by a different approach. Here the result does not depend on specific properties of Kerr holes (applying also to holes surrounded by massive disks, etc.), on the shape of the system, or on its motion (it may be subject to forces). We have already shown in paper I that the lower bound $\Delta\alpha = 2\mu b$ is actually attainable.

Combining (1) and (14) we get

$$\Delta S \quad \geq \mu b \hbar^{-1} \ln 2. \tag{15}$$

For a macroscopic system b, which is essentially an effective radius, must be enormous compared to the Compton lengths of the constituent particles, which themselves are large compared to $\hbar/\mu$. Therefore $\Delta S_{bh} \gg 1$. Thus, while the minimum increase in black-hole area [Eq. (14)] can be quite minute (10^{-19} cm^2 for the typical automobile), the corresponding increase in black-hole entropy [Eq. (15)] is always large in natural ($k=1$) units ($\simeq 10^{46}$ for the automobile). Failure to allow for this point on the part of Israel,[9] and Bardeen *et al.*,[6] vitiates their final conclusion. We now go on to compare ΔS_{bh} to the entropy carried by the system down the hole.

III. CHANGE IN GENERALIZED ENTROPY: THERMODYNAMIC ANALYSIS

We call the sum of S_{bh} and the common entropy in the black-hole exterior the generalized entropy S_g. For the process in question $\Delta S_g = \Delta S_{bh} - S$, where S is the common entropy carried by the system down the hole. The following demonstration that $\Delta S_g > 0$ applies to a closed macroscopic system describable by ordinary thermodynamics. We first define a characteristic temperature $T_b = \hbar(b \ln 2)^{-1}$. If our system is odd-shaped, we take for b the smallest "radius" of the system. Since for a macroscopic system b is large compared to the Compton length of any of its constituent nuclei or elementary particles, $T_b/m \ll 1$, where m is the rest mass of any such particle. Hence for absolute temperatures $T \leq T_b$, the thermal motions within the system are nonrelativistic. (The case of massless particles is treated later.)

We may express the entropy of the system at absolute temperature T as

$$S = S_0 + \int_0^T C(T')\, T'^{-1}\, dT', \tag{16}$$

where S_0 is its ground-state entropy (at $T=0$), and $C(T)>0$ is its heat capacity at constant volume and analogous parameters. Likewise, the rest-mass energy μ is just

$$\mu = E_0 + \int_0^T C(T')dT', \tag{17}$$

where E_0 is the ground-state energy (including rest masses).[13] From Eqs. (15)–(17) we have

$$\Delta S_g \geq E_0/T_b - S_0 + \int_0^T C(T')(T_b^{-1} - T'^{-1})dT'. \tag{18}$$

The last integral reaches its absolute minimum at $T=T_b$; thus for any T,

$$\Delta S_g \geq E_0/T_b - S_0 + \int_0^{T_b} C(T)(T_b^{-1} - T^{-1})dT. \tag{19}$$

This new integral involves only the range $0 \leq T \leq T_b$ for which the thermal motions are nonrelativistic.

We now show that S_0 is negligible compared to E_0/T_b. In general $S_0 = \ln g$, where g is the degree of degeneracy of the ground state; in practice one deals with (nuclear or particle) spin degeneracy. Suppose we have to do with N_n "nuclei" of spin $\frac{1}{2}\hbar$; then $g = N_n{}^2$ and $S_0 = N_n \ln 2$. If m_n is the mass of a nucleus, then $E_0 \geq N_n m_n$ since other particles, i.e., electrons, may be present.[14] As mentioned earlier $m_n/T_b \gg 1$; thus $S_0 \ll E_0/T_b$. If we try to circumvent this condition by considering nuclei of large spin $s\hbar$ so that $S_0 = N_n \ln(2s+1)$, we are soon thwarted. For $s \gg 1$ the spin is quasiclassical. We may then apply the result of Møller[15] that a spinning particle of mass m_n and spin $s\hbar$ has an effective radius no smaller than $s\hbar/m_n$. By implication $b \gg s\hbar/m_n$, $m_n/T_b \gg s$, and so $E_0/T_b \gg N_n s$. For large s, $s > \ln(2s+1)$ so that again $S_0 \ll E_0/T_b$. Thus we neglect S_0 in (19).

Now consider the integral in (19). For moderately high T, $C(T)$ varies slowly over each range of T corresponding to a given phase,[16] and is of the order of the number of effective degrees of freedom of the system at that T. This last number will never exceed the number of elementary particles in the system by more than a factor of order unity. For low T various degrees of freedom become ineffective or are "frozen out" (a strictly quantum effect), and $C(T)$ drops, ultimately vanishing as $T\to 0$ in harmony with the third law of thermodynamics.[17] It is consistent with all known results to assume that $C(T)$ vanishes at least as rapidly as T. It is then possible to define an "envelope" for C:

$$\mathfrak{C}(T)=\begin{cases} C_m T/T_c, & \text{for } T<T_c \\ C_m, & \text{for } T\geq T_c\,. \end{cases} \tag{20}$$

Here C_m is the largest value of C in $0\leq T\leq T_b$, and T_c is the largest parameter for which $\mathfrak{C}(T)\geq C(T)$ in that range. Depending on the value of T_b two basic cases are possible, one with $T_c=T_b$, the second with $0<T_c<T_b$. Both are illustrated in Fig. 1. Now the integrand in (19) is negative; hence by replacing $C(T)$ by $\mathfrak{C}(T)$ we obtain a lower bound for it. Thus

$$\Delta S_g\geq E_0/T_b-C_m[\tfrac{1}{2}T_c/T_b+\ln(T_b/T_c)]\,. \tag{21}$$

First we consider the case with $T_c=T_b$. Let $\overline{m}$ be the mean mass of an elementary particle defined by $E_0=N\overline{m}$, where N is the total number of such particles. As before $\overline{m}/T_b\gg 1$ so that $E_0/T_b\gg N$. We remarked that C/N is bounded from above by a number of order unity; likewise for C_m/N. Thus we see from (21) that $\Delta S_g>0$ since $E_0/T_b\gg\frac{1}{2}C_m$. The same argument will clearly go through if T_b/T_c exceeds 1 by a not very large factor (say up to 10; $\ln 10\approx 2.3$ only). This case is relevant when the graph of $C(T)$ is concave downward and T_b lies below the leveling-off point T_0.

Now consider the other cases under the label $T_c<T_b$. Just as above we see that $D\equiv\frac{1}{2}C_m T_c/T_b \ll E_0/T_b$, so we neglect D in (21). We now obtain an upper bound for T_b/T_c. Examples such as that in Fig. 1 show that T_c is either near the point T_0 from which $C(T)$ takes its last plunge to zero, or else is much larger. The final drop in $C(T)$ reflects the freezing out of the last group of degrees of freedom. It is well known that a degree of freedom is frozen out when T becomes comparable to or smaller than the energy spacing of its lower quantum states. Hence T_0 is no smaller than the energy spacing of the lower levels of the degrees of freedom in question. Let us estimate this spacing.

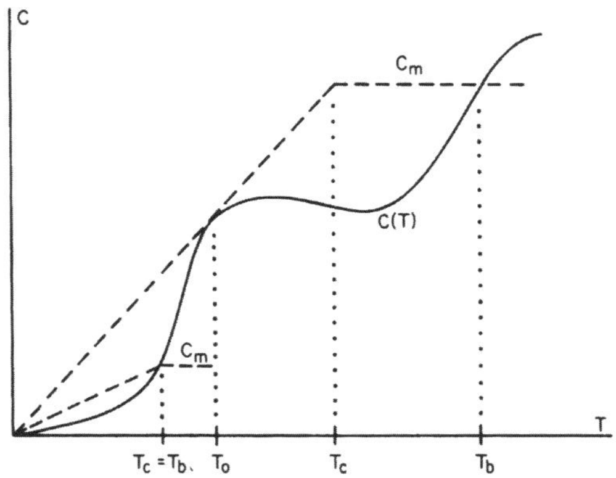

FIG. 1. Typical heat capacity $C(T)$ (solid line) and its "envelope" $\mathfrak{C}(T)$ (dashed lines) for two choices of T_b; T_0 is the point from which $C(T)$ starts its final drop to zero.

We may regard a degree of freedom as some motion involving an effective mass m_e and confined to a region of dimensions a. Order-of-magnitude estimates via WKB or uncertainty-principle arguments indicate that the energy spacing of the lower levels is of order $\hbar^2(m_e a^2)^{-1}$. Moreover, comparison with precise results shows that this estimate is correct for such varied degrees of freedom as translational motion, rotation and vibration of molecules, electronic motion, etc. It is clear that for our system $a\lesssim b$ so we may take $T_0 \gtrsim\hbar^2(m_e b^2)^{-1}$. Since $T_c\gtrsim T_0$, $T_b/T_c\lesssim(\ln 2)^{-2}m_e/T_b$. As usual $m_e/T_b\gg 1$ so that $C_m\ln(T_b/T_c)\lesssim(C_m/N) N\ln(m_e/T_b)$. This last term is seen to be small compared to $E_0/T_b=N\overline{m}/T_b$ if, as in ordinary systems, the effective masses m_e of the last effective degrees of freedom do not exceed $\overline{m}$ by more than a few orders of magnitude. Then according to (21) $\Delta S_g>0$.

What if $m_e/\overline{m}$ is sufficiently large to upset the last part in the previous argument? Consider a series of systems with fixed E_0, $\overline{m}$, and b, but with successively larger m_e. It is clear that the number of degrees of freedom in question, which roughly equals $C(T_0)$, must scale like m_e^{-1} if m_e gets really large. Then Fig. 1 shows that T_c/T_0 will scale as m_e. Since $T_0\propto m_e^{-1}$, T_c will be essentially unchanged as m_e is scaled up. It follows that increasing m_e cannot affect any term in (21) or our final conclusion that $\Delta S_g>0$.

Finally, we consider a system with $E_0=0$; the prototype is black-body radiation for which $S_0=0$ as well. Equation (18) seems to say that for $0<T\lesssim T_b$, $\Delta S_g<0$ in violation of the GSL. However, thermodynamic methods are actually inapplicable in this range due to the prevalence of quantum fluctuations.[4] Specifically, fluctuations are negligible and ordinary thermodynamics applicable to a system only if $T \gg \hbar/\tau$, where τ is a characteristic time for change if the system is disturbed from equilibrium.[17] For black-body radiation confined to dimensions b, $\tau \approx b$ and we need $T \gg T_b$ for thermodynamics to apply. If indeed $T \gg T_b$, ΔS_g is positive as is evident from (18) with $C(T) \propto T^3$. By contrast, if $T \lesssim T_b$, the entropy is ill defined[17] and the GSL is made meaningless by the fluctuations. To conclude, we have shown that for a system describable by thermodynamics, $\Delta S_g > 0$ in agreement with the GSL.

IV. CHANGE IN GENERALIZED ENTROPY: STATISTICAL ANALYSIS

Thus far our arguments have not depended critically on the choice $\frac{1}{2}\ln 2$ for the numerical coefficient in formula (1) for S_{bh}. Any similar but not much smaller number would have served just as well. We now show that the above choice is unique in allowing "generalized reversible" processes for which $\Delta S_g = 0$ as well as those with $\Delta S_g > 0$. For a system describable by thermodynamics $\Delta S_g = 0$ is not a real possibility: We made it clear that the positive term in ΔS_g far outweighs the negative ones. Hence we must consider microscopic systems, i.e., systems composed of a small number of quanta. We regard such a system as a small wave packet. The results of Sec. II will still apply if we think of $T_{B\gamma}$ as the expectation value of the corresponding normal-ordered quantum field operator.[18]

The lower bound for the ΔS_{bh} associated with the absorption of the system by the hole is given by (15). It should be attainable—at least in the case of a Kerr hole—as shown in paper I. We first suppose that the packet contains N identical bosons of mass m, all in the same quantum state. For some packets $b = \hbar/m$ should be a real possibility (it would correspond to a sort of minimum-uncertainty packet); otherwise $b > \hbar/m$. If we neglect interactions, $\mu = Nm$ so that according to (15)

$$\Delta S_{bh} \geq N \ln 2 . \tag{22}$$

Since m does not appear in (22), this relation should remain valid even for the case $m = 0$ to which our arguments are not strictly applicable. According to (22), ΔS_{bh} is not smaller than the equivalent of one "bit" per quantum. It was this feature which prompted us in paper I to choose the coefficient in (1) as we did.

Once the state of the quanta is agreed on, the packet is uniquely determined by N. We endow it with entropy by regarding N as a random variable subject to some probability distribution $\{p_N\}$. Clearly $p_0 = 0$ because no quanta corresponds to no system. The statistical mean of any function $f(N)$ is just $\langle f \rangle = \sum_N p_N f(N)$; the entropy associated with the distribution is just (see paper I)

$$S = -\sum_N p_N \ln p_N . \tag{23}$$

We are especially interested in that normalized distribution which maximizes S for given $\langle N \rangle$. It is easily found by the method of Lagrange multipliers to be

$$p_N = (e^\beta - 1)e^{-\beta N}, \quad N \geq 1 \tag{24}$$

with $\beta = \ln\langle N \rangle - \ln\langle N-1 \rangle$. From (23) it follows that

$$S_{max} = \langle N \rangle \ln\langle N \rangle - \langle N-1 \rangle \ln\langle N-1 \rangle . \tag{25}$$

One easily finds now that the quantity $\langle N \rangle \ln 2 - S_{max}$ is non-negative and vanishes only for $\langle N \rangle = 2$.

We conclude that $\langle \Delta S_g \rangle \equiv \langle \Delta S_{bh} \rangle - S \geq 0$ with the equality possible only if the lower bound in (15) is attained, if $b = \hbar/m$, and if the distribution (24) is realized with $\langle N \rangle = 2$. Note that any choice other than $\frac{1}{2}\ln 2$ in (1) would either result in $\langle \Delta S_g \rangle$ being negative sometimes, or would exclude the possibility $\langle \Delta S_g \rangle = 0$ altogether. Thus $\frac{1}{2}\ln 2$ is the "best" choice. The result $\langle \Delta S_g \rangle \geq 0$ means that the GSL is always satisfied in the mean (weak form of GSL). It need not always be satisfied in particular instances. For example, if the distribution (24) is realized with $\langle N \rangle = 2$ ($S_{max} = 2\ln 2$), the strong form of the GSL, $\Delta S_{bh} - S \geq 0$, can break down in an instance for which in fact $N = 1$, since $\Delta S_{bh} < 2\ln 2$ is then possible. Thus the GSL is a statistical law susceptible to violations due to large fluctuations. In this it is no different from the OSL. In the macroscopic limit of large $\langle N \rangle$ the strong form of the GSL becomes a statistical certainty. This is seen as follows. For large $\langle N \rangle$, $S_{max} \approx \ln\langle N \rangle$. Thus the probability p_v that $\Delta S_{bh} - S_{max} < 0$ is no greater than the sum of the p_N from $N = 1$ to the next larger integer to $(\ln\langle N \rangle)/\ln 2$. One easily finds that $p_v \lesssim (\ln 2)^{-1}(\ln\langle N \rangle)/\langle N \rangle$. Thus a violation of the strong form of the GSL becomes improbable for large $\langle N \rangle$. This again has a familiar ring to it.

When the packet contains different types of quanta in different states, the lower bound (22) for ΔS_{bh} will not be attainable in general since $b \geq \hbar/m$ for the smallest m while μ contains all the various m. In order for the system to be specified by the various occupation numbers, the available one-quantum states must be agreed on and the information conveyed with the system. The simplest way to do this is to always have at least one quantum occupying each available state, i.e., $p_0=0$ for each state. The entropy is maximized, first by taking the distributions for various states as statistically independent, and then by choosing (24) for each (boson) quantum state. The previous analysis can be applied and leads to the conclusion that $\langle\Delta S_g\rangle>0$ and that ΔS_g is almost always positive for large occupation numbers. In this way we again demonstrate the validity of the GSL independently of Sec. III.

V. MORE GENERAL SITUATIONS

One may be interested in situations more general than those discussed above. For example, one might ask whether the GSL holds for the case of some tenuous matter accreting onto a stationary black hole from all directions. One might think that the law would hold if each small element of matter causes an increase in black-hole area at least equivalent to its own entropy. But this condition may be neither necessary nor meaningful. First, it might be impossible to identify the increase in area ascribable to a given element; however, this identification is possible in the case of steady-state accretion. Second, the fluid description of the matter assumed above may be inappropriate: In some regime the smallest representative elements may be larger than the hole. And third, entropy may be generated in the blackhole exterior as an organic part of the whole process; this extra entropy cannot be ignored. We illustrate some of these points with an example.[3]

Consider a Kerr hole of mass M immersed in a black-body radiation "bath" within a cavity of temperature T. We assume that the cavity does not rotate so that the radiation has zero angular momentum, and that the process of accretion has reached a steady state. Then [see paper I, Eqs. (8), (16), and (18)] to radiation of energy E (as measured from infinity) going down the hole there corresponds $\Delta S_{bh}=E/T_{bh}$, where T_{bh} is a certain characteristic temperature. For given M, $T_{bh} \leq \frac{1}{4}\hbar(M\ln 2)^{-1}$. Now far from the hole the entropy-to-energy ratio of the radiation will be the usual one, $\frac{4}{3}T^{-1}$.[17] Since the accretion onto the hole takes place in steady state, it is plain that together with energy E, entropy $S=\frac{4}{3}ET^{-1}$ flows in. Thus for $T\geq\frac{4}{3}T_{bh}$, $\Delta S_{bh}-S\geq 0$ and the GSL is valid for each stage of the inflow.

One should not conclude that for $T<\frac{4}{3}T_{bh}$ the GSL is violated because in this regime a new feature must be taken into account. Indeed, for $T\lesssim\frac{4}{3}T_{bh}$, the typical wavelength in the radiation, $\hbar/T$, is already larger than the size of the hole $\leq 2M$. Thus the radiation no longer flows in as a fluid would, but rather most wavelengths in it must tunnel into the hole individually. It is clear that the rate of tunnelling is sensitive to the wavelength: The shorter the wavelength, the higher the rate. In steady state, the spectrum of the radiation must harmonize with this rate, and so will no longer be of Planck type as for $T\gg T_{bh}$ (it would be of Planck type only if the rate were wavelength-independent). This means that the radiation cannot be in thermodynamic equilibrium with the cavity, for if it were the spectrum would be of Planck type. We may conclude that irreversible processes will go on in the cavity material (and perhaps in the radiation also), which will generate additional entropy. This extra entropy must be allowed for in applying the GSL which, in the spirit of the second law, need hold only for a closed system: black hole, radiation, *and* cavity. To prove that the extra entropy suffices to make the GSL work in the regime in question would be a difficult problem in nonequilibrium statistical mechanics which we shall not pursue. But it is clear that the various physical factors conspire in favor of the GSL.[19]

One is also interested in whether the GSL holds for the process of black-hole formation. Since one must here give up the assumption of a stationary hole, this case is certainly the most difficult to discuss. What sort of framework would be required to construct a proof of the GSL for this, or for that matter, for any situation more complicated than those treated in Secs. II–IV? We have repeatedly seen that quantum effects play a prominent role in the operation of the law: The GSL is at its roots a quantum law. We have also seen that the GSL is a statistical law. It seems, therefore, that such a proof could be obtained only in the context of quantum-statistical mechanics in curved spacetime. Since this discipline does not yet exist, we shall simply have to assume, on the strength of the arguments of paper I, and in the absence of evidence to the contrary, that the GSL is valid generally.

VI. PREDICTING WITH THE GSL

One may make predictions with the GSL; two simple examples follow. Consider first a cold nonrotating assembly of N nucleons. If its mass exceeds a critical mass of the order of $2M_\odot$, it will collapse spontaneously, presumably to form a Schwarzschild black hole. However, in principle smaller masses can also be induced to collapse. One way to accomplish this[20] would be to artificially compress the central portion of the system into a "seed" black hole which would then gobble up the rest of the nucleons. The energy expended in "seeding" can be made negligible. The initial common entropy of the system is at least as large as the spin-degeneracy entropy of the spin-$\frac{1}{2}$ nuclei; thus $S \geq N \ln 2$ (see Sec. III). If the mass of the final Schwarzschild hole is M, then by (1)

$$S_{bh} = 2M^2 \hbar^{-1} \ln 2 . \tag{26}$$

The GSL predicts that $S_{bh} - S \geq 0$, or $M^2 \geq \frac{1}{2} N \hbar$. Because of the binding energy of the initial system and the possible mass loss to gravitational waves, we must have $M < Nm$, where m is the nucleon rest mass. These relations imply that $N > \frac{1}{2}\hbar/m^2$ and $2M > \hbar/m$.

Thus according to the GSL, induced collapse to a black hole is possible only if no fewer than $\frac{1}{2}\hbar/m^2 \approx 10^{38}$ nucleons participate. This corresponds to $M > 10^{-19} M_\odot$ (10^{14} g) and a gravitational radius exceeding the Compton length of the nucleon. This is reasonable result, for if $2M < \hbar/M$, the "seed" hole would necessarily be smaller than the Compton length, and the nucleons would have great difficulty in falling in.[21] Collapse could then take place only via quantum tunnelling, a very improbable process.[21] If such collapse did take place nevertheless, it could be classed as a statistical violation of the GSL.[22]

As a second example we consider a spherical thermal geon, that is, a spherical body of thermal radiation holding itself together by its own gravitational field.[23] The local temperature within it will not be strictly constant due to gravitational effects, but we shall replace it by an effective constant value T. The total radiation energy E is not a very precise concept when strong gravitational fields are present, but we shall use it nevertheless. Now the total entropy of the radiation should be approximately $S = \frac{4}{3} E T^{-1}$ (same as for ordinary thermal radiation). If our geon collapses to a Schwarzschild hole, its mass M will be conserved. The S_{bh} will again be given by (26), and the GSL now predicts that

$$M^2 \geq \tfrac{2}{3} \hbar E (T \ln 2)^{-1} = 0.96 \hbar E T^{-1} . \tag{27}$$

The geon's effective radius R must clearly exceed the characteristic wavelength in the *radiation* $\approx \hbar/T$. Hence (27) is seen to be a *necessary* condition for M^2/R, the geon's gravitational binding energy, to exceed E. That is, (27) is a necessary condition for the geon to be able to collapse rather than explode.

We have seen how the GSL makes very reasonable (if not astounding) predictions for processes of black-hole formation. This success supports the view that it must be generally valid for such processes.

Note added in proof. The argument leading to Eq. (14) is perfectly valid if the system's intrinsic angular momentum $\vec{s}$ is zero. If $\vec{s} \neq 0$ the position of the c.m. as computed in the inertial frame is at a distance $y = |\vec{v} \times \vec{s}| \mu^{-1}$ from that computed in the rest frame, where $\vec{v}$ is the relative velocity between the frames [C. W. Misner, K. S. Thorne, and J. A. Wheeler, *Gravitation* (Freeman, San Francisco, 1973), p. 161]. However, for macroscopic systems $y \ll b$, so in this case at least the effect of $\vec{s}$ is negligible. To see why $y \ll b$ consider first macroscopic rotation, which necessarily involves velocities much smaller than unity. Then $|\vec{s}| \ll \mu b$, so that $y \ll b$. If instead of macroscopic rotation we have all N spins aligned, $|\vec{s}| \approx \frac{1}{2} N \hbar$. But $b \gg \hbar/m \approx N\hbar/\mu$, so that again $y \ll b$.

---

*Work supported in part by National Science Foundation under Grant No. GP-32039.

[1]D. Christodoulou, Phys. Rev. Lett. 25, 1596 (1970); Ph.D. thesis, Princeton University, 1971 (unpublished).

[2]B. Carter, Nat. Phys. Sci. 238, 71 (1972).

[3]J. D. Bekenstein, Ph.D. thesis, Princeton University, 1972 (unpublished).

[4]J. D. Bekenstein, Phys. Rev. D 7, 2333 (1973).

[5]In *Black Holes*, edited by B. S. DeWitt and C. M. DeWitt (Gordon and Breach, New York, 1973).

[6]J. M. Bardeen, B. Carter, and S. W. Hawking, Commun. Math. Phys. 31, 161 (1973).

[7]S. W. Hawking, Phys. Rev. Lett. 26, 1344 (1971); Commun. Math. Phys. 25, 152 (1972), and Ref. 5.

[8]J. D. Bekenstein, Nuovo Cimento Lett. 4, 737 (1972).

[9]W. Israel, Nuovo Cimento Lett. 6, 267 (1973).

[10]S. W. Hawking and J. B. Hartle, Commun. Math. Phys. 27, 283 (1972).

[11]We ignore the increase for $v < 0$ which need not vanish.

[12]Generic tetrad indices are denoted by a or b, space tetrad indices by i or j, and time one by 0.

[13]Our implicit assumption of thermodynamic equilibrium is no restriction: For nonequilibrium S is smaller than at equilibrium. Thus if $\Delta S_g > 0$ for the equilibrium system, then automatically $\Delta S_g > 0$ for the nonequilibrium one.

[14]Ordinary binding energies are negligible; the possible exception is gravitational binding energy which we already assumed to be negligible.

[15]C. Møller, Commun. Dublin Inst. Adv. Stud. A 5 (1949).

[16]The specific heat at constant pressure is singular at phase transitions, but that at constant volume is not.

[17]L. D. Landau and E. M. Lifshitz, *Statistical Physics* (Addison-Wesley, Reading, Mass., 1969), 2nd edition, pp. 66, 344, 165.

[18]By calculating in an inertial frame we hope to bypass the thorny problem of quantization in curved spacetime.

[19]Another example involving an apparent violation of the second law was discussed in Ref. 8. The resolution given there was rightly criticized by L. Basano and

A. Morro, Nuovo Cimento Lett. 6, 193 (1973), and by Israel (Ref. 9). However, we have independently given in paper I a different resolution to which their objections are inapplicable.

[20]Ya. B. Zel'dovich and I. D. Novikov, *Stars and Relativity* (Univ. of Chicago Press, Chicago, 1971), p. 295.

[21]The existence of this "Compton barrier" has been demonstrated in detail by J. A. Wheeler, in *Cortona Symposium on Weak Interactions*, edited by L. Radicati (Accademia Nazionale dei Lincei, Rome, 1971).

[22]The existence of a nucleon number below which, classically, induced collapse will not occur has been inferred in a different way by B. K. Harrison, K. S. Thorne, M. Wakano, and J. A. Wheeler, *Gravitation Theory and Gravitational Collapse* (Univ. of Chicago Press, Chicago, 1965), p. 82.

[23]J. A. Wheeler and E. A. Power, Rev. Mod. Phys. 29, 480 (1957).

LETTERE AL NUOVO CIMENTO VOL. 11, N. 9 2 Novembre 1974

The Quantum Mass Spectrum of the Kerr Black Hole (*).

J. D. BEKENSTEIN

Center for Relativity Theory, University of Texas at Austin - Austin, Tex.

(ricevuto il 3 Aprile 1974)

Although the evolution of quantum fields in a black-hole background has been considered occasionally [1,2], a black hole has always been regarded as a classical entity devoid of quantum dynamics. Such a viewpoint is clearly only an approximation: a black hole must be subject to quantum laws. However, a rigorous quantum discussion of black holes must await the development of a complete quantum theory of gravitation. Nevertheless, it is possible to consider heuristically some quantum features of black holes by means of well-understood principles and arguments from ordinary quantum theory. By this approach we now discuss the possibility that the mass spectrum of the Kerr black hole may be discrete, and we infer a formula for the allowed mass levels. We also consider some applications of this formula.

The Kerr black hole is the most general stationary black hole with an exterior devoid of foreign matter; it is axisymmetric [3]. Classically it is completely parametrized by its mass-energy M, charge Q and angular momentum $\boldsymbol{J}$ [3]. Important quantities for any black hole are the surface area A [4] and the squared irreducible mass $M_{\text{ir}}^2 \equiv A/16\pi$ [5]. For the Kerr hole [5,6]

$$M^2 = M_{\text{ir}}^2(1 + \tfrac{1}{4}Q^2/M_{\text{ir}}^2)^2 + \tfrac{1}{4}\boldsymbol{J}^2/M_{\text{ir}}^2 \,. \tag{1}$$

In the quantum theory the operators representing $\boldsymbol{M}$, Q and the component of $\boldsymbol{J}$ along the symmetry axis J_z should play an important role. A stationary state is always an eigenstate of the energy operator [7]; hence the Kerr hole should be in an eigenstate of M. We expect the Kerr hole to have a definite charge; hence it should be in an eigen-

---

(*) Partially supported by N.S.F. grant GP 32039.

(1) W. UNRUH: to be published.

(2) S. W. HAWKING: *Nature*, in press.

(3) C. W. MISNER, K. S. THORNE and J. W. WHEELER: *Gravitation* (San Francisco, Cal., 1973).

(4) S. W. HAWKING: *Phys. Rev. Lett.*, **26**, 1344 (1971).

(5) D. CHRISTODOULOU: *Phys. Rev. Lett.*, **25**, 1596 (1970); D. CHRISTODOULOU and R. RUFFINI: *Phys. Rev. D*, **4**, 3552 (1971).

(6) We use units with $G = c = 1$.

(7) For example, see E. MERZBACHER: *Quantum Mechanics* (New York, N. Y., 1970).

state of the operator Q with the usual eigenvalues

$$Q = ke\,, \qquad k = 0, \pm 1, \dots, \tag{2}$$

where e is the proton charge. Finally, in quantum theory an axisymmetric state is always an eigenstate of the rotation generator J_z (7); hence the Kerr hole should be in an eigenstate of J_z with the usual eigenvalues

$$J_z = j\hbar\,, \qquad j = 0, \pm \tfrac{1}{2}, \dots . \tag{3}$$

Are other operators necessary for the complete description of a Kerr state? Classically $\boldsymbol{J}^2 = J_z^2$; this is no longer true for the quantum operators (7). Is the Kerr state also an eigenstate of the operator $\boldsymbol{J}^2$? We might be reluctant to admit this in view of the apparent lack of spherical symmetry of the system. However, the Kerr space-time possesses an unexpected subtle symmetry reflected in the separability of various wave equations in the Kerr background (8). Might the quantum version of this symmetry be the statement that $\boldsymbol{J}^2$ is a good quantum number for any Kerr state? Until this question is answered conclusively, it appears safest to assume that M, Q and J_z completely describe the generic Kerr state.

It has become clear by now that a black hole behaves in many repects like an ordinary bound system (3). Now in quantum theory a bound system generally possesses a discrete energy spectrum (7); thus by analogy we expect the mass spectrum of the Kerr hole to be discrete. If so, formula (1) with the substitutions $Q \to ke$ and $\boldsymbol{J}^2 \to (j\hbar)^2$ will define a quantity with a discrete spectrum M_{ir}^2, which in the classical limit is just Christodoulou's squared irreducible mass (5). The spectrum of M_{ir}^2 may be inferred as follows.

Classically M_{ir}^2 shares some properties with the action integral $\oint p dq$ of a periodic mechanical system. Such an integral is an adiabatic invariant (9). It is unchanged in an adiabatic process, one in which an external influence is exerted on the system very gradually (and with no particular relation to its natural periods) so that its parameters change slowly. An example is $\oint p\, dx$ for a harmonic oscillator, which essentially equals E/ω (E = energy, ω = frequency). A gradual change in the spring constant of the oscillator leaves E/ω unchanged (9).

Now M_{ir}^2 has the same dimensions as action, length squared (6). In addition, although it exhibits a tendency to increase irreversibly as shown by CHRISTODOULOU (5) and HAWKING (4), M_{ir}^2 does remain unchanged in processes analogous to the adiabatic processes of mechanics. For example, in Christodoulou's reversible process (5) a *point* particle can be injected into a Kerr hole without changing M_{ir}^2, but it must be injected tangentially to the horizon. Consequently, the process takes infinite time according to *local* observers; it is adiabatic in the sense of mechanics. As a second example consider a wave of frequency ω and azimuthal index m being scattered by a Kerr hole rotating with frequency Ω. In general the hole's properties will change due to absorption from, or reinforcement of, the wave. But in the limit $\omega - m\Omega \to 0$ the change in M_{ir}^2 becomes negligible compared to those of M and $\boldsymbol{J}$, changes which take place very slowly in that limit (10). Furthermore, a local observer at the horizon (necessarily dragged

(8) S. A. TEUKOLSKY: *Astrophys. Journ.*, **185**, 635 (1973).
(9) M. BORN: *The Mechanics of the Atom* (New York, N.Y., 1960).
(10) A. A. STAROBINSKII and S. M. CHURILOV: *Žhurn. Èksp. Teor. Fiz.*, **65**, 3 (1973) (English translation to appear in *Sov. Phys. JETP*).

around with frequency Ω) sees the wave to oscillate with frequency $\omega - m\Omega$. Thus as $\omega - m\Omega \to 0$ the external influence on the hole becomes very gradual. We thus have another example of an adiabatic process leaving M_{ir}^2 unchanged.

The above examples suggest that M_{ir}^2 may be a multiple of the black-hole analogue of an adiabatic invariant action integral in mechanics. In the old quantum theory a periodic system was quantized by equating its adiabatic invariant action integral to $2\pi\hbar n$, where $n = 0, 1, 2, \ldots$ (9). Proceeding by analogy, we equate M_{ir}^2 to $g\hbar n$, where g is some constant; we prescribe that $n = 1, 2, \ldots$, since the state with $n = 0$ does not represent a black hole. The value of g is suggested by the result (11) that (if one allows for the fact that the radius of a particle is never smaller than its Compton wavelength) the injection of a particle into any stationary black hole results in M_{ir}^2 increasing by no less that a universal minimum amount $\frac{1}{2}\hbar$. If this is identified with the smallest possible quantum jump of M_{ir}^2, then $g = \frac{1}{2}$. Hence we set

(4) $$M_{ir}^2 = \tfrac{1}{2}\hbar n\,, \qquad n = 1, 2, 3 \ldots .$$

Substitution of eqs. (2), (3) and (4) into (1) gives us the mass spectrum for the Kerr black hole

(5) $$M^2 = \tfrac{1}{2} M_P^2 [n(1 + \tfrac{1}{2} k^2 \alpha/n)^2 + j^2/n]\,,$$

where $M_P = \hbar^{\frac{1}{2}} = 2.18 \cdot 10^{-5}$ g is the Planck mass, and $\alpha = e^2/\hbar = 1/137$ is the fine-structure constant. Classically, the parameters of the Kerr blank hole are constrained by (5) $4J^2 + Q^4 = 16\, M_{ir}^4$. The quantum version of this constraint is

(6) $$j^2 + \tfrac{1}{4} k^4 \alpha^2 \leqslant n^2\,.$$

Let us see if formula (5) satisfies Bohr's correspondence principle. For simplicity consider a nearly Schwarzschild hole ($j^2 + \frac{1}{4}k^4\alpha^2 \ll n^2$). Then in the classical limit ($n \gg 1$) the energy spacing between states for which n differs by one is $\hbar/4M$. The corresponding frequency $1/4M$ is just what one would expect a classical frequency associated with the hole to be. Thus formula (5) is in the spirit of the correspondence principle.

It is well known that the quantization rule $\oint p\,dq = 2\pi\hbar n$ is generally accurate only for $n \gg 1$. (In fact, in latter days its accuracy was improved by replacing n by $n + \frac{1}{2}$.) However, there is a bit of evidence to suggest that the mass spectrum (5) may be correct even for small n. It is the observation that, if M is correctly given by (5), then for all allowed states the surface area of the hole exceeds $4\pi(\hbar/M)^2$, except for the ground state ($n = 1$, $j = k = 0$) for which $A = 4\pi(\hbar/M)^2$. This is a satisfying result: the Kerr black hole is always larger than its Compton wavelength, except in its ground state when it is of just the same size. The ground state may thus be regarded also as an elementary particle (12).

Will a Kerr hole not in its ground state spontaneously decay as would, say, an excited atom? If it is charged, its exterior electric field will produce electron-positron pairs at a low but nonzero rate (13). The hole will attract the oppositely charged particles and repel their antiparticles, thus slowly discharging itself, so that k tends to decay.

(11) J. D. BEKENSTEIN: *Phys. Rev. D*, **7**, 2333 (1973); **9**, 3292 (1974).

(12) This « elementary particle » with mass 10^{-5} g has often been discussed before, *i.e.* M. A. MARKOV: *Sov. Phys. JETP*, **24**, 584 (1967); L. MOTZ: *Nuovo Cimento*, **12** B, 239 (1972); J. SARFATT: *Nature Phys. Sc.*, **240**, 101 (1972).

(13) L. PARKER and J. TIOMNO: *Astrophys. Journ.*, **178**, 809 (1972).

For a rotating hole UNRUH (1) has shown that spontaneous creation of particles (*i.e.* neutrinos) in the hole's exterior will spin it down in a time scale $T = 10^{-43}(M/10^{-5}\text{ g})^3$ s. Thus j also tends to decay. What about n? A classical theorem of Hawking's (4) says that M_{ir}^2 cannot decrease, but the theorem can be violated in the quantum domain (14). In fact, a calculation by HAWKING himself (2) suggests that a Schwarzschild hole ($j = k = 0$) spontaneously decays by particle creation in the surrounding space (also with time scale T). Therefore, n may also tend to decay. Thus a Kerr hole behaves in some respects like an excited atom. Further, if our arguments about the mass spectrum are correct, the quanta « emitted » by the hole in its decay should have a discrete spectrum, at least when a single quantum is « emitted » during a quantum transition of the hole.

Other processes of interest are scattering of a quantum on a black hole leading to absorption, elastic scattering, inelastic scattering (with change in frequency), or stimulated emission of additional quanta. We can derive a selection rule for the allowed processes as follows. Let the frequency, azimuthal quantum number and charge number of the incident quantum be ω, m and ε ($\varepsilon = 0, \pm 1$), respectively, and those for outgoing quanta be ω', m' and ε'. Then by conservation of energy, angular momentum and charge, and by differentiation of formula (5) (for n, $|j|$ and $|k| \gg 1$) we obtain

$$(7)\qquad \omega - \sum \omega' = (2M)^{-1}\big[\tfrac{1}{2}(1 - j^2/n^2 - \tfrac{1}{4}k^4\alpha^2/n^2)\,\Delta n + (j/n)(m - \textstyle\sum m') + k\alpha(1 + \tfrac{1}{2}k^2\alpha/n)(\varepsilon - \textstyle\sum \varepsilon')\big]\,.$$

This formula with $\omega = m = \varepsilon = 0$ applies to spontaneous emission. In every case $\Delta n = 0, \pm 1, \pm 2 \ldots$. The various processes occur each with its own probability which we cannot calculate here, but we do know that the probability that $\Delta n < 0$ must be small in order that the average over many scatterings give back the classical result that $\Delta M_{\text{ir}}^2 > 0$.

As the final application of formula (5) consider the example of a *macroscopic* wave packet of energy E and frequency ω inpinging on a Schwarzschild hole of mass M ($n \gg 1$ assumed). According to formula (5) the energy treshold for excitation of the hole by $\Delta n \geqslant 1$ is $\hbar/4M$. Thus, if $E < \hbar/4M$, the wave will scatter without any absorption! This prediction seems to clash with the classical result (10) that some absorption will always occur. The paradox is resolved by taking account of the quantum nature of the wave.

Suppose first that $2M\omega$ is *not* very small compared to unity. Since the wave is macroscopic, $E \gg \hbar\omega$, and thus neccessarily $E \gg \hbar/4M$, hence the treshold is exceeded and there is no paradox. Now suppose that $2M\omega \ll 1$. Classically the energy absorbed here is (10)

$$(8)\qquad E_{\text{abs}} \approx 4(2M\omega)^{2l+2}\, C_{ls} E\,,$$

where l is the multipole order of the wave, s its spin ($s = 0$, scalar; $s = 1$, electromagnetic; $s = 2$, gravitational) and C_{ls} is a combinatorial factor never larger than unity. But the classical calculation of the absorption makes sense only if E_{abs} exceeds the energy of one quantum $\hbar\omega$. In this case it follows from (8) that again $E \gg \hbar/4M$. Thus whenever the classical result is valid, the treshold is exceeded and the paradox is removed. For $E \gtrsim \hbar/4M$ absorption will be of a statistical nature; the probabilities for the absorption of various amounts of energy from the wave cannot be calculated here. In the regime $E < \hbar/4M$ no absorption can take place.

---

(14) J. D. BEKENSTEIN: *Phys. Rev. D*, **7**, 949 (1973).

PHYSICAL REVIEW D VOLUME 23, NUMBER 2 15 JANUARY 1981

Universal upper bound on the entropy-to-energy ratio for bounded systems

Jacob D. Bekenstein*
Institute for Theoretical Physics, University of California, Santa Barbara, California 93106
(Received 7 July 1980; revised manuscript received 25 August 1980)

We present evidence for the existence of a universal upper bound of magnitude $2\pi R/\hbar c$ to the entropy-to-energy ratio S/E of an arbitrary system of effective radius R. For systems with negligible self-gravity, the bound follows from application of the second law of thermodynamics to a gedanken experiment involving a black hole. Direct statistical arguments are also discussed. A microcanonical approach of Gibbons illustrates for simple systems (gravitating and not) the reason behind the bound, and the connection of R with the longest dimension of the system. A more general approach establishes the bound for a relativistic field system contained in a cavity of arbitrary shape, or in a closed universe. Black holes also comply with the bound; in fact they actually attain it. Thus, as long suspected, black holes have the maximum entropy for given mass and size which is allowed by quantum theory and general relativity.

I. INTRODUCTION

Gravitational entropy is one of the most intriguing concepts that have emerged from much recent work on quantum fields in curved space-time and quantum gravity.[1,2] Its most striking manifestation occurs in Hawking's radiation process by black holes,[1] in which it is connected with the area of the event horizon. Even though this area behaves very much like entropy, two obstacles have stood in the way of attempts to understand the still mysterious connection between area, a geometrical quantity, and entropy, a thermodynamic one. First, since its very inception, black-hole entropy has seemed to be numerically much larger than the entropy of any ordinary system of like mass.[3] Thus, a solar-mass black hole has black-hole entropy 10^{20} times the sun's thermal entropy. Is it not preposterous to think there is a common denominator in two quantities so unlike in size? Second, even if the two entropies are of like origin, how can one hope to express black-hole entropy in statistical terms (the logarithm of a number of interior states or configurations) when that task evidently demands a full accounting of all that could possibly happen inside the hole?

In this paper we address only the first difficulty. We point out that it arises from the insistence in comparing black holes with nonrelativistic systems. When compared to relativistic systems of massless particles, black holes do not have inordinately large entropy. Rather, black-hole entropy is revealed as matching the maximal entropy for a given mass of more ordinary systems: There is no gap in magnitude between black-hole entropy and ordinary entropy. This comes about because of the existence of a hitherto unnoticed upper bound to the entropy-to-energy ratio of non-black-hole systems of given effective radius R (see Sec. II for definition):

$$S/E < 2\pi R \tag{1}$$

(in units where $k=\hbar=c=G=1$). Making this bound plausible is our main task in this paper. For systems with negligible self-gravity, inequality (1) keeps S from growing faster than E, a well-known property of ordinary bodies which is responsible for the seeming gap between this entropy and black-hole entropy (which grows as E^2). However, as one compresses a body to its gravitational radius, R becomes of order $2E$, and S can begin growing as E^2 thus "catching up" with black-hole entropy. The closing of the gap shows it is conceivable that black-hole entropy should be calculable in terms of the number of interior black-hole configurations.[3]

The plan of this paper is as follows. Relation (1) is formulated as a conjecture in Sec. II, where we also point out that for systems with negligible self-gravity it follows from the second law. Yet this way of deriving it gives no hint as to what physically limits the entropy. Thus other approaches, based on statistical physics, are imperative in understanding the limit, and in extending it to strongly gravitating systems. One such approach, based on the microcanonical ensemble, has been pioneered by Gibbons.[4] We describe this approach in Sec. II. It shows what keeps S/E from reaching arbitrarily large values. Its main disadvantage is that it can only be applied on a case-by-case basis. It does suggest a more general approach to the problem which we develop in Sec. III, and which is found to lead directly to the canonical ensemble. By this method one goes beyond examples to establish the upper bound on S/E for a broad class of systems with negligible self-gravity. The existence of the bound is found to be intimately connected with the vacuum energy of fields.

II. WHY A BOUND ON S/E?

Classically, entropy is a measure of the phase space available to the system in question. It is then not difficult to see why an upper bound might exist for S/E. Let the system have energy E, or alternatively, let its energy be no more than E. This amounts to a limitation on the momentum space available to the system's components provided the potential energy is bounded from below. If the system is also bounded in space, then its phase space is bounded, and so must its entropy. The bound evidently increases with E. But our simple argument cannot establish that it increases linearly and neither can it say anything about the proportionality constant. In fact, it would seem impossible to write down a concrete bound for S/E without going into details about the system. This may explain why such a bound has gone unnoticed by workers in statistical mechanics (but see Bremermann[5] for an information-theoretic analog).

In fact, black-hole physics yields a specific form for the upper bound on S/E for systems with negligible self-gravity. According to the generalized second law of thermodynamics,[1,3,6] the sum of the thermal entropy outside a black hole and the black-hole entropy ($\frac{1}{4}$ of the horizon's surface area) should never decrease. Now, it has long been known[3,6] that when a stationary hole absorbs a body with negligible self-gravity, energy E and effective radius R (for the precise definition see Ref. 6), the hole's surface area must increase by at least $8\pi ER$. Since one can arrange the absorption process so that this minimal increase can be attained,[3] the second law will be violated unless the body's entropy (what disappears from the hole's exterior) cannot exceed $2\pi ER$. Thus we obtain the bound (1) on S/E as applied to weakly gravitating bodies. Note that although the quoted minimum area increase is derived under the assumption that the body is small compared to the hole, nothing prevents us from choosing a hole as large as needed (this also makes the Hawking radiation entropy negligible). Hence (1) holds regardless of the scale of R.

The intriguing feature of the previous argument is that it uses a law whose very meaning stems from gravitation to derive a bound on S/E for systems in which gravitation is negligible, a bound which has nothing to do with gravity [written out fully, relation (1) would involve $\hbar$ and c, but not G]. This provides a striking illustration of the unity of physics, but it also throws a challenge at the theorist: Provide a proof for the bound independent of gravitational considerations. To attempt this will be the main task before us later on in this paper.

The question suggests itself, how dependent is the bound (1) on the assumption that the self-gravity of the system is negligible? To clarify matters, consider a Kerr black hole, a system with maximal gravitational effects. Let its energy be E; then its surface area will be

$$A = 4\pi\{[E + (E^2 - a^2 - Q^2)^{1/2}]^2 + a^2\}\,, \tag{2}$$

where a is the specific angular momentum and Q the charge. We define the effective radius R by $4\pi R^2 = A$. Then it is clear that $R < 2E$, the equality corresponding to $a = Q = 0$. Now since $S = A/4$, $S/ER = \pi R/E$; it then follows from the inequality that Kerr holes conform to the bound (1); the Schwarzschild hole actually attains the bound.

If systems with negligible self-gravity and black holes both obey the bound (1), it is reasonable to assume that systems in which gravity is of intermediate strength do also. Thus bound (1) appears to be of universal validity so long as R is appropriately defined. In fact, guided by the R for a Kerr black hole and that for an ordinary spherical object (the metric radius), we shall be assuming that R for an arbitrary system is given in terms of the area A of that (quasi) spherical surface which circumscribes the system by

$$R = (A/4\pi)^{1/2}\,. \tag{3}$$

The only exception to (3) will be for closed universes which, of course, lack any "circumscribing sphere". For them we define R in terms of the volume of the universe.

How can we begin to see *directly* that bound (1) actually holds for systems with negligible self-gravity? For systems composed of nonrelativistic particles this is easy and the basic idea has long been known.[6] Let the system's mass be M, and let m be the typical mass of the active constituent particles (stars, atoms, nuclei, and nucleons, depending on the kind of system). There are roughly M/m such particles. Thus the system's entropy will also be M/m in order of magnitude and its entropy-to-energy ratio $1/m$. On the other hand, the system's size will exceed, typically by many orders of magnitude, the Compton lengths of its constituent particles, which are of order $1/m$. Thus the bound (1) is obeyed, with orders of magnitude to spare.

Evidently, systems composed of nonrelativistic particles are not very interesting from the point of view of the bound. In them the bulk of the energy is tied up in rest masses, and does not partake of the degrees of freedom which are manifested in the entropy. To put our proposed bound to an interesting test, we must consider assemblies of massless particles. Consider, then, ordinary blackbody radiation confined to a cavity. The familiar formulas[7] tell us that

$$S/E = 4/3T\,, \tag{4}$$

where T is the temperature and E the thermodynamic energy of the radiation. Formula (4) seems to predict that S/E can be made as large as we please by lowering T sufficiently. But in fact the thermodynamic description of radiation on which (4) is based breaks down when T is no longer large compared to the reciprocal of the characteristic size of the system (typical wavelength not small compared to cavity size). Boundary effects make themselves felt. These can be expected to arrest the growth of S/E as T is lowered further. The transition in behavior comes just when S/E is of order of the bound set by (1), so that the present example comes close to challenging (1). This circumstance motivates a shift of interest to the S/E of massless quantum fields confined to a region of prescribed size and shape by a material cavity (or in the case of a closed universe by the topology of the curved spacetime).

Gibbons[4] has developed a microcanonical approach for determining the maximal S/E for such systems. He labels the quantum states of the field by the occupation numbers of the various eigenfrequencies and takes the energy E of each state to be the sum of the eigenfrequencies weighted by the appropriate occupation numbers. He then assumes the entropy at a given E to be given by the Boltzmann formula[7]

$$S_{MC}(E) = \ln N(E)\,, \tag{5}$$

where $N(E)$ is the number of distinct states with energy E. $N(E)$ can always be computed explicitly, at least for low E, if the eigenfrequency spectrum is known. But thus far no general method for doing this has emerged and $N(E)$ must be computed separately and laboriously for each spectrum. In all cases for which this has been done (i.e., scalar field confined to a one-dimensional cavity, to the flat interior of a two-sphere, or to an Einstein universe S^3), $\ln N(E)/E$ has a maximum at some moderately low E. The maxima all lie below the bound (1) (with R interpreted as the radius of the appropriate n-sphere) thus providing support for it.

These results also support our remarks about the relation between the bound on S/E and the breakdown of the thermodynamic (continuum) description of a field at low temperatures (energies). They suggest that the maximal S/E is a phenomenon of the low-excitation states of fields. This may explain why the existence of the bound (1) has not been widely noticed. When cavities having two or three very different scales of length are considered in the Gibbons formalism, it becomes clear that the maximal S/E is determined by the longest scale (that most clearly associated with the lowest eigenfrequencies). This feature is the primary motivation for the definition of effective radius (3): twice R is never less than the longest dimension of the system.

DeWitt has remarked[8] that the question of a universal bound for S/E is quite unique in that it can have no meaning except when a particular choice of the zero of energy is made: obviously (1) cannot hold for every such choice. Gibbons's procedure exemplifies one possible choice of zero: For every field system the zero of energy is the energy of the vacuum (no-particles) state. This choice is not unique; in fact there is an alternative one more in harmony with the lessons of quantum field theory. The (regularized) vacuum energy of a field in flat space confined by some boundaries is, in general, different from the vacuum energy of that field in unconfined Minkowski spacetime.[9] The most famous example of this is, of course, the Casimir effect[10,11] which has been verified experimentally. It thus seems most natural to take, for each type of field, the vacuum energy in unconfined Minkowski spacetime as the zero of energy. It is known that with this choice the energy of a given field state is its gravitating energy (active gravitational mass).[9] It is thus just that energy responsible for the growth in horizon area in the gedanken experiment just discussed. Thus, at least for systems with negligible self-gravity, the present definition of the zero of energy is the appropriate one for stating the bound on S/E in the form (1). By extension of this idea we shall take, for any field system, the energy of the no-quanta state to equal the vacuum energy as computed by an appropriate regularization scheme. Thus in the microcanonical approach just described, a constant (different for each system) must be added to every energy computed from occupation numbers alone.

The big question remaining is, how may the bound (1) be established by the microcanonical method without recourse to a case-by-case analysis which, in any case, could not proceed far because the eigenfrequency spectra of most field systems are not known? We will now replace this question by a more general one which, not only proves more tractable, but also turns out to be a more fundamental statement of the problem being considered. To describe this approach we first recall a few results from quantum statistics.[7] A quantum system is described by its Hamiltonian $\hat{H}$; its state is described by a density operator $\hat{\rho}$ satisfying the normalization condition

$$\mathrm{Tr}\hat{\rho} = 1\,. \tag{6}$$

The mean energy is

$$\bar{E} = \mathrm{Tr}(\hat{\rho}\hat{H}) \tag{7}$$

and the entropy

$$S = -\mathrm{Tr}(\hat{\rho}\ln\hat{\rho}) . \tag{8}$$

It is well known that the microcanonical entropy (5) is just the maximal value which S attains over all $\hat{\rho}$ which satisfy (6) and which assign nonvanishing probabilities only to states with energy E. Thus

$$\ln N(E)/E = \max_{E} (S/\bar{E}) , \tag{9}$$

with the maximization carried out under the same conditions as before. In particular, (9) holds for that E for which $\ln N(E)/E$ is largest. If we now remove the constraint that $\hat{\rho}$ is confined to a given E, $\max(S/\bar{E})$ can only increase. Therefore

$$\max[\ln N(E)/E] < \max(S/\bar{E}) , \tag{10}$$

where the maximization in the left-hand side is over E, while that in the right-hand side is over all $\hat{\rho}$ which fulfill (6). Thus the peak S_{MC}/E value is bounded by the maximum that $S/\bar{E}$ attains for an arbitrary normalized $\hat{\rho}$. If it can be shown that the last quantity is below the bound (1), then one will know that S_{MC}/E fulfills it also. As will become clear in Sec. III, this program can be implemented without recourse to a case-by-case investigation.

Most important, the new statement of the problem is important in its own right. Very often the state of a statistical system is not one of definite energy, but has a well-defined mean energy. The maximum possible value of $S/\bar{E}$ is then of interest and, as (9) shows, it cannot be inferred from microcanonical considerations. Thus from now on we shall concentrate on the question of what upper bound can be set on $S/\bar{E}$ for an arbitrary $\hat{\rho}$.

III. BOUND ON $S/\bar{E}$ FROM QUANTUM STATISTICS

For simplicity we consider only systems composed exclusively of massless fields: scalar, electromagnetic, and neutrino (for technical reasons we do not consider gauge or metric fields). We ignore interactions, except as needed in enforcing boundary conditions at cavity walls. We assume the systems are stationary; this is clearly not a great restriction, for stationarity often implies equilibrium and maximal S, other things being equal. We consider fields in flat spacetime confined by cavities of arbitrary shape, or fields confined in a model (Einstein) universe of constant space curvature. We do not, however, consider fields in flat spacetimes with non-Euclidean topologies (i.e., cube with opposite faces identified to make T^3 topology).

Let us ask which $\hat{\rho}$ makes $S/\bar{E}$ maximal? It is evidently given by the solution of the variational problem

$$\delta[-\mathrm{Tr}(\hat{\rho}\ln\hat{\rho})/\mathrm{Tr}(\hat{\rho}\hat{H}) - \lambda\,\mathrm{Tr}(\hat{\rho})] = 0 , \tag{11}$$

where λ is a Lagrange multiplier used to enforce the normalization condition (6). Varying $\hat{\rho}$, rearranging terms and normalizing, one finds

$$\hat{\rho} = Z(\beta_0)^{-1}\exp(-\beta_0\hat{H}) , \tag{12}$$

$$Z(\beta_0) \equiv \mathrm{Tr}[\exp(-\beta_0\hat{H})] , \tag{13}$$

$$\beta_0 = (S/\bar{E})_{\max} . \tag{14}$$

Thus the wanted distribution is the canonical one whose inverse temperature parameter is just the peak value of $S/\bar{E}$ for the system. To find this value one computes $S/\bar{E}$ with (12) and compares with (14) to get

$$\ln Z(\beta_0) = 0 . \tag{15}$$

The problem is thus superficially simple; the maximal $S/\bar{E}$ is just that β for which the partition function is unity. However, for the field systems we have in mind this prescription is not trivially applied. Each field mode has a ground-state energy $\pm\omega/2$ for bosons or fermions, respectively. If $\ln Z$ is calculated naively, the mode sum of ground energies will make it diverge unless the boson and fermion ground-state energies cancel miraculously. We thus need to regularize $\ln Z$. In principle, one knows how to do this.[12] But technical difficulties have limited explicit calculations of the finite $\ln Z$ to a few systems of high symmetry,[13,14] or to the high-[15,16] and low-[15] temperature limits. Since we are interested in the general β case, and wish to treat systems wholesale rather than on a case-by-case basis, it seems reasonable to adopt a pragmatic approach.

Let ω_i be the eigenfrequencies for our system and let them be g_i-fold degenerate. Then at inverse temperature β the mean energy in all species is given by

$$\bar{E} = E_0 + \sum_i g_i\omega_i(e^{\beta\omega_i} \mp 1)^{-1} , \tag{16}$$

where upper (lower) signs are for boson (fermion) modes. The E_0 is the sum of *regularized* vacuum energies, while the mode sum is just the usual thermal contribution. That the intuitive form (16) follows from detailed regularization of the finite-temperature quantum field theory has been amply demonstrated by Al'taie and Dowker[13] and Dowker and Kennedy.[16] Substituting (16) and (12) into (7) and integrating with respect to β, one obtains the *finite* expression

$$\ln Z = C - \beta E_0 + \sum_i \mp g_i \ln(1 \mp e^{-\beta\omega_i}) , \tag{17}$$

where C is a constant of integration. Now using (12) to calculate S, one finds that at zero temperature ($\beta \to \infty$) $S \to C$ provided no ω_i vanishes (see Appendix C). However, for the fields we have in mind the vacuum state (absence of quanta) is nondegenerate; thus we expect $S \to 0$ at zero temperature (third law of thermodynamics). Hence, we must set $C = 0$.

It is clear from (17) that, provided $E_0 > 0$, $\ln Z$ is a monotonic decreasing function of β. As $\beta \to 0$, $\ln Z \to \infty$, while $\ln Z \to -\infty$ for $\beta \to \infty$. Hence by continuity there exists some β for which (15) is satisfied and there exists a maximal $S/\bar{E}$. This is clearly not the case if $E_0 \leq 0$. Later we shall indicate how one copes with the case $E_0 = 0$ within our formalism, but we shall have to assume that *in nature* no complete system can have a negative vacuum energy. Actually, seeming counterexamples to this assumption can easily be found. The Casimir vacuum energies of the electromagnetic[10] and scalar fields[11] between two infinite parallel plates, or the scalar vacuum energies in some infinitely long rectangular pipes[16] are all negative. But these are infinite systems and for them (1) does not set a bound on S/E. The fact that $E_0 < 0$ and that no β_0 exists is, therefore, irrelevent here.

Fields in flat spaces with non-Euclidean topologies also frequently have $E_0 < 0$ even when the space is finite.[9] But though non-Euclidean topology is important in highly curved spacetimes, the physical relevance of flat non-Euclidean spaces is yet to be demonstrated. Presently they are mere mathematical models in which regularization calculations are tractable. For this reason we do not consider such spaces.[17] By contrast to the above, vacuum energies of massless fields in the Einstein (curved) universe with topology S^3 are all positive. Likewise, the electromagnetic vacuum energy in a Euclidean, flat, spherical space is positive.[18] The known facts are few but they do indicate that $E_0 > 0$ is a physically plausible assumption.

A. Case with $E_0 > 0$

Assume, as is probably the case, that the vacuum energy of every species of field confined to a finite cavity, or to a compact universe is positive; let E_{0k} be the vacuum energy for the kth field. Now solve the equation

$$E_{0k} = \beta_k^{-1} H_k(\beta_k) \equiv \mp \beta_k^{-1} \sum_i^{(k)} g_i \ln(1 \mp e^{-\beta_k \omega_i}) \qquad (18)$$

(which involves the sum over the kth field's modes only) for the parameter β_k. Repeat for every other species. One has then a series of distinct β_k. Now $E_0 = \sum_k E_{0k}$. Therefore, substituting the expressions (18) for all k into (17) evaluated at $\beta = \beta_0$ we get

$$\beta_0^{-1} \sum_k H_k(\beta_0) = \sum_k \beta_k^{-1} H_k(\beta_k) . \qquad (19)$$

Now, as direct differentiation shows, $\beta^{-1} H_k(\beta)$ is monotonic decreasing in β both for fermions and bosons. It is then fairly clear from (19) that β_0 is bracketed by the smallest and largest β_k. This leads to great simplification: *By treating field species separately one can find upper and lower bounds for β_0 for any conceivable mixture of species.*

1. Upper bound

Introduce $n_k(\omega)$, the number of *modes* (counting degeneracies) of species k with eigenfrequencies up to ω. Clearly at every ω which is an eigenvalue, $n_k(\omega)$ has a step discontinuity of strength $+g_i$. This (18) can be written as

$$E_{0k} = \mp \beta_k \int_0^\infty (dn_k/d\omega) \ln(1 \mp e^{-\beta_k \omega}) d\omega . \qquad (20)$$

Integration by parts gives

$$E_{0k} = \int_0^\infty n_k(\omega)(e^{\beta_k \omega} \mp 1)^{-1} d\omega \qquad (21)$$

as a condition for β_k. In most cases we know little about the precise form of n_k. However, one can easily obtain from (21) an upper bound for β_k which will suffice for our purpose—establishing that (1) holds generally.

One starts with the obvious inequality[4] (valid for $p > 0$)

$$n_k(\omega)\omega^{-p} < \sum^{\omega} g_i \omega_i^{-p} , \qquad (22)$$

where the sum includes all eigenfrequencies up to ω. The inequality holds (for any $p > 0$) since every $\omega_i < \omega$, while $n_k(\omega)$ is just the number of terms in the sum. The sense of the inequality is clearly preserved if the sum is extended to *all* eigenfrequencies in which case it becomes the well-known ζ function for the Hamiltonian of the field k[19,20] (see also Appendix B),

$$n_k(\omega) < \zeta_k(p)\omega^p . \qquad (23)$$

Since $\zeta_k(p)$ converges only for $p > 3$, we restrict all our following remarks to that range. We now use (23) to set an upper bound on the integral in (21):

$$E_{0k} < \beta_k^{-p-1} \zeta_k(p) \int_0^\infty x^p (e^x \mp 1)^{-1} dx . \qquad (24)$$

Evaluating the integral[21] in terms of Riemann's ζ function $\zeta_R(p)$ and isolating β_k, we get the bound

$$\beta_k < [\Gamma(1+p)\zeta_R(1+p)\zeta_k(p)E_{0k}^{-1}(1-\delta_k 2^{-p})]^{1/(1+p)}, \qquad (25)$$

where $\delta_k = 1$ for fermions and $\delta_k = 0$ for bosons.

For the scalar, electromagnetic, and neutrino fields in an Einstein universe one knows all quantities in (25).[13] For example, for the scalar field $E_{0k} = (240a)^{-1}$ and $\zeta_R(p) = a^p \zeta_R(p-2)$, where a is the radius of the universe. By trial and error one finds that $p = 6.4$ gives nearly the lowest bound on β_k. Inserting numbers[21] one gets

$$\beta_k(\text{scalar}; S^3) < 5.70a. \qquad (26)$$

This may be compared with the exact value $\beta = 4.13a$ which is obtained by finding (numerically) the zero of $\ln Z$, an exact expression for which may be inferred from the work of Dowker and Critchley[14] [Eqs. (30)-(33)]. In like manner, and still using $p = 6.4$, one finds

$$\beta_k(\text{neutrino}; S^3) < 4.25a, \qquad (27)$$

which again compares favorably with the exact result $\beta = 2.61a$ inferred from the $\ln Z$ implicit in Dowker and Al'taie.[12] Finally, for the electromagnetic field one has (also with $p = 6.4$)

$$\beta_k(\text{electromagnetic}; S^3) < 2.70a. \qquad (28)$$

In this case there exists no exact result for comparison. Evidently, if we identify a with the effective radius of the universe R, then all our β_k's are below $2\pi R$. Consequently, an Einstein universe filled with any mixture of photons, neutrinos and scalar particles obeys (1).

For fields in flat space confined to a cavity of arbitrary shape one does not generally know E_{0k}, or the spectrum out of which $\zeta_k(p)$ is constructed. To minimize the effect of our ignorance we work with large p; in the limit of large p, $\zeta_k(p) \to g_1\omega_1^{-p}$, so we need know only the first eigenfrequency and its degeneracy. Also, because of the exponent in (25), the dependence on E_{0k} is weak. Actually, we cannot take $p \to \infty$; because of the $\Gamma(1+p)$ the bound on β_k would blow up. But it will be sufficient to restrict attention to $p > 10$. Appendix B shows that for such p, approximating ζ_k by $g_1\omega_1^{-p}$ leads to fractional errors of order 10^{-3}–10^{-1} in the right-hand side (RHS) of (25) for all cases in which one knows ζ_k accurately. The replacement is thus a good approximation. Also for $p > 10$ one may neglect the 2^{-p} in (25) and replace $\zeta_R(1+p)$ by unity. The fractional errors incurred are only 10^{-4}. We thus get the bound

$$\beta_k\omega_1 < [\Gamma(1+p)\, g_1\omega_1/E_{0k}]^{1/(1+p)}. \qquad (29)$$

Calculations of E_{0k} are proverbial for their difficulty. Only for the electromagnetic field in a sphere,[18] and for all three fields in the Einstein universe,[13,14] are precise results known (and for the Casimir energies of some infinitely long systems not of relevance here[15,16]). In all these cases $\xi \equiv E_{0k}/g_1\omega_1$ is in the range 10^{-3}–10^{-2}. In fact, since ω_1 sets the energy scale in each case, and g_1 is of order unity, one expects, on order-of-magnitude grounds, that ξ will be within a few orders of magnitude of unity. We shall allow it the generous range 10^{-6}–10^{6}. Although one can conceive of E_{0k} being exactly zero (see below), it seems unlikely that it can fall below the range indicated and yet be nonvanishing.

Let us now find the tightest bound on $\beta_k\omega_1$ by differentiating the RHS of (29) with respect to p; it occurs for $p = p_c$ where p_c is determined by

$$\xi = \Gamma(1+p_c)\exp[-(1+p_c)\psi(1+p_c)] \qquad (30)$$

or its alternative form

$$[\Gamma(1+p_c)\xi^{-1}]^{1/(1+p_c)} = \exp[\psi(1+p_c)], \qquad (31)$$

where

$$\psi(p) \equiv d\ln\Gamma(p)/dp. \qquad (32)$$

By inserting in (30) a trial $p_c \geq 10$, one finds the ξ for which it gives the best upper bound for $\beta_k\omega_1$; that bound is obtained by simply evaluating the RHS of (31). Inserting numerical values,[21] one finds that p_c increases monotonically from 10 to 12.8 as ξ goes from 2.1×10^{-5} to 1.1×10^{-6}; simultaneously the bound increases from 10.50 to 13.30. For $\xi > 2.1\times10^{-5}$ the p_c would be below 10. In order to preserve the accuracy of our approximation, we simply use $p = 10$ in (29) for $\xi > 2.1\times10^{-5}$. Evidently the bound on $\beta_k\omega_1$ decreases as ξ increases beyond 2.1×10^{-5}. Hence for $10^{-6} < \xi < 10^{6}$ (or even larger)

$$\beta_k\omega_1 < 13.30. \qquad (33)$$

The only question now is what is the first eigenfrequency ω_1. As shown in Appendix A, for fields in a sphere of radius R, $\omega_1 = \pi/R$ for the scalar and neutrino fields and $\omega_1 = 2.082/R$ for the electromagnetic field. As shown in Appendix C, for a nonspherical cavity we have $\omega_1 > \pi/R$ for scalar and neutrino fields and $\omega_1 > 2.082/R$ for the electromagnetic field, where R is the radius of the sphere which circumscribes the cavity; this is just the cavity's effective radius by (3). Combining these values with (33) we finally get

$$\beta_k(\text{scalar or neutrino; cavity}) < 4.233R \qquad (34)$$

and

$$\beta_k(\text{electromagnetic; cavity}) < 6.388R. \qquad (35)$$

Thus for scalar and neutrino fields, $\beta_k < 2\pi R$. The RHS of (35) exceeds $2\pi R$ by only one percent. But one can be quite confident that for the electromagnetic field β_k is also below $2\pi R$. To have it otherwise would require an unlikely "conspiracy"[22]: $\xi \simeq 10^{-6}$, $\omega_1 \simeq 2.082/R$, and $n_k(\omega) \simeq \zeta_k(12.8)\omega^{12.8}$. The last condition cannot possibly hold for arbitrarily large ω. The second would be true only for a quasispherical cavity, but for a spherical cavity one knows[18] that $\xi = 7.39 \times 10^{-3}$, not 10^{-6}. Hence, we can see that an arbitrary mixture of scalar, electromagnetic, and neutrino fields would satisfy bound (1).

2. Lower bound

Thus far we have no inkling of how much our bounds for β_0 overestimate it. It is important to know this in order to support our claim in Sec. I that ordinary systems can approach bound (1) and thus close the gap between ordinary matter and black holes. There is no easier way to settle this than to obtain lower bounds for the β_k. Let us return to the defining relation (18). If one discards in the sum the terms beyond that with the lowest ω_i and solves

$$\bar{\beta}_k E_{0k} = \mp g_1 \ln(1 \mp e^{-\bar{\beta}_k \omega_1}) \qquad (36)$$

for $\bar{\beta}_k$, one will evidently get an underestimate for β_k. We only consider the case for which $E_{0k} \ll \omega_1$; examples[13,16,18] for which E_{0k} is known are of this type. Then (36) can be consistent only if $\exp(-\bar{\beta}_k\omega_1) \ll 1$ so that

$$\bar{\beta}_k E_{0k} \simeq g_1 e^{-\bar{\beta}_k \omega_1} \qquad (37)$$

or, in our previous notation,

$$\xi \bar{\beta}_k \omega_1 \simeq e^{-\bar{\beta}_k \omega_1}. \qquad (38)$$

Numerically we find that as ξ decreases from 10^{-2} to 10^{-4} to 10^{-6}, $\bar{\beta}_k\omega_1$ increases from 3.4 to 7.2 to 11.4. These values are consistent with (33). As we mentioned, $\omega_1 > 2.082/R$ for the electromagnetic field and $\omega_1 > \pi/R$ for the other two. Thus there could exist a cavity in which $\beta_k > 2.3R$ for a neutrino field if only the ξ in question had a rather modest value $\lesssim 10^{-4}$. Likewise, for the electromagnetic field in a spherical cavity (see Ref. 18 and Appendix A) $\xi = 7.39 \times 10^{-3}$, so that we know with confidence that $\bar{\beta}_k\omega_1 \simeq 3.62$; thus $\beta_k > 1.74R$.[23] There are, therefore, fields in cavities which are only a factor of 3 below the universal bound (1). As we mentioned earlier, a scalar-filled Einstein universe reaches $\frac{2}{3}$ of the bound (1). Hence there is no great gap between matter and black holes as far as $S/\bar{E}$ is concerned.

B. Case with $E_0 = 0$

If one or more E_{0k} vanish but still $E_0 > 0$, no change is required in our previous analysis. We may simply define fictitious positive E_{0k} for the fields which have none at the expense of those which have in such a way that the sum of E_{0k} is still the physical E_0. The fictitious E_{0k} are unlikely to have to be smaller than the minimum we require (corresponding to $\xi = 10^{-6}$). Thus our arguments go through. A problem of principle arises, however, if (and we emphasize if) *all* vacuum energies vanish. We may then argue that a vacuum (no particles) state with vanishing energy is impossible to detect and could well be discarded from $\hat{\rho}$ and Z without ill effect for the sort of question we have in mind. Since $Z = \sum e^{-\beta_0 E}$,[24] this amounts to subtracting unity from Z. Then the condition that the *new* Z is unity at β_0 implies [see (17)] that

$$\ln 2 = \sum_i \mp g_i \ln(1 \mp e^{-\beta_0 \omega_i}). \qquad (39)$$

Let there be ν species of massless fields present (including several neutrinos, etc.). The role of (18) is now taken over by

$$(\ln 2)/\nu = H_k(\beta_k). \qquad (40)$$

We can still conclude that β_0 is bracketed by the extreme β_k's and we can carry out an analysis analogous to the one leading to (29) to obtain

$$\beta_k \omega_1 < [\Gamma(1+p)\, g_1 \nu / \ln 2]^{1/p}. \qquad (41)$$

The slightly different power of p here has its root in the absence of a factor β_0 in (39) as compared with (17) at $\beta = \beta_0$. Implicit in (41) is the condition $p > 10$ introduced for the same reasons as before. In general, g_1 is a small integer while ν is evidently not large. We can with confidence assume that $g_1\nu/\ln 2 < 100$. Then with $p = 10$, (41) gives $\beta_k\omega_1 < 7.18$. In view of the above-mentioned lower bounds on ω_1, we see that no β_k may exceed $3.45R$. Thus an arbitrary mixture of scalar, electromagnetic, and neutrino fields with vanishing vacuum energy would also satisfy bound (1).

IV. ASSESSMENT

Resting as they do at three points on plausibility rather than on rigor, our arguments do not constitute an airtight proof that $S/\bar{E}$ complies with bound (1), but they do make this very plausible for mixtures of scalar, neutrino, and electromagnetic fields in cavities of arbitrary shapes. With the same degree of plausibility, they indicate that the microcanonical S_{MC}/E satisfies (1) for those same situations. These partial successes in "explaining" (1) underscore the need for deepening the rigor and broadening the scope of the arguments.

In particular, one would like to encompass in the scheme Dirac fields in interaction with gauge fields. Not only is this a case of great physical interest (are there thermodynamic constraints on the possible structure of hadrons?), but it also poses a great technical challenge: how to handle the zero-frequency Dirac modes that may appear in this context? One would also like to make bound (1) plausible for self-gravitating systems, not merely for fields in a prescribed gravitational background. This is perhaps the greatest challenge in connection with (1), relating as it does to the question of whether black-hole entropy can be calculated in terms of the same concepts used for calculating matter entropy.

Note added in proof. Recently K. A. Milton [Phys. Rev. D 22, 1444 (1980)] has demonstrated the positivity of the vacuum energy of a confined system of gauge fields and fermions in interaction.

ACKNOWLEDGMENTS

Many of the ideas reported were developed in the course of innumerable conversations with Gary Gibbons; they also owe much to the incisive criticism of Bryce DeWitt. Thanks are due to P. Candelas, D. Eardley, S. Fulling, U. Gerlach, J. Hartle, G. Horowitz, W. Kohn, G. Kennedy, F. Narcovitch, L. Parker, L. Smolin, W. Unruh, and J. Weber for their interest and suggestions, to W. Kohn and B. DeWitt for hospitality at the Institute for Theoretical Physics, and to the National Science Foundation for support (Grant No. PHY 77-27084).

APPENDIX A

Here we summarize the eigenfrequency spectra for scalar, electromagnetic, and neutrino fields in a spherical cavity of radius R and make some remarks about the choice of boundary conditions. For the neutrino case the well-known problem with prescribing boundary conditions is resolved. We regard spacetime as flat here and in Appendix C.

Solutions of the scalar equation which are harmonic in time may be found only for discrete eigenfrequencies ω_i which arise from the eigenvalue problem defined by

$$\nabla^2\phi = -\omega^2\phi\,, \tag{A1}$$

together with suitable boundary conditions for ϕ. We shall follow DeWitt in demanding $\phi=0$ on the boundary.[11] One advantage of this (Dirichlet) boundary condition is that it is conformally invariant (whereas the Neumann condition is not). Thus one can treat the conformal scalar field. The solutions of (A1) which are regular at the origin are $j_l(\omega r)Y_{lm}(\theta,\phi)e^{-i\omega t}$, where j_l is the standard spherical Bessel function of order l. The boundary condition then demands that ωR be a positive zero of j_l. Hence the spectrum is

$$\omega_{nl} = j_{n,l}R^{-1}\,, \quad n=1,2,\ldots;\quad l=0,1,\ldots\,, \tag{A2}$$

where $j_{n,l}$ is the nth positive zero of $j_l(x)$; the degeneracy is $2l+1$. The lowest eigenfrequency is $\omega_{10}=\pi/R$.

Sourceless electromagnetic fields which are harmonic in time obey the time-independent Maxwell equations

$$\vec\nabla\times\vec{\mathrm{E}} = i\omega\vec{\mathrm{B}}\,, \tag{A3}$$

$$\vec\nabla\times\vec{\mathrm{B}} = -i\omega\vec{\mathrm{E}}\,. \tag{A4}$$

The boundary conditions depend on the physical characteristics of the cavity. Intuitively one would like the fields to be confined to the cavity. Then one must assume it is a perfect conductor which implies that the tangential component of $\vec{\mathrm{E}}, \vec{\mathrm{E}}_t$ vanishes on the boundary. The "electric" solutions of (A3) and (A4) have[25] $\vec{\mathrm{E}} = j_l(\omega r)\vec{\mathrm{X}}_{lm}e^{-i\omega t}$, where $\vec{\mathrm{X}}_{lm}$ are the vector spherical harmonics obtained by operating on the Y_{lm} with the angular momentum differential operator. The $\vec{\mathrm{B}}$ may be inferred from (A3). Evidently $\vec{\mathrm{E}}_t$ vanishes on the surface only if j_l vanishes. Hence the spectrum of the electric modes is the same as the one for the scalar field (A2), except $l=1,2,\ldots$ because $\vec{\mathrm{X}}_{00}\equiv 0$. The magnetic modes have[25] $\vec{\mathrm{B}} = j_l(\omega r) \times \vec{\mathrm{X}}_{lm}e^{-i\omega t}$. By (A4) the $\vec{\mathrm{E}}$ field is proportional to the curl of this expression. Because $\vec{\mathrm{X}}_{lm}$ has only θ and ϕ components, $\vec{\mathrm{E}}_t$ can vanish on the surface only if the radial derivative of j_l vanishes. Thus the spectrum for the magnetic modes is

$$\omega_{nl} = j'_{n,l}R^{-1}\,, \quad n=1,2,\ldots;\quad l=1,2,\ldots\,, \tag{A5}$$

where $j'_{n,l}$ is the nth positive root of $j'_l(x)$; the degeneracy is $2l+1$. The lowest eigenfrequency is $\omega_{11}=2.082/R$.

The free neutrino field is described by the Weyl equation.[26] Since the neutrino is purely a negative-helicity particle, for a field with time dependence $\exp(-i\omega t)$ the equation reads

$$-\vec\sigma\cdot\vec{\mathrm{p}}\psi = \omega\psi\,, \tag{A6}$$

where $\vec\sigma$ is the triplet of Pauli matrices, ψ is the neutrino bispinor, and $\vec{\mathrm{p}}$ is the momentum operator $-i\vec\nabla$. One may introduce the operators for orbital angular momentum L^2 and total angular momentum J^2 and J^z. Their joint eigenfunctions are the spinorial spherical harmonics ${}_J\mathcal{Y}_L^M$ with eigenvalues $L(L+1)$, $J(J+1)$, and M for the three operators, respectively. One then has[27]

$$\vec\sigma\cdot\vec{\mathrm{p}} = -i\sigma^r r^{-1}[(\partial/\partial r)r - (J^2+\tfrac14 - L^2)]\,, \tag{A7}$$

where $\sigma^r = \vec\sigma\cdot\hat{\mathrm{r}}$; σ^r has the property[27]

$$\sigma^r\,{}_J\mathcal{Y}^M_{J\pm1/2} = -\,{}_J\mathcal{Y}^M_{J\mp1/2}\,. \tag{A8}$$

With the help of (A7) and (A8), as well as the relations[21]

$$j'_{J+1/2} = j_{J-1/2} - (J+\tfrac32)x^{-1}j_{J+1/2}\,, \tag{A9}$$

$$j'_{J-1/2} = -j_{J+1/2} + (J - \tfrac{1}{2})x^{-1}j_{J-1/2} \quad \text{(A10)}$$

satisfied by the spherical Bessel function $j_l(x)$, one verifies that

$$\psi = [j_{J+1/2}(\omega r)\,{}_J\mathcal{Y}^M_{J+1/2} - ij_{J-1/2}(\omega r)\,{}_J\mathcal{Y}^M_{J-1/2}]e^{-i\omega t} \quad \text{(A11)}$$

is a solution of (A6) (evidently $J \geq \frac{1}{2}$). Since (A11) is an eigenfunction of J^2, J^z, and the helicity $\omega^{-1}\vec{\sigma}\cdot p$ [eigenvalues $J(J+1)$, M, and -1, respectively], it is just the right mode function for the problem. The only thing missing is the spectrum of ω. To get it one must prescribe boundary conditions, and this is where the difficulty lies. One cannot demand $\psi = 0$ or $\hat{n}\cdot(\vec{\nabla}\psi) = 0$ on the boundary; the solution (A11) does not satisfy either condition essentially because the two spinorial harmonics are idependent. Thus traditional boundary conditions do not work. This is the well-known difficulty in posing the boundary conditions required to define a discrete spectrum for neutrino eigenfrequencies in a cavity.

We argue that the difficulty has its root in the fermionic nature of ψ; one should not impose boundary conditions on ψ as one does for a bosonic field (i.e., scalar). Rather, one should impose conditions on a bilinear expression in ψ, for only such a quantity is an observable. For example, one might require the normal component of the neutrino number current J^μ to vanish at the boundary (cavity material impermeable to neutrinos). In standard notation[26]

$$J^\mu = \psi^\dagger \tilde{\sigma} s^\mu \psi\,, \quad \text{(A12)}$$

where $\psi^\dagger$ is the Hermitian conjugate to ψ, s^μ is a four vector of 2×2 matrices whose anticommutator is twice the metric and $\tilde{\sigma}$ is a "Hermitizing" matrix. One may choose $\tilde{\sigma}$ and s^μ in such a way[26] that the space part of $\tilde{\sigma}s^\mu$ is $\vec{\sigma}$ while the time part is +1. Thus the radial component we require is just $\psi^\dagger\sigma^r\psi$. Substituting (A11) we have

$$J^r = -[j_{J+1/2}{}^2\,{}_J\mathcal{Y}^{M\dagger}_{J+1/2}\,{}_J\mathcal{Y}^M_{J-1/2} + (J+\tfrac{1}{2} \rightarrow J-\tfrac{1}{2})]$$
$$+ij_{J+1/2}j_{J-1/2}[{}_J\mathcal{Y}^{M\dagger}_{J+1/2}\,{}_J\mathcal{Y}^M_{J+1/2} - (J+\tfrac{1}{2} - J-\tfrac{1}{2})]\,. \quad \text{(A13)}$$

Evidently J^r must be real. It is obvious that the two angular terms in the last square brackets are real. Hence the second line in (A13) is pure imaginary. Its radial dependence is different from that of the first line, so it cannot be canceled by it. Hence the second square brackets in (A13) must vanish identically. By the same token, the two angular factors in the first square brackets (each other's complex conjugates) must be real (hence equal) since they are multiplied by different radial functions. Hence

$$J^r = -(j_{J+1/2}{}^2 + j_{J-1/2}{}^2)\,{}_J\mathcal{Y}^{M\dagger}_{J+1/2}\,{}_J\mathcal{Y}^M_{J-1/2}\,. \quad \text{(A14)}$$

Since the zeros of different j_l do not coincide, we see that J^r cannot vanish on a spherical boundary (although its angular integral does). Thus our presumed boundary condition is also inadequate.

According to Unruh,[28] the fact that J^r cannot be made to vanish at the boundary reflects a Klein-type paradox: If the cavity material is impermeable to neutrinos, it will create neutrino-antineutrino pairs and so defeat the boundary condition. Apparently for the same reasons, the condition that the normal component of the energy current vanish on the boundary fails also. However, a related, somewhat mysterious condition works.

From the four-spinor expression for the Dirac stress-energy tensor,[26] one may easily compute the energy flux for a fictitious *right*-handed neutrino field in terms of the appropriate bispinor ϕ:

$$\vec{P} = -\tfrac{1}{2}\omega\phi^\dagger\vec{\sigma}\phi + \tfrac{1}{4}i(\phi^\dagger\vec{\nabla}\phi - \vec{\nabla}\phi^\dagger\phi)\,. \quad \text{(A15)}$$

The boundary condition requires us to formally replace ϕ by the *left*-handed ψ and set the normal component of $\vec{P}$ to zero on the boundary:

$$-P^r = \tfrac{1}{2}\omega J^r - \tfrac{1}{4}i(\psi^\dagger\psi_{,r} - \psi^\dagger_{,r}\psi) = 0\,. \quad \text{(A16)}$$

Putting (A11) and (A14) into (A16), canceling terms, and taking account of the realness of the angular factor appearing in (A14) gives as boundary condition that at $r = R$,

$$j_{J+1/2}{}^2 + j_{J-1/2}{}^2 + j'_{J-1/2}j_{J+1/2} - j'_{J+1/2}j_{J-1/2} = 0\,. \quad \text{(A17)}$$

In view of (A9) and (A10) it now follows that

$$(J+1)j_{J+1/2}(\omega R)j_{J-1/2}(\omega R) = 0\,. \quad \text{(A18)}$$

Since $J \geq \frac{1}{2}$, we see that the allowed ωR are the positive zeros of either of the j_l in (A18). The neutrino eigenfrequency spectrum is then

$$\omega_{n\epsilon J} = j_{n,\,J+\epsilon}R^{-1}\,, \quad \text{(A19)}$$

where $\epsilon = \pm\frac{1}{2}$. These eigenvalues are $2J+1$ degenerate. In addition, half of the eigenvalues for given J ($\epsilon = +\frac{1}{2}$) are degenerate with half of the eigenvalues for $J+1$ ($\epsilon = -\frac{1}{2}$).

Thus far we have ignored the possibility of zero eigenfrequencies. We now show there are none in our problem. For $\omega = 0$ the scalar equation (A11) reduces to Laplace's equation. The nonsingular solutions have the form $\phi = r^l Y_{lm}(\theta,\phi)$; none vanish at $r = R$. Thus $\omega = 0$ is not an eigenfrequency. For $\omega = 0$, Maxwell's Eqs. (A3) and (A4) imply E and B are each the gradient of a different potential. The divergence equations then imply that the potentials solve Laplace's equation. The tangential derivative of the potential $r^l Y_{lm}$ cannot vanish at $r = R$. Hence no E field satisfying the boundary conditions for a perfectly conducting cavity exists for $\omega = 0$. The $\vec{B}$ field is uncoupled to $\vec{E}$; hence one must (for $\omega = 0$ only) specify boundary conditions for $\vec{B}$. It is simplest to imagine the cavity material as superconducting. It then excludes the $\vec{B}$ field; thus the normal component of $\vec{B}$ must vanish at the cavity boundary. But the radial derivative of $r^l Y_{lm}$ cannot vanish at $r=R$ unless $\vec{B}$ vanishes identically. Hence $\omega=0$ is not an electromagnetic eigenfrequency. An analysis of the Weyl equation along the lines of the one given above shows the regular solutions for $\omega=0$ to be of the form $\psi=r^{J-1/2}{}_J\mathcal{Y}^M_{J-1/2}$. Only that with $J=\frac{1}{2}$ satisfies the boundary condition and this only because P^r vanishes identically. Further, such a solution does not have negative helicity: $\vec{\sigma}\cdot\vec{p}$ operating on it gives zero. Hence we would not include the associated eigenfrequency in the mode sum (16). Thus, in the context of this paper, $\omega=0$ is not an eigenfrequency.

APPENDIX B

Here we show by example that for large p various ζ functions of interest are well approximated by their ground-state parts. For a spectrum ω_i with degeneracy g_i,

$$\zeta(p)=\sum_{i=1}^{\infty} g_i\omega_i^{-p}. \tag{B1}$$

The ground-state contribution is $g_1\omega_1^{-p}$. Evidently as $p\to\infty$, $\zeta(p)\to g_1\omega_1^{-p}$. But as we shall see, already for $p=10$ the approximation is adequate for the purposes of Sec. V. We first consider fields in an Einstein universe of radius a.

For the conformal scalar field[13] $\omega_i a=i$, $g_i=i^2$ with $i=1,2,\ldots$. Therefore $\zeta(p)=a^p\zeta_R(p-2)$, where ζ_R denotes the Riemann ζ function. Thus $\zeta(10)=1.004\omega_1^{-10}$. For the electromagnetic field,[12,29] $\omega_i a=i+1$, $g_i=2i(i+1)$ with $i=1,2,\ldots$. Thus $\zeta(p)=2a^p[\zeta_R(p-2)-\zeta_R(p)]$ and $\zeta(10)=1.05\times 6\omega_1^{-10}$. For the two-component neutrino field[13,30] $\omega_i a=i+\frac{1}{2}$, $g_i=2i(i+1)$ with $i=1,2,\ldots$. Thus[21]

$$\zeta(p)=\tfrac{1}{2}a^p[(2^p-4)\zeta_R(p-2)-(2^p-1)\zeta_R(p)],$$

so that $\zeta(10)=1.02\times 4\omega_1^{-10}$. Of interest in Sec. V is $\zeta(p)^{1/(1+p)}$. For $p=10$ the fractional errors incurred in this quantity by replacing any of the above ζ by $g_1\omega_1^{-p}$ are less than $\frac{1}{2}$ percent.

Let us now consider the various fields in a sphere of radius R. According to Appendix A, for the scalar field

$$\zeta(p)=R^p\sum_{l=0}^{\infty}(2l+1)\sum_{n=1}^{\infty}j_{n,l}^{-p}. \tag{B2}$$

For the neutrino field ζ is just twice this expression. A good approximation to these $\zeta(p)$ is obtained by summing only over the eigenvalues up to five times larger than $\omega_1=\pi/R$ (about 220 ω's). One finds $\zeta(10)\simeq 1.10\omega_1^{-10}$ for both the scalar and neutrino fields. For the electromagnetic field, $j_{n,l}+j'_{n,l}$ replaces $j_{n,l}$ in (B2). A good approximation to this $\zeta(p)$ may be obtained by summing over the eigenvalues up to five times larger than $\omega_1=2.082/R$ (about 140 ω's). One finds $\zeta(10)\simeq 3.33\times 3\omega_1^{-10}$. In all these cases the fractional errors incurred in $\zeta(p)^{1/(1+p)}$ by replacing $\zeta(p)$ by $g_1\omega_1^{-p}$ are 1–10% for $p=10$.

APPENDIX C

Here we establish lower bounds for the lowest eigenfrequencies of the scalar, electromagnetic, and neutrino fields in an odd-shaped cavity.[31] Let C denote the cavity's interior, and ∂C its boundary. Imagine C is circumscribed by a sphere of radius R. Let the sphere's interior be denoted by S, and its boundary by ∂S. Finally, let $\bar{C}$ denote the part of S outside C.

Let ϕ_1 be the normalized eigenfunction of the scalar equation in C which corresponds to the lowest eigenfrequency ω_1. We assume the boundary condition $\phi=0$ on ∂C. We now consider a test function $\bar{\phi}$ which equals ϕ_1 in C and vanishes in $\bar{C}$. Evidently $\bar{\phi}$ is normalized in S, obeys the boundary condition on ∂S, and is continuous throughout S. Thus by the Rayleigh-Ritz principle, as applied to Eq. (A1), we can set an upper bound on the lowest eigenfrequency for the scalar field in S, which we know (Appendix A) to be $\omega=\pi/R$:

$$-\int_S \bar{\phi}\nabla^2\bar{\phi}\,dV>\pi^2/R^2. \tag{C1}$$

Now the normal component of $\vec{\nabla}\bar{\phi}$ has a step discontinuity at ∂C, so $\nabla^2\bar{\phi}$ has a δ-function singularity there. However, $\nabla^2\bar{\phi}$ in (C1) appears multiplied by $\bar{\phi}$ which vanishes at ∂C. Hence the singularity does not contribute to the integral. Neither does the region $\bar{C}$ where $\bar{\phi}\equiv 0$. The contribution from C follows because by (A1) $\nabla^2\bar{\phi}=-\omega_1^2\bar{\phi}$ in C. Hence

$$\omega_1>\pi/R. \tag{C2}$$

In dealing with the electromagnetic field it proves convenient to write Maxwell's Eqs. (A3) and (A4) as

$$\omega F=\omega\begin{bmatrix}\vec{E}\\ \vec{B}\end{bmatrix}=\begin{bmatrix}0 & i\vec{\nabla}\times\\ -i\vec{\nabla}\times & 0\end{bmatrix}\begin{bmatrix}\vec{E}\\ \vec{B}\end{bmatrix}=MF. \tag{C3}$$

Let F_1 denote a solution of (C3) in C which obeys the boundary condition $\vec{E}_t=0$ on ∂C (conducting cavity) and corresponds to the lowest eigenfrequency ω_1. We assume it is normalized: $\int_C F_1^T\cdot F_1 dV=\int_C(E^2+B^2)dV=1$. As before, we define a test field $\bar{F}$ as coinciding with F_1 in C and vanishing in $\bar{C}$. Evidently $\bar{F}$ is normalized in S and obeys the boundary condition on ∂S. But in general it has step discontinuities in some of its components across ∂C. However, being piecewise continuous in S, it can be expanded in the eigensolutions of (C3) for S, and so a straightforward generalization of the Rayleigh-Ritz principle applies:

$$\int_S \tilde{F}^T \cdot M\tilde{F}\,dV > j_{1,1}/R\,, \tag{C4}$$

where $j_{1,1}/R$ is the lowest eigenfrequency in S (see Appendix A).

Now

$$\tilde{F}^T \cdot M\tilde{F} = i\tilde{\vec{E}} \cdot \vec{\nabla}\times\tilde{\vec{B}} - i\tilde{\vec{B}}\cdot\vec{\nabla}\times\tilde{\vec{E}}\,. \tag{C5}$$

Due to discontinuities in $\tilde{F}$ across ∂C there could appear δ-function contributions to (C5) which would then contribute to its integral in (C4). We show this does not occur. Evidently only derivatives of $\tilde{\vec{E}}$ or $\tilde{\vec{B}}$ normal to ∂C can generate δ functions. Since $\tilde{\vec{E}}_t$ is continuous across ∂C, $\vec{\nabla}\times\tilde{\vec{E}}$ does not have a singularity there. Now $\vec{\nabla}\times\tilde{\vec{B}}$ could have one, but it would occur only in the components tangential to ∂C. However, $\tilde{\vec{E}}_t = 0$ there, so $\tilde{\vec{E}}\cdot\vec{\nabla}\times\tilde{\vec{B}}$ cannot contribute to the integral right at ∂C. Neither can $\tilde{\vec{B}}\cdot\vec{\nabla}\times\tilde{\vec{E}}$ which, as we saw, does not have singularities. Thus only the integral over C contributes in (C4); it can be evaluated from (C3) and we get immediately

$$\omega_1 > j_{1,1}/R\,. \tag{C6}$$

We now turn to the neutrino field. Let ψ_1 be a solution of (A6) in C which obeys the boundary condition $P^n = 0$ (n for normal) at ∂C and which correspond to the lowest eigenfrequency ω_1 for the region C. Again we define a test field $\tilde{\psi}$ which coincides with ψ_1 in C and vanishes in $\bar{C}$. Evidently $\tilde{\psi}$ obeys the boundary condition at ∂S rather trivially. We normalize ψ_1 by $\int_C \psi_1^\dagger \psi_1\,dV = 1$; then $\tilde{\psi}$ is normalized in S. Since $\tilde{\psi}$ will have only a step discontinuity across ∂C, it can be expanded in eigensolutions for S and one can generalize the Rayleigh-Ritz principle to apply to the neutrino equation (A6). It gives

$$i\int_S \tilde{\psi}^\dagger \vec{\sigma}\cdot\vec{\nabla}\tilde{\psi}\,dV > \pi/R\,, \tag{C7}$$

where π/R is the lowest eigenfrequency for the neutrino in S (see Appendix A).

Of interest here is whether the δ-function singularity that arises in $\vec{\nabla}\tilde{\psi}$ from the jump of $\tilde{\psi}$ across ∂C contributes to the integral. We show it does not. If $\vec{n}$ denotes the outward normal to ∂C, then clearly

$$\vec{\nabla}\tilde{\psi} = -\psi_{1B}\,\delta(\partial C)\vec{n} + \cdots\,, \tag{C8}$$

where $\delta(\partial C)$ denotes a δ function in the coordinate running normal to ∂C, ψ_{1B} is ψ_1 at ∂C, and the additional terms $(\cdots)$ are tangential components with no δ functions. Hence the contribution from ∂C to the integral is

$$K = -i\int_{\partial C} \psi_{1B}^\dagger \sigma^n \psi_{1B}\,d\Sigma\,, \tag{C9}$$

where $\sigma^n = \vec{\sigma}\cdot\vec{n}$ and $d\Sigma$ is the element of area on ∂C. We shall now write down explicitly the boundary condition for ψ_1, namely $P^n = 0$, at ∂C by making use of (A15):

$$-i\psi_{1B}^\dagger\sigma^n\psi_{1B} + \tfrac{1}{2}\omega_1^{-1}(\vec{\nabla}\psi_{1B}^\dagger\psi_{1B} - \psi_{1B}^\dagger\vec{\nabla}\psi_{1B})\cdot\vec{n} = 0\,. \tag{C10}$$

Substituting this into (C9) and making use of Gauss's theorem it follows that

$$K = -\tfrac{1}{2}\omega_1^{-1}\sum_C (\nabla^2\psi_1^\dagger\psi_1 - \psi_1^\dagger\nabla^2\psi_1)\,dV\,. \tag{C11}$$

However, by iterating the neutrino equation (A6) one finds $\nabla^2\psi_1 = -\omega_1^2\psi_1$ and a like equation for $\psi_1^\dagger$. Hence the integrand in (C11) vanishes identically.

Thus only C contributes to the integral in (C7). By applying (A6) one has $i\vec{\sigma}\cdot\vec{\nabla}\tilde{\psi} = \omega_1\tilde{\psi}$ in C. Thus for neutrinos the bound for ω_1 is again given by (C2). We thus have lower bounds for the lowest eigenfrequencies of the three fields in terms of the effective radius R of the cavity (radius of circumscribing sphere). One can prove exactly analogous *upper* bounds for ω_1 in terms of the radius of the sphere *inscribed* in C.

---

*On sabbatical leave from Physics Department, Ben Gurion University, Beer Sheva, Israel.

[1]S. W. Hawking, Commun. Math. Phys. 43, 199 (1975).
[2]G. Gibbons and S. W. Hawking, Phys. Rev. D 15, 2738 (1977); 15, 2752 (1977).
[3]J. D. Bekenstein, Phys. Rev. D 7, 2333 (1973).
[4]G. Gibbons, private communication.
[5]H. J. Bremermann, in *The Encyclopedia of Ignorance*, edited by R. Duncan and M. Welton-Smith (Pergamon, Oxford, 1977).
[6]J. D. Bekenstein, Phys. Rev. D 12, 3292 (1974).
[7]L. D. Landau and E. M. Lifshiftz, *Statistical Physics* (Pergamon, Oxford, 1969).
[8]B. DeWitt, private communication.
[9]B. DeWitt, in *General Relativity*, edited by S. W. Hawking and W. Israel (Cambridge University Press, Cambridge, 1979).
[10]H. B. G. Casimir, Proc. K. Ned. Akad. Wet. 51, 793 (1948).
[11]B. DeWitt, Phys. Rep. 19C, 295 (1975).
[12]G. Gibbons, in *General Relativity*, edited by S. W. Hawking and W. Israel (Cambridge University Press, Cambridge, 1979).
[13]M. B. Al'taie and J. S. Dowker, Phys. Rev. D 18, 3557 (1978).
[14]J. S. Dowker and R. Critchley, Phys. Rev. D 15, 1484

(1977).

[15]R. Balian and B. Duplantier, Ann. Phys. (N.Y.) 112, 165 (1978).

[16]J. S. Dowker and G. Kennedy, J. Phys. A 11, 895 (1978).

[17]It is sometimes possible to avoid non-Euclidean topology while still employing the popular periodic boundary conditions. Consider a field in a finite cylindrical space. One requires the field to always have identical values at corresponding points of the two bases. By causality no physical mechanism can assure this if the bases are not identified. However, one may simply restrict attention to field configurations periodic along the axis. But of course these are infinitely long, and of no interest to our problem.

[18]T. Boyer, Phys. Rev. 174, 1764 (1968); K. A. Milton, L. L. DeRaad, and J. Schwinger, Ann. Phys. (N.Y.) 115, 388 (1978).

[19]J. S. Dowker and R. Critchley, Phys. Rev. D 13, 3224 (1976).

[20]S. W. Hawking, Commun. Math. Phys. 56, 133 (1977).

[21]*Handbook of Mathematical Functions*, edited by M. Abramowitz and I. A. Stegun (National Bureau of Standards, Washington, D.C., 1964).

[22]By "conspiracy" we mean that we are close to the equality limits of *all* inequalities used to get (35). The exponent 12.8 is just the p_c corresponding to $\xi = 10^{-6}$.

[23]In fact, by using (29) with $p = 10$ we have for the electromagnetic field in a spherical cavity $1.74\,R < (S/\bar{E})_{\max} < 2.96\,R$.

[24]This is just (13) evaluated by means of the energy representation.

[25]J. D. Jackson, *Classical Electrodynamics* (Wiley, New York, 1975).

[26]J. D. Brill and J. A. Wheeler, Rev. Mod. Phys. 29, 465 (1957).

[27]A. Messiah, *Quantum Mechanics* (Wiley, New York, 1960), Vol. II.

[28]W. Unruh, private communication.

[29]B. Mashhoon, Phys. Rev. D 8, 4297 (1973).

[30]L. H. Ford, Phys. Rev. D 14, 3304 (1976).

[31]Mathematicians have long been interested in lower bounds for the lowest nonzero eigenvalue of the Laplace operator in various spaces, but have not phrased such bounds in terms of the longest dimension of the space (our R). See, for example, J. Cheeger in *Problems in Analysis: A Symposium in Honor of Salomon Bochner*, edited by R. C. Gunning (Princeton University Press, Princeton, 1970); M. Berger, D. Gauduchon, and E. Mazet, *Lecture Notes in Mathematics, No. 194: Le Spectre d'une Variété Riemanniene* (Springer, New York, 1971).

PHYSICAL REVIEW D VOLUME 25, NUMBER 6 15 MARCH 1982

Fine-structure constant: Is it really a constant?

Jacob D. Bekenstein
Physics Department, Ben Gurion University of the Negev, Beer Sheva 84120, Israel
(Received 25 September 1981)

It is often claimed that the fine-structure "constant" α is shown to be strictly constant in time by a variety of astronomical and geophysical results. These constrain its fractional rate of change $\dot{\alpha}/\alpha$ to at least some orders of magnitude below the Hubble rate H_0. We argue that the conclusion is not as straightforward as claimed since there are good physical reasons to expect $\dot{\alpha}/\alpha << H_0$. We propose to decide the issue by constructing a framework for α variability based on very general assumptions: covariance, gauge invariance, causality, and time-reversal invariance of electromagnetism, as well as the idea that the Planck-Wheeler length (10^{-33} cm) is the shortest scale allowable in any theory. The framework endows α with well-defined dynamics, and entails a modification of Maxwell electrodynamics. It proves very difficult to rule it out with purely electromagnetic experiments. In a cosmological setting, the framework predicts an $\dot{\alpha}/\alpha$ which can be compatible with the astronomical constraints; hence, these are too insensitive to rule out α variability. There is marginal conflict with the geophysical constraints; however, no firm decision is possible because of uncertainty about various cosmological parameters. By contrast the framework's predictions for spatial gradients of α are in fatal conflict with the results of the Eötvös-Dicke-Braginsky experiments. Hence these tests of the equivalence principle rule out with confidence *spacetime* variability of α at any level.

I. INTRODUCTION

Early suggestions that the fine-structure constant α (or the elementary charge) varies with cosmological time were made by Dirac,[1] Teller,[2] and Jordan.[3] Dicke and Peebles[4] pointed out that the abundance ratios of Kr^{40} to Ar^{40} and of Rb^{87} to Sr^{87} in old ores and in meteorites constrain $\dot{\alpha}/\alpha$, the fractional rate of variation of α, to at least two orders of magnitude below the Hubble rate H_0.[5] They also suggested that better bounds should result from measurements of the Re^{187} to Os^{187} ratio, because the rate of the weak decay $Re^{187}\rightarrow Os^{187}$ is highly sensitive to the value of α. Later Dicke[6] showed that the Eötvös-Dicke experiments require any *spatial* variation of α caused by the Sun to be smaller, by orders of magnitude, than the corresponding variation in the solar gravitational potential. All this evidence against α variability was soon forgotten.

Interest in α variability reached a new peak following Gamow's proposal[7] (inspired by Dirac's large-numbers hypothesis) that α varies *at* the Hubble rate. Within weeks Gamow's $\alpha \propto t$ law had been newly ruled out. Peres[8] showed that it contradicts the excellent agreement between the list of nuclides found naturally and those expected to be β stable from nuclear mass systematics. Dyson[9] showed it to conflict with Re^{187} to Os^{187} ratios in old ores; these make sense only if $|\dot{\alpha}/\alpha| < 4\times10^{-4}H_0$ according to a later review.[10] Bahcall and Schmidt[11] called attention to the good agreement between fine-structure splittings in radiogalaxy spectra and their laboratory values; this implies $|\dot{\alpha}/\alpha| < 10^{-2}H_0$ in conflict with Gamow's suggestion. Later work has strengthened the trend. A laboratory intercomparison of cesium and superconducting cavity clocks by Turneaure and Stein[12] has given the only known bound on the instantaneous $\dot{\alpha}/\alpha$: $|\dot{\alpha}/\alpha| < 8\times10^{-2}H_0$. Wolfe, Brown, and Roberts[13] have extended the baseline of the astronomical test for $\dot{\alpha}$ by looking at a BL-Lacertae object at red-shift $z=0.52$; quasars make possible a further extension.[14] Finally, from an analysis of isotopic abundances of fission products in the Oklo "natural reactor" (a uranium vein in Gabon which spontaneously underwent a fission episode 1.8×10^9 years ago), Shlyakhter[15] derived the best constraint yet: $|\dot{\alpha}/\alpha| < 10^{-7}H_0$.

These impressive constraints are widely regarded as establishing perfect constancy of α. One often hears the argument that *if* α varied in time, it would be expected to vary roughly at the Hubble rate, which it does not; hence α is constant. There are several objections to such reasoning. It implicitly assumes a power law in time for α, with an exponent not widely different from unity. Yet other behaviors are conceivable. For example, α could exhibit an asymptotic approach to a finite value, with little present variation (this is the typical behavior of G in a large class of gravitational theories[16,17]). The extant constraints might not be sufficient to rule this out. Evidently, in the absence of a concrete dynamical equation for α, the mentioned argument essentially begs the question. Further, the idea that the Hubble rate is the only characteristic rate for variable quantities in an expanding universe is problematic. The Hubble expansion is driven (via the gravitation equations) by the *total* energy density in the universe. But α is an electromagnetic quantity; its temporal variation, if any, should be driven by the density of some electromagnetic quantity. Since various electromagnetic energy densities are orders of magnitude below the rest energy density (at least in recent times), one has good reasons to expect $|\dot{\alpha}/\alpha|$ to be *much smaller* than H_0.

All this underscores the need for a *general* dynamical framework for α. Only within the context of such can it be clear if the sensitivity of a given experimental constraint is sufficient to yield firm conclusions.[16] And only within such a framework can one combine constraints on spatial and on temporal variability of α into unified evidence. Without it one hardly knows how to compare the merits of the two, let alone integrate them. The main goal of the present paper is the construction of just such a framework based on a few uncontroversial assumptions, and its application to the question, Do the extant constraints rule out all sorts of α variability? We answer affirmatively.

Our assumptions are discussed in Sec. II. In Sec. III we construct the Lagrangian giving the dynamics of the electromagnetic field and of α. An unspecified scale of length l enters into it. From the observed scale invariance of electromagnetism $l < 10^{-15}$ cm; also $l > 10^{-33}$ cm because the Planck-Wheeler length is the shortest conceivable length in physics. In Sec. IV we show that purely electromagnetic experiments are incapable of ruling out α variability. Temporal variability of α in our expanding universe is taken up in Sec. V. The framework predicts that in recent epochs $|\dot{\alpha}/\alpha| \approx kH_0$ with k probably in the range $10^{-4}-10^{-6}$. This small value is due partly to the mentioned smallness of electromagnetic energies and partly to the appearance in the expression for $\dot{\alpha}/\alpha$ of an integration constant which can act to suppress $\dot{\alpha}/\alpha$. As a result the constraints from intercomparison of clocks[12] and from the spectra of radio galaxies and quasars[11,13,14] do not suffice to rule out α variability. The geophysical constraints[10,15] come rather close to ruling it out; however, uncertainties about various cosmological parameters prevent us from reaching a firm conclusion. Predictions of the framework for spatial α variability are the theme of Sec. VI. We confirm Dicke's conclusion[6] that tests of the equivalence principle provide very strong constraints on α variability. The Dicke[6] and Braginsky[18] versions of the Eötvös experiment strongly rule out any α variability, temporal or spatial. Our conclusions are discussed in Sec. VII.

II. POSTULATES OF THE FRAMEWORK

The fine-structure constant α is the low-energy limit of the (renormalized) electromagnetic coupling constant. Being interested primarily in macroscopic phenomena, we adopt a classical description of the electromagnetic field and its interaction with (classical or quantum) matter.

Once the possibility that α depends on space and time is admitted, modification of standard Maxwell electrodynamics is inevitable. For example, adopting the usual relation between α and elementary charge, $\alpha = e^2/\hbar c$, we see that α variability implies e variability in units for which $\hbar$ and c are constant. But e variability seems to clash with charge conservation (the electron's charge depends on position) which follows from Maxwell's equations. Evidently, something in the accepted picture of electrodynamics must give way. What we need, then, is a set of assumptions to guide us in modifying Maxwell electrodynamics in a reasonable way. These should allow a general, model-independent, framework for α variability; at the same time they should respect generally accepted physical principles. We adopt the following assumptions and discuss each in turn.

P1. *For constant α electromagnetism is Maxwellian and the coupling of the vector potential to matter is minimal.* This is a correspondence principle restraining us from introducing unnecessary modifications of standard physics.

P2. *Variations of α result from dynamics.* One must reject prescribed laws of variation. If α varies, the variation influences charged matter; charged matter should in turn influence α. Only dynamics for α can incorporate this important feature.

P3. *Dynamics of electromagnetism and α are derivable from an invariant action.* We know of no other way to consistently incorporate the principle "action equals reaction" than to derive dynamical equations from an action principle. And no better way offers itself to guarantee relativistic invariance than to start from an invariant action.

P4. *The action is locally gauge invariant.* The importance of the gauge principle in contemporary physics cannot be overstated. Thanks to it there is now a unified picture of all the microscopic interactions. If we gave up demanding full gauge invariance, we would be modifying Maxwell's theory in an unreasonable way.

P5. *Electromagnetism is causal.* There is no *experimental* evidence for causality violation. Theoretically it is sometimes discussed, but only in rather esoteric situations. In our rather mundane problem causality should be enforced: dynamical equations should be hyperbolic and of at most second order, in order to forestall runaway solutions and other noncausality.

P6. *The electromagnetic action is time reversal invariant.* This is taken for granted in contemporary particle physics where any violations of C, P, or T invariance are blamed on the weak interaction, or the superweak facet of it.[19] Concrete evidence for P and T invariance of electromagnetism are the very tight upper bounds on the electric dipole moments of the proton and neutron.[20] Of course such evidence establishes T invariance only to some finite precision. We assume it is exactly true and thereby accept a risk.

P7. *The shortest scale of length which can enter into physical theory is the Planck-Wheeler length $L_{PW} \equiv (G\hbar/c^3)^{1/2} \approx 10^{-33}$ cm.* It has been argued that at shorter scales spacetime itself is not smooth, or even simply connected.[21] If so, it makes no sense to talk of smooth fields obeying differential equations on scales shorter than $L_{\rm PW}$; introducing an even shorter length into a theory cannot have any palpable consequences. Another way of saying this is that to probe experimentally the predictions of a theory down the scale l requires use of particles of energy $E \approx \hbar c/l$. If $l < L_{\rm PW}$, E is so large that its gravitational radius $2GE/c^4$ exceeds l. So the particle probe would tend to fall into a black hole of its own making, thus frustrating any attempt to learn from its scattering about the structure of the theory at scales $< L_{\rm PW}$.

P8. *Gravitation is described by the metric of spacetime which satisfies Einstein's equations.* The aptness of the geometric description of gravitation is clear enough today. Einstein's equations are the simplest dynamics for the metric which have stood all experimental challenges to date. We adopt them, not out of blind confidence in their correctness, but because they allow a clean separation of the issue of α variability from that of G variability which exists in many competing gravitational theories. We do not assume the other half of general relativity—the strong equivalence principle—because it rules out α variability by definition.

III. DYNAMICS OF THE FRAMEWORK

A. Electrodynamics

We choose units of length, time, and mass so that $\hbar$, c, and G are spacetime constants; this can always be done. Since $\alpha = e^2/\hbar c$, where e is the electron charge, α variability means that e depends on the spacetime point. We expect the charges of all particle species to vary in exactly the same way. Otherwise charge ratios would vary, and there could not exist accurately neutral atoms (equal electron and proton charges) or hadrons (zero net quark charge), as observed. Thus every particle charge can be expressed in the form $e = e_0\epsilon(x^\mu)$, where e_0 is a constant characteristic of the particles and ϵ a dimensionless universal field. The particular split between e_0 and ϵ is our choice. We can rescale ϵ by a constant factor and make appropriate changes in all the e_0 so as to get the previous e's. Thus the theory governing ϵ should not fix its overall scale: it should be invariant under the change $\epsilon \rightarrow \text{constant} \times \epsilon$. We shall find this constraint very useful.

Consider now the classical dynamics of a charged particle of rest mass m and charge $e_0\epsilon$ in flat spacetime. We may start with the Lagrangian[22,23]

$$L = -mc(-u^\alpha u_\alpha)^{1/2} + (e_0\epsilon/c)u^\alpha A_\alpha \,, \tag{1}$$

where $u^\alpha = dx^\alpha/d\tau$ is the four-velocity and τ is the proper time. We notice that L is Lorentz invariant and involves mimimal coupling. The appropriate gauge transformation law for A_α is

$$\epsilon A_\alpha \rightarrow \epsilon A_\alpha + \chi_{,\alpha} \,, \tag{2}$$

where χ is an arbitrary function: under (2) L changes only by the perfect derivative

$(e_0/c)d\chi/d\tau$. The minimal coupling in (1) is required (P1) only in the $\epsilon\rightarrow$ const limit. One might thus be tempted to introduce other couplings consistent with this, such as $\epsilon_{,\alpha}A^{\alpha}$ or $\epsilon_{,\alpha}\epsilon_{,\beta}A^{\alpha}u^{\beta}$. However, no single rule such as (2) involving a truly arbitrary function guarantees gauge invariance of the dynamics. Thus we exclude these modifications by P4.

We notice that the curl of A_α is not invariant under (2); it cannot be the physical electromagnetic field $F_{\alpha\beta}$. To identify $F_{\mu\nu}$ we turn to the Lagrange equations for (1):

$$\frac{d}{d\tau}[mu_\alpha+(e_0/c)\epsilon A_\alpha]=-m_{,\alpha}c^2+(e_0/c)(\epsilon A_\beta)_{,\alpha}u^\beta . \quad (3)$$

In writing (3) we have already taken into account the normalization relation

$$u_\alpha u^\alpha=-c^2 . \quad (4)$$

In addition we have regarded m as $m(x^\alpha)$; since the particle is charged, part of its rest energy is electromagnetic in nature and should depend on position through ϵ.[6] Simplifying (3) we get

$$d(mu_\alpha)/d\tau=-m_{,\alpha}c^2+(e_0/c)[(\epsilon A_\beta)_{,\alpha}-(\epsilon A_\alpha)_{,\beta}]u^\beta . \quad (5)$$

The first term on the right-hand side is the anomalous force first discussed by Dicke,[6] due to variation of ϵ. The second is a Lorentz-type force. Identifying it with the usual expression $(e_0\epsilon/c)F_{\alpha\beta}u^\beta$ we arrive at the definition

$$F_{\alpha\beta}=\epsilon^{-1}[(\epsilon A_\beta)_{,\alpha}-(\epsilon A_\alpha)_{,\beta}] . \quad (6)$$

This agrees with the Maxwellian version only for $\epsilon\rightarrow$ const. We note that $F_{\alpha\beta}$ is gauge invariant and invariant under rescaling ϵ by a constant.

To build the electromagnetic Lagrangian (P3) we have at our disposal $F_{\alpha\beta}$ and its dual

$$^*F^{\mu\nu}=\tfrac{1}{2}\epsilon^{\mu\nu\alpha\beta}F_{\alpha\beta} , \quad (7)$$

where $\epsilon^{\mu\nu\alpha\beta}$ is the Levi-Civita tensor. $^*F_{\mu\nu}$ shares the invariances of $F_{\mu\nu}$. One possible electromagnetic action is

$$S_{\rm EM}=-(16\pi)^{-1}\int F^{\mu\nu}F_{\mu\nu}(-g)^{1/2}d^4x , \quad (8)$$

which reduces to the Maxwell action for constant ϵ. The other possible Lagrangian density is $^*F^{\mu\nu}F_{\mu\nu}$. When ϵ is constant this is just a perfect divergence (recall that $^*F^{\mu\nu}{}_{;\nu}\equiv0$). For this reason one never considers $^*F^{\mu\nu}F_{\mu\nu}$ in ordinary electromagnetism. When ϵ can vary, $^*F^{\mu\nu}{}_{;\nu}\neq0$ and $^*F^{\mu\nu}F_{\mu\nu}$ is not a perfect divergence; one could thus include it in the action. However, $^*F^{\mu\nu}F_{\mu\nu}$ changes sign under time reversal so it is excluded by P6. We did not include a mass term $m^2A_\mu A^\mu$ since it violates P4. Likewise, we excluded terms of fourth or higher order in $F_{\mu\nu}$ from $S_{\rm EM}$ because they do not vanish for constant ϵ and thus violate P1. We are thus left with (8) as the general electromagnetic action satisfying our assumptions.

B. Dynamics of α

$S_{\rm EM}$ is so simple in structure that variation of it with respect to ϵ does not give a proper dynamical equation for ϵ. By P2 and P3 we must thus introduce a separate action for ϵ. Evidently, the Lagrangian density should be constructed from the metric and the logarithmic gradient $\epsilon^{-1}\epsilon_{,\mu}$ which is invariant under rescaling of ϵ by a constant. One cannot introduce a function of ϵ by itself since it changes under rescaling. [For this reason also, we did not include such a function in (8).] Now, because ϵ is dimensionless, $\epsilon_{,\mu}\,\epsilon^{,\mu}\,\epsilon^{-2}$ does not have dimensions of a Lagrangian density (even if we introduce $\hbar$'s and c's at will). This led Meisels[24] to conclude that no satisfactory dynamics can be set up for a dimensionless coupling constant. Let us see if this pessimistic conclusion is unavoidable.

We can try the Lagrangian density $R\epsilon^{-2}\epsilon_{,\mu}\epsilon^{,\mu}$ or else $R^{\mu\nu}\epsilon^{-2}\,\epsilon_{,\nu}\epsilon_{,\mu}$, where $R^{\mu\nu}$ and R are, respectively, the Ricci tensor and scalar. These have, after multiplication by $\hbar c$, the correct dimensions. But they are problematical. The R or $R^{\mu\nu}$ are determined by Einstein's equations. The ϵ field is one of the sources of these equations, and up to third order derivatives of ϵ appear in the expressions for R or $R^{\mu\nu}$ (variation of R or $R^{\mu\nu}$ with respect to the metric brings two extra derivatives to bear on $\epsilon_{,\nu}\epsilon_{,\mu}$). Thus in its explicit form the differential equation for ϵ will contain third- and fourth-order derivatives as well as second. This contradicts P5 so we must exclude Lagrangians with curvature in them.

If a scale of length l is available, one can build a satisfactory action, namely,

$$S_\epsilon = -\tfrac{1}{2}\hbar c l^{-2} \int \epsilon^{-2}\epsilon_{,\mu}\epsilon^{,\mu}(-g)^{1/2}d^4x\ , \tag{9}$$

where the factor $-\frac{1}{2}$ is conventional. Of course, introduction of l must clash at some level with scale invariance, an experimentally established feature of electromagnetism. However, the electromagnetic interaction has been probed only up to energies in the GeV range, so by the usual argument one can vouch for scale invariance only down to 10^{-15} cm or so. Provided l is shorter than this, no contradiction need arise. In fact we shall see in Sec. IV that the electric field of a truly point charge described by $S_{\rm EM}$ and S_ϵ is accurately Coulombic down to a distance $\lesssim l$. Hence scattering experiments with the present energies cannot reveal the presence of a $l < 10^{-15}$ cm, say. One may also recall the current view that electromagnetism merges with the weak interaction at energies equivalent to a length $\approx 10^{-16}$ cm, and both merge with the strong interaction at energies equivalent to $\approx 10^{-29}$ cm. Whatever the details of this unification, one should not be surprised if a scale of length makes its way into dynamics of the electromagnetic *coupling*. We do not try to guess l, but merely require (P6)

$$l \geq L_{\rm PW} \equiv \left[\hbar G/c^3\right]^{1/2} \approx 10^{-33}\ {\rm cm}. \tag{10}$$

We did not consider the more general Lagrangian density

$$-\tfrac{1}{2}\hbar c l^{-4} f\left[l^2\epsilon^{-2}\epsilon_{,\mu}\epsilon^{,\mu}\right]$$

with f an arbitrary function because, as shown in the Appendix, any choice of f, except that in (9), leads to causality violations, and must be excluded by (P5). Thus a Lagrangian quadratic in $\epsilon^{-1}\epsilon_{,\mu}$ is forced on us.

C. Structure of the equations

In order to write the dynamical equations for $F_{\mu\nu}$ and ϵ, we must specify the matter action. Being interested in the macroscopic manifestations of ϵ variability, we adopt a model of point classical particles for the matter. By "point" we mean small by macroscopic standards, but large compared to l. Since $l < 10^{-15}$, the model is a good one for ordinary matter composed of atoms, nuclei, electrons, ... For many point particles with masses m coupling to A_μ with strengths $e_0\epsilon$, the action may be written as[22] [see (1)]

$$S_m = \sum \int [-mc^2 + (e_0\epsilon/c)u^\mu A_\mu] \times \gamma^{-1}\delta^3[x^i - x^i(\tau)]d^4x\ , \tag{11}$$

where $x^i(\tau)$ with $i=1,2,3$ describe the spatial track of the typical particle as a function of its proper time τ, $u^\mu = dx^\mu/d\tau$, while $\gamma = dx^0/cd\tau$ is the Lorentz factor. We regard each m as a function of ϵ because part of m is electromagnetic in nature and depends on the local strength of the electromagnetic interaction.

Variation of $S_{\rm EM} + S_\epsilon + S_m$ with respect to A_μ gives the electromagnetic field equations

$$(\epsilon^{-1}F^{\mu\nu})_{;\nu} = 4\pi j^\mu \tag{12}$$

with

$$j^\mu = \sum (e_0/c\gamma)u^\mu(-g)^{-1/2}\delta^3[x^i - x^i(\tau)]\ . \tag{13}$$

Because $\epsilon^{-1}F^{\mu\nu}$ is antisymmetric we have the identity $j^\mu{}_{;\mu}=0$. The conserved charge, the sum $\sum e_0$ over particles, is thus distinct from the sum of the "charges" $e_0\epsilon$ which couple to A_μ in the action. Thus is charge conservation reconciled with variability of ϵ (or α). We note, also from (12), that Gauss's theorem only permits the determination of $\sum e_0$ from the distant electric field of a system (provided ϵ is known at the field points).

There exists an alternative view of the electrodynamics just discussed. One can take the view that the Lorentz force in (5) is the force on a *constant* charge e_0 of the field $\epsilon F_{\alpha\beta}$ (a curl). If $\epsilon F_{\alpha\beta}$ is interpreted as the electromagnetic field, then Eqs. (12) have just the form of Maxwell's equations in a *material* medium with dielectric constant ϵ^{-2} and permeability ϵ^2. That is, if time-space components of $\epsilon F^{\alpha\beta}$ are identified with the electric field $\vec{\rm E} = \epsilon^2\vec{\rm D}$, while space-space components are identified with the magnetic induction $\vec{\rm B} = \epsilon^2\vec{\rm H}$, then $\vec{\rm D}$ and $\vec{\rm H}$ satisfy the usual "macroscopic" Maxwell equations with the conserved charge and current as sources. This alternative description of variable α was actually discussed long ago by Dicke [Eqs. (26)−(28), Appendix 4 of Ref. 6] who also pointed out that α is to be identified, not with $e_0{}^2/\hbar c$, but with $e_0{}^2\epsilon^2/\hbar c$ as in our formalism. Variation of α can thus be described in two languages (variable charges in a homogeneous vacuum or constant charges in an inhomogeneous medium) with the same physical content. The difference between our formalism and Dicke's is in the dynamics for ϵ. Dicke thinks of ϵ^{-2} as an ordinary scalar field;[6] our requirement that the overall scale of ϵ have no physical significance forces us to adopt scalar-field dynamics for $\ln\epsilon$ [Eq. (9)].

We know experimentally that ϵ varies little over "earthly" distances and times. Thus, for fields $F^{\mu\nu}$ which do vary on such scales, the factor ϵ^{-1} in (12) can be taken out from under the derivative, and we recover, to a good approximation, Maxwell's equations with the $e_0\epsilon$ as sources. At this point we use our prerogative to choose the overall scale of ϵ so as to make $\epsilon=1$ at our present cosmological epoch, and far away from strong sources of the field ϵ (see Sec. IV.). With this choice the e_0 coincide with the usual charges of particles—the sources of Maxwell's equations— *at the present epoch.*

The dynamical equation for ϵ is obtained by varying $S_{\rm EM}+S_\epsilon+S_m$ with respect to $\ln\epsilon$. With the notation

$$\sigma=\sum mc^2\gamma^{-1}(-g)^{-1/2}\delta^3[x^i-x^i(\tau)] , \qquad (14)$$

we have

$$\Box\ln\epsilon=\frac{l^2}{\hbar c}\left[\epsilon\frac{\partial\sigma}{\partial\epsilon}-\epsilon j^\mu A_\mu+\frac{1}{4\pi}\left(A_\mu F^{\mu\nu}\right)_{;\nu}\right] , \qquad (15)$$

where $\Box$ denotes the covariant D'Alembertian. Substituting j^μ from (12) into (15) we obtain the alternative form

$$\Box\ln\epsilon=\frac{l^2}{\hbar c}\left[\epsilon\frac{\partial\sigma}{\partial\epsilon}-\frac{1}{8\pi}F^{\mu\nu}F_{\mu\nu}\right] \qquad (16)$$

from which all reference to current has disappeared; this will be useful in Sec. V.

IV. ELECTROMAGNETIC TESTS

We now show that purely electrostatic laboratory experiments do not rule out the electrodynamics with variable α developed in Sec. III. One of the most accurate tests of Maxwell's equations is the classic Faraday "ice-pail" experiment. It is found that in the charge-free interior of a highly conducting cavity, there is no measurable electrostatic field, whatever the exterior situation. As is well known Gauss's law, together with the equipotential nature of the cavity walls predicts this vanishing of the electric field. A similar result holds in the variable α electrodynamics, as we now show.

Working in flat spacetime we identify F^{0i}, $i=1,2,3$ with the electric field vector $\vec{\rm E}$. Setting $\mu=0$ in (12) we get in the cavity interior

$$\vec\nabla\cdot(\epsilon^{-1}\vec{\rm E})=0 . \qquad (17)$$

Multiplying (17) by ϵA_0, integrating over the interior volume of the cavity, and integrating by parts we are left with the surface term

$$\oint \epsilon A_0\epsilon^{-1}\vec{\rm E}\cdot d\vec{\rm S} , \qquad (18)$$

where the integral extends over the walls. Since these conduct well, ϵA_0 must be strictly constant in (18); otherwise, $\epsilon\vec{\rm E}=\vec\nabla(\epsilon A_0)$ would not vanish in the walls, and a large current would flow. Thus, we may take ϵA_0 out of the integral (18). The remaining integral must vanish by (17) and Gauss's theorem. The volume integral is then

$$\int \vec{\rm E}^2 dV=0 \qquad (19)$$

which proves that $\vec{\rm E}$ vanishes identically inside the cavity, in harmony with the ice-pail experiment.

Does the proposed electrodynamics predict observable departure from Coulomb's law? To answer this we calculate the *exterior* field of a spherical distribution of charge in flat space. We exclude magnetic monopoles; thus we allow only a spherically symmetric radial electric field. Outside the charges (17) is solved by

$$\vec{\rm E}=\hat r\epsilon Q/r^2 , \qquad (20)$$

where Q is a constant. By Gauss's law one interprets Q as $\sum e_0$ over the particles in the charge distribution (see Sec. III). In order to calculate ϵ we substitute (20) into the right-hand side of (16). Again, outside the charges, for a static situation we get the equation

$$r^{-2}\partial_r[r^2(\partial_r\ln\epsilon)]=(4\pi\hbar c)^{-1}l^2Q^2\epsilon^2r^{-4} \qquad (21)$$

since $F^{\mu\nu}F_{\mu\nu}=-2\vec{\rm E}^2$.

By appropriate transformations we are able to solve (21). Here we only give the result:

$$\epsilon(r)=\sec[lQ(4\pi\hbar c)^{-1/2}r^{-1}] . \qquad (22)$$

This can be checked by direct substitution. We observe that as $r\to\infty$, $\epsilon\to1$ in harmony with our choice of scale. At large distances [when $r>>lQ(4\pi\hbar c)^{-1/2}$], $\vec{\rm E}$ has a purely Coulombic form: $\vec{\rm E}=\hat rQr^{-2}$. This accords with the success of Coulomb's law in experimental studies. If we are dealing with an elementary charge

$(Q/\sqrt{\hbar c} \cong 1/\sqrt{137})$, then we see that the Coulombic behavior is already accurate (argument of secant small compared to unity) for $r \gtrsim l$. Since leptons are known to have an accurately Coulombic field (apart from vacuum polarization corrections) down to distances 10^{-15} cm, we conclude that $l < 10^{-15}$ cm, in harmony with our earlier claim.

Let us now consider a spherical *macroscopic* distribution of charge of radius R. If $l < 10^{-15}$ cm, can one ever detect departures from the r^{-2} law outside such a distribution? This would require that $lQ(4\pi\hbar c)^{-1/2}R^{-1}$ be not much smaller than unity, i.e., that the *formal* electrostatic potential Q/R be no smaller than about $\sqrt{\hbar c}/l$. If $l = 10^{-16}$ cm, the required potential is at least 10^{10} V (1 esu = 300 V). For laboratory-sized charged objects, R is in the range $10^{-2}-10^{3}$ cm, so there would be fields of strength at least 10^7 V/cm present. Such fields exceed by far the critical field for air breakdown (3×10^4 V/cm), so that the required experiments would have to be carried out in high vacuum (which is not usually done). Even this option would prove useless if $l \lesssim 10^{-23}$ cm since the required fields could never be set up; they would exceed the vacuum's breakdown field (10^{14} V/cm). As it is, potentials of 10^7 V obtainable in Van de Graaf accelerators are about as large as can be had in the laboratory, but they fall short of our requirements.

Departures from spherical symmetry do not mend matters. For a collection of point charges Q_i, the solution of (17) is

$$\vec{E} = \epsilon \sum Q_i \vec{r}_i r_i^{-3}, \tag{23}$$

where $\vec{r}_i$ is the vector from Q_i to the field point. Instead of (21) we now have

$$\nabla^2 \ln\epsilon = (4\pi\hbar c)^{-1} l^2 \vec{E}^2 \tag{24}$$

outside the sources. As direct substitution and use of (17) shows, the solution of (23)–(24) is

$$\epsilon = \sec\left[l(4\pi\hbar c)^{-1/2} \sum Q_i r_i^{-1}\right]. \tag{25}$$

Thus despite the nonlinearity of (23)–(24), there is a sense in which the effects of separate charges can be superimposed. For $\vec{E}$ to depart from a superposition of Coulomb fields, ϵ must depart from unity. From (25) it is clear that this requires the potential $\sum Q_i/r_i$ to be large in our earlier sense. For the reasons already mentioned we again conclude that electrostatic experiments have trouble distinguishing between the variable α electrodynamics and Maxwell electrodynamics.

V. TEMPORAL VARIATIONS OF α

We now wish to calculate temporal variations of α or ϵ in a cosmological setting. We assume the universe to be homogeneous and isotropic in the large over the epochs of interest. Adopting a Robertson-Walker metric with expansion factor $a(t)$, we write the equation (16) for ϵ as

$$(a^3\dot{\epsilon}/\epsilon)^{\cdot} = -\frac{a^3 l^2}{\hbar c}\left[\epsilon\frac{\partial\sigma}{\partial\epsilon} + \frac{1}{4\pi}(\vec{E}^2 - \vec{B}^2)\right], \tag{26}$$

where a dot stands for a time derivative, and $\vec{E}$ and $\vec{B}$ are the fields measured by a commoving observer. Since we are not interested in spatial variations of ϵ resulting from graininess in the contents of the Universe, the source term in the right-hand side of (26) is to be understood as averaged over a large volume.

The Universe is filled with electromagnetic radiation, but this does not contribute to the source of ϵ in (26). We recall that for a plane electromagnetic wave the invariant $\vec{E}^2 - \vec{B}^2$ vanishes.[25] An incoherent distribution of waves (i.e., blackbody radiation) is a linear superposition of plane waves with no definite phase relations. Because of this last point, a space average of $\vec{E}^2 - \vec{B}^2$ lacks any cross terms between different waves. Thus, incoherent radiation does not contribute to (26). The incoherence also eliminates cross terms between fields of the waves and field of material particles. Thus *only matter* is a source of ϵ.

Both terms in the square brackets in (26) are Lorentz invariants. In evaluating the contribution of a particle to the source, one can thus pass to its rest frame.[26] Averaging such a contribution over a *large* volume V_0 in the rest frame of the particle gives, by (14), the source term

$$\left[\epsilon(\partial m/\partial\epsilon)c^2 + (4\pi)^{-1}\int_{V_0}(\vec{E}^2 - \vec{B}^2)dV\right]V_0^{-1}. \tag{27}$$

Now, one can represent the electromagnetic part of the energy mc^2, $m_{\rm EM}c^2$, by

$$m_{\rm EM}c^2 = (8\pi)^{-1}\int_{V_0}(\vec{E}^2 + \vec{B}^2)dV \tag{28}$$

so that the source can be written as

$$\left[\epsilon(\partial m/\partial\epsilon)c^2+2m_{\rm EM}c^2-(2\pi)^{-1}\int_{V_0}\vec{B}^2dV\right]V_0^{-1}\,. \tag{29}$$

Since magnetic energy is a very tiny fraction of the total electromagnetic energy in ordinary nonrelativistic matter, the magnetic term in (29) can be neglected. Further, the bulk of $m_{\rm EM}c^2$ is Coulombic in nature, so to a good approximation $m_{\rm EM}\propto\epsilon^2$. It follows that $\epsilon(\partial m/\partial\epsilon)c^2\approx 2m_{\rm EM}c^2$. Thus the quantity in square brackets in (26) is essentially *four* times the density of Coulomb energy.

The bulk of rest mass in the Universe is in the form of protons and neutrons, free and bound in nuclei; these are the main contributors to the Coulomb energy density. Electrons, because they are so light, contribute negligibly. So do atomic and interionic interactions. The Coulomb energy of the proton can be estimated in two ways. If we extrapolate the nuclear mass formula[27,28] to $Z=1$, $A=1$ we get $m_{\rm EM}c^2\approx 0.7$ MeV. If we model the proton by a uniformly charged sphere of radius 1 fm, we get $m_{\rm EM}c^2\approx 0.86$ MeV. Both estimates should be on the small side since the charge in the proton is strongly concentrated.[27] But even if it were concentrated within the proton's Compton length (which it is not), we should have $m_{\rm EM}c^2\approx\alpha mc^2\approx 7$ MeV. Thus, to within a factor of 2, $m_{\rm EM}c^2\approx 3$ MeV $\approx 3\times 10^{-3}mc^2$. The neutron's rest energy exceeds the proton's by 1.29 MeV. It is generally agreed that this splitting is purely electromagnetic in origin.[27] Thus we estimate for the neutron $m_{\rm EM}c^2\approx 4.3$ MeV.

Most neutrons in the Universe have been incorporated into nuclei, primarily into He^4 which makes up about a quarter by mass of matter in the Universe. The mass formula indicates that the Coulomb contribution to the *binding* energy of the He^4 nucleus is ≈ 1.8 MeV, or 0.45 MeV per nucleon. Thus a nucleon in He^4 contributes only $\approx 15\%$ more Coulomb energy than a free nucleon. Nucleons in other abundant nuclei do better. For example, in O^{16} and Fe^{56} they contribute 40% and 70% more Coulomb energy than when free. However, all nuclei apart from H^1 and He^4 account for only 2% of all matter by mass.[28,29] Thus, when weighted by the appropriate abundances, the average nucleon's Coulomb energy exceeds the free proton contributions by only 10%. We may thus rewrite (26) as

$$(a^3\dot{\epsilon}/\epsilon)^{\cdot}=-a^3\zeta(l^2/\hbar c)\rho_m c^2\,, \tag{30}$$

where ρ_m is the total rest mass density of matter, and $\zeta\approx 1.3\times 10^{-2}$.

We now argue that ζ may be taken as constant in (30). For it to change significantly there must be a significant change in the relative abundance of H^1 and other nuclei. It might be argued that nucleosynthesis in stars has been accomplishing just such a change. However, as we now show, the effect is negligible. All transformation of H^1 into heavier nuclei proceeds via the process $4H^1\to He^4$ which releases 2×10^{-5} erg per proton converted. This energy is the principal source of optical luminosity of galaxies. A recent estimate[30] sets the present mean luminosity density at 1.6×10^{-32} erg sec^{-1} cm^{-3} corresponding to a proton conversion rate of 8×10^{-28} sec^{-1} cm^{-3}. A *minimal* estimate for the nucleon number density in the Universe follows from considerations of primordial He^4 synthesis[31]: 1×10^{-7} cm^{-3}. Some 90% of these are protons.[28] Thus the time scale for proton conversion is $\gtrsim 10^{20}$ sec; that for significant changes in ζ is similar. But the time scale for changes of ρ_m or a^3 in (30) is $(3H_0)^{-1}\approx 10^{17}$ sec. Thus in (30) ζ is highly constant when compared with the other quantities. It changes $\lesssim 0.3\%$ per Hubble time $H_0{}^{-1}$. However, galaxies, when young, should have been much brighter than today.[32] Thus, constancy of ζ can safely be assumed only as far back as $z\approx 1$. Before then galaxies may have converted protons much faster than our estimate suggests, and ζ might have varied substantially.

Since ρ varies as a^{-3} in an expanding Universe, we can immediately integrate (30); assuming ζ constant

$$\dot{\epsilon}/\epsilon=-\zeta(l^2c/\hbar)\rho_m(t-t_C)\,, \tag{31}$$

where t_C is an integration constant. To estimate $\dot{\epsilon}/\epsilon$ at the present time t_0, we may as usual express ρ_m as a multiple Ω_{m0} of the critical density $\rho_C\equiv(3/8\pi G)H_0{}^2$, where G is the gravitation constant. Since $(G\hbar/c^3)^{1/2}$ is just the Planck-Wheeler length $L_{\rm PW}$, we get the suggestive relation

$$(\dot{\epsilon}/\epsilon)_{t=t_0}=-1.6\times 10^{-3}(l/L_{\rm PW})^2\Omega_{m0}H_0{}^2\tau\,, \tag{32}$$

where $t\equiv t_0-t_C$ and we have set $\zeta=1.3\times 10^{-2}$. An upper bound on $\dot{\epsilon}/\epsilon$ in terms of the unknowns l and τ follows by taking for H_0 the largest value discussed today, $H_0=100$ km/Mpc$=1.02\times 10^{-10}$ yr^{-1}, and for Ω_{m0} the large value 0.2 obtained by assuming that all gravitating matter is nucleonic (an exaggeration), and that the specific luminosity of great clusters of galaxies equals that of the typical galaxy.[31] Both assumptions tend to overestimate Ω_{m0}. Thus

$$|(\dot{\epsilon}/\epsilon_0)_{t=t_0}| \leq 3.2\times 10^{-24}(l/L_{\rm PW})^2|\tau|y^{-2}\,. \qquad (33)$$

Turneaure and Stein[12] have set the laboratory bound $|\dot{\epsilon}/\epsilon| = \frac{1}{2}|\dot{\alpha}/\alpha| < 5\times 10^{-12}\ \text{yr}^{-1}$. We do not know τ but since our equation is valid for times of order t_0, it would be strange indeed if $|\tau|$ were orders of magnitude larger than t_0. If $|\tau| < t_0$ and $l/L_{\rm PW} < 10$, or if $|\tau| < 0.1H_0^{-1} \approx 0.1\ t_0$ and $(l/L_{\rm PW}) < 40$ there is no contradiction between (33) and the experimental bound. Evidently, in the absence of a concrete value for l, the experiment *cannot* rule out α variability.

We now turn to the evolution of ϵ in the past. Since $\rho_m \propto (1+z)^3$, where z is the red-shift, we may integrate (31) by using the standard relation between "lookback time" $\Delta t \equiv t_0 - t$ and red-shift[33]

$$z = H_0\Delta t + (1+\tfrac{1}{2}q_0)H_0^2\Delta t^2 + O(\Delta t^3)\,, \qquad (34)$$

where q_0 is the deceleration parameter. We get

$$\rho_m(t) = (3/8\pi G)\Omega_{m0}H_0^2[1+3H_0\Delta t+3(2+\tfrac{1}{2}q_0)H_0^2\Delta t^2+O(\Delta t^3)]\,. \qquad (35)$$

Recalling our convention $\epsilon(t_0)=1$ and assuming $|\epsilon-1|$ to be small we get

$$\epsilon(t)-1 = -1.6\times 10^{-3}(l/L_{\rm PW})^2\Omega_{m0}H_0^2\Delta t F(\Delta t,\tau)\,, \qquad (36)$$

$$F(\Delta t,\tau) \equiv \tau[1+\tfrac{3}{2}H_0\Delta t+(2+\tfrac{1}{2}q_0)H_0^2\Delta t^2] - \tfrac{1}{2}\Delta t - H_0\Delta t^2 + O(\Delta t^3)\,. \qquad (37)$$

From observations of fine-structure splitting in radio galaxies, Bahcall and Schmidt[11] concluded that at red-shift $z=0.2$, $|\epsilon(t)-1| = \frac{1}{2}|\alpha(t)/\alpha(t_0)-1| < 1\times 10^{-3}$. For the currently discussed values, $0.05 < q_0 < 0.5$, $z=0.2$ corresponds to $H_0\Delta t \approx 0.17$. Since $\Omega_{m0} \leq 0.2$, (36) gives

$$|\epsilon(z=0.2)-1| \leq 5.4\times 10^{-5}(l/L_{\rm PW})^2|1.32H_0\tau-0.11|\,. \qquad (38)$$

Thus, if $|\tau| < t_0$ while $l/L_{\rm PW} < 3$, or if $|\tau| < 0.1H_0^{-1} \approx 0.1t_0$ while $l/L_{\rm PW} < 8$, the prediction is below the observational bound. From observations of fine-structure splitting in a BL-Lac object, Wolfe, Brown, and Roberts[13] deduce that at $z=0.52$, $|\epsilon-1| < 0.03$. Now $z=0.52$ corresponds to $H_0\Delta t \approx 0.37$ for $0.05 < q_0 < 0.5$. With $\Omega_{m0} \leq 0.2$ (36) gives

$$|\epsilon(z=0.52)-1| \leq 1.1\times 10^{-4}(l/L_{\rm PW})^2(1.86H_0\tau-0.32)\,. \qquad (39)$$

Thus, if $|\tau| < t_0$ and $l/L_{\rm PW} < 10$ or if $|\tau| < 0.1H_0^{-1}$ while $l/L_{\rm PW} < 20$, the prediction is again below the observational bound. Thus the astronomical observations do *not* rule out α variability unless one is willing to assume that the shortest conceivable scale is more than an order-of-magnitude larger than $L_{\rm PW}$.

From the Re^{187}, Os^{187} abundances in ores, Dyson[9,10] deduced that 4.5×10^9 yr ago $|\epsilon-1| < 5\times 10^{-5}$. With $H_0 \approx 1.02\times 10^{-10}\ \text{yr}^{-1}$ and $\Omega_{m0} \approx 0.2$, (36) predicts

$$|\epsilon(\Delta t=4.5\times 10^9\text{yr})-1| \approx 1.5\times 10^{-4}(l/L_{\rm PW})^2|2.2H_0\tau-0.44|\,. \qquad (40)$$

Even for $l \approx L_{\rm PW}$ this is consistent with Dyson's constraint only if $\tau = H_0^{-1}(0.2\pm 0.15)$. Shlyakhter[15] deduced from the isotopic abundances in the Oklo "natural reactor" that 1.8×10^9 yr ago $|\epsilon-1| < 5\times 10^{-9}$. The appropriate prediction is made by (38) since $\Delta t = 1.8\times 10^9$ yr corresponds closely to $H_0\Delta t = 0.17$. With $\Omega_{m0}=0.2$ and $l \approx L_{\rm PW}$ it is consistent with Shlyakhter's constraint only if $\tau = H_0^{-1}(0.08\pm 7\times 10^{-5})$. The two determinations of τ are consistent; however, this is no longer so if $l/L_{\rm PW} \geq 1.3$. Thus Dyson and Shlyakhter constraints *together* force l down to near its minimum allowable value; in this sense they *marginally* rule out α variability. However, this conclusion is strongly conditional on the assumed cosmological parameters. Consideration of He^4 synthesis in the big bang suggests[31] that values of $\Omega_{m0}H_0^2$ a factor 25 times smaller than we have assumed are quite possible (recall, Ω_{m0} refers to nuclear matter only). With these low values $l/L_{\rm PW}$ need no longer be near unity. Also, small $\Omega_{m0}H_0^2$ allows the very narrow range of values for

τ deduced from Shlyakhter's constraint to broaden considerably. Thus the suggestive implication that Shlyakhter's result requires the present time t_0 to be a very special one ($t_0 - t_C$ very precisely determined), or α not to vary, looses much of its force. We conclude that only when Ω_{m0} is determined more accurately will it be clear if the geophysical constraints provide a strong case against α variability.

VI. SPATIAL VARIATIONS OF α

In the laboratory there should be spatial gradients of α due to nearby matter with electromagnetic structure. The relevant equation is (16). If one is interested only in macroscopic gradients, one may average its source over volumes containing many atoms. As shown in Sec. V, one can then replace the quantity in parentheses in (16) by $\zeta \rho_m c^2$, where ζ is a number of order 10^{-2}. Actually ζ varies somewhat with the composition of the source. Thus the ζ appropriate to the sun (predominantly *H*) is somewhat smaller than that appropriate to the earth (mostly heavy elements). In a static situation (16) becomes

$$\nabla^2 \ln\epsilon = \zeta (l/L_{\rm PW})^2 G c^{-2} \rho_m \tag{41}$$

[recall $L_{\rm PW} = (G\hbar/c^3)^{1/2}$].

Our convention for ϵ is that $\epsilon \to 1$ at infinity. Hence ln ϵ obeys the same boundary condition as gravitational potential ϕ. In fact, by comparing (41) with the Poisson equation, we have

$$\ln\epsilon = (4\pi c^2)^{-1} \zeta (l/L_{\rm PW})^2 \phi \ . \tag{42}$$

That any fractional variation of α (or 2 ln ϵ for $\epsilon \approx 1$) should be proportional to ϕ was conjectured by Dicke.[34,6] He estimated the proportionality constant as α/c^2. Since we know now that $l \sim L_{\rm PW}$, we see that this is about right.

As discussed by Dicke,[6] gradients of α give rise to "anomalous" acceleration. Basically, the total force acting on a *neutral* body of mass *M* bearing electromagnetic energy $E_{\rm EM}$ is, by energy conservation,

$$\begin{aligned} \vec{F} &= -M\vec{\nabla}\phi - \vec{\nabla} E_{\rm EM} \\ &= -M\vec{\nabla}\phi - (\partial E_{\rm EM}/\partial\epsilon)\vec{\nabla}\epsilon \ . \end{aligned} \tag{43}$$

Here we have neglected a fractional correction of order ϕ/c^2 to the last term. Since $E_{\rm EM}$ is for the most part Coulombic, one can replace $\partial E_{\rm EM}/\partial E$ by $2E_{\rm EM}\epsilon^{-1}$ (see Sec. V). By substituting (42) into (43) we get for the acceleration $\vec{F}/M$

$$\vec{a} = -[1 + (2\pi)^{-1}\zeta(l/L_{\rm PW})^2 (E_{\rm EM}/Mc^2)]\vec{\nabla}\phi \ . \tag{44}$$

The second term in the square brackets varies from body to body, giving rise to an anomalous composition-dependent acceleration. Tests of the equivalence principle search for just such a term. In the typical experiment the accelerations of two substances are compared and a bound is set on the relative difference, i.e., on

$$D = (2\pi)^{-1}\zeta(l/L_{\rm PW})^2 \Delta(E_{\rm EM}/Mc^2) \ . \tag{45}$$

Bessel[35] used a pendulum to establish that $|D| < 2\times10^{-5}$ for various pairs of substances. The source in this case was the earth (Fe, Si, C, O, $\cdots$) for which we estimate roughly $\zeta \approx 1\times10^{-2}$. We also estimate $\Delta(E_{\rm EM}/Mc^2) \approx 1\times10^{-3}$ (see below). Bessel's constraint then tells us that $l/L_{\rm PW} < 3.5$. Thus, this ancient test of the equivalence principle sets a bound on l as tight as any of the astronomical bounds (Sec. V), but one free of ambiguity related to the uncertain values of Ω_{m0} and t_C. The 1922 torsion balance test by Eötvös, Pekár, and Fekete[36] gave $|D| < 5\times10^{-9}$ for various pairs. Using again $\zeta \approx 1\times10^{-2}$ and $\Delta(E_{\rm EM}/Mc^2) \approx 1\times10^{-3}$ we infer that $l/L_{\rm PW} < 10^{-1}$. This result argues against α variability. Evidently, experiments sensitive to spatial variation of α are the only ones sensitive enough to decide the question. In order to make the most of the data from modern tests of the equivalence principle, we now turn to some fine points.

In his pioneering analysis of the relevance of the experiments for α variability,[6] Dicke calculated $E_{\rm EM}$ from the Coulomb interaction energies between nucleons alone. Evidently atomic and lattice energies can safely be neglected by comparison. But at first sight it seems that self-energies of nucleons cannot, so let us include them. If ϵ_e, ϵ_p, and ϵ_n denote the self-electromagnetic energies of electron, proton, and neutron, respectively, we should have for an atom of atomic and mass numbers *Z* and *A*

$$\begin{aligned} E_{\rm EM}(Z,A) = {} & Z(\epsilon_e + \epsilon_p) + (A - Z)\epsilon_n \\ & + 7.7\times10^{-4}\bar{m}c^2 Z^2 A^{-1/3} \ , \end{aligned} \tag{46}$$

where $\bar{m}$ is the atomic mass unit ($\frac{1}{12}$ of the mass of the C^{12} atom); the last term in (46) is the Coulomb term from the nuclear mass formula.[28] To a sufficiently good approximation $M(Z,A) = A\bar{m}$ (the maximal error is 0.3%). Hence, for nuclei specified by Z_1, A_1 and Z_2, A_2,

$$\left[\frac{E_{\rm EM}}{Mc^2}\right]_1-\left[\frac{E_{\rm EM}}{Mc^2}\right]_2=\left[\frac{Z_2}{A_2}-\frac{Z_1}{A_1}\right]\frac{\Delta}{\bar{m}c^2}+7.7\times10^{-4}\left[\frac{Z_1{}^2}{A_1{}^{4/3}}-\frac{Z_2{}^2}{A_2{}^{4/3}}\right], \tag{47}$$

where $\Delta\equiv\epsilon_n-\epsilon_p-\epsilon_e$. Evidently $\epsilon_e<m_ec^2=0.51$ MeV. Also, the 1.29 MeV mass splitting between neutron and proton is known to come exclusively from electromagnetic effects: $\epsilon_n-\epsilon_p=1.29$ MeV. Thus $8.4\times10^{-4}<\Delta/\bar{m}c^2<1.4\times10^{-3}$. In most cases the two terms in (47) are opposite in sign. When one nucleus is heavy and the other light, as in the actual experiments, the correction due to self-electromagnetic energies is small (typically 5%); this justifies Dicke's approximation.

Roll, Krotkov, and Dicke[37,6] showed that the sun accelerates gold and aluminum masses equally to great accuracy. Their formal result was $D=(0.96\pm1.04)\times10^{-11}$ (1σ interval). Thus at the 95% confidence level $D<2.7\times10^{-11}$. For gold and aluminum the difference (47) is 2.7×10^{-3}. We get a definite lower bound on ζ of the sun by regarding it as composed only of hydrogen, and taking the minimal value 0.7 MeV for ϵ_p (see Sec. V). This gives $\zeta>3\times10^{-3}$. Putting all these into (45) we come up with the 95% confidence bound $l/L_{\rm PW}<5\times10^{-3}$. Braginsky and Panov[18] established with 95% confidence that $D=(0.3\pm0.9)\times10^{-12}$ for platinum and aluminum. For these the difference in (47) is also about 2.7×10^{-3}. With our earlier bound on ζ we get (95% confidence) $l/L_{\rm PW}<1\times10^{-3}$. Thus, the Eötvös-Dicke-Braginsky experiments strongly rule out the framework for α variability developed here from assumptions P1–P8. Because these are reasonable assumptions we now know with some confidence that α is a parameter, not a dynamical variable. It undergoes *no* spatial or temporal changes whatsoever.

VI. CONCLUSIONS AND CAVEATS

Experimental constraints on variation of α cannot by themselves rule out variability. A theoretical framework capable of making specific predictions is required to judge the relevance of any particular constraint. Just such a framework has been developed under very general assumptions; it necessarily entails a modification of Maxwellian electrodynamics. A characteristic length l enters into it. An experimental constraint rules out α variability of any kind if it is in clear conflict with predictions of the framework for l no shorter than the fundamental length 10^{-33} cm. By this criterion neither the astronomical constraints nor those from laboratory intercomparison of clocks are sufficiently sensitive to rule out variability. The geophysical constraints are marginally sensitive, but uncertainty about various cosmological parameters precludes firm conclusions. The constraints from laboratory tests of the equivalence principle are very sensitive and strongly rule out all α variability.

Our approach divorces electromagnetism from the weak and strong interactions. Yet, it is suspected that all three are unified at high energies. Thus it would be more logical to introduce a spacetime variable coupling constant into the unified gauge theory of the three interactions and recover electromagnetism with a variable α after symmetry breaking. What guarantee have we that our scheme would emerge? Although only a full treatment can decide the question, we would argue that the very generality of our assumptions precludes other outcomes *if,* in fact, electromagnetism separates cleanly from the other interactions in the presence of a variable coupling constant. A full treatment would be worthwhile, not only to verify this point, but because it would make it clear whether our conclusions about α carry through to the constants of the weak and strong interactions, or whether further experiments are needed to establish their strict constancy.[38]

ACKNOWLEDGMENTS

The author is grateful to Freeman Dyson, Amnon Meisels, and Jim Peebles for helpful discussions and to Robert Dicke for drawing attention to his earlier formulation of the variable-α problem. This work was supported by the Rector's Fund, Ben Gurion University, Israel.

APPENDIX

Consider the Lagrangian density

$$\mathscr{L}=-\tfrac{1}{2}f(\psi_{,\mu}\psi^{,\mu}), \tag{A1}$$

where $f(x)$ is an arbitrary function. The factor $-\frac{1}{2}$ is conventional; ψ represents $l\ln\epsilon$, and for simplicity we have dropped l. For what f's is (A1) consistent with causality? We show that only for $f(x) \propto x$.

The wave equation corresponding to (A1) is

$$f' g^{\mu\nu} + 2f'' \psi^{,\mu} \psi^{,\nu} \psi_{;\mu\nu} = \text{source} , \quad \text{(A2)}$$

where f' and f'' denote first and second derivatives with respect to the *argument*. We consider the evolution of small disturbances $\delta\psi$ on the background $(g_{\mu\nu}, \psi)$. The characteristics $S(x^\alpha) = \text{const}$ of the corresponding linearized equation obey

$$f' g^{\mu\nu} S_{,\mu} S_{,\nu} + 2f'' (\psi^{,\mu} S_{,\mu})^2 = 0 . \quad \text{(A3)}$$

Now, by causality characteristics cannot be spacelike surfaces (information transmitted outside the light cone) so $g^{\mu\nu} S_{,\mu} S_{,\nu} \geq 0$. Consider the option $g^{\mu\nu} S_{,\mu} S_{,\nu} > 0$. It follows from (A3) that f' and f'' always have opposite signs. This rules out maxima, minima, or inflection points of $f(x)$. Hence, either (a) $f' > 0$ and $f'' < 0$ for all x, or else (b) $f' < 0$ and $f'' > 0$ for all x. We also want f' and f'' to be well behaved in order that the wave equation be well defined. This is especially true near $x = 0$ because for easily realizable gradients $| l^2 \epsilon^{-2} \epsilon_{,\alpha} \epsilon^{,\alpha} |$ should be small if, as we know, l is a very short length. Thus for small $|x|$ we must have $f' \cong \alpha + \beta x^\gamma$ for constant α, β and γ.[39] For case (a) above we need $\alpha, \beta > 0$ ($f' > 0$ for $x > 0$), γ = ratio of even to odd integers (f' real and positive for $x < 0$), and $\gamma < 0$ ($f'' < 0$ for $x > 0$). But then $f'' > 0$ for $x < 0$ and $f', f'' \to \infty$ at $x = 0$, so this alternative is excluded. Case (b) is excluded in a similar manner.

We return to the alternative $g^{\mu\nu} S_{,\mu} S_{,\nu} = 0$ (null characteristics). The field $\psi_{,\mu}$ can be quite complicated and need not lie in the null surfaces of the spacetime. In fact if $\psi_{,\mu}$ is timelike ($\epsilon_{,\mu}$ in an expanding Universe), it cannot lie in null surfaces. Hence, in general $\psi^{,\mu} S_{,\mu} \neq 0$. It follows from (A3) that $f'' = 0$. Now, $\psi^{,\mu} \psi_{,\mu}$ need not be constant on a characteristic. The implication is that $f(x) = ax + b$ for constant a, b. Of course b does not generate any dynamics for ψ and so can be dropped.

---

[1]P. A. M. Dirac, Nature 139, 323 (1937); Proc. R. Soc. London A165, 199 (1938).
[2]E. Teller, Phys. Rev. 73, 801 (1948).
[3]P. Jordan, *Schwerkraft and Weltall* (Vieweg, Braunschweig, 1955); Z. Phys. 157, 112 (1959).
[4]R. H. Dicke and P. J. E. Peebles, Phys. Rev. 128, 2006 (1962).
[5]Whenever necessary we adopt the value $H_0 = 100$ km/Mpc $= 1.02 \times 10^{-10}$ yr^{-1}.
[6]R. H. Dicke, *The Theoretical Significance of Experimental Relativity* (Gordon and Breach, New York, 1965).
[7]G. Gamow, Phys. Rev. Lett. 19, 759 (1967).
[8]A. Peres, Phys. Rev. Lett. 19, 1293 (1967).
[9]F. J. Dyson, Phys. Rev. Lett. 19, 1291 (1967).
[10]F. J. Dyson, in *Aspects of Quantum Theory,* edited by A. Salam and E. Wigner (Cambridge University Press, London, 1972).
[11]J. N. Bahcall and M. Schmidt, Phys. Rev. Lett. 19, 1294 (1967).
[12]J. P. Turneaure and S. R. Stein, in *Atomic Masses and Fundamental Constants, Vol. 5,* edited by J. H. Sanders and A. H. Wapstra (Plenum, New York, 1976).
[13]A. M. Wolfe, R. L. Brown, and M. S. Roberts, Phys. Rev. Lett. 37, 179 (1976).
[14]M. P. Savedoff, Nature 178, 689 (1956); J. Bahcall, W. Sargent, and M. Schmidt, Astrophys. J Lett. 149, L11 (1967); A. D. Tubbs and A. M. Wolfe, *ibid.* 236, L105 (1980).
[15]A. I. Shlyakhter, Nature 264, 340 (1976).
[16]J. Bekenstein, Comments Astrophys. 8, 89 (1979).
[17]J. Bekenstein and A. Meisels, Phys. Rev. D 22, 1313 (1980); Astrophys. J. 237, 342 (1980).
[18]V. B. Braginsky and V. I. Panov, Zh. Eksp. Teor. Fiz. 61, 873 (1972) [Sov. Phys.—JETP 34, 463 (1972)].
[19]D. C. Cheng and G. K. O'Neill, *Elementary Particle Physics* (Addison-Wesley, Reading, 1979).
[20]I. S. Altarev *et al.*, Nucl. Phys. A341, 269 (1980); E. A. Hinds and P. G. H. Sanders, Phys. Rev. A 21, 480 (1980); W. B. Dress, P. D. Miller, J. M. Pendlebury, D. Perrin, and N. F. Ramsey, Phys. Rev. D 15, 9 (1977).
[21]C. W. Misner, K. S. Thorne, and J. A. Wheeler, *Gravitation* (Freeman, San Francisco, 1973), p. 1191.
[22]L. D. Landau and E. M. Lifshitz, *Classical Theory of Fields* (Pergamon, Oxford, 1962), 2nd edition.
[23]Our conventions are signature $-+++$; Greek (Latin) indices are spacetime (space) indices; the summation convention applies; semicolon denotes covariant derivative; g denotes determinant of metric $g_{\mu\nu}$.
[24]A Meisels, dissertation, Ben Gurion University, 1979 (unpublished).
[25]This Maxwellian theorem applies to an excellent approximation because ϵ varies little; see Sec. III.
[26]For several particles this approach still gives the correct answer if interactions are negligible. If not there will be an error of order v^2/c^2 times the interac-

tion energy density.

[27]E. Segre, *Nuclei and Particles* (Benjamin, New York, 1965).

[28]K. R. Lang, *Astrophysical Formulae* (Springer, New York, 1974).

[29]L. Spitzer, *Physical Processes in the Interstellar Medium* (Wiley, New York, 1978).

[30]M. Davis, M. J. Geller, and J. Huchra, Astrophys. J. 221, 1 (1978).

[31]K. A. Olive, D. N. Schramm, G. Steigman, M. S. Turner, and J. Yang, Astrophys. J. 264, 557 (1981).

[32]B. Tinsley, in *Physical Cosmology,* edited by R. Balian, J. Adouze, and D. N. Schramm (North-Holland, Amsterdam, 1980).

[33]S. Weinberg, *Gravitation and Cosmology* (Wiley, New York, 1972), p. 442.

[34]R. H. Dicke, Science 129, 621 (1959).

[35]F. W. Bessel, Poggendorff's Ann. 25, 401 (1832).

[36]R. V. Eötvös, V. Pekár, and E. Fekete, Ann. Phys. (Leipzig) 68, 11 (1922).

[37]P. G. Roll, R. Krotkov, and R. H. Dicke, Ann. Phys. (N.Y.) 26, 442 (1964).

[38]For bounds on their variability see C. M. Will, in *General Relativity,* edited by S. W. Hawking and W. Israel (Cambridge University Press, Cambridge, 1979).

[39]Other conceivable behaviors such as $\ln|x|$ and $\exp(-1/x^2)$ are singular and lead to varying signs for f' and f''.

PHYSICAL REVIEW D VOLUME 49, NUMBER 4 15 FEBRUARY 1994

Entropy bounds and black hole remnants

Jacob D. Bekenstein*

Department of Physics, University of California at Santa Barbara, Santa Barbara, California 93106
and The Racah Institute of Physics, Hebrew University of Jerusalem, Givat Ram, Jerusalem 91904, Israel†

(Received 14 June 1993)

We rederive the universal bound on entropy with the help of black holes while allowing for Unruh-Wald buoyancy. We consider a box full of entropy lowered toward and then dropped into a Reissner-Nordström black hole in equilibrium with thermal radiation. We avoid the approximation that the buoyant pressure varies slowly across the box, and compute the buoyant force exactly. We find, in agreement with independent investigations, that the neutral point generically lies very near the horizon. A consequence is that in the generic case the Unruh-Wald entropy restriction is neither necessary nor sufficient for enforcement of the generalized second law. Another consequence is that generically the buoyancy makes only a negligible contribution to the energy bookkeeping, so that the original entropy bound is recovered if the generalized second law is assumed to hold. The number of particle species does not figure in the entropy bound, a point that has caused some perplexity. We demonstrate by explicit calculation that, for an arbitrarily large number of particle species, the bound is indeed satisfied by cavity thermal radiation in the thermodynamic regime, provided vacuum energies are included. We also show directly that thermal radiation in a cavity in D-dimensional space also respects the bound regardless of the value of D. As an application of the bound we show that it strongly restricts the information capacity of the posited black hole remnants, so that they cannot serve to resolve the information paradox.

PACS number(s): 04.70.Dy, 04.62.+v, 95.30.Tg, 97.60.Lf

I. INTRODUCTION

The generalized second law (GSL) of thermodynamics for black holes [1,2] states that when entropy flows into a black hole, the sum of black hole entropy and ordinary entropy outside the hole does not decrease. Arguing from the GSL, Bekenstein [3] has proposed the existence of a universal bound on the entropy S of any object of maximal radius R and total energy E:

$$S \leq \frac{2\pi R E}{\hbar c}. \tag{1}$$

This bound was inferred from the requirement that the GSL be respected when a box containing entropy is deposited with no radial motion next to the horizon of a Schwarzschild black hole (for a Kerr black hole the condition is more complex), and then allowed to fall in. The box's entropy disappears but an increase in black hole entropy occurs. The second law is respected provided S is bounded as in Eq. (1). Other derivations of the bound based on black holes have been given by Zaslavskii [4,5] and by Li and Liu [6]. Bound (1) can also be interpreted as a bound on the information capacity of any object with total energy E and circumscribing radius R [7,8].

Bound (1) has been checked directly for quantum fields enclosed in boxes of various shapes, when S is interpreted as the logarithm of the number of field quantum states up to energy E above the ground state (for a review see Ref. [9]). Numerical checks have been made for free scalar, electromagnetic, and massless spinor fields enclosed in rectangular or spherical boxes [8]. And an analytic proof of the bound for those same free fields valid for boxes of arbitrary shape and topology has been provided [10]. A couple of checks exist for self-interacting fields [9,11]. All the above demonstrations can be supplemented by the observation that if the box the fields are enclosed in is reckoned as part of the system, the bound is even more strongly satisfied because E is thus augmented while S is hardly changed. Black holes play no part in any of the above considerations. Therefore, bound (1) is known to be true independently of black hole physics for a variety of systems in which gravity is negligible.

By contrast, just a few pieces of evidence exist concerning the validity of bound (1) for self-gravitating systems [4,12,13]. To these we may add that black holes themselves comply with bound (1) if R in the formula is interpreted as $(\mathcal{A}/4\pi)^{1/2}$, where $\mathcal{A}$ is the horizon area [3]. Thus our confidence in bound (1) for gravitating systems rests almost entirely on the black hole arguments of Refs. [3–6].

In Ref. [3] it was assumed that the energy at infinity added to the black hole with the box (which determines the increase in black hole entropy) is that which may be inferred from the redshift factor at the deposition point. However, Unruh and Wald (UW) [14,15] have pointed out that when the deposition is attempted by lowering the box from far away, buoyancy in the radiative black hole environs will prevent lowering the box down to the

*Electronic address: bekenste@vms.huji.ac.il
†Permanent address.

horizon (if one does not wish to invest energy by pushing it in), and the box will "float" at some neutral point. The total energy at infinity added to the hole after the box has been dropped from the neutral point is larger than the redshifted proper energy of the box. Accordingly, UW concluded that the GSL is respected when the box is dropped, provided only its entropy is restricted by

$$S \leq V\, s(E/V), \tag{2}$$

where $s(e)$ is the entropy density as function of energy density e of *unconfined* thermal radiation. UW thus concluded that bound (1) is unecessary for the proper functioning of the GSL. They further argued that that bound (1) could not possibly remain correct when the number of particle species in nature is arbitrarily large.

Although it seems reasonable that thermal radiation maximizes entropy as a function of energy density, the UW entropy restriction (2) can easily fail for a system in which surface effects are sizable, because such a system cannot be described entirely in terms of extensive or intensive variables such as V or E/V, respectively. In effect, the shape of the system is also a variable. For example, in a rectangular box with dimensions $d \times d \times 0.1d$, radiation in the energy range $[\hbar/d, 80\hbar/d]$ explicitly exceeds the UW restriction on entropy [16]. [A more dramatic violation occurs for a black hole; Eq. (2) predicts a bound on S that rises like $E^{3/2}$ whereas black hole entropy grows like E^2.] In light of this, UW [15] conjectured that the entropy restriction (2) applies only for box and contents together. In this revised form the restriction may be generally correct for nongravitating systems; however, it did not play a role in UW's subsequent discussion of the functioning of the GSL in an alternative *gedanken* experiment (box emptied into a black hole and withdrawn open) [15], and its status has remained unclear. And because UW's rescue of the GSL by an appeal to buoyancy [14] in the original *gedanken* experiment [3] relied on the UW entropy restriction, that experiment has continued to be problematic for the GSL.

In light of the above, we retrace in Sec. II A UW's analysis of the original *gedanken* experiment in which a box full of entropy is lowered towards and then dropped into a black hole from the neutral point. For later convenience we perform the *gedanken* experiment with a Reissner-Nordström (RN) black hole. We studiously avoid UW's approximation that the ambient buoyant pressure varies slowly across the box because it has become clear [17,5,6] that the neutral point lies very near the horizon where this approximation must fail because of the large gradients. Our exact treatment shows that the UW entropy restriction (2) is neither sufficient nor necessary for the satisfaction of the GSL.

This situation motivates our rederivation of entropy bound (1) which takes full account of UW buoyancy. Our new analysis closely parallels UW's original one, much more so than alternative ones proposed in the wake of UW's paper [17,16,5,6]; we pinpoint the stage at which the analysis departs from UW's. In Sec. II B we determine the position of the neutral point, confirming that it lies very near the horizon. As a result, the buoyancy makes only a small change in the energy bookkeeping. The original entropy bound, Eq. (1), is recovered in Sec. II C.

As mentioned, the independence of bound (1) on the number of particle species has been regarded as a sign that it must fail when the number of species is large [14]. In Sec. III A we show by explicit calculation of thermal radiation entropy in the thermodynamic limit that the bound is respected for a box containing an arbitrarily large number of massless particle species, provided care is taken to include the vacuum energy in the energy of the full system. We also show, in Sec. III B, that cavity thermal radiation in the thermodynamic regime in D-dimensional space also respects the bound regardless of the value of D.

Finally, in Sec. IV we show that the bound strongly restricts the information capacity of the posited black hole remnants, so that they cannot serve to resolve the information paradox.

Henceforth we use units with $G = c = k_{\rm Boltzmann} = 1$, but continue to display $\hbar$.

II. DERIVATION OF THE ENTROPY BOUND

A. Critique of the Unruh-Wald analysis

The system of interest is a macroscopic rectangular box of total energy E which holds entropy S. We label the horizontal crossectional area of the box A and its height b. The box is to be lowered towards a black hole with a face facing it in a standard orientation which is defined by two technically convenient conditions. First, we require that b *not* be very small compared to $A^{1/2}$. If this turns out to be the case for the initial orientation, the box is to be rotated by 90° about a horizontal axis so that a longer edge is brought to vertical orientation, and that edge is to be labeled by b. Further, we require that the center of mass (c.m.) of the box lie initially on the centroid plane (horizontal plane halfway up the box), or below it. If this condition is not satisfied, the box is to be turned upside down (which respects the previous arrangement), and lowered in that orientation. We denote by R the circumscribing radius of the box.

For convenience we carry out the *gedanken* experiment with a RN black hole of mass M and (not necessarily electric) charge Q. The exterior metric may be written as

$$ds^2 = -\chi^2\, dt^2 + \chi^{-2} dr^2 + r^2\,(d\theta^2 + \sin\theta^2\, d\phi^2), \tag{3}$$

where r denotes the usual Schwarzschild radial coordinate and

$$\chi^2 = \frac{(r - r_+)(r - r_-)}{r^2}, \tag{4}$$

with $r_\pm \equiv M \pm \sqrt{M^2 - Q^2}$. The event horizon lies at $r = r_+$ and has area $\mathcal{A} = 4\pi r_+^2$ and entropy $S_{\rm BH} = \mathcal{A}/(4\hbar)$. In what follows we also employ the radial proper length measured from the event horizon, l, defined by $dl = \chi^{-1}\, dr$, and use the notation $\Delta = r_+ - r_-$. In order that the box may ultimately be dropped into the hole we assume that

$$A \ll r_+^2 \qquad b \ll r_+ \,. \tag{5}$$

We assume that the box and its contents are "transparent" to the gauge field of the hole, i.e., noninteracting with it. This is certainly possible if the gauge field is distinct from the electromagnetic one and the box is made of ordinary matter. Thus we do not worry about buoyancy due to stresses of this field.

Following UW we assume the black hole has reached equilibrium with its own Hawking radiation, the whole system being enclosed in a large cavity. We assume the cavity to be made of ordinary matter, like the UW box, so that it does not interact with the hole's gauge field. One consequence of such a setup is that superradiance of the charges of the gauge field does not lead to an unstable situation because any such stray particles escape from the cavity and do not get multiplied by multiple reflection onto the hole. The black hole temperature $T_{\rm BH}$ and the local temperature T are related by

$$T = \frac{T_{\rm BH}}{\chi} = \frac{\hbar\Delta}{4\pi r_+^2\chi} \,. \tag{6}$$

The redshift factor χ enters here as in any equilibrium situation in a gravitational field.

Applying the first law of thermodynamics to a parcel of equilibrium radiation, and assuming that its proper energy density e, pressure P, and proper entropy density s are all functions only of T, UW derived the relation

$$e + P - Ts = 0 \,, \tag{7}$$

which merely says that the Gibbs free energy of the parcel vanishes, as befits a collection of photons or a mixture of equal numbers of neutrinos and antineutrinos. Differentiation of Eq. (7), use of $T = de/ds$, and simplification with help of Eqs. (6) and (7) gives

$$d(P\chi) = -e\, d\chi \,, \tag{8}$$

which is equivalent to the condition of hydrostatic equilibrium [14].

The buoyant force acting on the box, as measured by an observer at infinity, is the difference of the redshifted local forces acting on the upper and lower faces [14]:

$$f_{\rm buoy}(l) = A[(P\chi)_{l-b/2} - (P\chi)_{l+b/2}], \tag{9}$$

where l is the proper height of the centroid plane of the box above the horizon. UW approximated the difference in this equation by the first term in a Taylor series. An exact expression follows from using Eq. (8) to convert Eq. (9) to

$$f_{\rm buoy}(l) = A\int_{l-b/2}^{l+b/2} e\frac{d\chi}{dl'}dl' \,. \tag{10}$$

Let us now express the gravitational force acting on the box and contents. The four-acceleration of a point with four-velocity u^α is defined as $a^\alpha = u^\alpha{}_{;\beta}u^\beta$. A simple calculation shows that a point stationary in Schwarzschild coordinates has invariant acceleration

$$a \equiv (a^\alpha a_\alpha)^{1/2} = \chi', \tag{11}$$

where a prime denotes a derivative with respect to r. We can thus write the "gravitational force" acting on the box as measured at infinity in the form

$$f_{\rm grav}(l) = -A\int_{l-b/2}^{l+b/2} \rho\frac{d\chi}{dl'}dl' \,. \tag{12}$$

Here ρ denotes the proper energy density of the box and contents, and $d\chi/dl = a\chi$ is the local acceleration as measured at infinity. The minus sign reminds us that gravity and buoyancy act in opposite senses.

Putting together Eqs. (10) and (12) we write the work done by the box on the agent lowering it from infinity down to proper height l above the horizon as

$$W(l) = \int_\infty^l (f_{\rm buoy} + f_{\rm grav})dl' \,. \tag{13}$$

$W(l)$ is maximized when the box's centroid plane reaches the neutral point, $l = l_0$. Setting $dW/dl = 0$ we obtain the condition determining l_0:

$$f_{\rm buoy}(l_0) + f_{\rm grav}(l_0) = A\int_{l_0-b/2}^{l_0+b/2} (e-\rho)\frac{d\chi}{dl}dl = 0 \,. \tag{14}$$

Henceforth, we shall adopt the following notation for integrals such as those appearing in Eqs. (10), (12), and (14):

$$A\int_{l_0-b/2}^{l_0+b/2} F dl \Longleftrightarrow \int_V F dV \,, \tag{15}$$

where dV stands for the element of box volume $A\,dl$. Note that if e and $d\chi/dl$ are nearly constant across the box, the condition (14) may be *approximated* by UW's form [cf., Eq. (2.2) of Ref. [14];

$$eV = E \equiv \int_V \rho dV \,, \tag{16}$$

where $V = Ab$ is the box's volume.

Since $W(l)$ is maximum at $l = l_0$, the mass increment δM of the black hole is minimal if the box is dropped in from the neutral point. Evidently $(\delta M)_{\rm min} = E - W(l_0)$. Using Eq. (13) we may reexpress this as

$$(\delta M)_{\rm min} = \int_V \rho\chi dV + \int_V P\chi dV \,. \tag{17}$$

The first integral is just the energy at infinity of the box, with each parcel properly redshifted. This integral equals $E - \int_\infty^{l_0} f_{\rm grav}dl$. The second integral is just $-\int_\infty^{l_0} f_{\rm buoy}dl$ with $f_{\rm buoy}(l)$ in the form (9). As a result of the cancellation of the work done by buoyant forces on top and bottom of the box over the range $[l_0 + b/2, \infty]$, the buoyant contribution to $(\delta M)_{\rm min}$ only depends on the distribution of $P\chi$ over the height of the box at the neutral point.

UW's version of Eq. (17), namely,

$$(\delta M)_{\rm min} = (E + PV)\chi \ , \tag{18}$$

would follow from Eqs. (16) and (17) if P and χ separately varied little across the box. Because the neutral point turns out to be so near the horizon (see Sec. II B), these quantities actually vary a lot across the box in the generic case, so that the approximations leading to Eq. (18) are questionable.

Instead, let us replace P in Eq. (17) by means of the identity (7). Taking cognizance of Eq. (6) we get

$$(\delta M)_{\rm min} = \int_V (\rho - e)\chi dV + T_{\rm BH}\int_V s dV. \tag{19}$$

Were χ to be nearly constant across the box, the first integral here would vanish by virtue of Eq. (16), and we would be left with UW's expression (2.22) of Ref. [14]. Since taking χ as nearly constant is an unwarranted approximation in view of the closeness of the neutral point to the horizon, let us instead use Eqs. (19) and (6) to compute the overall entropy change of the world, $(\delta S)_{\rm tot} \equiv (\delta M)_{\rm min}/T_{\rm BH} - S$, when the box is dropped from the neutral point. We find

$$(\delta S)_{\rm tot} = T_{\rm BH}^{-1}\int_V (\rho - e)\chi dV + \int_V s dV - S \ . \tag{20}$$

In obtaining this equation we have ignored any entropy emitted by the black hole. We have already mentioned that, by virtue of the cavity it is in, the black hole is in equilibrium with thermal radiation, so that Hawking radiance entropy is not produced. The entropy produced in spontaneous Unruh emission corresponding to the *superradiant* modes, and the concomitant black hole entropy *increase*, can be made negligible by taking the black hole as very massive, which we do in this section.

In UW's discussion, the first integral in Eq. (20) dropped out in wake of the indicated approximations. It may be seen by comparing with Eq. (14) that, in fact, its magnitude and even its sign depend on the distribution of e and ρ across the box. Thus, in contrast with UW's discussion, we must conclude that validity of the UW entropy restriction, Eq. (2), does not by itself guarantee that the GSL will be obeyed. And, conversely, assuming that the GSL is satisfied in the *gedanken* experiment in question does not allow us to derive the UW entropy restriction for the box. Unless supplemented by detailed information about the box, the entropy restriction is neither a necessary nor a sufficient condition for the operation of the GSL. Therefore, it seems best to calculate in detail the sum of integrals in Eq. (20).

B. Neutral point

As the first step we make an approximate determination of the location of the neutral point by using UW's criterion Eq. (16) and a model of radiation as a mixture of noninteracting gases of massless particles, one for each species in nature. According to Boltzmann, for such radiation

$$P = e/3 = \frac{N\pi^2 T^4}{45\hbar^3} \ , \tag{21}$$

where N is the effective number of particle species (photon and graviton contribute one to N, while each neutrino and antineutrino species counts as 7/16).

With Eq. (6) for T, criterion (16) amounts to

$$\chi(l_0) = 0.0717 N^{1/4}(Ab\hbar/E)^{1/4}\Delta r_+^{-2} \ . \tag{22}$$

Since we assume the box to be macroscopic, and hence large compared to its Compton wavelength, $\hbar/E \ll b$. This together with inequalities (5) tells us that for realistic values of N (in Sec. III A we shall reconsider the situation where N is arbitrarily large), $\chi(l_0) \ll \Delta r_+^{-1}$. A look at Eq. (4) shows that the neutral point must lie in the region $r - r_+ \ll \Delta$. But in this region, a good approximation to the proper distance from the horizon to a point r is

$$l \approx 2r_+(r - r_+)^{1/2}\Delta^{-1/2}. \tag{23}$$

It is clear that in that region $l \ll r_+$. In the same approximation Eq. (4) can be written

$$\chi(l) \approx \frac{1}{2}\Delta r_+^{-2} l. \tag{24}$$

Two important consequences follow from the linear form of χ in the near horizon region where the neutral point lies. First, from Eq. (6),

$$T = \hbar/(2\pi l) \ , \tag{25}$$

so that, as a function of proper distance from the horizon, T does not depend on the black hole's parameters. This universal form for T may be understood by writing down the Unruh temperature [18] for an (accelerated) stationary observer, $T_U = \hbar a/(2\pi)$, and using Eqs. (11) and (24) to recast it into a form identical to Eq. (25). The equality $T = T_U$ in the near horizon region allows one to interpret the thermal radiation in equilibrium with the hole as observer dependent Unruh acceleration radiation seen by the stationary observers. One may thus follow UW in asserting that the large e and P corresponding to T at the neutral point do not generate strong curvature, i.e., that the metric (3) remains an excellent approximation.

The second consequence of the $\chi \propto l$ form is that condition (14) can be rewritten as

$$\int_V e dV = E \ , \tag{26}$$

which, unlike the UW condition (16), is accurate in the near horizon region, and indeed becomes the more accurate the nearer to the horizon since Eq. (24) is asymptotically exact there.

Let us now use Eq. (26) to accurately determine l_0. Substituting e from Eq. (21) and T from Eq. (25), and remembering that the range of l in the integral is $[l_0 - b/2, l_0 + b/2]$, the condition leads to

$$\frac{(l_0^2 - b^2/4)^3}{3l_0^2 b^4 + b^6/4} = \eta^3 \equiv \frac{N\hbar A}{720\pi^2 E b^3} \ . \tag{27}$$

In our world where the number of particle species that would be excited by a massive black hole is limited, $\eta \ll 1$ because $\hbar/E \ll b$ and because in the standard box orientation, b^2 cannot be very small compared to A. In Sec. III A we shall argue that $\eta \ll 1$ remains true even when N is arbitrarily large.

When $\eta \ll 1$, it follows from Eq. (27) to lowest order in η that

$$l_0^2 \approx (1/4 + \eta)b^2. \tag{28}$$

Therefore, at the neutral point, the centroid plane of the box is just a little farther from the horizon than half the box's height, which means the box floats almost touching the horizon. This phenomenon has been noted earlier [17,5,6]. Since the neutral point does occur near the horizon, the approximations involved in Eqs. (23) and (24) are seen to be justified *a posteriori*. The finding here serves to emphasize that the UW neutral point condition (16) cannot be accurate because of the strong gradient of e near the horizon, and should properly be replaced by condition (26).

Of course, all the above hinges on the correctness of the Boltzmann model, Eq. (21). However, three effects may introduce corrections to it in the horizon's vicinity: vacuum polarization in the gravitational field, the increase in the number N of species present as the local temperature rises upon approaching the horizon, and interactions between constituents of the radiation. We now discuss each in turn.

The importance of vacuum polarization is gauged by the expectation values of components of the energy-momentum tensor for the appropriate (Hartle-Hawking) vacuum. According to calculations by Page [19], the mixed components $\langle T_\mu^\nu \rangle$ are all finite and small near the horizon. By comparison, the energy density and stresses in the acceleration radiation rise as l^{-4} near the horizon. It is thus clear that vacuum polarization is a minor effect.

The appearance of new species as T rises is well understood (c.f. early cosmology), and is known not to cause a significant departure from the Boltzmann "equation of state," provided N is regarded as slowly rising with T. Very often, as in Eq. (22), an important quantity depends on a low power of N. Thus the slow variation of N is rather innocuous.

In contrast with the above, interactions between radiation constituents may cause a substantial departure from the simple relation (21). Interactions, when not weak, prevent the radiation from behaving like a gas of noninteracting quanta as postulated in the Boltzmann model. Departures from the simple Boltzmann model will set in according to the relation between the relevant lengthscales of the problem: r_+, the size of the hole; λ, the typical locally measured wavelength in the radiation; l, the proper distance from the horizon and the scale over which radiation properties change significantly; and some length scale L characteristic of the interaction physics. Now, according to Eq. (25), $\lambda \sim l$ over a broad region near the horizon, so that λ is not an independent variable. We are interested in a large black hole, and so we take it for granted that $L \ll r_+$. Under these circumstances departures from the Boltzmann model are not expected for $l > r_+$. The size of the black hole thus becomes an irrelevant variable in that only the ratio l/L decides whether departures are expected. Clearly such departure will set in when l becomes comparable with L. We reach the same conclusion if we stipulate that interactions become important when T becomes comparable to the temperature $\hbar/L$ associated with the interaction length scale.

How big is L? We know that $L < 10^{-12}$ cm because Eq. (21) works well in cosmology all the way back to the lepton era. If the strong interaction is the one that spoils the Boltzmann model, we may expect $L \leq 10^{-13}$ cm. At any rate, it is clear that L is much smaller than the dimensions of a macroscopic box.

A departure from Boltzmann's model affects our discussion only if at the neutral point the lower side of the box has penetrated to within $l < L$, so that it is exposed to temperatures at which interactions are important. But, of course, since the box is much larger than L, this means that at the neutral point the box must nearly touch the horizon, with its centroid plane at proper height $\approx b/2$, just as we found by relying on the Boltzmann model alone. Thus, as far as the location of the neutral point is concerned, complications beyond the Boltzmann model are of no practical consequence.

C. Entropy bound

Proceeding with our program, we now compute explicitly $\delta S_{\rm BH}$ by going back to Eq. (17). Because of the linearity of χ with l, the integral over ρ, the energy density in the box, may be expressed in terms of the proper height of its c.m. above the horizon, $l_{\rm c.m.}$, defined by $\frac{1}{2}\Delta r_+^{-2} l_{\rm cm} = \overline{\chi}|_{l_0}$ (average with respect to ρ), and E as defined by Eq. (16). And the integral over P may be worked out with help of Boltzmann's model, Eq. (21), and Eqs. (23)–(25). The result is

$$(\delta M)_{\rm min} = \frac{1}{2}\Delta r_+^{-2} E\, [l_{\rm c.m.} + \eta^3 l_0 b^4 (l_0^2 - b^2/4)^{-2}]. \tag{29}$$

A look at Eq. (28) shows that the square parentheses in Eq. (29) amounts to $l_{\rm c.m.} + \eta l_0$; the second term comes from the buoyancy. We recall now that in the standard box orientation employed here, the c.m. of the box cannot lie above the centroid plane ($l_{\rm cm} \leq l_0$ initially, and the c.m. can only slip down as the box enters strong gravitational fields). And because $l_0 \approx b/2 + \eta b$ [see Eq. (28)], the square parentheses in Eq. (29) cannot exceed $(1 + 3\eta)b/2$. The buoyancy corrections of $O(\eta)$ are evidently negligible under the same assumptions that led us to conclude that the neutral point is near the horizon. We thus obtain

$$(\delta M)_{\rm min} \leq \frac{1}{4}\Delta r_+^{-2} b E\ , \tag{30}$$

which is another version of Eq. (19).

From Eq. (30) we may compute, with help of the first law for black holes, Eq. (6) and the obvious constraint

$b < 2R$ that, when the box is dropped from the neutral point,

$$(\delta S)_{\rm tot} < 2\pi RE/\hbar - S \ , \tag{31}$$

which replaces Eq. (20). It is evident from this inequality that in order for the GSL to be satisfied $[(\delta S)_{\rm tot} \geq 0]$, the box entropy S must satisfy the bound on entropy (1).

The above argument requires modification if at the neutral point the box already penetrates into the region with T beyond the transition temperature at which the Boltzmann model gets modified. We first note that whatever the true relation between e, P, and T, the inequality $P < e$ must be satisfied. The reason is that causality demands that $dP/de \leq 1$ (speed of sound in the radiation subluminal) for all T. Integrating the inequality $dP - de \leq 0$ from low T where $P = e/3$ certainly applies, we see that $P < e$ for any T, however large. Consider now the identity

$$d(P\chi l) = l d(P\chi) + P\chi\, dl \ . \tag{32}$$

If we substitute in it from Eq. (8), and take into account that in the near horizon region $l\,d\chi = \chi\, dl$, we see that $P < e$ implies that $d(P\chi l)/dl < 0$ near the horizon. This means that $P(l)$ drops off with l faster than l^{-2}, which means that whatever the modification to the Boltzmann model, $P(T)$ must grow faster than T^2.

In light of the above, consider the buoyant contribution to $(\delta M)_{\rm min}$ in Eq. (17), written as

$$\int_V P\chi dV = \frac{\int_V P\chi dV}{\int_V PdV}\int_V PdV \ . \tag{33}$$

By the causality constraint, $\int_V PdV < E$ [see Eq. (26)]. The ratio of integrals is evidently largest when P decreases slowest with l. We may thus bound from above that ratio by using the limiting form $P \propto l^{-2}$ to compute it. Recalling Eq. (15) we thus have

$$\int_V P\chi dV < \frac{1}{2}\Delta r_+^{-2} E(l_0 + b/2)\frac{\ln y}{y-1} \ , \tag{34}$$

where $y \equiv (l_0 + b/2)(l_0 - b/2)^{-1}$. The closeness of the box's bottom to the horizon at the floating point means that $l_0 \approx b/2$; therefore, $l_0 + b/2 \approx 2l_0$ and $y \gg 1$. Hence,

$$\int_V P\chi dV = \varpi \Delta r_+^{-2} l_0 E \ , \tag{35}$$

where $\varpi < \ln y/(y-1) \ll 1$.

Inserting this result in Eq. (17) for $(\delta M)_{\rm min}$ and recalling that the integral over ρ is already evaluated in our previous result (29), we have

$$(\delta M)_{\rm min} = \frac{1}{2}\Delta r_+^{-2} E\ [l_{\rm c.m.} + 2\varpi l_0] \ . \tag{36}$$

Again, because the box is standardly oriented and very close to the horizon, $l_{\rm c.m.} \leq l_0 \approx b/2 \leq R$. And because $\varpi \ll 1$, the buoyant term evidently cannot make the square brackets larger than R. We thus find that Eqs. (30) and (31) apply again, and the entropy bound (1) follows from the assumed validity of the GSL, as in the case when the Boltzmann model could be used down to the neutral point.

In all arguments in this subsection, a slightly tighter bound on entropy would follow if we worked throughout in terms of b, rather than appealing to the inequality $b < 2R$. The impression we get from this, that the maximal entropy of a thin box decreases with decreasing thickness b, is supported by numerical computations [8] if b is not too small. Those calculations do not support the Li-Liu conclusion [6] that the entropy bound is unconditionally set by the smallest box dimension. After all, in three-dimensional space a two-dimensional box filled with massless quanta can hold nonvanishing entropy. In our approach here the bound could not be derived in terms of b for arbitrarily small b because the condition $\eta \ll 1$ will fail when b gets sufficiently small [see Eq. (27)].

III. EXTENSIONS

A. Irrelevance of species number

Our derivation of the bound in Sec. II assumed $\eta \ll 1$, which condition would seem to fail in a world where N is very large. Indeed, UW contended that no bound of type (1), which is independent of the number of species, could possibly retain its validity as N becomes large because it is known that the more species there are, the larger the number of states (entropy) accessible with given energy. Their own entropy restriction, Eq. (2), scales as $N^{1/4}$, that being the dependence of Boltzmann's formula for blackbody radiation entropy at given energy on the number of species. UW conceived of buoyancy, with its intrinsic dependence on the number of species in the radiation, as nature's exclusive way to defend the GSL against a violation when a body made up of many species (large entropy with moderate energy) is lost down a black hole.

That this view must have restricted validity is clear from the following argument demonstrating that an N-independent entropy bound must exist in order for the GSL to function, even in situations where buoyancy cannot play a role.

Consider a box with energy E and entropy S dropped freely *from far away* into an *exactly extremal* RN black hole of mass M in empty space. Because the fall is free, i.e., geodesic, the box does not feel Unruh radiation. And because the hole is extremal, thermal radiant pressure is absent. Hence there is no buoyancy to complicate the energy bookkeeping. Initially $S_{\rm BH} = \pi M^2/\hbar$. After the black hole has assimilated the box, the mass has gone up to $M + E$ and $S_{\rm BH} = \pi(M + E + \sqrt{2ME + E^2})^2/\hbar$ (because Q is unchanged). We regard E as a small quantity. Then to lowest order in E,

$$\delta S_{\rm BH} = \pi(2M)^{3/2}E^{1/2}/\hbar. \tag{37}$$

Coherent gravitational and electromagnetic radiation may be emitted due to the infall. The energy in these radiations is expected to be of order E^2/M [21]. Its subtraction from the final M will produce terms in Eq. (37) of the same order as those we have already discarded.

Thus the infall radiations prove negligible, and in any case they would act to reduce $\delta S_{\rm BH}$. There may also be some radiation entropy emitted by the Hawking process as the box is assimilated and the black hole departs from exact extremality. Once the black hole is not extreme, it should radiate entropy at a rate a bit higher than a blackbody of temperature $T_{\rm BH}$ and area $\mathcal{A}$ [20]. This entropy emission should be somewhat larger than the induced decrease in $S_{\rm BH}$. Therefore, the overall entropy growth rate would be

$$dS_{\rm BH}/dt + dS_{\rm rad}/dt \sim (180)^{-1}\Delta^3 r_+^{-4}\ . \tag{38}$$

In our case, after the assimilation, $\Delta = 2(2ME)^{1/2}$. Therefore, over a period of coordinate time $\sim M$, which is of the order of the time required for the effective disappearance of the box, the just perturbed black hole should generate entropy $\sim (4\sqrt{2}/45)(E/M)^{3/2}$, which is certainly much smaller than $\delta S_{\rm BH}$ because M has to be large compared to the Planck mass.

Thus we have accounted for all entropy contributions. In order for the GSL not to be violated, the box's entropy S must be bounded by $\pi(2M)^{3/2}E^{1/2}/\hbar$. Of course, for large M this is a much larger bound than Eq. (1). However, the point is that the new bound, though derived without recourse to lowering into the black hole, is also independent of N. This shows that there must exist an N-independent bound on the entropy of a bounded system, arguments involving buoyancy notwithstanding.

This said, we still face a paradox: By Boltzmann's formulas it seems that for large enough N, a quantity of radiation with given energy should surpass any N-independent entropy bound. The resolution we develop here depends on recognizing that the bounds on entropy derivable with the help of black holes must always refer to an entire system, not to part of one. In particular, bound (1) refers to the entropy and energy of the box *and* its contents. Therefore, when comparing thermal radiation entropy in a box with the bound, one would like to restate the Boltzmann formula for entropy in a form that takes cognizance of the existence of the box and those of its properties which are responsible for confining the radiation. Then a comparison can be made.

We shall perform the concrete calculations for a spherical box of radius R and total energy E. According to Boltzmann, at temperature T thermal radiation in the box has energy

$$E_{\rm rad} = (4\pi^3/45)NR^3T^4\hbar^{-3} \tag{39}$$

and entropy

$$S_{\rm rad} = (16\pi^3/135)NR^3T^3\hbar^{-3}. \tag{40}$$

If E_0 stands for the energy of the empty box (E at $T=0$), then upon eliminating T we have

$$S = (4\sqrt{2}/135)N^{1/4}[45\pi R(E-E_0)/\hbar]^{3/4}\ . \tag{41}$$

How small can E_0 be? Even when empty, the box's energy receives a contribution from vacuum (Casimir) energy of those fields it can entrap (or keep out). On dimensional grounds each species contributes vacuum energy $\varepsilon = \alpha\hbar/R$ where α is dimensionless. For the electromagnetic field in a sphere Boyer [22] showed that $\alpha = 0.045$. For all species together we write $\varepsilon = \bar{\alpha}N\hbar/R$, where $\bar{\alpha}$ is the suitable average. If positive, the vacuum energy sets a lower bound on E_0.

Even when the vacuum energy is negative [23,24], the total box mass must end up being positive. The physical mechanism is the suction on the box's wall that must accompany negative vacuum energy. In fact, from the rate at which ε gets more negative as the box's radius decreases, it follows that the wall sustains a negative pressure $P = -\bar{\alpha}N\hbar/4\pi R^3$. To resist this suction the wall must maintain a surface pressure (force per unit length) of the same order [17]. But unless the wall's surface mass density is bigger than the surface pressure, the speed of sound in the wall would be superluminal. The conclusion is that the wall must have a mass comparable to the magnitude of ε, or larger. This will make E_0 positive. Since an exact cancellation between wall mass and vacuum energy is unlikely, we expect E_0 to be of order $|\bar{\alpha}|\hbar/R$, or larger. Thus, henceforth, we simply write $E_0 = \alpha N\hbar/R$ and assume $\alpha > 0$.

The formula for entropy, Eq. (41), is now

$$S_{\rm rad} = (4\sqrt{2}/135)N^{1/4}[45\pi(RE/\hbar - \alpha N)]^{3/4}. \tag{42}$$

Now the function $f(x) = (x-a)^{3/4}/x$ has a maximum value of $\frac{1}{4}3^{3/4}a^{-1/4}$. We may, therefore, deduce from Eq. (42) the inequality

$$S_{\rm rad} \le \frac{\sqrt{2}\,\pi^{3/4}}{(135\alpha)^{1/4}}\frac{RE}{\hbar}. \tag{43}$$

This bound for blackbody radiation in a box is of the same form as bound (1). Indeed, since typically $\alpha \sim 10^{-3} - 10^{-2}$ for a single field [23,24], the numerical coefficient here is $\sim 3 - 5.5$ in harmony with bound (1). Most important, the bound here obtained from statistical physics is independent of N, just as (1) is. The physical mechanism is that some of the energy E is inert energy whose magnitude depends on N: $E > E_0 = \alpha N\hbar/R$. Thus the factor $RE/\hbar$ actually grows with N. But the point is that when we state the entropy bound in terms of R and E, we do not have to worry about how big N is; the bound will take care of that automatically.

At this junction we return to the issue of the allowed range of η. We have already seen that when N is a few, and the box is standardly oriented, $\eta \ll 1$. Let us now analyze the case when N is large for a spherical box of radius R. Then we may use our result that $E > E_0 = \alpha N\hbar/R$ in the definition (27) by taking $A = 4\pi R^2$ and $b = 2R$. We get

$$\eta^3 = \frac{N\hbar}{1440\pi ER} < \frac{1}{1440\pi\alpha}, \tag{44}$$

which shows that η does not grow indefinitely with N. Furthermore, unless α is much smaller than the typical value for separate fields, $\sim 10^{-2} - 10^{-3}$, η is indeed small compared to unity, thus allowing the arguments following Eq. (27) to yield the entropy bound from an appeal to black hole physics. And this bound is, of course, consistent with Eq. (43).

The case of a nonspherical box is much harder to analyze because E_0 depends not only on the typical dimension of the box, but also on axis ratios [23]. However, it seems plausible that η will also be small in those cases provided the various box dimensions are not too different.

B. Irrelevance of many dimensions

Not only does the proliferation of particle species increase the entropy for given energy, proliferation of spatial dimensions has the same effect. Evidently the more the dimensions, the more ways there are to split up the energy, so that a higher entropy is obtained. We might thus naively expect a bound such as (1) to be violated as the number of dimensions increases without bound. As we now show, this conclusion would be premature. Consider in D flat spatial dimensions a spherical space of radius R into which we dump energy E. What can we say about the entropy $S(E)$ as D grows? Evidently maximal $S(E)$ corresponds to the excitation of all existing field species with like inverse temperature β. The description of our fixed energy system in terms of temperature (thermodynamic regime) is tenable provided E is large enough that energy fluctuations in the canonical ensemble for β are small. In practice this means $\beta\hbar/R \ll 1$ (many wavelengths small compared to R are thermally excited), which we assume to be true. We shall simplify matters by ignoring massive species. This corresponds to the case that β times any of the rest masses is large.

The volume of a sphere of radius r in D dimensions is [25]

$$V_D(r) = \frac{2\pi^{D/2} r^D}{D\Gamma(D/2)}, \tag{45}$$

where Γ denotes the Euler gamma function. Consequently, the volume in frequency space of the shell $(\omega, \omega + d\omega)$ is

$$dV_D(\omega) = D[V_D(\omega)/\omega]d\omega. \tag{46}$$

The mean thermal energy in the sphere from one helicity degree of freedom is

$$E = V_D(R) \int_0^\infty \frac{\hbar\omega \, dV_D(\omega)}{(e^{\beta\hbar\omega} \mp 1)(2\pi)^D}, \tag{47}$$

where upper (lower) signs correspond to boson (fermion) fields. Using Eqs. (45) and (46), and

$$\int_0^\infty \frac{x^D \, dx}{e^x \mp 1} = \zeta(D+1)\Gamma(D+1) \times \begin{cases} 1 & \text{for bosons,} \\ 1 - 2^{-D} & \text{for fermions,} \end{cases} \tag{48}$$

where $\zeta(z)$ is Riemann's zeta function, we can cast the mean energy of all massless species in the form

$$E = \frac{N\zeta(D+1)\Gamma(D)R^D}{2^{D-3}[\Gamma(D/2)]^2 \beta^{D+1}\hbar^D}, \tag{49}$$

where N is the number of massless species (massless nonscalar bosons contribute one to N, while massless fermions contribute $2^{-1} - 2^{-(D+1)}$).

Likewise, we can write the thermal entropy of one helicity degree of freedom as

$$S = V_D(R) \int_0^\infty \left[\mp \ln(1 \mp e^{-\beta\hbar\omega}) + \frac{\beta\hbar\omega}{e^{\beta\hbar\omega} \mp 1}\right] \frac{dV_D(\omega)}{(2\pi)^D}. \tag{50}$$

After integration by parts, the first term in square brackets is seen to reduce, by virtue of Eq. (46), to D^{-1} times the second. Comparing the result for the entropy of all helicity degrees of freedom with Eq. (47) we see that

$$S_N = (1 + 1/D)\beta E. \tag{51}$$

The special case for $D = 3$ of this equation is well known.

As mentioned, the condition $\beta\hbar/R \ll 1$ must be respected in order for the above continuum treatment to make sense. Solving Eq. (49) for the relevant quantity we have

$$\beta\hbar/R = C_D(N\hbar/RE)^{\frac{1}{D+1}}, \tag{52}$$

with

$$C_D \equiv \left[\frac{\zeta(D+1)\Gamma(D)}{2^{D-3}[\Gamma(D/2)]^2}\right]^{\frac{1}{D+1}}. \tag{53}$$

Numerically it is found that C_D decreases with D: $C_D = (1.34, 1.29, 1.25, \ldots, 1)$ for $D = (1, 2, 3, \ldots, \infty)$. Hence the condition on β amounts to $RE/N\hbar \gg 1$. So, just as in Sec. III A, we find here that N constrains the value of RE. Substituting Eqs. (52) and (53) into Eq. (51) we get

$$S_N = C_D(1 + 1/D)N^{\frac{1}{D+1}}(RE/\hbar)^{\frac{D}{D+1}}. \tag{54}$$

The case $D = 3$ of this should be compared with Eq. (42); the only real difference is that here E does not include vacuum energy.

Let us now take the limit $D \to \infty$ while keeping N fixed. In this limit $\zeta(D+1) \to 1$ and $\Gamma(D+1) \to \sqrt{2\pi}e^{-D}D^D$, and so we find rigorously that

$$\lim_{D\to\infty} S_N = RE/\hbar. \tag{55}$$

Thus despite the many dimensions, and the arbitrary number of species N, this entropy is consistent with bound (1). As in Sec. III A, the dependence on N has dropped out except for the fact that our end result is valid only if E is large enough on a scale set by N. And contrary to intuition, in the thermodynamic limit the large phase space that opens up because of the multiplicity of dimensions does not help the entropy to surpass the entropy bound (1).

What happens for D finite? We rewrite Eq. (54) as

$$S_N = C_D(1 + 1/D)(N\hbar/RE)^{\frac{1}{D+1}} RE/\hbar. \tag{56}$$

Numerically $C_D(1 + 1/D)$ decreases monotonically from 2.89 at $D = 1$ to unity as $D \to \infty$. Because $N\hbar/RE \ll 1$ in the thermodynamic regime, it is clear that the entropy

S_N conforms to bound (1) for any D. We must stress that this result, like the previous one, is not guaranteed if the condition for the thermodynamic regime fails. In that case one cannot rely on continuum formulas such as Eqs. (50) and (51), and must take recourse to numerical calculations of the energy distribution of quantum states in a cavity. As mentioned, for $D = 3$ such calculations have been made [8], and fully confirm bound (1). When generalizing these to arbitrary D and N, care should be taken to include vacuum energies.

IV. INFORMATION CAPACITY OF REMNANTS

Remnants of black holes have been suggested as a resolution of the black hole information paradox (for reviews see Ref. [26]). In one version of this idea [27], the black hole stops evaporating when its dimension approaches the Planck scale. It is hypothesized that this remnant quiescent object retains the large information that went down the initial black hole upon its formation. With a notable exception [28], the arguments for and against Planck scale remnants have not confronted *quantitatively* the question of how much information can really be contained in a space which looks so small from the outside, and has a correspondingly tiny externally measured mass energy.

Now, if a system contains information I (natural units), then it must have at its disposal at least $n = e^I$ internal states. When looked at in coarse graining, such a system will be ascribed entropy $S = \max(\ln n) \geq I$. Now, the entropy bound (1) must also bound the entropy of a gravitating system since we never specified in Sec. II details about the box's interior. Applied to a remnant the bound would predict that for radius $\ell_{\rm Planck}$ and mass $\hbar/\ell_{\rm Planck}$ its entropy is 2π at most. The remnants information content is thus just a few bits. This makes Planck-size remnants irrelevant for the information paradox (the system that collapsed to the black hole which fathered the remnant may have been specified by 10^{12} of bits).

Giddings introduced the idea of large remnants [28] to overcome problems such as this. The idea is that each black hole decays to a remnant of a different size, which is deemed capable of retaining the initial information. Now, for a Schwarzschild black hole of initial mass M_0, the initial black hole entropy $S_{\rm BH} = 4\pi M_0^2$ measures the total information that becomes hidden at the moment of collapse. As argued above, Eq. (1) should bound the information that can be held by the remnant whose mass and *external* radius are E and R. We assume $R \sim 2E$ because the remnant must be a strongly gravitating object. Thus for the remnant to succesfully retain the initial information, we need $E \geq M_0$ which is impossible since some evaporation must have taken place. The point is that black holes saturate the entropy (information) bound so that the equivalent amount of information cannot be held by a lighter strongly gravitating object.

Clearly the role of remnants, Planck size or larger, might be rescued if they were not to respect the entropy bound (1). We have argued that the bound must apply to any object that can be lowered to the horizon of a black hole. Perhaps a remnant, by virtue of its black hole nature, cannot be supported in the requisite way to be lowered. However that may be, any remnant can still be dropped freely into an extreme RN black hole of mass M provided the black hole radius M is larger than the remnant's external radius $R \sim E$. Repeating the argument at the beginning of Sec. III A, and arguing that the information-coding states of the remnant contribute towards coarse-grained entropy, we see that the procedure bounds the entropy, and hence the information capacity, of the remnant by $\pi(2M)^{3/2}E^{1/2}$ (otherwise the GSL gets violated in the remnant–black-hole merger).

Now M has to be bigger than E so that the remnant can be absorbed, but it does not have to be arbitrarily large. Thus this independent argument bounds the remnant's information capacity by γE^2 with γ perhaps a few times $2^{3/2}\pi$. If the remnant has descended from a neutral black hole of initial mass M_0, it must retain information $16\pi M_0^2$ to do its job as information repository in the resolution of the information paradox. This information will conform to the bound we have just set provided $E > 4(\pi/\gamma)^{1/2}M_0$. We see that the remnant cannot be much lighter than M_0. However, there seems to be no reason for black hole evaporation to turn off as soon as the black hole has lost a moderate fraction of its mass. This problem dramatizes the difficulty the remnant scenario has in resolving the black hole information paradox.

The usual argument for remnants as information repositories is that they can retain large information despite their small external dimension because they contain a very large internal space in the form of a throat or horn. According to this widespread argument, a bound such as (1), which is stated in terms of external remnant dimensions alone, cannot apply. But we have just seen that if a remnant had an information capacity well above that indicated by bound (1) in terms of its external dimension, it would cause a violation of the GSL, were it dropped into a black hole sufficiently large to contain its outer dimension, the only one relevant for assimilation. Thus we must conclude that the postulated large internal space of a remnant does not allow it to hold a lot of information: The usual intuition that a large space can store a lot of information must thus fail for these strongly gravitating objects.

This last conclusion might be countered if the infall of the remnant into a black hole produces a singularity, thus making the GSL moot. However, in that case the whole role of remnants as stable information repositories is put in doubt. The bottom line is that black remnants, as discussed heretofore, cannot resolve the information paradox.

ACKNOWLEDGMENTS

I thank Ted Jacobson, Don Page, and Bob Wald for a number of remarks, Leonard Parker for suggestions, Jim Hartle for his interest in the subject, Hartle and Gary Horowitz for hospitality in Santa Barbara, and the Aspen Center for Physics where the draft of this paper was completed.

[1] J. D. Bekenstein, Lett. Nuovo Cimento **4**, 737 (1972); Phys. Rev. D **7**, 2333 (1973); S. W. Hawking, *ibid.* **13**, 2460 (1976).
[2] J. D. Bekenstein, Phys. Rev. D **9**, 3292 (1974).
[3] J. D. Bekenstein, Phys. Rev. D **23**, 287 (1981).
[4] O. B. Zaslavskii, Phys. Lett. A **160**, 339 (1991); Gen. Relativ. Gravit. **24**, 973 (1992).
[5] O. B. Zaslavskii, "Bekenstein entropy upper bound and laws of thermodynamics," Kharkov University report, 1993 (unpublished).
[6] L. X. Li and L. Liu, Phys. Rev. D **46**, 3296 (1992).
[7] J. D. Bekenstein, Phys. Rev. Lett. **46**, 623 (1980).
[8] J. D. Bekenstein, Phys. Rev. D **30**, 1669 (1984).
[9] J. D. Bekenstein and M. Schiffer, Int. J. Mod. Phys. C **1**, 355 (1990).
[10] M. Schiffer and J. D. Bekenstein, Phys. Rev. D **39**, 1109 (1989).
[11] J. D. Bekenstein and E. I. Guendelman, Phys. Rev. D **35**, 716 (1987).
[12] R. Sorkin, R. M. Wald, and Z. Z. Jiu, Gen. Relativ. Gravit. **13**, 1127 (1981).
[13] W. H. Zurek and D. N. Page, Phys. Rev. D **29**, 628 (1984).
[14] W. G. Unruh and R. M. Wald, Phys. Rev. D **25**, 942 (1982).
[15] W. G. Unruh and R. M. Wald, Phys. Rev. D **27**, 2271 (1983).
[16] J. D. Bekenstein, Phys. Rev. D **27**, 2262 (1983).
[17] J. D. Bekenstein, Phys. Rev. D **26**, 950 (1982).
[18] W. G. Unruh, Phys. Rev. D **14**, 870 (1976).
[19] D. N. Page, Phys. Rev. D **25**, 1499 (1982).
[20] D. N. Page, Phys. Rev. D **14**, 3260 (1976).
[21] M. Davis, R. Ruffini, and J. Tiomno, Phys. Rev. D **5**, 2932 (1972).
[22] T. H. Boyer, Phys. Rev. **174**, 1764 (1968).
[23] W. Lukosz, Z. Phys. **262**, 327 (1975).
[24] S. D. Unwin, Phys. Rev. D **26**, 944 (1982).
[25] K. Huang, *Statistical Mechanics* (Wiley, New York, 1963).
[26] J. Preskill, "Do black holes destroy information?," Report No. hep-th/9209058 (unpublished); J. Harvey and A. Strominger, "Quantum aspects of black holes," Report No. hep-th/9209055 (unpublished); S. Giddings, "Toy models for black hole evaporation," Report No. hep-th/9209113 (unpublished).
[27] L. Susskind and L. Thorlacius, Nucl. Phys. **B382**, 123 (1992); T. Banks, A. Dabholkar, M. R. Douglas, and M. O'Loughlin, Phys. Rev. D **45**, 3607 (1992); T. Banks, M. O'Loughlin, and A. Strominger, *ibid.* **47**, 4476 (1993).
[28] S. Giddings, Phys. Rev. D **46**, 1347 (1992).